# Solid Mechanics and Its Applications

Volume 210

*Series editor*

G. M. L. Gladwell, Waterloo, Canada

For further volumes:
http://www.springer.com/series/6557

## Aims and Scope of the Series

The fundamental questions arising in mechanics are: *Why?*, *How?*, and *How much?* The aim of this series is to provide lucid accounts written by authoritative researchers giving vision and insight in answering these questions on the subject of mechanics as it relates to solids.

The scope of the series covers the entire spectrum of solid mechanics. Thus it includes the foundation of mechanics; variational formulations; computational mechanics; statics, kinematics and dynamics of rigid and elastic bodies: vibrations of solids and structures; dynamical systems and chaos; the theories of elasticity, plasticity and viscoelasticity; composite materials; rods, beams, shells and membranes; structural control and stability; soils, rocks and geomechanics; fracture; tribology; experimental mechanics; biomechanics and machine design.

The median level of presentation is to the first year graduate student. Some texts are monographs defining the current state of the field; others are accessible to final year undergraduates; but essentially the emphasis is on readability and clarity.

Wolfgang H. Müller

# An Expedition to Continuum Theory

Springer

Wolfgang H. Müller
Institute of Mechanics
Technical University of Berlin
Berlin
Germany

ISSN 0925-0042 ISSN 2214-7764 (electronic)
ISBN 978-94-007-7798-9 ISBN 978-94-007-7799-6 (eBook)
DOI 10.1007/978-94-007-7799-6
Springer Dordrecht Heidelberg New York London

Library of Congress Control Number: 2013951120



Printed on acid-free paper

Springer is part of Springer Science+Business Media (www.springer.com)

*For S.A.R.*

*Durch Sturm und bösen Wind verschlagen,*
*irr' auf den Wassern ich umher –*
*wie lange, weiß ich kaum zu sagen;*
*Schon zähl' ich nicht die Jahre mehr.*
*Unmöglich dünkt mich', daß ich nenne*
*die Länder alle, die ich fand:*
*das eine nur, nach dem ich brenne,*
*ich find' es nicht, mein Heimatland!*

(Der fliegende Holländer, Richard Wagner)

# Preface

Just to make it clear from the very start: This is not a monograph for the specialists who are looking for a compendium on continuum theory. Rather I attempt to fill a niche with this book, which started to widen after our government in all their infinite wisdom decided to terminate the diploma syllabi at German Universities and replace them by bachelor and master modules instead:

True students of engineering have a notorious aversion to mathematical abstraction. They prefer a quick and dirty way of solution, in particular when modeling the behavior of advanced materials in complex technical systems. The usual, highly formalized textbooks on continuum theory do not really support such an approach. Even their last resorts, for example consulting the user manuals of the all-time-favorite finite element codes, are blocked, because the same cryptic symbols are lurking there. On the other hand, students of physics face another problem, which is due to the way physics is commonly taught. They hear much about discrete systems, in particular in mechanics and in thermodynamics. The concept of fields is usually not presented before they attend classes on electrodynamics or quantum mechanics. Finally, the education of both groups of students has in common that usually no difference is made between the laws of nature (the balances of mass, momentum, energy, etc.) and constitutive equations. Both are usually well mixed to form a hodge-podge of recipes. The best examples are the NAVIER–STOKES equations. Moreover, every subject of physics, i.e., mechanics, thermodynamics, electrodynamics, etc., is usually taught separately without emphasizing the connections and the similarities. This is definitely not what we need when developing modern technologies, which only thrive because of their multiphysics interaction.

This is where continuum theory can help. It provides a bridge between the various subjects, by working out a common structure and by emphasizing the common roots. In this context constitutive equations form a most essential joint.

This is where this book sets in. It is the result of two teaching modules of four contact hours per week each. These modules are currently taught at TU Berlin, in particular in the master course Physical Engineering Science. The exercises compiled in this book play an important role in the teaching: Approximately two of the four hours per week are reserved for a seminar, where each student presents one of the various problems. Moreover, further problems are worked out in written form every week. Thus the students study the subject matter on a continuous basis, throughout the semester. They are forced to learn for their future job and not just for a final examination, after which they may forget everything immediately. Another side effect of this hard way of learning is to make sure that the students do not study singular aspects of continuum theory and ignore the rest. And, finally, the students learn from each other so that collectivism can be good for something after all.

Another important aspect of this book is to educate students in tensor calculus both in abstract as well as in index notation. Future engineers and physicists must be able to perform calculations in their daily practice, which is why the latter method is particularly important.

Finally, I would like to thank all those who helped me with this book. First of all, Prof. Ingo Müller and Priv.-Doz. Dr. Wolf Weiss, who contributed to various sections that were not included in the German edition of this book. Further credits are due to Messrs. B. Emek Abali, M.Sc., Dipl.-Ing. Benjamin Schmorl, and Felix Reich, M.Sc. The latter contributed in particular Exercises 13.2.3, 13.4.1, and 13.4.2. Several students found typos in the book: Messrs. Heinrich Grümmer, Wilhelm Hübner, Andre Klunker, Christian Seidel, and Oliver Stahn. Finally, I would like to thank Springer Engineering in the Netherlands, in particular Ms. Nathalie Jacobs and Ms. Cynthia Feenstra.

And now we start

Berlin, July 2013 Wolfgang H. Müller

# Contents

# Chapter 1
# Prologue

**Abstract** This introductory chapter explains the scope and structure of this book and why it may be useful for the reader. Some basic notions and concepts of continuum theory are introduced. As a motivation for studying continuum theory several engineering problems are presented, which will eventually be solved in the following chapters.

> *Relations between pure and applied mathematicians are based on trust and understanding.*
> *Namely, pure mathematicians do not trust applied mathematicians,*
> *and applied mathematicians do not understand pure mathematicians.*
>
> Albert Einstein

## 1.1 Some Remarks Regarding the Purpose of this Book

"What!? Yet another book on continuum theory?," one may ask. "How is this one different from the established literature?" In this context we must first of all explain what this book does not want to be: It certainly *does not* cover all there is to know in continuum theory, not even rudimentarily. Rather the intention is to give a first impression of what continuum theory can do for the engineer who has to solve technical problems involving solids, fluids, or gases. It is by no means a compendium for a specialist or an advanced student. Its clientele are beginners instead. The various fields of application of continuum theory will be illustrated to suit their needs, and we will explain to them how continuum theory "works," at least in principle and, to a certain degree, also in detail. One of our intentions is to create awareness for certain notions, such as universal balance equations in contrast to material-specific constitutive equations and, by making this distinction, show how relatively simple problems for solids, fluids, and gases can be solved analytically.

W. H. Müller, *An Expedition to Continuum Theory*, Solid Mechanics and Its Applications 210, DOI: 10.1007/978-94-007-7799-6_1,

Robert HOOKE was born on July 18, 1635 in Freshwater on the Isle of Wight. He died on March 3, 1703 in London. He was a contemporary of NEWTON. However, other than Sir Isaac, he was interested in physics *and* in biology. His experiments with springs led him to more or less verbal forms of the law named after him and which he summarized in an anagram as follows: CEIIINOSSSTTUU. Deciphered this means "ut tensio sic vis," what we may translate as "like the distension so the force." (painting by Rita GREER, Oxford University)

For a quantitative investigation of such problems suitable engineering mathematics is required. For continuum theory this is predominantly the *tensor calculus* which, normally, is not part of the mathematics syllabus in engineering education. It is for this reason that we start in Chap. 2 with *tensor algebra* and the representation of tensors in arbitrary coordinate systems, before we elaborate on continuum theory any further. The emphasis in this book is on *practical* mathematics and not on formalisms. Therefore our preferred way of writing tensors will be in index notation. However, in order to be able to read more advanced textbooks as well as the current scientific literature the index notation will always be juxtaposed to the symbolic tensor calculus, albeit by sacrificing mathematical stringency.

Chapter 3 is dedicated to *balances*, in particular to the balances of *mass*, *momentum*, and *energy*, in other words to the conservation laws of classical physics. The balances will be stated in integral form—for a material volume—as well as locally—in regular and singular points of the continuum. The effects of a changing *total* energy or, in other words, non-isothermal processes, are broadly covered in this book. By doing so it goes well beyond traditional continuum *mechanics* and we may speak of an introduction to continuum *physics* instead.

Claude Louis Marie Henri NAVIER was born on February 10, 1785 in Dijon and died on August 21, 1836 in Paris. In 1819 he became a professor of mechanics at the École des Ponts et Chaussées. In 1831 he succeeded CAUCHY at the École Polytechnique. He worked in various fields relevant to mechanical engineering, such as elasticity theory and fluid mechanics. Being an "offspring" of FOURIER he contributed much to the theory of FOURIER series and suggested an equation for fluid friction, the NAVIER-STOKES-model.

In order to evaluate balances in arbitrary as well as technically important coordinate systems (such as cylindrical, spherical, or elliptical coordinates) it is necessary to learn more about spatial derivatives of tensor fields. This will be addressed in Chap. 4, which is dedicated to elements of *tensor analysis*. The formalism will then be applied to balance equations in Chap. 5.

George Gabriel STOKES was born on August 13, 1819 in Skreen, County Sligo, Ireland and died on February 1, 1903 in Cambridge, England. After a detour to Bristol College he starts studying mathematics in Cambridge at Pembroke College. He graduates in 1841 in the Mathematical Tripos Examination with highest honors as a Senior Wrangler. In 1849 STOKES becomes one of the successors to NEWTON's famous Lucasian Chair at the University of Cambridge.

In Chap. 6 we substantiate the notion of *constitutive equations*. However, we will not present a stringent form of materials theory where constitutive equations follow as a consequence of superior principles. Rather we will argue in an informal, engineering manner, i.e., cast the constitutive equations into mathematical form after they have been motivated and illustrate their usage in arbitrary coordinate systems. In this context we will discuss HOOKE's law for the anisotropic and for the isotropic linear solid, the NAVIER-STOKES law for viscous fluids, the thermal and the caloric equations of state for the ideal gas, the connection between the specific heats and the internal energy for simple solids according to DULONG-PETIT, and FOURIER's law for the heat flux vector.

After balance equations and (simple) constitutive relations have been introduced, they will be combined in Chap. 7 to result in *field equations*. These are used to pose and to solve initial-boundary value problems for simple geometries and to reach the primary goal of continuum theory, which is the determination of the five fields for mass density, velocity, and temperature in each point of a material body and at all times.

Pierre Louis DULONG was born on February 12, 1785 in Rouen and died on July 19, 1838 in Paris. He was a French physico-chemist. He also studied at the famous École Polytechnique in Paris and became a professor at the school in 1820. Moreover, he became a member of the Académie des Sciences in 1823 and assumed the position of a secretary in 1832. Chemists live dangerously: While working with nitrogen-trichloride DULONG tragically lost his eye and three fingers.

Alexis Thérèse PETIT was born on October 2, 1791 in Vesoul and died on June 21, 1820 in Paris. He was a French physicist and studied, just like DULONG, at the École Polytechnique where he became successor to Pierre-Simon LAPLACE. He discovered "his law" in 1819 together with DULONG. Moreover, being a French patriot, he was also a strong supporter of the wavy nature of light and strongly opposed NEWTON's belief of a corpuscular theory.

Chapter 8 deals with a relatively abstract question: Do balance laws and constitutive equations keep their form if transferred from an observer at rest to an arbitrarily moving one? In mathematical terms this problem can be analyzed by using so-called *Euclidean transformations*. We will touch upon an almost philosophical principle according to which a genuine law of nature must keep its form, independently of the frame of reference.

The geometrician Euclid was born around the year 360 B.C., presumably in Athens. He died around 280 B.C., presumably in Alexandria. His great heritage is a set of textbooks, the so-called "elements" in which he compiled the knowledge about mathematics of his times. The most famous anecdote about Euclid tells us about his encounter with the pharaoh Ptolemy I who, being a lazy politician, asked him if there was a shortcut to geometry other than the tedious lore of the "elements." Euclid answered laconically that there is no royal road to knowledge.

Chapters 9–11 give a first, rudimentary impression of three large fields of continuum physics: *linear elasticity*, *mechanics of frictional fluids*, and *time-independent plasticity*, where the latter provides a bridge between solids and fluids.

Carl Henry Eckart was born on May 4, 1902 in St. Louis, Missouri. He died on October 23, 1973 in La Jolla, California. He was a multi-talented American scientist who contributed to theoretical physics as well to geology and oceanography. For example, in quantum mechanics he provided proof that Heisenberg's and Schrödinger's point of view are equivalent to each other. He can also be considered to be one of the fathers of irreversible and relativistic thermodynamics. However, on top of that he was a competent administrator of academic affairs which shows that, surprisingly, science and business can mix after all. (Photo public domain, credit SIO Archives/UCSD)

Chapter 12 is dedicated to a delicate subject: if one intends to read textbooks on materials theory it is necessary to have some basic knowledge of the concept of *entropy*. This is presented here: Entropy is introduced as just another field obeying a balance law. Thus entropy is directly in line with other additive quantities that were used previously in this book, namely mass, momentum, and energy. Moreover, the potential paranoia over entropy is reduced by computing it for a few simple but illuminating cases: We shall understand why it is a measure of the state of disorder of a system. Similarly we shall calculate the entropy production in simple cases and understand that it is a measure of irreversibility, or of how difficult it is to reverse a "natural" process. At the end of the chapter we present an introduction to the theory of irreversible processes according to Eckart. We shall rediscover the constitutive equations of Navier-Stokes-Fourier. This may serve as a starting point for further studies of entropy principles.

In the final Chap. 13 we shall learn how the realm of continuum physics can be extended to *electromagnetic fields*. The emphasis is on a rational presentation of fundamental principles: What are the foundations of MAXWELL's equations, how can the occurring fields be measured, at least in principle, and how are they linked to each other? Moreover, the question regarding the frame indifference of the equations and the transformation properties of the electromagnetic fields will be posed, which had already been answered before in context with the thermo-mechanical fields. This will lead us to the beginnings of relativistic field theories. At the very end of the chapter we will present simple constitutive equations and couple electro-magnetic to mechanical phenomena.

In order to consolidate the acquired knowledge many exercises have been added to the text. In part they pick up and examine statements of the surrounding text in detail because they had not been proven before. However, some of them also require deeper thinking since their solution requires a broader understanding of the subject matter. Clearly, the reader will gain maximum profit by solving *all* of the exercises. However, a more superficial reading is also possible by recognizing the meaning and relevance of the given solution. Moreover, at the end of each chapter an overview to additional literature on the current topics is presented together with a preliminary discussion of each reference: *Would you like to know more?*

James Clerk MAXWELL was born on June 13, 1831 in Edinburgh, Scotland and died on November 5, 1879 in Cambridge, Cambridgeshire, England. MAXWELL coined 19th century theoretical physics like no one else. He laid the foundations to various fields, starting from mechanics, kinetic theory of gases and thermodynamics to electrodynamics. Notoriously famous is his heated-polemic controversy about entropy and recurrence with his Austrian colleague BOLTZMANN.

Finally, for the entertainment, relaxation, and inspiration of the reader pictures and interesting details of the life of each scientist who was mentioned in the text have been added.

In what follows we present some motivation of what makes learning continuum theory worth while for the engineer.

Jean Baptiste Joseph Baron de FOURIER was born on March 21, 1768 in Auxerre (Bourgogne, France). He dies on May 16, 1830 in Paris. After having survived the chaos of the French revolution (luckily FOURIER was not born as a nobleman) FOURIER starts as a student at the École Normale in Paris. His teacher, the great LAGRANGE, considers him among the top scientists of Europe at that time. It is, therefore, not surprising that he becomes successor to LAGRANGE's chair for analysis and mechanics at the École Polytechnique in 1797. He is also one of the scientists who accompanied Napoleon during his military campaign to Egypt. This acquaintance certainly helped getting him the title of a baron.

## 1.2 A Reminder of Scalars, Vectors, and Tensors

Many fields of continuum theory are no scalars and cannot simply be described by a single "number." Rather they have a "direction" and show an "orientation." For their quantification more than just one number is required. Such quantities are known as *vectors* or, even more general, as *tensors*. Typical vectors mentioned in a beginner's class on engineering mechanics are the position, $\boldsymbol{x}$, its time derivatives, the velocity, $\boldsymbol{\upsilon}$, and the acceleration, $\boldsymbol{a}$, or the force *vector*, $\boldsymbol{F}$, as well as the *specific force*,[1] $\boldsymbol{f}$. Other mechanics-related vectors are the angular velocity, $\boldsymbol{\omega}$, or the torque, $\boldsymbol{M}$. It is intuitively clear that the latter vectors are associated with some rotation or, in other words, they have both a direction and an orientation—in contrast to the vectors previously mentioned which had only a direction. Therefore, such objects are called *axial* or *polar vectors*. Moreover, in thermodynamics we encounter the *heat flux*, $\boldsymbol{q}$, and the *temperature gradient*, grad $T$. And in electrical engineering use is made of the *electric field*, $\boldsymbol{E}$, the *dielectric displacement*, $\boldsymbol{D}$, the *polarization*, $\boldsymbol{P}$, or the *magnetic fields*, $\boldsymbol{H}$ and $\boldsymbol{B}$. As in mechanics some of these vectors are of a polar nature. However, due to lack of intuition and everyday experience it is hard to tell which ones. Finally, we should also think of vectors related to the geometry of a body, such as the *unit surface normal*, $\boldsymbol{n}$, the directed *surface element*, $\mathrm{d}\boldsymbol{A}$, the *unit tangent*, $\boldsymbol{\tau}$, or the *directed line element*, $\mathrm{d}\boldsymbol{s}$. In a more general manner of speech we may also refer to vectors as first order tensors. Scalar quantities, like *mass density*, $\rho$, *temperature*, $T$, or the *electric potential*, $U$, may analogously be called tensors of zeroth order. We shall later add some precision to this nomenclature, which for the time being is purely heuristic.

Well known mechanical tensors of second order, or "tensors" for short, are the *stress*,$\boldsymbol{\sigma}$, the linear *strain*, $\boldsymbol{\varepsilon}$, or the *velocity gradient*, grad $\boldsymbol{\upsilon}$. In HOOKE's law for anisotropic materials we even encounter a tensor of fourth order—the *stiffness matrix*, $\boldsymbol{C}$. Constitutive equations of electromagnetism even use third order tensors with a physical meaning, for example the *piezo-electric* matrix of *coefficients*, $\boldsymbol{e}$. However, as we shall see, tensors can also be used to shed a different light on a well-established mathematical concept like the vector-product. As we shall see it can be rewritten in terms of an axial third order tensor, the *LEVI-CIVITA-symbol*.

Note that whenever we wish to emphasize the absolute vector or tensor character of a physical quantity, which is independent of a coordinate system, the coordinate base or, even more general, the observer, we denote the corresponding quantity by a bold letter, e.g., by $\boldsymbol{F}$ in the case of a force. This is quite customary in the contemporary scientific literature and we have already followed this custom in the text above. Other symbolic ways of writing is to underline[2] the symbol, e.g., by writing $\underline{F}$, or to use an arrow, i.e., $\vec{F}$. However, for the solution of a specific engineering problem it is in most cases necessary to refer to a suitable set of

[1] The adjective "specific" refers to a quantity per unit mass. An example for a specific force is the well-known gravitational acceleration, $\boldsymbol{g}$.

[2] Second order tensors are sometimes underlined twice.

coordinates, i.e., to specify the corresponding *vector base* and to represent all physical quantities in this very base by means of *components*. Therefore in this book the reader will, first, be introduced to important notions and concepts of continuum theory and, second, learn the appropriate engineering mathematics for the solution of technical problems. The following examples were taken from daily engineering practice and illustrate the importance of a deeper understanding of how to evaluate vector and tensor relations in arbitrary coordinate systems.

### *1.2.1 Example I: Glass Fiber Optics—Pressing of a Spherical Lens into a Bushing*

For an optical link a spherical lens (∅ 1 mm) made of glass or sapphire is pressed into a cylindrical bushing of a slightly smaller diameter (∅ 0.995 mm) (cf., Fig. 1.1). This leads to building up a pressure along the equators of the sphere and the corresponding deformation will lead to tensile stresses in its interior.

However, brittle materials such as glass or sapphire react extremely sensitive to tensile stresses. Even if these stresses are not high enough to crush the sphere immediately, the phenomenon of *subcritical crack growth* will lead to failure eventually: Under the influence of tensile stress water vapor, which is always present in the environment, will preferably diffuse to the tips of microcracks in the glass. There it will loosen the atomic bonds and stepwise increase the crack length. In the very moment where the first microcrack reaches a critical size, the sphere of glass will fail. Indeed, it was shown experimentally that fracture starts in the equatorial region of the sphere. This is where one would intuitively expect the highest tensile stresses: Fig. 1.2. The engineer must now provide an answer to the

**Fig. 1.1** Inserting a sphere of glass into a bushing made of metal

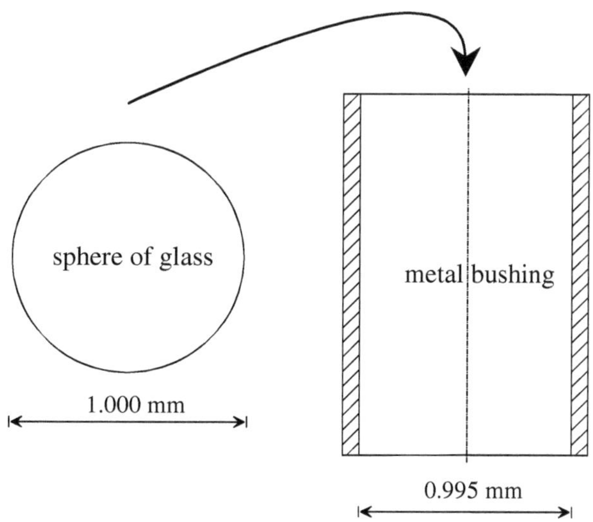

**Fig. 1.2** A broken spherical glass lens after insertion into a bushing

following question: How is the misfit in diameters or, in other words, how is the clearing tolerance related to an acceptable malfunction rate in view of the lifetime warranty of the whole system?

In order to answer this question the stress distribution in the sphere must be known first. Basically it can be calculated as follows. In a first step a suitable differential equation for the stresses needs to be established. As we shall explain in more detail below this equation can be obtained from the static *balance of momentum*:

$$\operatorname{div}\boldsymbol{\sigma} = \mathbf{0} \tag{1.2.1}$$

The symbol "div" denotes a differential operator known as *divergence*. We shall see later that it entails a certain differentiation with respect to position. Eq. (1.2.1) now needs to be complemented by a suitable constitutive law for the stress tensor. For brittle matter HOOKE's law may serve as a suitable model. It connects the *stress* $\boldsymbol{\sigma}$ and the *strain* $\boldsymbol{\varepsilon}$ linearly via the *stiffness tensor* $\boldsymbol{C}$:

$$\boldsymbol{\sigma} = \boldsymbol{C} \cdot\cdot\, \boldsymbol{\varepsilon}. \tag{1.2.2}$$

The symbol "$\cdot\cdot$" stands for a double scalar product. We shall see later what its precise meaning is. At this point it suffices to say that Eq. (1.2.2) is rather general and holds for arbitrary anisotropic, linear elastic materials. In the case of an isotropic linear-elastic material the above-mentioned equation simplifies considerably: The ominous double scalar product after the stiffness matrix can be evaluated and rewritten in terms of the *LAMÉ* parameters $\lambda$ and $\mu$:

$$\boldsymbol{\sigma} = \lambda \operatorname{tr}(\boldsymbol{\varepsilon})\mathbf{1} + 2\mu\boldsymbol{\varepsilon}. \tag{1.2.3}$$

Gabriel LAMÉ was born on July 22, 1795 in Tours (France). He died on May 1, 1870 in Paris. In 1813 he enters the École Polytechnique in Paris and graduates from that school in 1817. This is followed by further studies at the famous École des Mines, which he finishes with another degree in 1820. In the same year LAMÉ moves to Russia and becomes a professor and engineer at the Institut et Corps du Génie des Voies de Communication in St. Petersburg. In 1832 he returns to Paris, founds an engineering firm together with French colleagues and finally accepts a chair in physics at his first *alma mater*.

In this equation "**1**" denotes the *unit tensor* and "tr" is an operator known as the *trace* of a tensor. We shall see later that the LAMÉ parameters, $\lambda$ and $\mu$, are related to *Young's modulus, E*, and to the *shear modulus, G*, as follows:

$$\lambda = \frac{G(E-2G)}{3G-E}, \quad \mu = G. \tag{1.2.4}$$

If the stress $\boldsymbol{\sigma}$ in Eq. (1.2.1) is substituted by (1.2.3) a set of partial differential equations of second order in space for the *displacement vector*, $\boldsymbol{u}$, results. This is because the strain tensor, $\boldsymbol{\varepsilon}$, contains first order derivatives in $\boldsymbol{u}$. For a concrete solution it is necessary to specify *boundary conditions*. For example: The normal stress in the equatorial region of the sphere must be equal to the pressure, $p_0$, due to the misfit. Moreover, the stresses in the interior of the sphere must remain finite.

Thomas YOUNG was born on June 13, 1773 in Milverton, Somersetshire and died on May 10, 1829 in London. In Anglo-Saxon countries his name is inseparably connected to the modulus of elasticity. It is highly illuminating to read YOUNG's own words when he introduces the modulus: "The modulus of elasticity of any substance is a column of the same substance, capable of producing a pressure on its base which is to the weight causing a certain degree of compression as the length of the substance is to the diminution of its length." His commissioner—the British Admiralty—responded immediately: "Though science is much respected by their Lordships and your paper is much esteemed, it is too learned … in short it is not understood." (cited after [2])

Thus the problem is well-posed and a mathematician could turn to the proof of a unique solution. However, this would be insufficient for the engineer. For an explicit calculation of the stresses it is first of all necessary to choose a suitable coordinate system. In the present case *spherical coordinates* $r$, $\varphi$, $\vartheta$ would do nicely. The coordinate system is placed in the center of the sphere of glass (cf., Fig. 1.3). The letter $r$ denotes the *radial distance* in the sphere of glass; $\varphi$ and $\vartheta$ refer to the so-called *azimuthal* and *polar angles,* respectively.

Before a solution can be obtained, Eqs. (1.2.1) and (1.2.2) need to be rewritten in spherical coordinates. This requires two things. First, one must be able to express physical quantities, such as stress, strain, and displacements correctly. Second, it is required to perform derivatives in spherical coordinates, for example

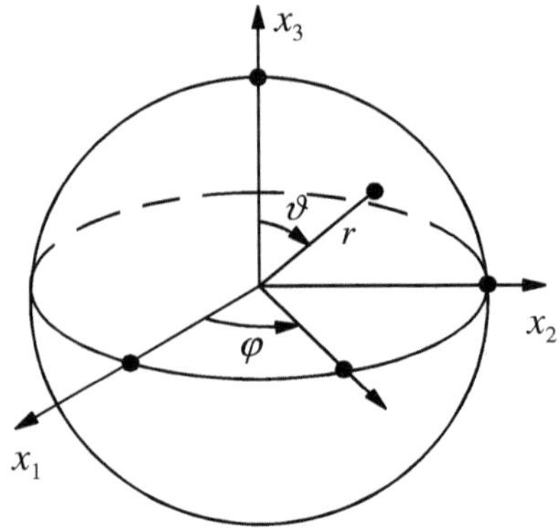


**Fig. 1.3** Spherical coordinates

to specify the abstract operation "div," as well as the relation between strain and displacement which, as indicated above, also contains a spatial derivative. How to do this will be described in Chaps. 2–4.

Adrien Marie LEGENDRE was born on September 18, 1752 in Paris where he also died on January 10, 1833. His life falls into the heydays of the French Revolution. In fact he strongly supported the new ideals that originated during those days of change. For example, he was engaged in the definition of rational units independent of the weight or the length of the currently acting monarch. In 1791 he becomes a member of the corresponding committees in the French Academy and starts producing logarithmic tables in 1792. Note that he did not do this all by himself. At certain times he had more than 80 (!) assistants, which clearly shows that revolutions can be good for something.

Once the partial differential equation has successfully been formulated in spherical coordinates it needs to be solved by using the afore-mentioned boundary conditions. By separating the variables $r$ and $\vartheta$ the radial part can be converted into an ordinary differential equation of the LEGENDRE type. We will get back to that in Chap. 9.

### *1.2.2 Example II: Composites—Thermal Stresses Around Fibers*

Materials reinforced with (for example) carbon or silicon carbide fibers are used more and more frequently in advanced lightweight engineering constructions. Fig. 1.4 shows a typical hexagonal unit cell of such a composite material. A potential reliability problem originates from the thermal stresses in these materials, which are due to the Coefficients of Thermal Expansion (CTEs) of the various matrices and fibers that are being used and which can be considerably different. The resulting stresses (also known as thermal eigenstresses) can be considerable

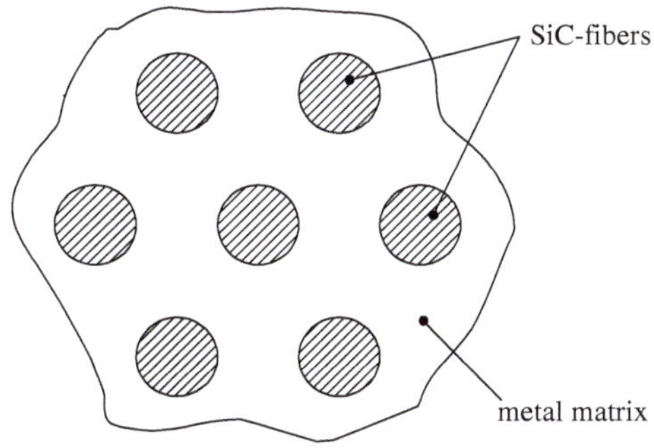


**Fig. 1.4** A hexagonal fiber arrangement

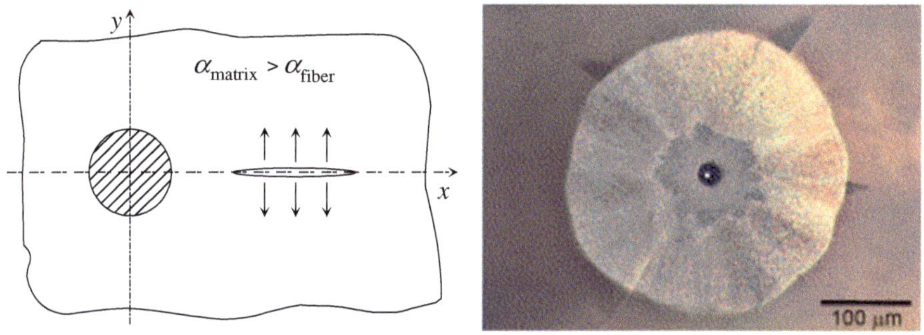


**Fig. 1.5** *Left* generating radial cracks in fiber-matrix composition, e.g., by thermal stresses; *right* optical micrograph of radial cracks generated by a lithium niobate-lithium disilicate double crystal in a glass matrix after Serbena and Zanzotto [3]

and will eventually lead to cracking of the material. For example, if the CTE of the matrix is greater than the CTE of the fiber the matrix will shrink upon the fiber during cooling and tensile stresses will be generated in the matrix perpendicularly to the radial direction (cf., Fig. 1.5, left). These stresses can lead to the formation and growth of radial cracks (see Fig. 1.5, right). On the other hand, if the CTE of the matrix is smaller, radial stresses along the circumference of the fiber will form and the matrix will debond from the fiber (see Fig. 1.6).

In order to quantify the thermal stresses the compound is first idealized by the following system shown in Fig. 1.7: A cylinder "1" (the fiber) is inserted in a hollow cylinder "2" (the matrix). Both cylinders have different LAMÉ parameters $\lambda_1$, $\mu_1$ and $\lambda_2$, $\mu_2$, respectively. Their CTEs are also different, $\alpha_1$ and $\alpha_2$. An interaction between the various fibers of the composite is neglected in this simple model.[3]

---

[3] A more careful investigation shows that this assumption is reasonable up to a fiber volume of 40 %.

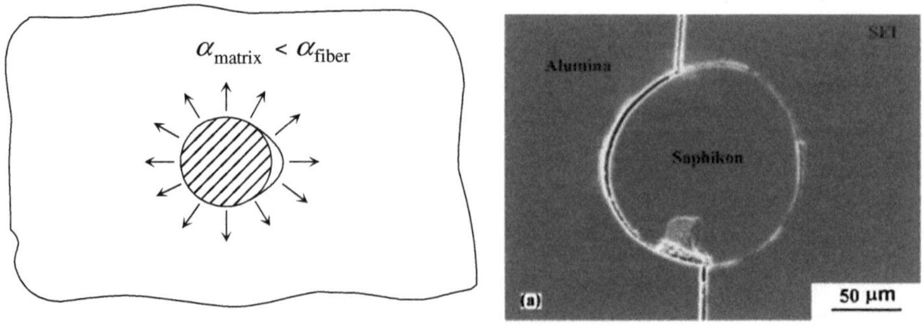


**Fig. 1.6** *Left* detachment of a fiber from the matrix by thermal stresses; *right* Interfacial debonding at a monazite/fiber interface after Chawla et al. [1]

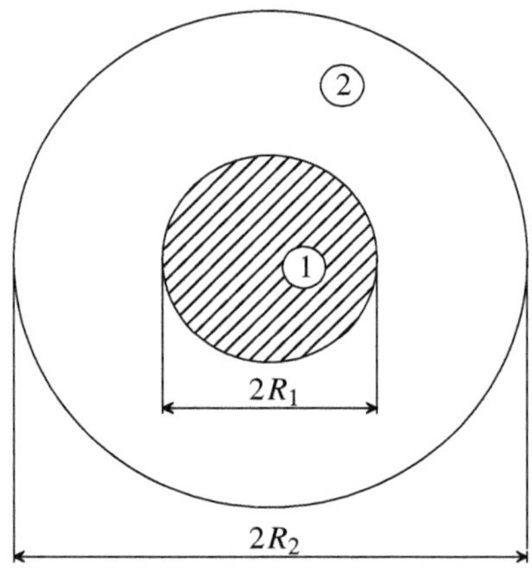


**Fig. 1.7** A simple cylinder model of a fiber-matrix composite

As in the previous case the computation of the stresses relies upon the balance of momentum of elastostatics, Eq. (1.2.1). In the present case it is suitably evaluated in *polar coordinatesr*, $\vartheta$, and $z$. As a constitutive equation we will once more use Hooke's law, although it needs to be extended to cover thermal expansion. According to Duhamel and Neumann we write:

$$\boldsymbol{\sigma} = -(3\lambda + 2\mu)\alpha[T - T_{\mathrm{R}}]\mathbf{1} + \lambda\operatorname{tr}(\boldsymbol{\varepsilon})\mathbf{1} + 2\mu\boldsymbol{\varepsilon}. \tag{1.2.5}$$

In this equation $T$ and $T_{\mathrm{R}}$ denote the current and the reference temperature, respectively. The latter can be interpreted as the manufacturing temperature at which the composite is typically free of stress.

Note that we have to distinguish between two regions filled with different materials and a borderline at which both materials meet. Clearly, there must be a transition between the mathematical solutions in both regions. More specifically, we proceed as follows. The differential equations resulting by combination of Eqs. (1.2.1) and (1.2.5) are first solved for a hollow cylinder. We shall see that in the resulting expressions two constants of integration will occur. This solution is then applied to each of the two regions separately so that four constants of integration result. These are determined from suitable boundary conditions. One can argue from first principles that the radial stresses at all boundaries must be continuous as well as finite in all points of the continuum. Moreover, the displacement at the inner borderline must be continuous, at least as long as the fiber and the matrix does not debond.

Jean Marie Constant DUHAMEL was born on February 5, 1797 in St. Malo. He died on April 29, 1872 in Paris. In 1814 he became a student at the École Polytechnique in Paris from where he graduates in 1816 in mathematics. After that he goes to Rennes to study and practice law. In the end, however, he goes back to Paris and teaches mathematics at various higher schools. In 1830 he becomes the successor to CORIOLIS who taught calculus at DUHAMEL's old school.

Moreover, the same remarks as in the case of spherical coordinates hold, which were made in context with the sphere of glass in a bushing. More explicitly, in order to obtain an explicit solution it is now necessary to represent operators like the divergence as well as all occurring tensors in cylindrical coordinates.

Franz Ernst NEUMANN was born on September 11, 1798 in Joachimsthal (Brandenburg, Germany). He died on May 23, 1895 in Königsberg (now Kaliningrad, East Prussia). His youth falls into the final days of NAPOLEON when Prussia tried to liberate herself. Consequently, he is almost beaten to death when campaigning with Marshal BLÜCHER. However, he recovers and finishes high school in 1817, enrolls as a student at Berlin University and studies theology, law, and science. After an assistantship with the mineralogist E.C. WEISS his scientific career starts blooming. His son, however, turns completely to mathematics to give us what we now know as NEUMANN boundary conditions.

In Chap. 2 we will first turn to the problem of general coordinate transformations. The resulting rules and formulae will then be applied to cylindrical, spherical, and other coordinates of interest.

### *1.2.3 Example III: Cracks*

In the previous examples the notion of cracks has been mentioned already. From a mechanics point of view we may, more generally, talk about *stress concentrators* instead. The first two-dimensional mathematical model of such a concentrator and, in the limit, of a sharp crack in a brittle solid was suggested and detailed by the Englishman A.A. GRIFFITH.

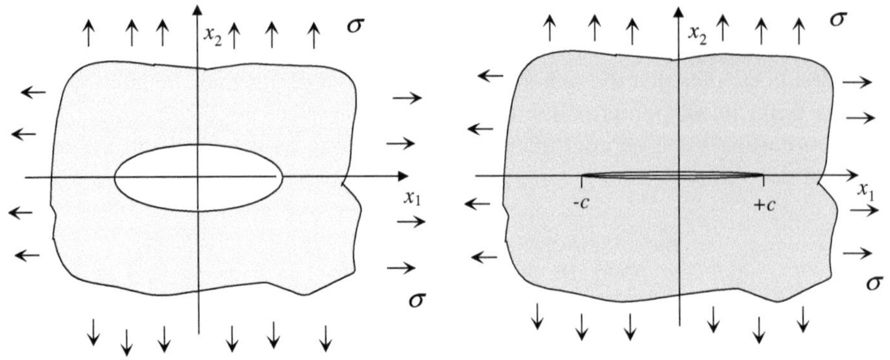


**Fig. 1.8** GRIFFITH's model of an elliptic crack

Alan Arnold GRIFFITH was born on June 13, 1893 in London and died on October 13, 1963. He graduated from the University of Liverpool. From his early days on he was very much interested in avionics and the problems of material fatigue and fracture encountered therein. In 1920 he publishes his famous paper on brittle fracture which already contains the fracture criterion that later carried his name. However, it contained a faulty factor, which Griffith corrected soon—but without providing details. It took a long time until in the sixties another paper was published which proved his correction.

As indicated in Fig. 1.8 GRIFFITH considered a sequence of ellipses of smaller and smaller minor axes. Elliptic coordinates are used to tackle the GRIFFITH crack problem (cf., Chaps. 2 and 9). As we shall see, such coordinates are given by perpendicularly intersecting lines of ellipses and hyperbolae. It is not surprising that the equations of linear elasticity will take a mathematically particularly simple structure for the case of an elliptical hole becoming increasingly flatter.

## References

1. Chawla KK, Liu H, Janczak-Rusch J, Sambasivan S (2000) Microstructure and properties of monazite (LaPO4) coated saphikon fiber/alumina matrix composites. J Eur Ceram Soc 20(5):551–559
2. Gordon JE (1978) Structures or why things don't fall down. Penguin Books, London
3. Serbena FC, Zanotto ED (2012) Internal residual stresses in glass-ceramics: a review. J Non-Cryst Solids 358(6–7):975–984

# Chapter 2
# Coordinate Transformations

**Abstract** For explicit solutions of engineering problems it is necessary to choose an appropriate coordinate system. In fact, the choice of coordinates should depict the geometry of the problem so that the corresponding mathematical formulation simplifies as much as possible. How to rewrite the equations of continuum theory in arbitrary coordinates is subject of *tensor calculus*, which is usually no part of the mathematics syllabus in engineering education. For this reason we start in this chapter with *tensor algebra* and the representation of tensors in arbitrary coordinate systems, before we elaborate on continuum theory any further. As the emphasis in this book is on practical mathematics, the preferred way of writing tensors will be in index notation. However, many textbooks and the scientific literature use symbolic notation instead. Therefore we will always present both, although at the expense of mathematical stringency.

*Capt. Terrell: Chekov, are you sure these are the correct coordinates?*
*Chekov: Captain, this is the garden spot of Ceti Alpha Six!*

Star Trek II, The Wrath of Khan

## 2.1 A Personal Remark

Typically scientists working in continuum theory split into two extremely hostile ideological factions: the supporters of symbolic tensor calculus and the friends of index notation. The former emphasize the absolute character laws of nature and constitutive equations should have. In other words these laws should be stated independently of an observer and, therefore, independently of a coordinate system. As to whether this is possible, at least in principle, or, in other words, as to whether the tensorial relations that force the laws of nature and constitutive relations into a

W. H. Müller, *An Expedition to Continuum Theory*, Solid Mechanics and Its Applications 210, DOI: 10.1007/978-94-007-7799-6_2,

mathematical form are correct, is a highly philosophical question which, in the end, can only be decided experimentally. However, for this purpose measurements in space and time need to be performed (by an observer). This is exactly what is implicitly emphasized by index notation. Certainly symbolic notation is aesthetically pleasing. This becomes immediately visible if it is juxtaposed to the cumbersome form of the index calculus. However, if the objective is to solve concrete engineering problems the symbolic way of writing is only of limited use. The situation is similar to the world of fashion: A tuxedo with a top hat may be appropriate at the Met but it is hopelessly unsuitable for gardening.

Indeed, the engineer-to-be should be capable to read the pertinent technical literature as well as use the results therein to perform further calculations. Therefore we shall present both lines of reasoning in this book and learn to appreciate the corresponding advantages—but, hopefully, without turning into ideologists.

## 2.2 First Definitions and Notions in Index Notation

We will initially start from an index point-of-view that circumvents the notion of a (unit) vector base. In this spirit we consider a three-dimensional *Cartesian* coordinate grid consisting of three straight lines, $x_1$, $x_2$, $x_3$, which are orthogonal to each other. For brevity we shall denote them by $x_k$ and note that a Latin case index runs from 1 to 3. Moreover, we consider other three-dimensional coordinate lines which may be curvilinear and denote them by $z^i$. Clearly points in space should be identifiable by either set of lines. In other words invertible relations of the following kind must hold:

$$z^i = z^i(x_1, x_2, x_3) \equiv z^i(x_k) \quad \text{and} \quad x_k = x_k\left(z^1, z^2, z^3\right) \equiv x_k\left(z^i\right). \tag{2.2.1}$$

Mathematically speaking such invertible relations are also known as *isomorphisms*. The distance $s$ between two points (1) and (2) with the corresponding Cartesian coordinates $\overset{(1)}{x_i}$ and $\overset{(2)}{x_i}$ can easily be calculated following PYTHAGORAS:

$$s = \sqrt{\left(\overset{(1)}{x_1} - \overset{(2)}{x_1}\right)^2 + \left(\overset{(1)}{x_2} - \overset{(2)}{x_2}\right)^2 + \left(\overset{(1)}{x_3} - \overset{(2)}{x_3}\right)^2}. \tag{2.2.2}$$

PYTHAGORAS of Samos lived around 580–500 BC. He was a versatile man engaged in philosophy, mysticism, mathematics, astronomy, music, healing arts, wrestling, and politics. In 532 BC he leaves Samos, flees from the local tyrant, and moves to southern Italy. In Croton he founds his famous philosophical-religious school: the so-called Pythagoreans. Surely he was not the first to know about the Pythagorean theorem but, maybe, he was one of the first interested in its proof.

Note that, in general, the corresponding formula does not hold in curvilinear coordinates, $z^i$:

$$s \neq \sqrt{\left(\overset{(1)}{z^1} - \overset{(2)}{z^1}\right)^2 + \left(\overset{(1)}{z^2} - \overset{(2)}{z^2}\right)^2 + \left(\overset{(1)}{z^3} - \overset{(2)}{z^3}\right)^2}. \tag{2.2.3}$$

However, if the coordinate points (1) and (2) are infinitesimally close, it becomes possible to derive a relation similar to Eq. (2.2.2). We first define:

$$\mathrm{d}x_i = \overset{(1)}{x_i} - \overset{(2)}{x_i} \quad \text{or} \quad \mathrm{d}z^k = \overset{(1)}{z^k} - \overset{(2)}{z^k} \tag{2.2.4}$$

and obtain for the total differential by using Eq. $(2.2.1)_2$:

$$\mathrm{d}x_i = \frac{\partial x_i}{\partial z^1}\mathrm{d}z^1 + \frac{\partial x_i}{\partial z^2}\mathrm{d}z^2 + \frac{\partial x_i}{\partial z^3}\mathrm{d}z^3. \tag{2.2.5}$$

This can be expressed in shorthand notation:

$$\mathrm{d}x_i = \frac{\partial x_i}{\partial z^k}\mathrm{d}z^k, \tag{2.2.6}$$

if we agree *to sum up automatically from 1 to 3 (or up to 2 for planar problems) whenever an index appears twice in a product.* In the present case this concerns the index $k$. This is known as *EINSTEIN's summation rule in the literature.* We also refer to *$k$ as a bound index* in contrast to *free indices*, i.e., those that do not appear twice (in the present case the letter "$i$"). The infinitesimal distance d$s$ between the two infinitesimally close points (1) and (2) can now be calculated if we insert Eqs. (2.2.4) and (2.2.6) into (2.2.2):

$$\mathrm{d}s = \sqrt{\mathrm{d}x_i\mathrm{d}x_i} = \sqrt{\frac{\partial x_i}{\partial z^k}\frac{\partial x_i}{\partial z^j}\mathrm{d}z^k\mathrm{d}z^j}. \tag{2.2.7}$$

Albert EINSTEIN was born on March 14, 1879 in Ulm (South Germany) and died on April 18th, 1955 in Princeton. He is certainly the most eminent scientist of the twentieth century. Similar to NEWTON or MAXWELL he enriched physics by many fundamental discoveries from different fields. The development of General Relativity and the tensor calculus that was used therein for describing space–time is probably his most popular contribution. However, this did not win him the NOBEL price. On the contrary: This award was given to him for something much less "obscure," namely for his interpretation of the photo-electric effect.

**Exercise 2.2.1:** ***Line element***

Go through the proof of Eq. (2.2.7). By doing so learn to understand the meaning of the different indices by writing down all terms according to EINSTEIN's summation rule. Realize that a double summation occurs in Eq. (2.2.7).

Obviously it is "almost" possible to write the infinitesimal distance in terms of products in $\mathrm{d}z^k$. Well, almost, since Eq. (2.2.7) teaches us that one has to multiply these products by derivatives of the Cartesian coordinates $x_i$ w.r.t. the curvilinear coordinates $z^k$. These derivatives have a special name. They are known as the (covariant) components of the *metric tensor* **g** and defined as follows:

$$g_{kj} = \frac{\partial x_i}{\partial z^k}\frac{\partial x_i}{\partial z^j}. \tag{2.2.8}$$

The word *metric* stems from the Greek word μετρώ for measuring. This makes immediate sense because this quantity renders it possible to determine distances, cf., Eq. (2.2.7), provided the differences of coordinates $\mathrm{d}z^k$ are known. We now rewrite Eq. (2.2.7) in the following compact form:

$$\mathrm{d}s = \sqrt{g_{kj}\mathrm{d}z^k\mathrm{d}z^j}. \tag{2.2.9}$$

Note that in order to determine $g_{kj}$ for a particular coordinate transformation it is sufficient to determine only "half of it," for example, for the sequence of indices $(k, j)$ = (1,1), (1,2), (1,3), (2,2), (2,3), (3,3). This is due to the fact that the metric tensor is symmetric:

$$g_{kj} = \frac{\partial x_i}{\partial z^k}\frac{\partial x_i}{\partial z^j} = \frac{\partial x_i}{\partial z^j}\frac{\partial x_i}{\partial z^k} = g_{jk}. \tag{2.2.10}$$

As a typical example for curvilinear coordinates, the corresponding metric tensor, and the line element we consider the case of cylindrical coordinates: Here a point in space is characterized by its radial distance, $r$, the polar angle, $\vartheta$, and the height, $z$:$(r, \vartheta, z)$, $r \in [0,\infty)$, $\vartheta \in [0,\pi)$, $z \in (-\infty,+\infty)$. Just like Cartesian coordinate lines cylindrical ones are orthogonal to each other. However, in contrast to those not all of them are "straight." In particular the lines describing a constant radial distance are of circular shape (cf., Fig. 2.1). The following relations between Cartesian and cylindrical coordinates follow by simple geometric considerations from Fig. 2.1:

$$\begin{aligned} z^1 &\equiv r = \sqrt{x_1^2 + x_2^2}, & x_1 &= r\cos\vartheta = z^1\cos z^2, \\ z^2 &\equiv \vartheta = \arctan\frac{x_2}{x_1}, & x_2 &= r\sin\vartheta = z^1\sin z^2, \\ z^3 &= z = x_3, & x_3 &= z = z^3. \end{aligned} \tag{2.2.11}$$

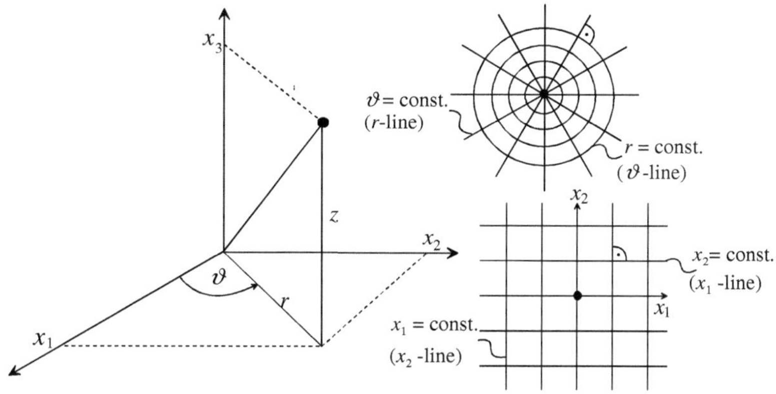


**Fig. 2.1** Cylindrical coordinates

This in turn yields (for example):

$$\begin{aligned} g_{11} &= \frac{\partial x_1}{\partial z^1}\frac{\partial x_1}{\partial z^1} + \frac{\partial x_2}{\partial z^1}\frac{\partial x_2}{\partial z^1} + \frac{\partial x_3}{\partial z^1}\frac{\partial x_3}{\partial z^1} = \cos^2 z^2 + \sin^2 z^2 + 0 = 1, \\ g_{12} &= \frac{\partial x_1}{\partial z^1}\frac{\partial x_1}{\partial z^2} + \frac{\partial x_2}{\partial z^1}\frac{\partial x_2}{\partial z^2} + \frac{\partial x_3}{\partial z^1}\frac{\partial x_3}{\partial z^2} \\ &= -z^1 \sin z^2 \cos z^2 + z^1 \sin z^2 \cos z^2 = 0. \end{aligned} \tag{2.2.12}$$

and in the same manner we show that:

$$g_{kj} = \begin{pmatrix} 1 & 0 & 0 \\ 0 & r^2 & 0 \\ 0 & 0 & 1 \end{pmatrix}. \tag{2.2.13}$$

By means of Eqs. (2.2.9/2.2.13) it becomes now easily possible to obtain the line element:

$$\begin{aligned} \mathrm{d}s &= \sqrt{g_{kj}\mathrm{d}z^k\mathrm{d}z^j} \\ &= \sqrt{(\,\mathrm{d}r \quad \mathrm{d}\vartheta \quad \mathrm{d}z\,) \cdot \begin{pmatrix} 1 & 0 & 0 \\ 0 & r^2 & 0 \\ 0 & 0 & 1 \end{pmatrix} \cdot \begin{pmatrix} \mathrm{d}r \\ \mathrm{d}\vartheta \\ \mathrm{d}z \end{pmatrix}} = \sqrt{\mathrm{d}r^2 + r^2\mathrm{d}\vartheta^2 + \mathrm{d}z^2}. \end{aligned} \tag{2.2.14}$$

In this context we have used the rules of matrix multiplication. Note that this was done for practical reasons and not because of necessity. The same result can be obtained by "expanding" the double sum, i.e., by writing down each term and suitable combination. The latter is always possible, even if sums of order higher than two are concerned, i.e., triple summations, etc. However, in general, the beautiful matrix notation is then no longer possible.

**Exercise 2.2.2:** ***Metric tensor for spherical coordinates***

Use geometric arguments in context with Fig. 1.3 to show that the following relations hold between Cartesian and spherical coordinates, $(x_1, x_2, x_3)$ and $(r, \vartheta, \varphi)$, $r \in [0, \infty)$, $\vartheta \in [0, \pi)$, $\varphi \in [0, 2\pi)$, respectively:

$$\begin{aligned}
&z^1 \equiv r = \sqrt{x_1^2 + x_2^2 + x_3^2}, \quad x_1 = r \sin\vartheta \cos\varphi = z^1 \sin z^2 \cos z^3, \\
&z^2 \equiv \vartheta = \arccos \frac{x_3}{\sqrt{x_1^2 + x_2^2 + x_3^2}}, \quad x_2 = r \sin\vartheta \sin\varphi = z^1 \sin z^2 \sin z^3, \\
&z^3 \equiv \varphi = \arctan \frac{x_2}{x_1}, \quad x_3 = r \cos\vartheta = z^1 \cos z^2.
\end{aligned} \tag{2.2.15}$$

Discuss the shape of coordinate lines of a constant radial distance, azimuthal as well as polar angle. Show that these lines are perpendicular to each other. Moreover, show that the components of the metric tensor read:

$$g_{kj} = \begin{pmatrix} 1 & 0 & 0 \\ 0 & r^2 & 0 \\ 0 & 0 & r^2 \sin^2\vartheta \end{pmatrix}. \tag{2.2.16}$$

As a trivial example for an application of the formula for the line element (2.2.9) we calculate the circumferential length $U$ of a circle $C_R$ of radius $R$. In this case we have $\mathrm{d}r = 0$ and $\mathrm{d}z = 0$ and obtain:

$$U = \oint_{C_R} \mathrm{d}s = \int_0^{2\pi} R\mathrm{d}\vartheta = 2\pi R. \tag{2.2.17}$$

As a more complex example for Eq. (2.2.14) we consider the situation shown in Fig. 2.2: The objective is to determine the length $L$ of the diagonal in a rectangle of height $H$ and width $2\pi R$. Obviously this can be obtained by using the Pythagorean theorem:

$$L = \sqrt{H^2 + (2\pi R)^2}, \tag{2.2.18}$$

which has nothing to do with Eq. (2.2.14). However, now transform the rectangle into a three-dimensional object, namely the mantle of a cylinder, as shown in Fig. 2.2. This way the former diagonal is also transformed into a three-dimensional curve.

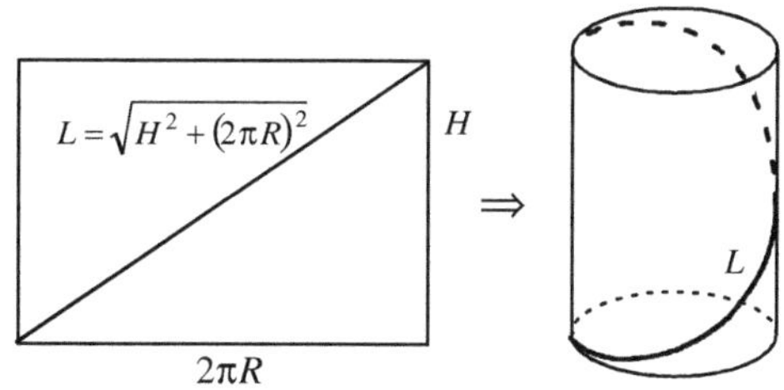


**Fig. 2.2** Transformation of a straight line in the plane into a three-dimensional curve

It makes sense to calculate the length of the curve with cylindrical coordinates. We first note that the radial distance of the curve on the mantle does not change:

$$r = R = \text{const.} \tag{2.2.19}$$

Consequently Eq. (2.2.14) becomes:

$$\mathrm{d}r = 0 \Rightarrow \mathrm{d}s = \sqrt{R^2\mathrm{d}\vartheta^2 + \mathrm{d}z^2}. \tag{2.2.20}$$

Now we assume that the height $z$ changes linearly with the polar angle $\vartheta$:

$$z = A\vartheta + B. \tag{2.2.21}$$

Of course we have:

$$z(\vartheta = 0) = 0 \quad \text{and} \quad z(\vartheta = 2\pi) = H, \tag{2.2.22}$$

and therefore the constants $A$ and $B$ become:

$$A = \frac{H}{2\pi} \quad \text{and} \quad B = 0. \tag{2.2.23}$$

If we insert this in Eq. (2.2.20) we obtain:

$$\mathrm{d}z = \frac{H}{2\pi}\mathrm{d}\vartheta \Rightarrow s = \int\limits_{\vartheta=0}^{\vartheta=2\pi} \sqrt{R^2 + \left(\frac{H}{2\pi}\right)^2}\,\mathrm{d}\vartheta = \sqrt{R^2 + \left(\frac{H}{2\pi}\right)^2}\,2\pi, \tag{2.2.24}$$

in other words, a result identical to Eq. (2.2.18).

**Exercise 2.2.3:** ***Line element in spherical coordinates***

Show with the result for the metric tensor from Exercise 2.2.2 that the line element d$s$ in spherical coordinates is given by:

$$\mathrm{d}s = \sqrt{\mathrm{d}r^2 + r^2\sin^2\vartheta\mathrm{d}\varphi^2 + r^2\mathrm{d}\vartheta^2}. \tag{2.2.25}$$

Use the equation to show that the length of the equatorial circumference as well as the length of any great circle of a sphere of radius $R$ is given by:

$$U = 2\pi R. \tag{2.2.26}$$

**Exercise 2.2.4: *Decoration***

A producer of hats for the Loge of the Metric Tensor Fetishists faces the problem of providing enough string for the decoration of length $L$ on a piece of felt of quarter-circle-shape. The felt is then rolled into a fashionable cone hat as shown in Fig. 2.3. By doing so the radial distance $r$ of the string grows proportionally with the polar angle $\vartheta$ starting from zero to the radius $R$ of the quarter circle.

Calculate the length $L$ of the string

(a) in plane polar coordinates (Fig. 2.3, left)

(b) in three-dimensional cylindrical coordinates (Fig. 2.3, right) and show that in both cases $L \approx 1.324R$.

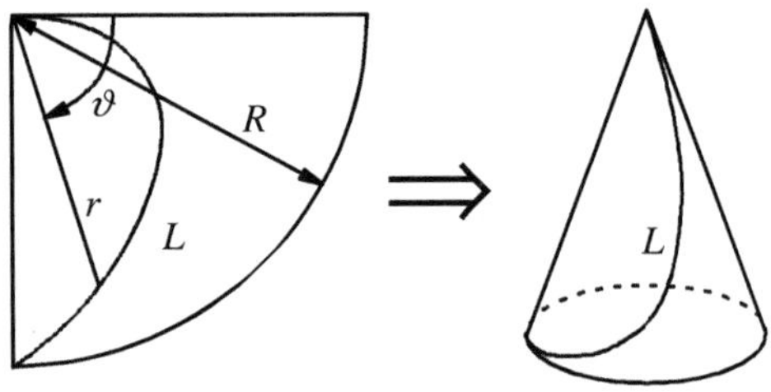


**Fig. 2.3** Making of a decoration

Clearly the same length must result, independently of the choice of coordinates. Discuss the pros and cons of both methods, also by comparison with the previously discussed problem of the diagonal.

We summarize what we have learned so far: Cartesian coordinates are characterized by straight lines and form a square-like mesh in the plane and a cube-like mesh in three dimensions. Polar coordinates (or in other words cylindrical coordinates of the plane), however, form a spider-like web such that the lines of constant radius are given by concentric circles around the origin, whereas the lines of constant polar angle are straight and run through the origin. So one set of lines is curved, but both sets are *perpendicular* to each other, just like the Cartesian case. However, in general, coordinate meshes do not have to intersect at a 90° angle. An example is shown in Fig. 2.4: Next to the traditional Cartesian system denoted by $x$ a scissors-like coordinate system is drawn and denoted by $z$. It obviously takes two angles, $\alpha$ and $\beta$, in order to characterize the orientation of the $z$ system. In Exercise 2.3.1 it will be shown that the corresponding metric tensor is *not* diagonal unlike the previous cases of polar, cylindrical, and spherical coordinates. This is due to the fact that these were *orthogonal* coordinate systems whereas the scissors-like system is not. In fact, we shall see shortly that components of the metric tensor can be interpreted as scalar products between vectors pointing in the direction of the coordinate lines. If these are orthogonal to each other their scalar products must vanish.

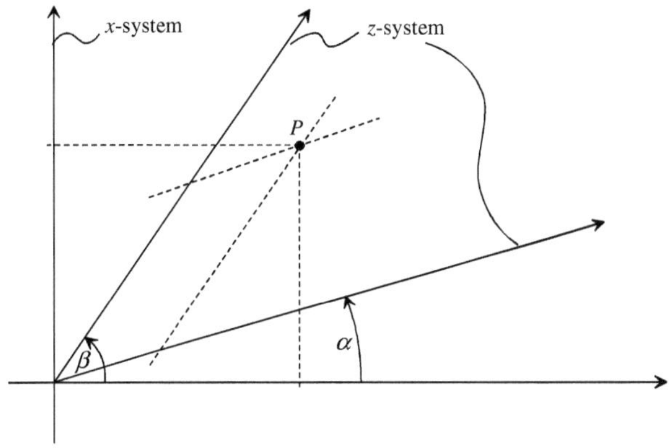


**Fig. 2.4** Skew coordinate system

## 2.3 Vector Interpretation of the Metric

We now return to the problem of calculating the distance between two infinitesimally close points, which was already investigated component-wise in Eq. (2.2.6). In the absolute language of vectors we denote the infinitesimal distance between both points by $\mathrm{d}\boldsymbol{x}$. Moreover, $\boldsymbol{e}_1, \boldsymbol{e}_2$, and $\boldsymbol{e}_3$ (or short $\boldsymbol{e}_i,\ i = 1,\ 2,\ 3$) stand for three Cartesian unit vectors satisfying so-called conditions of *orthonormality*:

$$\boldsymbol{e}_i \cdot \boldsymbol{e}_j = \delta_{ij}, \tag{2.3.1}$$

where the so-called *Kronecker symbol* a.k.a. *unit tensor* has been used:

$$\delta_{ij} = \begin{cases} 1, & \text{if } i = j \\ 0, & \text{if } i \neq j. \end{cases} \tag{2.3.2}$$

If we now observe the relations (2.2.1) between the coordinates $x_i$ and $z^k$ we may write:

$$\mathrm{d}\boldsymbol{x} = \mathrm{d}x_i \boldsymbol{e}_i = \frac{\partial x_i}{\partial z^k} \mathrm{d}z^k \boldsymbol{e}_i, \tag{2.3.3}$$

where Eq. (2.2.6) has been used again and Einstein's summation rule has been extended to expressions related to coordinate lines and vectors. Reshuffling terms in (2.3.3) yields:

$$\mathrm{d}\boldsymbol{x} = \mathrm{d}z^k \boldsymbol{g}_k, \quad \boldsymbol{g}_k = \frac{\partial x_i}{\partial z^k} \boldsymbol{e}_i. \tag{2.3.4}$$

Leopold Kronecker was born on December 7, 1823 in Liegnitz (Silesia) and died on December 29, 1891 in Berlin. In the spring of 1841 Kronecker starts to study mathematics at the University of Berlin. There he attends lectures by the famous mathematicians Dirichlet, Jacobi, and Steiner. However, he is also interested in philosophy and participates in the lectures offered by Schelling. Another famous mathematician, Ernst Eduard Kummer, becomes his teacher and mentor. It is therefore not too surprising that Kronecker inherits Kummer's chair in 1883. It is curious to note that during a mathematical debate on the *infinite* he insisted on proofs of related theorems within a *finite* number of steps, i.e., by avoiding the concept of quantities of size epsilon. This is why David Hilbert referred to him as *Verbotsdiktator* (forbidding dictator). This name, however, would also suit to characterize the effect of the symbol named after him.

Figure 2.5 illustrates for the planar case (which is simpler to draw) how we have to interpret the new defined vectors $\boldsymbol{g}_k$. The figure shows how a tangent vector to the line $z^1 = \text{const.}$, namely $\boldsymbol{g}_2$, is obtained:

$$\begin{aligned} \mathrm{d}\boldsymbol{x} &= \lim_{\Delta\boldsymbol{x}\to 0} \Delta\boldsymbol{x} = \lim_{\Delta x_i \to 0} \Delta x_i \boldsymbol{e}_i \\ &= \lim_{\Delta\boldsymbol{x}\to 0} \frac{x_i(z^1, z^2 + \Delta z^2, z^3) - x_i(z^1, z^2, z^3)}{\Delta z^2} \Delta z^2 \boldsymbol{e}_i = \frac{\partial x_i}{\partial z^2} \boldsymbol{e}_i \mathrm{d} z^2. \end{aligned} \tag{2.3.5}$$

Consequently, for this case we have:

$$\frac{\mathrm{d}\boldsymbol{x}}{\mathrm{d}z^2} = \frac{\partial x_i}{\partial z^2} \boldsymbol{e}_i \equiv \boldsymbol{g}_2. \tag{2.3.6}$$

It is slightly irritating that $\boldsymbol{g}_2$ (and not $\boldsymbol{g}_1$) denotes the tangent vector to the lines $z^1 = \text{const.}$ and *vice versa*. A solution to this puzzle is offered in Fig. 2.1: In the Cartesian case lines for which $x_1 = \text{const.}$ holds (say) are parallel to the abscissa $x_2$ and we also call them $x_2$-lines for short. Consequently, (see again Fig. 2.1) lines for which $r = \text{const.}$ are $\vartheta$-lines and *vice versa*. However, words are always vague and confusing. What is important, though, is the equation that tells us how to calculate the new vector base $\boldsymbol{g}_k$, i.e., Eq. $(2.3.4)_2$. The terminology used for these vectors is, in the end, arbitrary.

**Fig. 2.5** Tangent vectors to the lines of skew/curvilinear coordinates

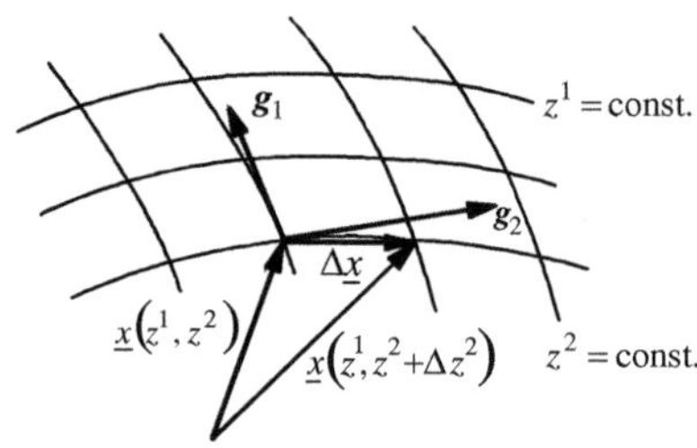

For the case of three dimensions all arguments hold analogously. Therefore we may say that, in general, $\boldsymbol{g}_k$ denotes tangent vectors to the lines $z^j = \text{const.}$ However, note that these are *not* necessarily unit vectors. We now use them to calculate the following scalar product and rearrange terms slightly:

$$\begin{aligned} \boldsymbol{g}_k \cdot \boldsymbol{g}_l &= \left(\frac{\partial x_i}{\partial z^k}\boldsymbol{e}_i\right) \cdot \left(\frac{\partial x_j}{\partial z^l}\boldsymbol{e}_j\right) = \frac{\partial x_i}{\partial z^k}\frac{\partial x_j}{\partial z^l}\boldsymbol{e}_i \cdot \boldsymbol{e}_j \\ &= \frac{\partial x_i}{\partial z^k}\frac{\partial x_j}{\partial z^l}\delta_{ij} = \frac{\partial x_i}{\partial z^k}\frac{\partial x_i}{\partial z^l} \equiv g_{kl}, \end{aligned} \tag{2.3.7}$$

where the Kronecker symbol of Eq. (2.3.2) has been used. Note that the effect of the Kronecker symbol consists of replacing one of its bound indices in the remaining expression of a product with its other index. In order to prove this statement all sums must be expanded first. Then all of the occurring terms can be simplified by observing Eq. (2.3.2).

We conclude that the scalar product between the two tangent vectors yields the components of the metric tensor. The basic definition of the scalar product of two vectors involves the cosine of the angle they enclose. Therefore non-diagonal components of the metric must vanish, if the curvilinear coordinates are *orthogonal* as, for example, in the case of cylindrical or spherical transformations. Consequently, their explicit calculation is unnecessary, albeit possible, as for example demonstrated in Eq. $(2.2.12)_2$.

As a specific example we consider the case of plane polar coordinates for which the tangent vectors to the coordinate lines can be calculated explicitly. We use Eqs. (2.2.11) in context with $(2.3.4)_2$ to obtain:

$$\begin{aligned} \boldsymbol{g}_1 &= \frac{\partial x_1}{\partial z^1}\boldsymbol{e}_1 + \frac{\partial x_2}{\partial z^1}\boldsymbol{e}_2 = \cos\vartheta\,\boldsymbol{e}_1 + \sin\vartheta\,\boldsymbol{e}_2 \equiv \boldsymbol{e}_r, \\ \boldsymbol{g}_2 &= \frac{\partial x_1}{\partial z^2}\boldsymbol{e}_1 + \frac{\partial x_2}{\partial z^2}\boldsymbol{e}_2 = -r\sin\vartheta\,\boldsymbol{e}_1 + r\cos\vartheta\,\boldsymbol{e}_2 \equiv r\boldsymbol{e}_\vartheta. \end{aligned} \tag{2.3.8}$$

In this equation we have introduced the commonly used unit vectors $\boldsymbol{e}_r$ and $\boldsymbol{e}_\vartheta$ of polar coordinates. They are shown in Fig. 2.6. In particular, the second chain of equations shows that the tangent vectors $\boldsymbol{g}_i$ do not necessarily need to be normalized.

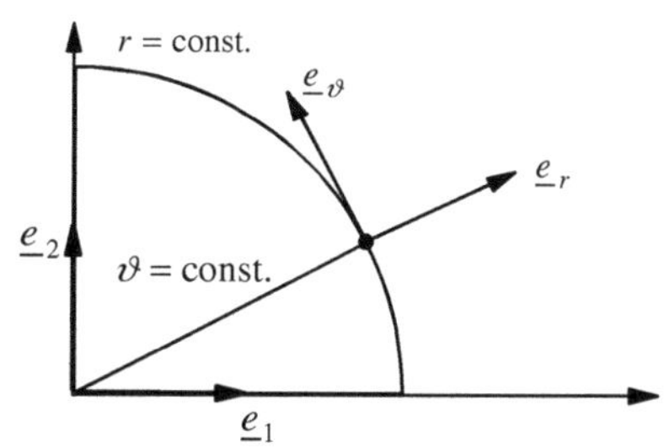


**Fig. 2.6** Plane polar coordinates and the corresponding (unit) tangent vectors

**Exercise 2.3.1:** ***Metric of a plane skew coordinate system***

Use geometric arguments to show that for the scissor-like coordinates of Fig. 2.4 the following relations hold:

$$x_1 = \cos\alpha\, z^1 + \cos\beta\, z^2,\; x_2 = \sin\alpha\, z^1 + \sin\beta\, z^2, \tag{2.3.9}$$

and the metric tensor reads:

$$g_{kj} = \begin{pmatrix} 1 & \cos(\alpha-\beta) \\ \cos(\alpha-\beta) & 1 \end{pmatrix}. \tag{2.3.10}$$

Discuss and interpret the case $\alpha - \beta = \pi/2$. Show that the tangent vectors are given by:

$$\boldsymbol{g}_1 = \cos\alpha\,\boldsymbol{e}_1 + \sin\alpha\,\boldsymbol{e}_2,\; \boldsymbol{g}_2 = \cos\beta\,\boldsymbol{e}_1 + \sin\beta\,\boldsymbol{e}_2. \tag{2.3.11}$$

Now use this result in context with the interpretation of metric components as scalar products and rederive Eq. (2.3.10).

**Exercise 2.3.2:** ***Metric for spherical coordinates in vector notation***

Recall the equations for spherical coordinates discussed in Exercise 2.2.2 and use them in context with Eq. $(2.3.4)_2$ to show that the tangent vectors read:

$$\begin{aligned} \boldsymbol{g}_1 &= \sin\vartheta\,\cos\varphi\,\boldsymbol{e}_1 + \sin\vartheta\,\sin\varphi\,\boldsymbol{e}_2 + \cos\vartheta\,\boldsymbol{e}_3, \\ \boldsymbol{g}_2 &= r\cos\vartheta\,\cos\varphi\,\boldsymbol{e}_1 + r\cos\vartheta\,\sin\varphi\,\boldsymbol{e}_2 - r\sin\vartheta\,\boldsymbol{e}_3, \\ \boldsymbol{g}_3 &= -r\sin\vartheta\,\sin\varphi\,\boldsymbol{e}_1 + r\sin\vartheta\,\cos\varphi\,\boldsymbol{e}_2. \end{aligned} \tag{2.3.12}$$

Use these results to reconfirm the expression for the metric tensor shown in Eq. (2.2.16). Also confirm that the tangent vectors can be linked to the unit vectors $\boldsymbol{e}_r$, $\boldsymbol{e}_\vartheta$, and $\boldsymbol{e}_\varphi$ of Fig. 1.3 as follows:

$$\boldsymbol{g}_1 = \boldsymbol{e}_r, \quad \boldsymbol{g}_2 = r\,\boldsymbol{e}_\vartheta, \quad \boldsymbol{g}_3 = r\sin\vartheta\,\boldsymbol{e}_\varphi. \tag{2.3.13}$$

**Exercise 2.3.3: *Elliptic coordinates***

The following relations hold between the so-called elliptic coordinates $z^1$, $z^2$ and planar Cartesian coordinates $x_1, x_2$:

$$x_1 = c \cosh z^1 \cos z^2, \ x_2 = c \sinh z^1 \sin z^2 , \\ z^1 \in [0, \infty), \ z^2 \in [0, 2\pi) . \tag{2.3.14}$$

Rearrange these equations to show that of constant values of $z^1$ and $z^2$ can be interpreted as confocal ellipses and hyperbolae. Identify both axes of the ellipses in terms of the parameter $c$ shown in Eq. (2.3.14). Discuss the limit for which the ellipse degenerates into a crack. Determine the length of the crack. Show that the metric is given by:

$$g_{ij} = \frac{c^2}{2} \begin{pmatrix} \cosh(2z^1) - \cos(2z^2) & 0 \\ 0 & \cosh(2z^1) - \cos(2z^2) \end{pmatrix}, \tag{2.3.15}$$

and confirm that the tangential vectors can be written as:

$$\boldsymbol{g}_1 = c \sinh z^1 \cos z^2 \boldsymbol{e}_1 + c \cosh z^1 \sin z^2 \boldsymbol{e}_2, \\ \boldsymbol{g}_2 = -c \cosh z^1 \sin z^2 \boldsymbol{e}_1 + c \sinh z^1 \cos z^2 \boldsymbol{e}_2. \tag{2.3.16}$$

Use the interpretation of metric coefficients as scalar products to rederive Eq. (2.3.15).

## 2.4 Co- and Contravariant Components

In this section we introduce the notions of *co-* and *contravariant* components which are important in context with the representation of vectors and tensors in arbitrary curvilinear coordinate systems. For this purpose we consider the situation shown in Fig. 2.7: The vector $\boldsymbol{A}$ is first decomposed w.r.t. a Cartesian coordinate system called $x$. In this frame it is characterized by the components $\underset{(x)}{A_1}$ and $\underset{(x)}{A_2}$. In order to point out that these are components w.r.t. a Cartesian frame they carry the additional suffix "($x$)."

Besides the Cartesian system a skew coordinate system called $z$ is depicted. The vector $\boldsymbol{A}$ can be decomposed in two ways w.r.t. that system. One possibility is to project the vector *parallel* to the coordinate lines $z^1$ and $z^2$: Fig. 2.7 (right). This way we obtain the components $\underset{(z)}{A^1}$ and $\underset{(z)}{A^2}$. They are components w.r.t. the skew coordinate system, and this is emphasized by a suffix "($z$)." They are also known as *contravariant* components or, in other words, we speak of the contravariant representation of the vector $\boldsymbol{A}$ in the $z$-system. This way of representation is characterized by *upper indices* at the vector symbol. Indeed, without knowing, we have already used this notation in context with the coordinate lines $z^i$ from Sect. 2.2.

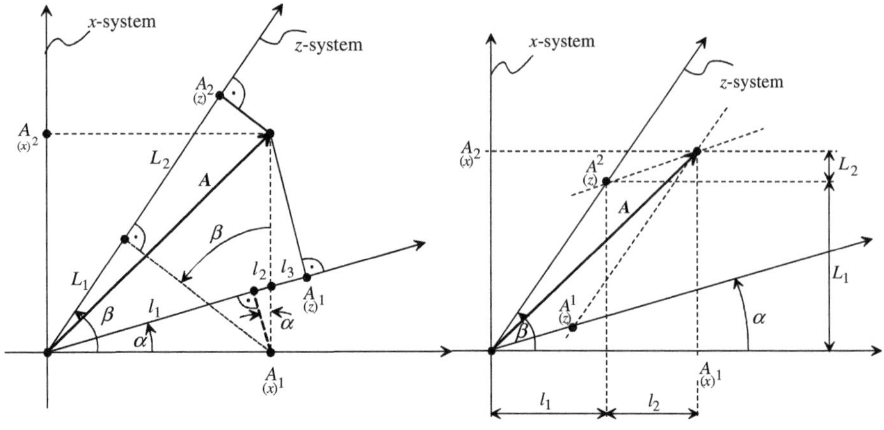


**Fig. 2.7** Co- and contravariant components

However, there is a second possibility of how to represent vectors in skew curvilinear coordinate systems: The vector $\boldsymbol{A}$ can also be projected *perpendicularly* to the $z$-axes. This is indicated in Fig. 2.7 (left). In this manner we obtain $\underset{(z)}{A_1}$ and $\underset{(z)}{A_2}$, which are called *covariant* components and identified by lower indices.

Note that in Cartesian coordinate systems, i.e., in the $x$-system, it is impossible to distinguish co- and contravariant components. It is for that reason that we have used $x_i$ for the components of the position vector above. However, the notation $x^i$ would be equally appropriate.

Next we shall prove a formula that allows us to transform a certain set of coordinates—co-, contravariant, or Cartesian—into another one. The proof will be illustrated for the special case of two dimensions. However, the formula also holds for the 3D-case: The components $\underset{(x)}{A_i}$ of a vector $\boldsymbol{A}$ w.r.t. a Cartesian coordinate system, $x$, can be converted into co- and contravariant components, $\underset{(z)}{A_i}$ or $\underset{(z)}{A^i}$, w.r.t. a skew or curvilinear coordinate system, $z$, by differentiation of the coordinate transformation from Eq. (2.2.1):

$$\underset{(z)}{A^i} = \frac{\partial z^i}{\partial x_k} \underset{(x)}{A_k}, \quad \underset{(z)}{A_i} = \frac{\partial x_k}{\partial z^i} \underset{(x)}{A_k}. \tag{2.4.1}$$

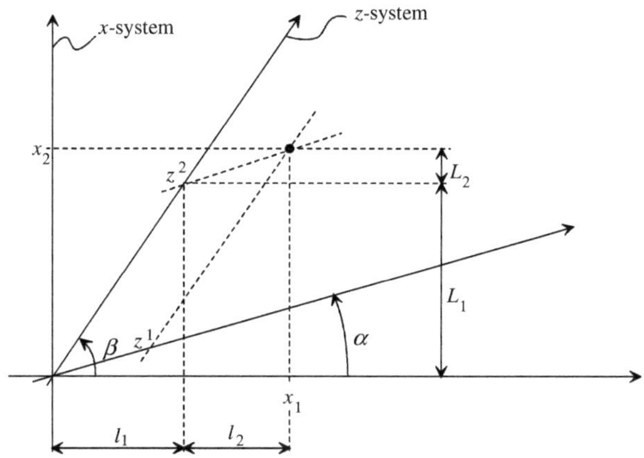


**Fig. 2.8** A position represented in a Cartesian and a skew coordinate system

For the proof we consider the systems $x$ and $z$ in Fig. 2.8 and conclude that the coordinate transformation of Eq. (2.2.1) can be written explicitly as (also see Exercise 2.3.1):

$$
\begin{aligned}
x_1 &= \cos\alpha\, z^1 + \cos\beta\, z^2, \quad z^1 = H\,[\sin\beta\, x_1 - \cos\beta\, x_2],\\
x_2 &= \sin\alpha\, z^1 + \sin\beta\, z^2, \quad z^2 = H\,[-\sin\alpha\, x_1 + \cos\alpha\, x_2],\\
H &= [\cos\alpha\,\sin\beta - \sin\alpha\,\cos\beta]^{-1} \equiv [\sin(\beta-\alpha)]^{-1}.
\end{aligned}
\tag{2.4.2}
$$

A closer examination of Fig. 2.7 shows that the contravariant components can be written as:

$$
\underset{(z)}{A^1} = H\left[\sin\beta\,\underset{(x)}{A_1} - \cos\beta\,\underset{(x)}{A_2}\right], \quad \underset{(z)}{A^2} = H\left[-\sin\alpha\,\underset{(x)}{A_1} + \cos\alpha\,\underset{(x)}{A_2}\right],
\tag{2.4.3}
$$

whereas we find for the covariant ones:

$$
\underset{(z)}{A_1} = \cos\alpha\,\underset{(x)}{A_1} + \sin\alpha\,\underset{(x)}{A_2}, \quad \underset{(z)}{A_2} = \cos\beta\,\underset{(x)}{A_1} + \sin\beta\,\underset{(x)}{A_2}\,.
\tag{2.4.4}
$$

By differentiation of Eq. (2.4.2) the validity of Eq. (2.4.1) is easily established. Of course, being a planar problem, the summation runs only from 1 to 2.

**Exercise 2.4.1:** ***Transformation formulae from a plane Cartesian to a skew coordinate system***

Use the auxiliary quantities $l_1$, $l_2$, $l_3$, $L_1$, $L_2$ in Fig. 2.7 in order to confirm Eqs. (2.4.3), (2.4.4), and (2.4.1).

Next we multiply in Eq. (2.4.1) $\underset{(z)}{A^i}$ by the metric $g_{ni}$, observe Eq. (2.2.8) and obtain:

$$g_{ni} \underset{(z)}{A^i} = \frac{\partial x_l}{\partial z^n} \frac{\partial x_l}{\partial z^i} \frac{\partial z^i}{\partial x_k} \underset{(x)}{A_k} = \frac{\partial x_l}{\partial z^n} \delta_{lk} \underset{(x)}{A_k} = \frac{\partial x_k}{\partial z^n} \underset{(x)}{A_k} = \underset{(z)}{A_n} \,. \tag{2.4.5}$$

Here we have used the *chain rule* after the second equality sign (or, figuratively speaking, "cancelled out" $\partial z^i$). This generates a KRONECKER symbol, $\delta_{lk}$, first and then Eq. $(2.4.1)_2$ was observed. Of course, the KRONECKER symbol is nothing else but the unit matrix in component form, i.e., we may write:

$$\frac{\partial x_l}{\partial z^n} \delta_{lk} \underset{(x)}{A_k} = \frac{\partial x_k}{\partial z^n} \underset{(x)}{A_k} \quad \text{or} \quad \delta_{lk} \underset{(x)}{A_k} = \underset{(x)}{A_l}, \tag{2.4.6}$$

which, consequently, transforms the index $l$ in Eq. (2.4.5) into the index $k$, or vice versa. Eq. (2.4.1) was applied once more after the last equality sign of Eq. (2.4.5). This time, however, for the covariant components $\underset{(z)}{A_n}$.

We conclude that by means of the covariant components $g_{lk}$ of the metric tensor it becomes possible to convert the contravariant index $k$ into a covariant one, $l$. This process is also known as *contraction* in textbooks on tensors: Multiplication with the covariant metric components $g_{lk}$ *lowers* the index $k$. However, it is also possible to *raise* indices. To this end we now introduce the inverse to $g_{lk}$ by:

$$g^{lk} = \frac{\partial z^l}{\partial x_p} \frac{\partial z^k}{\partial x_p}, \tag{2.4.7}$$

and may write:

$$\underset{(z)}{A^i} = g^{ij} \underset{(z)}{A_j} \,, \quad \underset{(z)}{A_i} = g_{ij} \underset{(z)}{A^j} \,. \tag{2.4.8}$$

The first equation shows that contraction of an expression with $g^{ij}$ will raise the covariant index $j$ to a contravariant index $i$.

**Exercise 2.4.2: *Contravariant metric components***

Convince yourself by explicit expansion and use of the chain rule that the contravariant components $g^{lk}$ shown in Eq. (2.4.7) truly constitute the inverse of the covariant form $g_{lk}$.

**Exercise 2.4.3: *Transforming covariant into contravariant components***

Follow the arguments of the text and show analogously that covariant components can be converted into contravariant ones according to $(2.4.8)_1$.

In fact, we may manipulate the partial derivatives in Eq. (2.4.1) as if they were fractions, i.e., write:

$$\underset{(x)}{A_k} = \frac{\partial x_k}{\partial z^i} \underset{(z)}{A^i}, \quad \underset{(x)}{A_k} = \frac{\partial z^i}{\partial x_k} \underset{(z)}{A_i} \,. \tag{2.4.9}$$

All of this is a consequence of the chain rule. If, for example, we multiply Eq. $(2.4.1)_1$ by the expression $\partial x_m/\partial z^i$, we obtain:

$$\frac{\partial x_m}{\partial z^i} \underset{(z)}{A^i} = \frac{\partial x_m}{\partial z^i} \frac{\partial z^i}{\partial x_k} \underset{(x)}{A_k} = \frac{\partial x_m}{\partial x_k} \underset{(x)}{A_k} = \delta_{mk} \underset{(x)}{A_k} = \underset{(x)}{A_m}, \tag{2.4.10}$$

i.e., by renaming the indices $m \to k$ exactly Eq. $(2.4.9)_1$. Equation $(2.4.9)_2$ can be validated in an analogous manner.

It was mentioned in context with Eq. (2.2.9) that the metric tensor allows to calculate distances with curvilinear coordinates. We will now show that it can also be used to determine the length of a vector. For this purpose we start from the basic definition for the length of a vector, namely with the scalar product. If the vector is represented by Cartesian components we may write:

$$\begin{aligned} A = \sqrt{\boldsymbol{A} \cdot \boldsymbol{A}} &= \sqrt{(\underset{(x)}{A_i} \boldsymbol{e}_i) \cdot (\underset{(x)}{A_j} \boldsymbol{e}_j)} = \sqrt{\underset{(x)}{A_i} \underset{(x)}{A_j} \boldsymbol{e}_i \cdot \boldsymbol{e}_j} \\ &= \sqrt{\underset{(x)}{A_i} \underset{(x)}{A_j} \delta_{ij}} = \sqrt{\underset{(x)}{A_i} \underset{(x)}{A_i}}, \end{aligned} \tag{2.4.11}$$

where use was made of Eqs. (2.3.1/2.3.2) and the properties of the Kronecker symbol. By observing Eq. $(2.4.9)_1$ and the basic definition of the metric tensor (2.2.8) this yields:

$$A = \sqrt{\frac{\partial x_i}{\partial z^k} \underset{(z)}{A^k} \frac{\partial x_i}{\partial z^l} \underset{(z)}{A^l}} = \sqrt{\frac{\partial x_i}{\partial z^k} \frac{\partial x_i}{\partial z^l} \underset{(z)}{A^k} \underset{(z)}{A^l}} = \sqrt{g_{kl} \underset{(z)}{A^k} \underset{(z)}{A^l}}, \tag{2.4.12}$$

or by $(2.4.9)_2$ and Eq. (2.4.7) for the inverse metric:

$$A = \sqrt{\frac{\partial z^k}{\partial x_i} \underset{(z)}{A_k} \frac{\partial z^l}{\partial x_i} \underset{(z)}{A_l}} = \sqrt{\frac{\partial z^k}{\partial x_i} \frac{\partial z^l}{\partial x_i} \underset{(z)}{A_k} \underset{(z)}{A_l}} = \sqrt{g^{kl} \underset{(z)}{A_k} \underset{(z)}{A_l}}. \tag{2.4.13}$$

Note that bound indices in the sense of the summation convention can be used only once (consequently we have to distinguish between $k$ and $l$). Moreover, the index calculus allows us to check easily if the summation convention has been applied correctly: Bound indices in a tensor equation always have to appear in pairs, i.e., one of them is covariant and the other one is contravariant (see, for example, the index $k$ in $g^{kl}$ connecting to $\underset{(z)}{A_k}$). This property becomes also evident in the following third alternative equation for the length of a vector:

$$\begin{aligned} A &= \sqrt{\frac{\partial x_i}{\partial z^k}\underset{(z)}{A^k}\frac{\partial z^l}{\partial x_i}\underset{(z)}{A_l}} = \sqrt{\frac{\partial x_i}{\partial z^k}\frac{\partial z^l}{\partial x_i}\underset{(z)}{A^k}\,\underset{(z)}{A_l}} \\ &= \sqrt{\delta^l_k\,\underset{(z)}{A^k}\,\underset{(z)}{A_l}} = \sqrt{\underset{(z)}{A^l}\,\underset{(z)}{A_l}}. \end{aligned} \tag{2.4.14}$$

Analogously to the case of vectors co- and contravariant components can also be introduced for tensors. If we consider the absolute tensor quantity $\boldsymbol{B}$, which could represent the stress tensor $\boldsymbol{\sigma}$ or the strain tensor $\boldsymbol{\varepsilon}$ (say), we can write analogously to Eq. (2.4.1):

$$\begin{aligned} \underset{(z)}{B^{ij}} &= \frac{\partial z^i}{\partial x_k}\frac{\partial z^j}{\partial x_l}\underset{(x)}{B_{kl}}, \quad \underset{(z)}{B_{ij}} = \frac{\partial x_k}{\partial z^i}\frac{\partial x_l}{\partial z^j}\underset{(x)}{B_{kl}}, \\ \underset{(z)}{B^i{}_j} &= \frac{\partial z^i}{\partial x_k}\frac{\partial x_l}{\partial z^j}\underset{(x)}{B_{kl}}, \quad \underset{(z)}{B^j{}_i} = \frac{\partial x_k}{\partial z^i}\frac{\partial z^j}{\partial x_l}\underset{(x)}{B_{kl}}. \end{aligned} \tag{2.4.15}$$

For obvious reasons the components in the last two Eq. of (2.4.15) are also known as *mixed components* of the tensor $\boldsymbol{B}$. As in the case of vectors all indices can be raised and lowered by means of the co- and contravariant metric components of Eq. (2.4.8). For example:

$$\underset{(z)}{B^{ij}} = g^{ik}g^{jl}\,\underset{(z)}{B_{kl}}, \; \underset{(z)}{B_{ij}} = g_{ik}g_{jl}\,\underset{(z)}{B^{kl}}. \tag{2.4.16}$$

And we may treat the derivatives like ordinary fractions following Eq. (2.4.9):

$$\begin{aligned} \underset{(x)}{B_{kl}} &= \frac{\partial x_k}{\partial z^i}\frac{\partial x_l}{\partial z^j}\underset{(z)}{B^{ij}}, \quad \underset{(x)}{B_{kl}} = \frac{\partial z^i}{\partial x_k}\frac{\partial z^j}{\partial x_l}\underset{(z)}{B_{ij}}, \\ \underset{(x)}{B_{kl}} &= \frac{\partial x_k}{\partial z^i}\frac{\partial z^j}{\partial x_l}\underset{(z)}{B^i{}_j}, \quad \underset{(x)}{B_{kl}} = \frac{\partial z^i}{\partial x_k}\frac{\partial x_l}{\partial z^j}\underset{(z)}{B^j{}_i}. \end{aligned} \tag{2.4.17}$$

Again the proof is based on successive application of the chain rule. Observe that bound indices always appear in co-/contravariant pairs.

**Exercise 2.4.4:** ***The components of the metric tensor as co-/contravariant components of the unit tensor***

Show by using the definitions (2.2.8), (2.4.7) and by application of the chain rule to $\delta^i_{\ j} = \partial z^i / \partial z^j$ that the following relations hold for the KRONECKER symbol (which was originally defines in a Cartesian frame):

$$g^{ij} = \frac{\partial z^i}{{}_k} \frac{\partial z^j}{\partial x_l} \delta_{kl}, \quad g_{ij} = \frac{\partial x_k}{\partial z^i} \frac{\partial x_l}{\partial z^j} \delta_{kl},$$
$$\delta^i_{\ j} = \frac{\partial z^i}{\partial x_k} \frac{\partial x_l}{\partial z^j} \delta_{kl}, \quad \delta^j_i = \frac{\partial x_k}{\partial z^i} \frac{\partial z^j}{\partial x_l} \delta_{kl}. \tag{2.4.18}$$

Interpret these equations by using (2.4.15).

**Exercise 2.4.5:** ***The co-/contravariant components of the position vector in cylindrical and spherical coordinates***

Show by using the Eqs. (2.4.9), (2.2.11), and (2.2.15) that the following relations hold for the co-contravariant components of the position vector, $\boldsymbol{x}$, in cylindrical and spherical coordinates, respectively:

$$\underset{(z)}{x^1} = r\,, \ \underset{(z)}{x^2} = 0\,, \ \underset{(z)}{x^3} = z; \quad \underset{(z)}{x_1} = r, \ \underset{(z)}{x_2} = 0, \ \underset{(z)}{x_3} = 0. \tag{2.4.19}$$

Explain the difference between coordinate lines and the position vector.

**Exercise 2.4.6:** ***The co-/contravariant components of the stress tensor in cylindrical coordinates***

Use Eqs. $(2.4.15)_1$ and (2.2.11) to show that the contravariant components of the stress tensor in polar coordinates (i.e., plane cylindrical coordinates) read:

$$\underset{(z)}{\sigma^{11}} = \cos^2\vartheta\ \sigma_{xx} + 2\sin\vartheta\cos\vartheta\ \sigma_{xy} + \sin^2\vartheta\ \sigma_{yy},$$
$$\underset{(z)}{\sigma^{12}} = \frac{1}{r}\big[-\sin\vartheta\cos\vartheta\ \sigma_{xx} + \big(\cos^2\vartheta - \sin^2\vartheta\big)\ \sigma_{xy} + \sin\vartheta\cos\vartheta\ \sigma_{yy}\big], \tag{2.4.20}$$
$$\underset{(z)}{\sigma^{22}} = \frac{1}{r^2}\big(\sin^2\vartheta\ \sigma_{xx} - 2\sin\vartheta\cos\vartheta\ \sigma_{xy} + \cos^2\vartheta\ \sigma_{yy}\big).$$

The symbols $\sigma_{xx}, \sigma_{xy}, \sigma_{xz}, \sigma_{yy}, \sigma_{yz}, \sigma_{zz}$ denote the components for plane stress in Cartesian coordinates $x_1, x_2, x_3 \equiv x, y, z$. Moreover, use Eqs. $(2.4.15)_2$ and (2.2.11) and derive corresponding expressions for the covariant components of the stress tensor in cylindrical coordinates. How do the co-/contravariant expressions for the stress tensor fit into the world of MOHR's circle in 2D ?

**Exercise 2.4.7**: ***The co-/contravariant components of the stress tensor in spherical coordinates***

Use Eqs. (2.2.15) and (2.4.15) and derive expressions for the co-/contravariant components of the stress tensor in spherical coordinates for given Cartesian stress components $\sigma_{xx}$, $\sigma_{xy}$, $\sigma_{xz}$, $\sigma_{yy}$, $\sigma_{yz}$, and $\sigma_{zz}$, radial distance $r$, and the two angles $\varphi$ and $\vartheta$.

Christian Otto Mohr was born on October 8, 1835 in Wesselburen in Holstein (Germany) and died on October 2, 1918 in Dresden. Despite being a full-blood engineer he also showed a certain appreciation to useful mathematical concepts. Examples of this are MOHR's circles, which allowed for an intuitive representation of the components of the various stress tensors without knowing tensor calculus. Another one is MOHR's analogy, a graphical method for solving the differential equation of deflective beams for geometrically difficult cases. MOHR held chairs for technical mechanics in Stuttgart and Dresden and proved himself to be a didactically sensitive professor whenever teaching strength of materials.

## 2.5 Co- and Contravariant from the Perspective of Vectors

By using the base vectors $\boldsymbol{g}_k$ from Eq. $(2.3.4)_2$ we may write for an arbitrary vector $\boldsymbol{A}$:

$$\boldsymbol{A} = \underset{(z)}{A^k} \boldsymbol{g}_k = \underset{(z)}{A^k} \frac{\partial x_i}{\partial z^k} \boldsymbol{e}_i. \tag{2.5.1}$$

On the other hand we have:

$$\boldsymbol{A} = \underset{(x)}{A_j} \boldsymbol{e}_j. \tag{2.5.2}$$

Consequently we conclude that:

$$\underset{(x)}{A_i} = \underset{(z)}{A^k} \frac{\partial x_i}{\partial z^k} \Rightarrow \underset{(z)}{A^k} = \frac{\partial z^k}{\partial x_i} \underset{(x)}{A_i}. \tag{2.5.3}$$

In other words: In the representation $\boldsymbol{A} = \underset{(z)}{A^k} \boldsymbol{g}_k$ the quantities $\underset{(z)}{A^k}$ must be interpreted as the contravariant components of the vector $\boldsymbol{A}$. We now define another base $\boldsymbol{g}^l$ (a.k.a. as the *dual base*) according to:

$$\boldsymbol{g}^l = \frac{\partial z^l}{\partial x_j} \boldsymbol{e}_j \tag{2.5.4}$$

and use it for decomposing the vector $\boldsymbol{A}$:

$$\boldsymbol{A} = \boldsymbol{g}^l \underset{(z)}{A_l} = \frac{\partial z^l}{\partial x_j} \boldsymbol{e}_j \underset{(z)}{A_l} . \tag{2.5.5}$$

By comparison with Eq. (2.5.2) we obtain:

$$\underset{(x)}{A_i} = \frac{\partial z^l}{\partial x_i} \underset{(z)}{A_l} \quad \Rightarrow \quad \underset{(z)}{A_l} = \frac{\partial x_i}{\partial z^l} \underset{(x)}{A_i} . \tag{2.5.6}$$

Consequently the quantities $\underset{(z)}{A_l}$ are truly the covariant components of $\boldsymbol{A}$ and we may write:

$$\boldsymbol{A} = \underset{(z)}{A_l}\, \boldsymbol{g}^l = \underset{(z)}{A^k}\, \boldsymbol{g}_k . \tag{2.5.7}$$

For the scalar product of the two sets of base vectors, $\boldsymbol{g}^l$ and $\boldsymbol{g}_k$, we find:

$$\begin{aligned} \boldsymbol{g}^l \cdot \boldsymbol{g}_k &= \left(\frac{\partial z^l}{\partial x_j} \boldsymbol{e}_j\right) \cdot \left(\frac{\partial x_i}{\partial z^k} \boldsymbol{e}_i\right) = \frac{\partial z^l}{\partial x_j} \frac{\partial x_i}{\partial z^k} \boldsymbol{e}_j \cdot \boldsymbol{e}_i \\ &= \frac{\partial z^l}{\partial x_j} \frac{\partial x_i}{\partial z^k} \delta_{ji} = \frac{\partial z^l}{\partial x_i} \frac{\partial x_i}{\partial z^k} = \frac{\partial z^l}{\partial z^k} = \delta^l{}_k . \end{aligned} \tag{2.5.8}$$

If we recall Eq. (2.3.7), i.e., the relation $\boldsymbol{g}_k \cdot \boldsymbol{g}_l = g_{kl}$ for the scalar product and the analogous condition:

$$\begin{aligned} \boldsymbol{g}^l \cdot \boldsymbol{g}^k &= \left(\frac{\partial z^l}{\partial x_j} \boldsymbol{e}_j\right) \cdot \left(\frac{\partial z^k}{\partial x_i} \boldsymbol{e}_i\right) = \frac{\partial z^l}{\partial x_j} \frac{\partial z^k}{\partial x_i} \boldsymbol{e}_j \cdot \boldsymbol{e}_i \\ &= \frac{\partial z^l}{\partial x_j} \frac{\partial z^k}{\partial x_i} \delta_{ji} = \frac{\partial z^l}{\partial x_j} \frac{\partial z^k}{\partial x_j} \equiv g^{lk} , \end{aligned} \tag{2.5.9}$$

we find by using Eq. (2.5.7) after scalar multiplication by $\boldsymbol{g}^m$:

$$\boldsymbol{g}^m \cdot \left(\boldsymbol{g}^l \underset{(z)}{A_l}\right) = \boldsymbol{g}^m \cdot \left(\underset{(z)}{A^k} \boldsymbol{g}_k\right) \quad \Rightarrow \quad \underset{(z)}{A^m} = g^{ml} \underset{(z)}{A_l} , \tag{2.5.10}$$

or by $\boldsymbol{g}_m$:

$$\boldsymbol{g}_m \cdot \left(\underset{(z)}{A_l} \boldsymbol{g}^l\right) = \boldsymbol{g}_m \cdot \left(\underset{(z)}{A^k} \boldsymbol{g}_k\right) \quad \Rightarrow \quad \underset{(z)}{A_m} = g_{mk} \underset{(z)}{A^k} . \tag{2.5.11}$$

Note that we have run across these formulae before in Eq. (2.4.8).

**Exercise 2.5.1:** ***Base vectors for skew coordinates***

Recall the results of Exercise 2.3.1. Use the corresponding coordinate transformation $z^k = z^k(x_i)$ and calculate the base vectors $\boldsymbol{g}^l$. Verify the orthogonality conditions $\boldsymbol{g}^l \cdot \boldsymbol{g}_k = \delta^l_{\ k}$ and depict the $\boldsymbol{g}$-vectors in the $z$-coordinate system.

Thus the vector $\boldsymbol{A}$ can be represented in the Cartesian base $\boldsymbol{e}_j$ (see Eq. 2.5.2) as well as in the skew-curvilinear bases $\boldsymbol{g}_k$ and $\boldsymbol{g}^l$, which are not normalized (Eq. 2.5.7). The same holds for tensorial quantities. As an example we consider the tensor of second order of Eq. (2.4.15), $\boldsymbol{B}$. The following representation is valid in the Cartesian base:

$$\boldsymbol{B} = \underset{(x)}{B_{kl}}\, \boldsymbol{e}_k \otimes \boldsymbol{e}_l \tag{2.5.12}$$

and the following ones in the skew-curvilinear case:

$$\begin{aligned} \boldsymbol{B} &= \underset{(z)}{B^{kl}}\, \boldsymbol{g}_k \otimes \boldsymbol{g}_l,\ \boldsymbol{B} = \underset{(z)}{B_{kl}}\, \boldsymbol{g}^k \otimes \boldsymbol{g}^l, \\ \boldsymbol{B} &= \underset{(z)_l}{B^k}{}_l\, \boldsymbol{g}_k \otimes \boldsymbol{g}^l,\ \boldsymbol{B} = \underset{(z)}{B_k{}^l}\, \boldsymbol{g}^k \otimes \boldsymbol{g}_l.. \end{aligned} \tag{2.5.13}$$

We conclude that the number of possible ways of representation in skew-curvilinear bases increases significantly with the order of the tensors. A second order tensor offers already four different possibilities. However, a second order tensor is by no means the highest type encountered in continuum theory. For example, the stiffness tensor $\boldsymbol{C}$ of linear elasticity is of fourth order and allows for sixteen co-/contravariant representations. Indeed, it looks quite innocent in a purely Cartesian frame:

$$\boldsymbol{C} = \underset{(x)}{C}{}_{klmn}\, \boldsymbol{e}_k \otimes \boldsymbol{e}_l \otimes \boldsymbol{e}_m \otimes \boldsymbol{e}_n. \tag{2.5.14}$$

The notion of the *tensor product* or *dyad*, i.e., the symbol "$\otimes$" deserves an explanation. Even though it is as necessary as a Mercedes star it is customarily used in the literature, the main reason being to distinguish a product between vectors (or tensors) in absolute notation from the scalar and the vector product, which are identifiable by the symbols "$\cdot$" and "$\times$". Mathematically speaking, two vectors (first order tensors), $\boldsymbol{A}$ and $\boldsymbol{B}$, are mapped onto a number (zeroth order tensor) by writing $\boldsymbol{A} \cdot \boldsymbol{B}$, onto another (axial) vector by $\boldsymbol{A} \times \boldsymbol{B}$, and onto a second order tensor by $\boldsymbol{A} \otimes \boldsymbol{B}$. Just like the scalar or vector product "$\otimes$" can also be introduced axiomatically by defining a corresponding algebra. However, this will not be detailed any further in this book and the reader is referred to the more mathematically oriented literature cited below.

**Exercise 2.5.2:** ***The trace of a second order tensor***

The *trace* of a tensor of second order is defined in Cartesian coordinates as follows:

$$\operatorname{tr} \boldsymbol{B} = \underset{(x)}{B_{kk}} \,. \tag{2.5.15}$$

Use Eq. (2.4.17) to show that:

$$\operatorname{tr} \boldsymbol{B} = g^{ij} \underset{(z)}{B_{ij}} = g_{ij} \underset{(z)}{B^{ij}} = \underset{(z)}{B^{i}}_{\,i} = \underset{(z)}{B_{j}^{j}} \,. \tag{2.5.16}$$

Interpret the last two expressions in terms of bound co-/contravariant indices.

## 2.6 Physical Components of Vectors and Tensors

Exercises 2.4.6 and 2.4.7 have shown that the co-/contravariant components of vectors and tensors do not necessarily all have the same physical dimension, for example that of a stress. We have to concede that in engineering terms co-/contravariant components of a vector or a tensor are, in general, rather unphysical quantities.

However, in the case of orthogonal coordinate transformations whose coordinate lines are perpendicular to each other it becomes possible to recover the physical notion of Cartesian components by introducing so called *physical components*. The key to their definition lies in the fact that the metric tensor of orthogonal coordinate transformations is diagonal:

$$\boldsymbol{g}_m \cdot \boldsymbol{g}_l = \begin{bmatrix} g_{11} & 0 & 0 \\ 0 & g_{22} & 0 \\ 0 & 0 & g_{33} \end{bmatrix}, \quad \boldsymbol{g}^m \cdot \boldsymbol{g}^l = \begin{bmatrix} g^{11} & 0 & 0 \\ 0 & g^{22} & 0 \\ 0 & 0 & g^{33} \end{bmatrix}. \tag{2.6.1}$$

**Exercise 2.6.1:** ***Diagonal property of the metric for orthogonal coordinate transformations***

Prove Eq. (2.6.1) and show that the metric tensor for coordinate transformations with perpendicularly oriented coordinate lines has only components along its diagonal.

To do so use the definition (2.2.8) for the metric tensor and recall that the derivatives $\partial x_i / \partial z_k$ are the components of tangent vectors to the coordinate lines $z_k$: Eq. (2.3.4).

In that case the length of a vector (cf., Eqs. 2.4.12/2.4.13) can be expressed as a sum of quadratic terms:

$$A = \sqrt{\sum_{i=1}^{3} g_{\underline{ii}} (\underset{(z)}{A^i})^2} = \sqrt{\sum_{i=1}^{3} g^{\underline{ii}} (\underset{(z)}{A_i})^2}. \tag{2.6.2}$$

In order not to break with EINSTEIN's summation convention we had to use an explicit summation sign. Hence the expressions $g_{\underline{ii}}$ and $g^{\underline{ii}}$ do not consist of three terms. In fact they stand for only one diagonal element of the co-/contravariant components of the metric tensor, namely no. "$i$". We have emphasized the exception to the summation convention by an underscore, and agree that underlined indices appearing twice in a product will not be summed up.We now define so-called *physical vector components*, $A_{\langle i \rangle}$, by:

$$A_{\langle i \rangle} = \sqrt{g_{\underline{ii}}}\, \underset{(z)}{A^i} = \sqrt{g^{\underline{ii}}}\, \underset{(z)}{A_i}\,. \tag{2.6.3}$$

If the square of these components is summed up we immediately obtain the length of the vector $\boldsymbol{A}$:

$$A = \sqrt{A_{\langle i \rangle} A_{\langle i \rangle}}\,. \tag{2.6.4}$$

Note that physical components can also be defined for tensors of second and higher order:

$$B_{\langle ij \rangle} = \sqrt{g_{\underline{ii}}} \sqrt{g_{\underline{jj}}}\, \underset{(z)}{B^{ij}} = \sqrt{g^{\underline{ii}}} \sqrt{g^{\underline{jj}}}\, \underset{(z)}{B_{ij}}\,. \tag{2.6.5}$$

**Exercise 2.6.2: *The length of a vector in physical components***

Combine Eqs. (2.6.1) and (2.6.2) to prove Eq. (2.6.4). Clarify the meaning of underlined indices. Also try to calculate the length through scalar products according to Eqs. (2.6.2) and (2.5.7), which need to be specialized to orthogonal coordinate transformations before:

$$\begin{aligned} \boldsymbol{A} \cdot \boldsymbol{A} &= \left( \boldsymbol{g}_l \underset{(z)}{A^l} \right) \cdot \left( \boldsymbol{g}_k \underset{(z)}{A^k} \right) = \underset{(z)}{A^k} \underset{(z)}{A^k} g_{\underline{kk}}, \\ \boldsymbol{A} \cdot \boldsymbol{A} &= \left( \boldsymbol{g}^l \underset{(z)}{A_l} \right) \cdot \left( \boldsymbol{g}^k \underset{(z)}{A_k} \right) = \underset{(z)}{A_k} \underset{(z)}{A_k} g^{\underline{kk}}. \end{aligned} \tag{2.6.6}$$

**Exercise 2.6.3:** ***Physical components of the stress* tensor *in cylindrical coordinates***

Use the results from Exercise 2.4.6 and derive the following expressions for the physical components of the stress tensor in cylindrical coordinates. In particular, verify that all physical components have the physical dimension of stress:

$$\begin{aligned}
\sigma_{\langle rr\rangle} &= \cos^2\vartheta\ \sigma_{xx} + 2\sin\vartheta\cos\vartheta\ \sigma_{xy} + \sin^2\vartheta\ \sigma_{yy},\\
\sigma_{\langle r\vartheta\rangle} &= -\sin\vartheta\cos\vartheta\ \sigma_{xx} + \left(\cos^2\vartheta - \sin^2\vartheta\right)\sigma_{xy} + \sin\vartheta\cos\vartheta\ \sigma_{yy},\\
\sigma_{\langle rz\rangle} &= \cos\vartheta\ \sigma_{xz} + \sin\vartheta\ \sigma_{yz}\ ,\\
\sigma_{\langle\vartheta\vartheta\rangle} &= \sin^2\vartheta\ \sigma_{xx} - 2\sin\vartheta\cos\vartheta\ \sigma_{xy} + \cos^2\vartheta\ \sigma_{yy},\\
\sigma_{\langle\vartheta z\rangle} &= -\sin\vartheta\ \sigma_{xz} + \cos\vartheta\ \sigma_{yz},\\
\sigma_{\langle zz\rangle} &= \sigma_{zz}\ .
\end{aligned} \tag{2.6.7}$$

Recall the problem of equilibrium of forces shown in Fig. 2.9. The analysis will lead to the following equations for MOHR's circle:

$$\begin{aligned}
\sigma_s &= \frac{1}{2}\left(\sigma_{xx} + \sigma_{yy}\right) - \frac{1}{2}\left(\sigma_{yy} - \sigma_{xx}\right)\cos(2\vartheta) + \sigma_{xy}\sin(2\vartheta),\\
\tau_s &= \frac{1}{2}\left(\sigma_{yy} - \sigma_{xx}\right)\sin(2\vartheta) + \sigma_{xy}\cos(2\vartheta).
\end{aligned} \tag{2.6.8}$$

**Fig. 2.9** A reminder of MOHR's circle in 2D

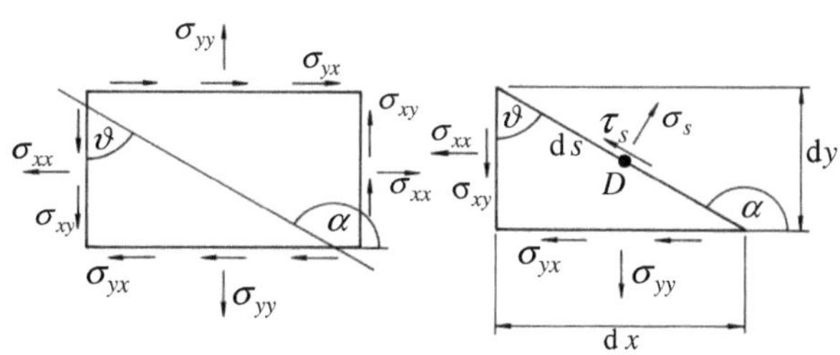


How is this result related to Eq. (2.6.7)? Use the following trigonometric theorems to answer this question:

$$\begin{aligned}
&\sin^2\vartheta = \frac{1}{2}[1 - \cos(2\vartheta)],\quad \cos^2\vartheta = \frac{1}{2}[1 + \cos(2\vartheta)],\\
&\sin(2\vartheta) = 2\sin\vartheta\cos\vartheta.
\end{aligned} \tag{2.6.9}$$

**Exercise 2.6.4:** ***Physical components of the metric tensors***

Show that for orthogonal coordinates systems the physical components of the metric tensor read:

$$g_{\langle ij\rangle} = \begin{bmatrix} 1 & 0 & 0 \\ 0 & 1 & 0 \\ 0 & 0 & 1 \end{bmatrix}. \tag{2.6.10}$$

**Exercise 2.6.5:** ***The* VON MISES *flow rule***

In Cartesian coordinates the *stress deviator* $\underset{(x)}{\Sigma}_{ij}$ is defined as follows:

$$\underset{(x)}{\Sigma}_{ij} = \underset{(x)}{\sigma}_{ij} - \frac{1}{3}\underset{(x)}{\sigma}_{kk}\,\delta_{ij}. \tag{2.6.11}$$

Show that this quantity is trace-free:

$$\underset{(x)}{\Sigma}_{ll} = 0. \tag{2.6.12}$$

Recall the postulate by VON MISES according to which a metal starts flowing if the following threshold value, $\sigma_{\mathrm{y}}$, is reached:

$$\sigma_{\mathrm{y}}^2 = \frac{3}{2}\underset{(x)}{\Sigma}_{ij}\,\underset{(x)}{\Sigma}_{ij}. \tag{2.6.13}$$

$\sigma_{\mathrm{y}}$ is a material specific quantity and known as the *yield stress*. Explain why it makes sense to remove the trace of state of stress from a strength criterion, at least for polycrystalline metals. Show that in the case of plane stress it is possible to write:

$$\sigma_{\mathrm{y}}^2 = \sigma_{xx}^2 + \sigma_{yy}^2 - \sigma_{xx}\sigma_{yy} + 3\sigma_{xy}^2. \tag{2.6.14}$$

Moreover, specialize the VON MISES criterion to a 1D tensile bar with a tensile stress $\sigma$ as well as to a block subjected to a shear load $\tau$ and show that:

$$\tau = \frac{1}{\sqrt{3}}\sigma. \tag{2.6.15}$$

Transform from the Cartesian system $x$ to an arbitrary system $z$ to show that:

$$\underset{(x)}{\sigma}_{kk} = g_{ij}\,\underset{(z)}{\sigma}^{ji} = g^{ji}\,\underset{(z)}{\sigma}_{ij} = \underset{(z)}{\sigma}_{i}{}^{i} \tag{2.6.16}$$

and, in a similar manner, that we may also write:

$$\underset{(z)}{\Sigma}_{ij} = \underset{(z)}{\sigma}_{ij} - \frac{1}{3}\underset{(z)}{\sigma}{}^k_k\, g_{ij}, \quad \underset{(z)}{\Sigma}{}^{ij} = \underset{(z)}{\sigma}{}^{ij} - \frac{1}{3}\underset{(z)}{\sigma}{}^k_k\, g^{ij} \tag{2.6.17}$$

and:

$$\sigma_{\mathrm{y}}^2 = \frac{3}{2}\underset{(z)}{\Sigma}_{ij}\underset{(z)}{\Sigma}{}^{ji} = \frac{3}{2} g_{ir} g_{js} \underset{(z)}{\Sigma}{}^{rs}\underset{(z)}{\Sigma}{}^{ji} = \frac{3}{2} g^{ir} g^{js} \underset{(z)}{\Sigma}_{ij}\underset{(z)}{\Sigma}_{rs}. \tag{2.6.18}$$

Prove the following relations by assuming orthogonal coordinates:

$$\underset{(x)}{\sigma}_{kk} = \underset{(z)}{\sigma}_{\langle kk\rangle}, \quad \underset{(z)}{\Sigma}_{\langle ij\rangle} = \underset{(z)}{\sigma}_{\langle ij\rangle} - \frac{1}{3}\underset{(z)}{\sigma}_{\langle kk\rangle}\delta_{\langle ij\rangle} \tag{2.6.19}$$

and:

$$\sigma_{\mathrm{y}}^2 = \frac{3}{2}\underset{(z)}{\Sigma}_{\langle ij\rangle}\underset{(z)}{\Sigma}_{\langle ij\rangle}. \tag{2.6.20}$$

Finally specialize to plane polar coordinates and show that:

$$\sigma_{\mathrm{y}}^2 = \sigma_{\langle rr\rangle}^2 + \sigma_{\langle\vartheta\vartheta\rangle}^2 - \sigma_{\langle rr\rangle}\sigma_{\langle\vartheta\vartheta\rangle} + 3\sigma_{\langle r\vartheta\rangle}^2. \tag{2.6.21}$$

In context with physical coordinates a few remarks or rules for computing combinations between scalar and tensor products are in order. For simplicity we assume that the following vectors, $\boldsymbol{A}$ and $\boldsymbol{B}$, and the (second order) tensors, $\boldsymbol{C}$ and $\boldsymbol{D}$, are expressed in an orthogonal coordinate system with unit vectors $\boldsymbol{e}_i = \sqrt{g^{\underline{ii}}}\boldsymbol{g}_i$ so that physical coordinates can be used. Then by definition of the scalar product we may write:

$$\begin{aligned} &\boldsymbol{e}_i \cdot (\boldsymbol{e}_j \otimes \boldsymbol{e}_k) = \delta_{ij}\boldsymbol{e}_k, \quad (\boldsymbol{e}_i \otimes \boldsymbol{e}_j) \cdot \boldsymbol{e}_k = \boldsymbol{e}_i\delta_{jk}, \\ &\boldsymbol{e}_i \cdot (\boldsymbol{e}_j \otimes \boldsymbol{e}_k) \cdot \boldsymbol{e}_l = \delta_{ij}\delta_{kl}, \quad (\boldsymbol{e}_i \otimes \boldsymbol{e}_j) : (\boldsymbol{e}_k \otimes \boldsymbol{e}_l) = \delta_{il}\delta_{jk}. \end{aligned} \tag{2.6.22}$$

Richard Edler VON MISES was born on April 19, 1883 in Lemberg (now Ukraine) and died on July 14, 1953 in Boston. From 1909 to 1918 he was a professor of applied mathematics in Straßburg (now France) where he investigated problems of solid and fluid mechanics, aerodynamics, statistics, and probability theory. During WW I he was on the Austrian side where he built and flew his own battle plane. After the war he went to Berlin until the Nazis forced him into exile in 1933, first to Istanbul and then on to Harvard.

It should be noted that some authors use the following relation for the double scalar product instead:

$$(\boldsymbol{e}_i \otimes \boldsymbol{e}_j)..(\boldsymbol{e}_k \otimes \boldsymbol{e}_l) = \delta_{ik}\delta_{jl}. \tag{2.6.23}$$

Also note that all of these complications can be avoided if the index notation is used exclusively from the very start.

**Exercise 2.6.6:** ***Scalar and double scalar products in vector notation***

Use Eq. (2.6.22) to show that:

$$\begin{gathered}\boldsymbol{A} \cdot \boldsymbol{C} = A_{\langle i\rangle} C_{\langle ik\rangle} \boldsymbol{e}_k, \ \boldsymbol{C} \cdot \boldsymbol{B} = C_{\langle ij\rangle} B_{\langle j\rangle} \boldsymbol{e}_i, \\ \boldsymbol{A} \cdot \boldsymbol{C} \cdot \boldsymbol{B} = A_{\langle i\rangle} C_{\langle ij\rangle} B_{\langle j\rangle}, \ \boldsymbol{C} : \boldsymbol{D} = C_{\langle ij\rangle} D_{\langle ji\rangle}.\end{gathered} \tag{2.6.24}$$

Moreover, show that for four vectors $\boldsymbol{A}$, $\boldsymbol{B}$, $\boldsymbol{E}$, and $\boldsymbol{F}$ we may write:

$$(\boldsymbol{A} \otimes \boldsymbol{B}) : (\boldsymbol{E} \otimes \boldsymbol{F}) = (\boldsymbol{A} \cdot \boldsymbol{F})\,(\boldsymbol{B} \cdot \boldsymbol{E}). \tag{2.6.25}$$

## 2.7 A Touch of Differential Geometry

The EINSTEIN convention requires summation from 1 to 2 for plane problems or from 1 to 3 for spatial ones. So far corresponding tensor equations could be applied to both cases. However, in what follows we will deliberately concentrate on two-dimensional situations. Our intention is to capture the *geometry of curved surfaces* mathematically. Such a tool will later be of great importance when these surfaces have a real meaning in terms of continuum physics. For example, they can represent the boundary between a liquid (or gaseous) and a solid region, such as the wall of pressure vessel or the skin of a soap bubble. In general such surfaces are curved and by no means planar. In order to become familiar with the mathematics pertinent to their characterization (i.e., differential geometry) some training is required. However, as we shall see, it is just a straightforward continuation of our remarks on metric tensors and tangent vectors.

In order to quantify the amount of curvature of a surface in three-dimensional space two curvilinear surface coordinates are used, called $z^\alpha$, $\alpha = 1{,}2$. They behave analogously to the curvilinear coordinates $z^i$ from Eq. (2.2.1) the only difference being that there are only two of them. We emphasize this by small Greek indices, for example, $\alpha$. A position vector $\boldsymbol{x}$ on the surface is described by its three Cartesian coordinates $x_i$ which, in turn, are functions of the surface coordinates:

$$x_i = x_i(z^1, z^2) \equiv x_i(z^\alpha), \quad i = 1,\ 2,\ 3. \tag{2.7.1}$$

Two tangent vectors, $\boldsymbol{\tau}_1$ and $\boldsymbol{\tau}_2$, can now be defined analogously to Eq. $(2.3.4)_2$:

$$\boldsymbol{\tau}_\alpha = \frac{\partial x_i(z^\gamma)}{\partial z^\alpha}\boldsymbol{e}_i, \quad \alpha = 1,\ 2. \tag{2.7.2}$$

Obviously their components in (three-dimensional) Cartesian coordinates are then given by:

$$\tau^i_\alpha = \frac{\partial x_i(z^\gamma)}{\partial z^\alpha}. \tag{2.7.3}$$

In this case it does not matter if we write the index $i$ at the top or at the bottom of the symbol $\tau$ because it refers to a Cartesian representation. However, the positioning of the index $\alpha$ does matter: Just as in the case of index $k$ from Eq. $(2.3.4)_2$ it signals covariance.

Note that just like the case of $\boldsymbol{g}_k$ the tangent vectors $\boldsymbol{\tau}_\alpha$ are no unit vectors. They still need to be normalized in order to define a unit vector $\boldsymbol{n}$ normal to the surface:

$$\boldsymbol{n} = \frac{\boldsymbol{\tau}_1 \times \boldsymbol{\tau}_2}{|\boldsymbol{\tau}_1 \times \boldsymbol{\tau}_2|}. \tag{2.7.4}$$

Moreover, just like in Eq. (2.5.4), it is possible to define dual tangent vectors $\boldsymbol{\tau}^\beta$ by:

$$\boldsymbol{\tau}^\beta = \frac{\partial z^\beta}{\partial x_j}\boldsymbol{e}_j. \tag{2.7.5}$$

Because of the chain rule the following orthogonality conditions hold (cf., (2.5.8)):

$$\boldsymbol{\tau}^\beta \cdot \boldsymbol{\tau}_\alpha = \delta^\beta_\alpha. \tag{2.7.6}$$

We are now in a position to define co- and contravariant surface metrics in complete analogy to the Eqs. (2.2.10), (2.3.7), (2.4.7), and (2.5.8/2.5.9):

$$g_{\alpha\beta} = \frac{\partial x_i}{\partial z^\alpha}\frac{\partial x_i}{\partial z^\beta} \equiv \boldsymbol{\tau}_\alpha \cdot \boldsymbol{\tau}_\beta, \quad g^{\alpha\beta} = \frac{\partial z^\alpha}{\partial x_j}\frac{\partial z^\beta}{\partial x_j} \equiv \boldsymbol{\tau}^\alpha \cdot \boldsymbol{\tau}^\beta. \tag{2.7.7}$$

In order to obtain a measure for the local curvature of the surface we first define the so-called *curvature tensor* (with the Cartesian components $n_i$ of the normal $\boldsymbol{n}$ from Eq. (2.7.4)):

$$b_{\alpha\beta} = \frac{\partial^2 x_i}{\partial z^\alpha \partial z^\beta} n_i. \tag{2.7.8}$$

This definition deserves a comment: Calculus teaches us that the extreme values of a curve $y = y(x)$ within the plane are governed by the second derivative $\mathrm{d}^2 y/\mathrm{d}^2 x$, i.e., the local curvature in a point $x$. This explains the second derivatives in Eq. (2.7.8). Moreover, a plane surface is, of course, not curved. Therefore we expect that the curvature tensor vanishes. And, indeed, the scalar product (note the summation implied by the index $i$) in Eq. (2.7.8) or, in other words the projection of the curvature onto the normal vector, becomes zero in this case. Moreover, note that by virtue of Eq. (2.7.3) we may write for the curvature in Eq. (2.7.8) as well:

$$\frac{\partial^2 x_i}{\partial z^\alpha \partial z^\beta} = \frac{\partial \tau^i_\alpha}{\partial z^\beta} \Rightarrow \frac{\partial^2 x_i}{\partial z^\alpha \partial z^\beta} \boldsymbol{e}_i = \frac{\partial \tau^i_\alpha}{\partial z^\beta} \boldsymbol{e}_i \equiv \frac{\partial \tau^i_\alpha \boldsymbol{e}_i}{\partial z^\beta} = \frac{\partial \boldsymbol{\tau}_\alpha}{\partial z^\beta}. \tag{2.7.9}$$

Note that because of their constancy the Cartesian unit vectors $\boldsymbol{e}_i$ are not affected by partial differentiation. The curvature may also be interpreted as the change of tangent vectors with the lines of coordinates, which is less intuitive. In vector notation the covariant components of the curvature tensor can therefore be written as:

$$b_{\alpha\beta} = \frac{\partial \boldsymbol{\tau}_\alpha}{\partial z^\beta} \cdot \boldsymbol{n}. \tag{2.7.10}$$

And, if it pleases, the curvature tensor can also be written in complete invariant form:

$$\boldsymbol{b} = b_{\alpha\beta} \boldsymbol{\tau}^\alpha \otimes \boldsymbol{\tau}^\beta = \frac{\partial \boldsymbol{\tau}_\alpha}{\partial z^\beta} \cdot \boldsymbol{n}\, \boldsymbol{\tau}^\alpha \otimes \boldsymbol{\tau}^\beta. \tag{2.7.11}$$

It is customary to define the *mean curvature* through the mean trace of the curvature tensors (cf., Exercise 2.5.2):

$$K_\mathrm{m} = \frac{1}{2} \operatorname{tr} \boldsymbol{b} \Rightarrow K_\mathrm{m} = \frac{1}{2} b^\alpha_\alpha = \frac{1}{2} g^{\alpha\beta} b_{\alpha\beta}. \tag{2.7.12}$$

Obviously this is an invariant, a scalar, since we have seen in Exercise 2.5.2 that, independently of the coordinate system, a trace will always yield the same value.

**Exercise 2.7.1:** ***Differential geometry of a spherical surface***

Investigate the surface of a sphere of radius $R$. To this end identify as surface coordinates $z^1 = \vartheta$, $z^2 = \varphi$ and recall the transformation rules for spherical coordinates from Exercise 2.2.2. Use the definition (2.7.3) and calculate the components of the tangent vectors, $\tau^i_\vartheta$ and $\tau^i_\varphi$, w.r.t. a Cartesian base in the center of the sphere:

$$\begin{aligned} \tau^1_\vartheta &= R\cos\vartheta\, \cos\varphi\,, \quad \tau^2_\vartheta = R\cos\vartheta\, \sin\varphi\,, \quad \tau^3_\vartheta = -R\sin\vartheta, \\ \tau^1_\varphi &= -R\sin\vartheta\, \sin\varphi\,, \quad \tau^2_\varphi = R\sin\vartheta\, \cos\varphi\,, \quad \tau^3_\varphi = 0. \end{aligned} \tag{2.7.13}$$

Show that both vectors are orthogonal to each other. Are they related to the base vectors $\boldsymbol{g}_1, \boldsymbol{g}_2, \boldsymbol{g}_3$ of Exercise 2.3.1? Depict them on the surface of a sphere together with surface coordinate lines. Are the tangent vectors normalized? Show by using Eq. (2.7.7) that the surface metric $g_{\alpha\beta}$ is given by:

$$g_{\alpha\beta} = \begin{pmatrix} R^2 & 0 \\ 0 & R^2 \sin^2 \vartheta \end{pmatrix}. \qquad (2.7.14)$$

Calculate its inverse $g^{\alpha\beta}$. Show by using Eq. (2.7.4) that the Cartesian components of the unit normal $n_i$ to the sphere are given by:

$$n_1 = \sin\vartheta \, \cos\varphi \, , \quad n_2 = \sin\vartheta \, \sin\varphi \, , \quad n_3 = \cos\vartheta. \qquad (2.7.15)$$

Use them to calculate the curvature tensor based on the definition shown in Eq. (2.7.8). Finally show that the mean curvature is given by $K_{\rm m} = -1/R$. Try to interpret the minus sign by using terms like "convex" or "concave."

**Exercise 2.7.2:** ***Differential geometry of a cylindrical surface***

Investigate now the mantle surface of a circular cylinder of radius $R$. For this purpose choose $z^1 = \vartheta$, $z^2 = z$ as surface coordinates. Recall the transformations for cylindrical coordinates from Sect. 2.2 and follow the procedures of Exercise 2.7.1. Show first that the tangent vectors are given by:

$$\begin{aligned} &\tau_\vartheta^1 = -R\sin\vartheta, \quad \tau_\vartheta^2 = R\cos\vartheta, \quad \tau_\vartheta^3 = 0, \\ &\tau_z^1 = 0, \quad \tau_z^2 = 0, \quad \tau_z^3 = 1. \end{aligned} \qquad (2.7.16)$$

Use them and calculate the surface metric:

$$g_{\alpha\beta} = \begin{pmatrix} R^2 & 0 \\ 0 & 1 \end{pmatrix}. \qquad (2.7.17)$$

Show that the unit normal in Cartesian coordinates is given by:

$$n_1 = \cos\vartheta \, , \quad n_2 = \sin\vartheta \, , \quad n_3 = 0. \qquad (2.7.18)$$

Use the results and prove that the mean curvature is given by $K_{\rm m} = -1/(2R)$. Interpret the factor $\frac{1}{2}$ and compare it to the result for a spherical surface.

## 2.8 Would You Like to Know More?

The book by Schade and Neemann [1] is a real treasure chest of mathematical formulae (which makes it easier to read since it is written in German) for true disciples of the index calculus. In particular one should look at Sects. 4.2.2 and

4.2.4 for the concepts of "metric," and "co-/contravariance." The books by Itskov [2] and Bertram [3] insist on a mathematically more stringent approach and emphasize the absolute tensor calculus. Particularly worth reading in context with the present section are Chap. 1 and Sect. 3.2 (for differential geometry) in the first and Sects. 1.1 and 1.2 in the second book. Tensor algebra and tensor analysis are also treated in concise form in Irgens [4], Chap. 12, in index as well as in absolute notation, and also in Liu [5], Appendix A.1.

In general the tensor concepts presented so far have been known for a long time. Consequently it is also worth while to study the "classics." In this context the article by Ericksen [6] in the Encyclopedia of Physics, Sects. I, II, and (in parts) III should be mentioned first. Moreover, the book by Green and Zerna [7] is to be recommended, in particular Sects. 1.1 to 1.10. Finally Chaps. 1, 12 and, notably, Sect. 13 (with many exotic coordinate transformations) in the book by Flügge [8] should be pointed out.

Several notions, such as Mohr's circle or yield stress, were used in this section without further explanations. In this context it may be useful to study textbooks on strength of materials, e.g., Hibbeler [9], Sects. 10.3 and 10.7, or Gross et al. [10], Sects. 2.2.3 and 3.3.

## References

1. Schade H, Neemann N (2009) Tensor analysis. 3. überarbeitete Auflage. de Gruyter, Berlin
2. Itskov M (2007) Tensor algebra and tensor analysis for engineers with applications to continuum mechanics. Springer, Berlin
3. Bertram A (2008) Elasticity and plasticity of large deformations, 2nd edn. Springer, Berlin
4. Irgens F (2008) Continuum mechanics. Springer, Berlin
5. Liu I-S (2010) Continuum mechanics. Springer, Berlin
6. Ericksen JL (1960) Appendix. Tensor fields. In: Flügge S (ed) Encyclopedia of physics, vol III/1. Principles of classical mechanics and field theory. Springer
7. Green AE, Zerna W (1968) Theoretical elasticity, 2nd edn. Dover Publications Inc., New York
8. Flügge W (1972) Tensor analysis and continuum mechanics. Springer, New York
9. Hibbeler RC (2005) Mechanics of materials, 6th edn. Pearson Education Inc., Upper Saddle River (07458)
10. Gross D, Hauger W, Schröder J, Wall WA, Bonet J (2011) Engineering mechanics 2, Mechanics of materials. Springer, Berlin

# Chapter 3
# Balances (in Particular in Cartesian Systems)

**Abstract** In this chapter we introduce the concept of *balances* in particular the balances of *mass momentum* angular momentum *energy* or in other words the conservation laws of classical physics. The balances will be stated in integral form—for a material volume—as well as locally—in regular singular points of the continuum. In particular the effects of changes of non-mechanical energy in time i.e. non-isothermal processes are broadly covered in later sections of this book. This way we go well beyond the scope of traditional continuum *mechanics* may speak of an introduction to continuum *physics* instead.

*We all must have a balance in our life.*

John D. ROCKEFELLER

## 3.1 Preliminary Remarks

In the previous chapter we have already specified a few important notions of tensor calculus, namely the metric tensor and co-/contravariant components including their intuitive interpretation as parallel and orthogonal projections onto the coordinate axes. In the Chap. 4 we will consider *derivatives with respect to skew and curvilinear coordinates*. We will explain how these can be introduced quite naturally as a part of tensor calculus. However, before that we will motivate how derivatives of fields in arbitrary coordinate systems arise. For that reason we present in this chapter the first "ingredient" for the solution of continuum related problems, the so-called *balance equations*, where spatial derivatives play an essential role.

Our primary objective is the following one: In *continuum thermo-mechanics* we wish to determine *five* fields, namely mass density, $\rho = \hat{\rho}(\boldsymbol{x}, t)$,[1] the three

[1] We distinguish a function of current position and time from its value by a circumflex.

W. H. Müller, *An Expedition to Continuum Theory*, Solid Mechanics and Its Applications 210, DOI: 10.1007/978-94-007-7799-6_3,

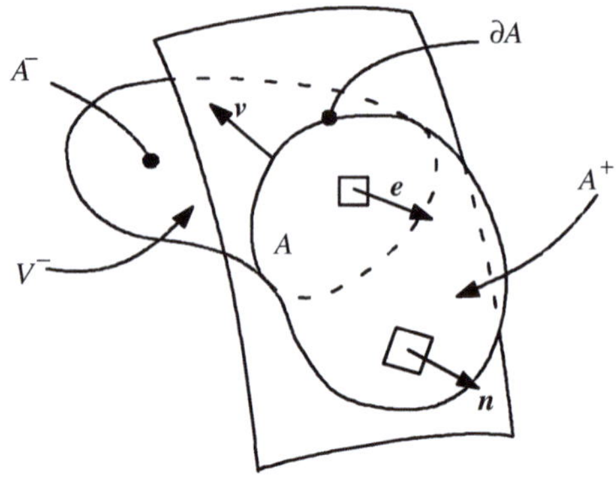


**Fig. 3.1** Balance volume with a singular surface

components of the velocity vector, $\boldsymbol{v} = \hat{\boldsymbol{v}}(\boldsymbol{x}, t)$, and temperature, $T = \hat{T}(\boldsymbol{x}, t)$, in all material points, $\boldsymbol{x}$, of a body, $V(t)$, and at all times, $t$. Initially we will assume that these five fields are described in a Cartesian frame $\boldsymbol{x} = x_i \boldsymbol{e}_i$. From an atomistic point of view the three fields can be interpreted as follows: The sum of the mass of all the molecules, $\mathrm{d}m$, divided by the volume of the material point, $\mathrm{d}V$, gives the mass density, $\rho = \hat{\rho}(\boldsymbol{x}, t)$. Similarly, if we sum up the momenta of all the molecules in $\mathrm{d}V$, i.e., the product of their individual masses and velocities, and divide the result by the total mass, we obtain the macroscopically observable velocity field, $\boldsymbol{v} = \hat{\boldsymbol{v}}(\boldsymbol{x}, t)$, of that very "point." And, finally, by computing the "average" kinetic energy of all the molecules in the volume we obtain a measure of the intensity of the macroscopically invisible erratic atomic movement. This is the (imprecise) kinetic interpretation of temperature, $T = \hat{T}(\boldsymbol{x}, t)$.

Anders Jonas ÅNGSTRÖM was born on August 13, 1814 in Medelpad (Sweden) and died on June 21, 1874 in Uppsala. He studied physics at Uppsala University where he also became a lecturer in 1839. In 1842 he joined the observatory in Stockholm and became the Keeper of Uppsala Observatory the year after. In 1858 he finally obtained the Chair of Physics at Uppsala University. His work touches on various branches of physics: Magnetism, spectroscopy, optics. Clearly the wavelength of light made his name immortal whenever the quantification of small length scales is required.

The notion of a *material point of a continuum* deserves some comments. In mathematics a point in $\mathbb{R}^3$, $\boldsymbol{x}$, has no extension at all, and it is without structure. Moreover, a continuous mathematical function, $f(\boldsymbol{x})$, defined on a subset of $\mathbb{R}^3$ can be evaluated and differentiated in any neighborhood of $\boldsymbol{x}$, as small as this may be. In contrast to that, the material point of continuum theory does have a size as well as an internal structure. It can be visualized as an assembly of a "sufficiently large" number of atoms or molecules which is of the order of the AVOGADRO number, despite the fact that we denote its size by a differentially small volume, $\mathrm{d}V$. The number of molecules must be high enough so that fluctuations within the "point" can be neglected. Only then the concept of fields becomes *physically* relevant: If we consider real volumes that are only a few Ångströms wide, very few atoms will be inside and, what is more, they will leave and enter the volume over extremely short time periods. Thus it does not make sense to compute the local mass density as outlined above. In fact, one must ask, how small the real volume element can be so that it can be treated on the basis of continuum theory. Note that, just like space, appropriate smallness of time is another big issue in continuum theory. Scientists love arguing about the "limits of the continuum approach," for example, when they talk about nanostructures. However, they rarely give a satisfying general answer, maybe, because this question can only be answered on a case-by-case basis.

Interestingly the fields of continuum theory are treated as if they were mathematical functions in space and time: They are integrated with respect to an arbitrarily small environment, they are differentiated without hesitation, etc. In what follows we shall learn about the corresponding mathematical tools. However, we must realize that mathematics has nothing to do with reality and we should never forget that it is nothing more but a neat "language" conceived by the human brain to model reality.

Obviously, in order to determine the five fields, *five* equations are required, the so-called *field equations*. The foundation of these equations are the *balance of mass* (a scalar relation), the *balance of momentum* (a vector equation with three components), and the *balance of energy* (a scalar equation).

Moreover, we have to acknowledge the possibility that the fields are discontinuous within the body. In other words, they *jump* when crossing a certain boundary. The corresponding situation is illustrated in Fig. 3.1: A *material body* (i.e., a region that always consists of the same material particles) is split into two halves, denoted by $V^{\pm}$, by an open surface, $A$. Note that these quantities can be time-dependent. Very often we assume this implicitly and do not acknowledge it in the symbol. So we just write $A$ and not $A(t)$. Moreover, we use the symbol $A^{\pm}$ for the (open) surfaces of $V^{\pm}$. In general these are time-dependent as well. Their outward normal unit vectors are denoted by $\boldsymbol{n}$ whereas, for the sake of distinction, the normal on the separating surface $A$ is characterized by the symbol $\boldsymbol{e}$. Moreover, the symbol $L = \partial A$ is used for the closed line bordering the surface $A$. Finally, the outward normal to that line, and which is also a tangent to the surface $A$, was denoted by $\boldsymbol{v}$.

Osborne REYNOLDS was born on August 23, 1842 in Belfast (Ireland) and died on February 21, 1912 in Watchet, Somerset (England). He graduated from Cambridge University in mathematics. In 1868 he became the very first Professor of Engineering at the University of Manchester. He stayed there until his retirement in 1905. He is mostly known because of his pioneering work in fluid mechanics. Particularly noteworthy among his findings is the REYNOLDS' number, which characterizes the transition between laminar and turbulent flow and made him immortal.

Note that the singular surface, $A$, can move at its own speed, independently of the movement of the surrounding material particles. In other words, $A$ is not necessarily a material surface. Consequently, the two volumes $V^{\pm}$ separated by $A$ are not necessarily material and may consist of different particles as time goes on. Recall that at the singular surface the primary fields and their derivates will "jump," i.e., they are discontinuous. As an example of such a situation consider the interface between liquid water and its vapor. If the temperature rises the liquid water will vaporize: Water molecules will leave the liquid and enter the gaseous region, i.e., the matter in both subvolumes is clearly not conserved. They are open systems. Moreover, the interface between a fiber and the surrounding matrix of a composite material may be considered as a singular surface: Although both regions do not show macroscopically visible motion, there are density gradients. The matter in these systems does not change, they are material systems. It is a judgment call to decide as to whether the interface in both examples is a material singular surface or not. This depends on the physical accuracy our modeling requires. For example there could be very thin coatings on the fibers or a "chemical reaction" takes place in a thin transition region between the fiber and the matrix. We may wish to assign intrinsic properties to a material singular surface to model such regions. With a grain of salt we may say that our model is somewhere between two- and three-dimensions. An example of an immaterial moving singular surface through which matter is not transported is a shock wave. Behind and in front of a shock wave the velocity of the material particles is zero. However, the mass densities, the pressure fields, and, if the shock wave moves through a solid, certain components of the stress tensor show a huge gradient.

In what follows we shall specialize on material bodies, i.e., $V(t) = V^{+} \cup V^{-} \cup A$ will always consist of the same material particles. However, since we do not treat open systems explicitly, this is a deficit only on first glance: It is true that for open systems relative convective flows across the open surface of the to-be-balanced quantity need to be added to global balance equations. Moreover, the velocities in Reynolds' transport theorem as stated below need to be changed from material velocities to non-material ones of the surface of the open system. However, we shall see that the global balances are only a vehicle leading us to the final result namely *local* equations: In regular points of the continuum we will obtain partial differential equations for the fields and their derivatives. They hold in a material point and, therefore, they will no longer carry information regarding the movement of the system boundaries, open or not. Moreover, in singular points, i.e., points on the singular surface we will obtain so-called jump conditions. These do carry information about the state of the singular surface, including its movement.

In the next chapters we will first concentrate on bodies without a singular surface and, in particular, state the global balances of mass and momentum for these. We will discuss the general structure common to all global balances and derive a general field formulation. This will allow us to derive local balances in regular points, in particular for mass and momentum. The local balance of momentum in regular points will then be used to derive, first, the local and, second, the global balance of kinetic energy. We shall realize that this quantity is not conserved since it contains a production term. This in mind we shall postulate the balance of total energy, which is a conserved quantity. The local balance of momentum will finally also be used to obtain local and global balances for the moment of momentum. All balances will then be formulated for bodies with a singular surface.

## 3.2 Balances of Mass and Momentum

The total mass, $M$, of a material body, $V(t)$, follows by adding the masses, $\mathrm{d}M$, of all material points within small sub-volumes, $\mathrm{d}V$. Moreover, the ratio $\mathrm{d}M/\mathrm{d}V$ represents the mass density, $\rho = \hat{\rho}(\boldsymbol{x}, t)$, of the material point, which can change in space and time. Consequently we may write:

$$M = \int_M \mathrm{d}M = \iiint_{V(t)} \rho \, \mathrm{d}V. \tag{3.2.1}$$

The total mass of a material body is (by definition) conserved, i.e., it does not change with time:

$$\frac{\mathrm{d}M}{\mathrm{d}t} = 0 \Rightarrow \frac{\mathrm{d}}{\mathrm{d}t} \iiint_{V(t)} \hat{\rho}(\boldsymbol{x}, t) \, \mathrm{d}V = 0. \tag{3.2.2}$$

Rewriting the last expression any further is not possible yet: We have to clarify how to perform the time derivative since the time variable is present in the integration boundaries *and* in the integrand as well. At this point we only note that Eq. (3.2.2) is the *global balance of mass*. Note again that mass is a conserved quantity, it cannot be produced and it obeys a *conservation law*. In classical physics mass cannot be created out of nothing nor can it vanish into nowhere. Moreover, a material system does not allow that mass is transported across its boundaries, i.e., mass cannot leave nor enter through the system's boundaries. This is why the right hand side of Eq. (3.2.2) is simply zero.

Isaac NEWTON was born on January 4, 1643 in Woolsthorpe-by-Colsterworth in Lincolnshire and died on March 31, 1727 in Kensington. Undoubtedly he was one of the greatest scientific minds of all times. In his most famous book entitled *Philosophiae Naturalis Principia Mathematica* he establishes the all-time principles of classical mechanics, nowadays called NEWTON's laws. However, he had also fundamental things to say about many other branches of physics, for example on optics. As a human being, however, SIR ISAAC was no NEWTON. He had a strong tendency to secret-mongering as far as his scientific discoveries were concerned. On the other hand, he insisted adamantly and vengefully on his priority rights, whenever one of "his" discoveries appeared in someone else's work. A notorious example of this attitude is his feud about the discovery of calculus with LEIBNIZ.

Just like mass the momentum of mass is an additive quantity, too. Thus the total momentum can be obtained by summing up the contributions $\mathrm{d}\boldsymbol{P} = \mathrm{d}M\boldsymbol{\upsilon} = \rho\boldsymbol{\upsilon}\,\mathrm{d}V$ of all material points. However, unlike mass momentum is a vector, due to the velocity field $\boldsymbol{\upsilon} = \hat{\boldsymbol{\upsilon}}(\boldsymbol{x}, t)$ it contains. Consequently, we obtain for the total momentum of a body:

$$\boldsymbol{P} = \iiint\limits_{V(t)} \rho\,\boldsymbol{\upsilon}\,\mathrm{d}V. \tag{3.2.3}$$

It is NEWTON's merit to have realized that forces, $\boldsymbol{K}$, cause a change of momentum in time. Therefore we write:

$$\frac{\mathrm{d}\boldsymbol{P}}{\mathrm{d}t} = \boldsymbol{K} \equiv \boldsymbol{T} + \boldsymbol{F}. \tag{3.2.4}$$

Note that it is customary in continuum mechanics to split forces additively into long and short range contributions. The latter—identified by the symbol $\boldsymbol{T}$—are transferred across the surface of the body by contact with other bodies. The former ones, however,—denoted by the symbol $\boldsymbol{F}$—reach directly into the interior of the body, for example gravity or electromagnetic fields. In order to rewrite both types of forces in Eq. (3.2.4) in terms of field quantities we refer $\boldsymbol{T}$ to unit surface and $\boldsymbol{F}$ to unit mass. In other words, we postulate that additive fields $\boldsymbol{t} = \hat{\boldsymbol{t}}(\boldsymbol{x}, t; \boldsymbol{n})$ and $\boldsymbol{f} = \hat{\boldsymbol{f}}(\boldsymbol{x}, t)$ exist so that:

$$\boldsymbol{T} = \oiint\limits_{\partial V(t)} \boldsymbol{t}\,\mathrm{d}A, \quad \boldsymbol{F} = \iiint\limits_{V(t)} \rho\,\boldsymbol{f}\,\mathrm{d}V. \tag{3.2.5}$$

Baron Augustin Louis CAUCHY was born on August 21, 1789 in Paris. He died on May 23, 1857 in Sceaux near Paris. In his early years he was a scientist in NAPOLEON's army, just like FOURIER. At the age of 26 he was already a professor at the École Polytechnique, where he soon established himself as one of the leading French mathematicians of his time. More than 780 publications are attributed to him. In the same context, however, rumor has it that he was not free of committing plagiarism. True or not, in any case it got him the dubious nickname "cochon" in academic circles.

Note that the so-called *traction*, $\boldsymbol{t}$, does not only depend on position and time, but also on the unit normal, $\boldsymbol{n}$, of the surface element, $\mathrm{d}A$. CAUCHY showed by using the so-called *tetrahedron argument* (cf., Exercise 3.2.1) that the dependence is linear and the following relation holds to the field of the *stress tensor*, $\boldsymbol{\sigma}$:

$$\boldsymbol{t} = \boldsymbol{n} \cdot \boldsymbol{\sigma}. \tag{3.2.6}$$

**Exercise 3.2.1: CAUCHY'S *formula***

Motivate in a first step what prompts us to write $\boldsymbol{t} = \hat{\boldsymbol{t}}(\boldsymbol{x}, t; \boldsymbol{n})$. To this end use differently oriented surfaces to show that the effect of a given force vector $\boldsymbol{t}\,\mathrm{d}A$ in terms of tension, shear or pressure on a surface $\mathrm{d}A$ with the unit normal vector $\boldsymbol{n}$ depends on the direction of that very normal.

Study now the equilibrium of forces on the tetrahedron shown in Fig. 3.2. Its four surfaces, $\mathrm{d}A$, $\mathrm{d}A_1$, $\mathrm{d}A_2$, and $\mathrm{d}A_3$ are loaded with forces $\boldsymbol{t}\,\mathrm{d}A$, $\boldsymbol{t}^1\mathrm{d}A_1$, $\boldsymbol{t}^2\mathrm{d}A_2$, and $\boldsymbol{t}^2\mathrm{d}A_3$, respectively. Show that w.r.t. the Cartesian frame of the figure we may write:

$$\hat{t}_i(\boldsymbol{x}, t; \boldsymbol{n}) = n_j \sigma_{ji} \tag{3.2.7}$$

with the abbreviation:

$$\sigma_{ji} = \hat{t}^i_j(\boldsymbol{x}, t; -\boldsymbol{e}_i), \tag{3.2.8}$$

in which we recognize the Cartesian components of the stress tensor. For the proof also apply and comment on the reaction-principle in the form:

$$\hat{\boldsymbol{t}}(\boldsymbol{x}, t; \boldsymbol{n}) = -\hat{\boldsymbol{t}}(\boldsymbol{x}, t; -\boldsymbol{n}). \tag{3.2.9}$$

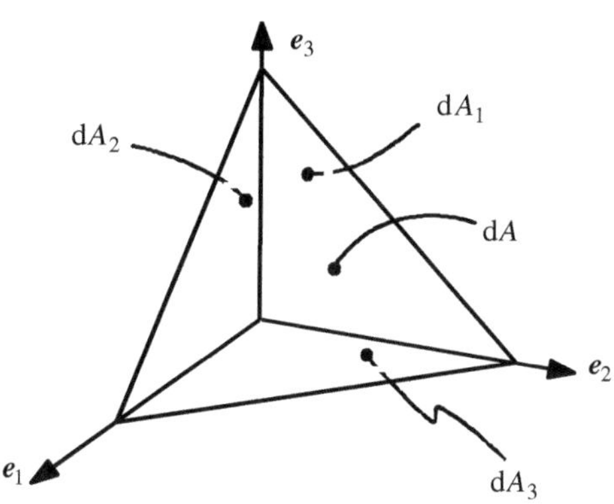


**Fig. 3.2** Equilibrium of forces on a tetrahedron

Also interpret Eq. (3.2.7) in terms of a (left sided) scalar product between the vector $\boldsymbol{n} = n_k \boldsymbol{e}_k$ and the tensor $\boldsymbol{\sigma} = \sigma_{ji} \boldsymbol{e}_j \otimes \boldsymbol{e}_i$ so that the general vector relation (3.2.6) results.

In honor of its inventor the quantity $\boldsymbol{\sigma}$ is also referred to as the CAUCHY *stress tensor*. Sometimes it is also called the *current* or *true stress tensor*, since it relates current forces to current surfaces.

Consequently we obtain the following *global balance of momentum* (NEWTON'S *Lex Secunda*):

$$\frac{\mathrm{d}}{\mathrm{d}t} \iiint\limits_{V(t)} \hat{\rho}(\boldsymbol{x},t)\, \hat{\upsilon}(\boldsymbol{x},t)\, \mathrm{d}V = \oint\!\!\oint\limits_{\partial V(t)} \boldsymbol{n} \cdot \boldsymbol{\sigma}\, \mathrm{d}A + \iiint\limits_{V(t)} \rho \boldsymbol{f}\, \mathrm{d}V. \tag{3.2.10}$$

As before, the left hand side of this equation cannot be transformed any further yet, because of the complication involved with the time differentiation. Once again we note that this equation represents a global balance, this time for the momentum. Momentum is a conserved quantity. There is no production of momentum. However, there is a flux of momentum across the boundary, namely $\boldsymbol{T}$, and a volume supply of momentum, namely $\boldsymbol{F}$. One may ask why is $\boldsymbol{F}$ a supply and not a production? The answer is hidden in the following argument: In principle supplies can be shielded and, therefore, be controlled. This is not possible with productions. They develop inside the system and cannot be influenced by an outside observer, even in principle. Gravity is a concrete example of a supply of momentum. It can be switched off if we move the system far away from all other masses. This should be possible, at least in the mind.

Now that we have learned about two specific balances it is time to generalize and then face the problem of how to perform the time derivatives on their left hand side.

## 3.3 General Global Balances

We wish to study the change in time of an arbitrary physical quantity $\psi$ assigned to the body shown in Fig. 3.1. This quantity is supposed to be additive w.r.t. the matter in that body or, more precisely, w.r.t. the unit volume or unit surface, respectively. Consequently, densities $\underset{V}{\psi}$ and $\underset{A}{\psi}$ exist so that:

$$\psi = \iiint\limits_{V^{+} \cup V^{-}} \underset{V}{\psi}\, \mathrm{d}V + \iint\limits_{A} \underset{A}{\psi}\, \mathrm{d}A. \tag{3.3.1}$$

Examples of such additive quantities are mass, momentum, and—as we shall see later—angular momentum, energy, entropy, and electric charge. Below we will detail the corresponding volume and surface densities $\underset{V}{\psi}$ and $\underset{A}{\psi}$. Nevertheless, in general we may say that the temporal change of $\psi$ is determined by a *flux* across the surface, $F$, a *production*, $P$, and a *supply*, $S$. The latter two are defined within the interior of the body. Consequently we write:

$$\frac{\mathrm{d}\psi}{\mathrm{d}t} = F + P + S. \tag{3.3.2}$$

In its present form this equation is not very helpful yet. We must represent $F$, $P$, and $S$ by fields, i.e., densities that vary in space and time. The flux $F$ occurs across the surfaces $A^{\pm}$ and across the periphery of the singular surface $A$, i.e., the line $L \equiv \partial A$, all of which are depicted in the figure. By using the *vector field densities* $\underset{A}{\boldsymbol{\phi}}$ and $\underset{L}{\boldsymbol{\phi}}$, which are with reference to unit surface and unit length, respectively, we write:

$$F = -\iint\limits_{A^+ \cup A^-} \underset{A}{\boldsymbol{\phi}} \cdot \boldsymbol{n}\, \mathrm{d}A - \oint\limits_{L} \underset{L}{\boldsymbol{\phi}} \cdot \boldsymbol{\nu}\, \mathrm{d}l. \tag{3.3.3}$$

The two minus signs are a matter of convention: If the vectors $\underset{A}{\boldsymbol{\phi}}$ and $\underset{L}{\boldsymbol{\phi}}$ point into the body their scalar product with the outward normals $\boldsymbol{n}$ and $\boldsymbol{\nu}$ becomes negative. In this case, however, we expect $\psi$ to grow. This will be guaranteed by the minus signs. As far as $P$ is concerned, we note that production is possible within the volumes $V^{\pm}$ as well as in the (open) surface $A$. Therefore we write with the corresponding volume and surface densities of production $\underset{V}{p}$ and $\underset{A}{p}$:

$$P = \iiint\limits_{V^+ \cup V^-} \underset{V}{p}\, \mathrm{d}V + \iint\limits_{A} \underset{A}{p}\, \mathrm{d}A. \tag{3.3.4}$$

Just like the production the supply can be described by volume and surface densities $\underset{V}{s}$ and $\underset{A}{s}$:

$$S = \iiint\limits_{V^+ \cup V^-} \underset{V}{s}\, \mathrm{d}V + \iint\limits_{A} \underset{A}{s}\, \mathrm{d}A. \tag{3.3.5}$$

At this point we may ask again why it is useful to distinguish between productions and supplies which, from a mathematical point of view, are completely alike. As already mentioned the reason for the distinction is a physical one: Supplies can be controlled and suppressed by the experimenter, whereas productions cannot. As an example of a supply, gravity has already been mentioned, which can be switched off by performing the experiment in outer space (say). Another example (in context with the energy balance, see this section below) is radiation, which can be prevented to enter the system by suitable shielding. An example of a production is the dissipated power of the stresses within a volume due to velocity gradients, profanely known as internal friction. Friction cannot simply be switched off. On the contrary, it will adjust itself during the process. Thus we obtain by combination of Eqs. (3.3.1–3.3.5):

$$\begin{aligned}\frac{\mathrm{d}}{\mathrm{d}t}\iiint\limits_{V^+ \cup V^-} \underset{V}{\psi}\, \mathrm{d}V + \frac{\mathrm{d}}{\mathrm{d}t}\iint\limits_{A} \underset{A}{\psi}\, \mathrm{d}A = & -\iint\limits_{A^+ \cup A^-} \underset{A}{\boldsymbol{\phi}} \cdot \boldsymbol{n}\, \mathrm{d}A - \oint\limits_{L} \underset{L}{\boldsymbol{\phi}} \cdot \boldsymbol{\nu}\, \mathrm{d}l \\ & + \iiint\limits_{V^+ \cup V^-} \left(\underset{V}{p} + \underset{V}{s}\right) \mathrm{d}V + \iint\limits_{A} \left(\underset{A}{p} + \underset{A}{s}\right) \mathrm{d}A.\end{aligned} \tag{3.3.6}$$

This equation is much more detailed than Eq. (3.3.2). However, it is comparatively unwieldy to use during calculations. The integrals pose a potential problem. They are often hard to evaluate. Mathematically "easier" are *localbalances*. In these the various densities are related differentially. As we shall see they can be turned into partial differential equations for the aforementioned primary fields, so that well-established solution methods can be applied. In fact we have to distinguish between *localbalances inregular* and *in singularpoints* of the body, respectively, or in other words, points within the (material) volumes, $V^+$ and $V^-$, and points on the singular surface, $A$. To obtain them we first of all have to deal with the time derivatives on the left side of Eq. (3.3.6) and pull them beneath the integrals, so to speak. For this purpose we need so-called transport theorems which will be discussed—for volume integrals—in the Sect. 3.4. Finally, it is worth mentioning that Eq. (3.3.6) simplifies considerably for the case of a material volume $V$ with the closed surface $\partial V$ that does not contain a singular surface:

$$\frac{\mathrm{d}}{\mathrm{d}t}\iiint\limits_{V} \underset{V}{\psi}\,\mathrm{d}V = -\oint\!\!\!\!\oint\limits_{\partial V} \underset{A}{\boldsymbol{\varphi}}\cdot\boldsymbol{n}\,\mathrm{d}A + \iiint\limits_{V}\left(\underset{V}{p} + \underset{V}{s}\right)\mathrm{d}V. \tag{3.3.7}$$

## 3.4 Transport Theorem for Volumes

Executing the time derivatives on the left hand side of Eqs. (3.3.6 / 3.3.7) is problematic because the integrand as well as the integration boundaries are both time-dependent. In this section we concentrate on the case of the volume integral shown in Eqs. (3.3.6 / 3.3.7) and anticipate the result. It turns out that in Cartesian coordinates we may write:

$$\frac{\mathrm{d}}{\mathrm{d}t}\iiint\limits_{V^+\cup V^-} \underset{V}{\psi}\,\mathrm{d}V = \iiint\limits_{V^+\cup V^-}\left[\frac{\partial \underset{V}{\psi}}{\partial t} + \frac{\partial}{\partial x_i}\left(\underset{V}{\psi}\, \upsilon_i\right)\right]\mathrm{d}V. \tag{3.4.1}$$

Johann Carl Friedrich Gauss was born on April 30 1777 in Braunschweig and died on February 23, 1855 in Göttingen. Just like Newton he was an eminent scientist and rather nasty human character. As far as his professional side was concerned he was interested in applied as well as pure mathematics. However, he also contributed to the solution of real world problems, such as electricity, surveying, or the setup of a telegraph line near Göttingen. Hence it is not surprising that he was a professor of mathematics (1807) at Göttingen University as well as the director of the observatory in Göttingen (1821).

Note that the same equation holds also for a volume without a singular surface. In this case we simply replace $V^+ \cup V^-$ by $V$. Also note that we have chosen the Cartesian representation for the velocity field, mainly for computational reasons as we shall see shortly. However, before we turn to the proof of this relation we will rewrite the second integral. This is where GAUSS' theorem comes in.

**Exercise 3.4.1:Mostly GAUSS' theorem**

This exercise is dedicated to GAUSS' theorem, which allows for a conversion of volume integrals over a continuous field $g = g(\boldsymbol{x})$ defined within the region $V$ with the closed hull $\partial V$ to surface integrals as follows:

$$\iiint\limits_V \frac{\partial g}{\partial x_i}\,\mathrm{d}V = \oiint\limits_{\partial V} g\, n_i\,\mathrm{d}A. \tag{3.4.2}$$

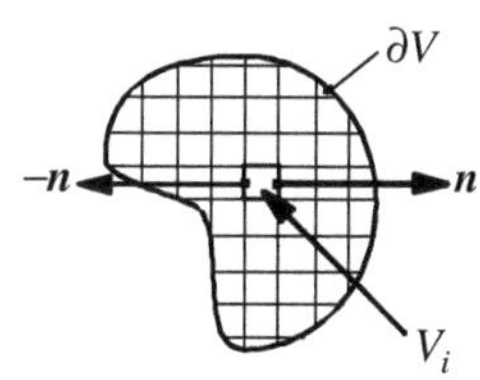


**Fig. 3.3** Proof of GAUSS' theorem

In Cartesian coordinates the field $g$ can, but does not have to be, a component of a vector or a tensor. We proceed to sketch a proof of the theorem in Cartesian coordinates and will extend it later to arbitrary coordinates (see Sect. 4.5). To this end consider the sketch of Fig. 3.3. Start with the surface integral on the right hand side of Eq. (3.4.2) and apply it suitably to the six surfaces of the small cubes shown in the figure.

Now extend the surfaces of each cube to cover its whole volume. Combine adjacent surfaces of the cube and generate partial derivatives while observing the mean value theorem for integrals. Consider now the limit case of infinitesimally small cubes and arrive at Eq. (3.4.2). Reconsider each step of the sequence of mathematical operations and explain why Eq. (3.4.2) is only valid for continuous fields. Explain that the equation can be generalized as follows in order to cover also the case of a singular surface, $A$, dividing the volume as shown in Fig. 3.1:

$$\iiint\limits_{V^+\cup V^-} \frac{\partial g}{\partial x_i}\,\mathrm{d}V = \iint\limits_{A^+\cup A^-} g\, n_i\,\mathrm{d}A - \iint\limits_{A} [\![g]\!]\, e_i\,\mathrm{d}A, \tag{3.4.3}$$

where double brackets refer to the *jump* at the surface:

$$[\![g]\!] = g^+ - g^-. \tag{3.4.4}$$

$g^+$ and $g^-$ denote the right- and left-sided limits of the field $g$ when approaching the singular surface.

Now define the so-called *del operator* or *nabla symbol* by using the Cartesian base of unit vectors, $\boldsymbol{e}_i$:

$$\nabla = \frac{\partial(\cdot)}{\partial x_i}\boldsymbol{e}_i. \tag{3.4.5}$$

Show that Eqs. (3.4.2 / 3.4.3) can be written in the following symbolic, system independent form:

$$\begin{aligned}
&\iiint\limits_V \nabla g \, \mathrm{d}V = \oiint\limits_{\partial V} g\,\boldsymbol{n}\,\mathrm{d}A,\\
&\iiint\limits_{V^+ \cup V^-} \nabla g \, \mathrm{d}V = \iint\limits_{A^+ \cup A^-} g\,\boldsymbol{n}\,\mathrm{d}A - \iint\limits_{A} [\![g]\!]\,\boldsymbol{e}\,\mathrm{d}A
\end{aligned} \tag{3.4.6}$$

Now suppose that $g$ is a vector field $\boldsymbol{g} = g_j\boldsymbol{e}_j$ given in Cartesian representation. Use Eq. (3.4.6) to prove the so-called *divergence theorem*:

$$\begin{aligned}
&\iiint\limits_V \frac{\partial g_i}{\partial x_i}\,\mathrm{d}V = \oiint\limits_{\partial V} g_i n_i\,\mathrm{d}A,\\
&\iiint\limits_{V^+ \cup V^-} \frac{\partial g_i}{\partial x_i}\,\mathrm{d}V = \iint\limits_{A^+ \cup A^-} g_i n_i\,\mathrm{d}A - \iint\limits_{A} [\![g_i]\!]\,e_i\,\mathrm{d}A
\end{aligned} \tag{3.4.7}$$

or:

$$\begin{aligned}
&\iiint\limits_V \nabla\cdot\boldsymbol{g}\,\mathrm{d}V = \oiint\limits_{\partial V} \boldsymbol{g}\cdot\boldsymbol{n}\,\mathrm{d}A,\\
&\iiint\limits_{V^+ \cup V^-} \nabla\cdot\boldsymbol{g}\,\mathrm{d}V = \iint\limits_{A^+ \cup A^-} \boldsymbol{g}\cdot\boldsymbol{n}\,\mathrm{d}A - \iint\limits_{A} [\![\boldsymbol{g}]\!]\cdot\boldsymbol{e}\,\mathrm{d}A.
\end{aligned} \tag{3.4.8}$$

By using the result from Eq. (3.4.3) we can now rewrite Eq. (3.4.1) as follows:

$$\begin{aligned}
\frac{\mathrm{d}}{\mathrm{d}t}\iiint\limits_{V^+ \cup V^-} \psi_V\,\mathrm{d}V = &\iiint\limits_{V^+ \cup V^-} \frac{\partial \psi_V}{\partial t}\,\mathrm{d}V\\
&+ \iint\limits_{A^+ \cup A^-} \psi_V\,\upsilon_i n_i\,\mathrm{d}A - \iint\limits_{A} [\![\psi_V \upsilon_i]\!]\,e_i\,\mathrm{d}A.
\end{aligned} \tag{3.4.9}$$

Note that during the proof of this equations we have implicitly assumed (namely during the application of the extended GAUSS' theorem with discontinuities) that $V^+$ and $V^-$ are *material* volumes, in which the fields are *continuous*.

We write for the jump $[\![\psi_V \upsilon_i]\!] = \underset{V}{\psi}^+ \upsilon_i^+ - \underset{V}{\psi}^- \upsilon_i^- = \left(\underset{V}{\psi}^+ - \underset{V}{\psi}^-\right) \underset{A}{\upsilon}_i$. After the last equality sign we have assumed that $\underset{A}{\upsilon}_i = \upsilon_i^+ = \upsilon_i^-$, where $\underset{A}{\upsilon}_i$ denotes the local velocity of the singular surface. By doing so we obviously assume that the singular surface is a material one, which can only move along with the adjacent particles. This, however, must not necessarily always be the case. In general, the velocity $\underset{A}{\upsilon}_i$ of the singular surface $A$ is independent of the velocities $\upsilon_i^+$ and $\upsilon_i^-$ of the material particles on both sides and these do not necessarily need to be equal. Then an additional amount of $\psi$ will be swept into the volumes $V^+$ and $V^-$, due to the relative motion of $A$ against the left- and right-sided particle velocities $\upsilon_i^-$ and $\upsilon_i^+$, namely $-\iint_A \underset{V}{\psi}^+ \left(\underset{A}{\upsilon}_i - \upsilon_i^+\right) e_i \, dA + \iint_A \underset{V}{\psi}^- \left(\underset{A}{\upsilon}_i - \upsilon_i^-\right) e_i \, dA$: This must be added to Eq. (3.4.9). Consequently, the most general transport theorem for a region divided by a singular surface reads:

$$\frac{d}{dt} \iiint_{V^+ \cup V^-} \underset{V}{\psi} \, dV = \iiint_{V^+ \cup V^-} \frac{\partial \underset{V}{\psi}}{\partial t} dV + \iint_{A^+ \cup A^-} \underset{V}{\psi} \boldsymbol{\upsilon} \cdot \boldsymbol{n} \, dA - \iint_A [\![\psi_V]\!] \underset{A}{\upsilon}_\perp \, dA, \tag{3.4.10}$$

where for the sake of brevity the normal component of the velocity of the immaterial surface $A$ has been introduced:

$$\underset{A}{\upsilon}_\perp = \underset{A}{\upsilon}_i e_i \equiv \underset{A}{\boldsymbol{\upsilon}} \cdot \boldsymbol{e}. \tag{3.4.11}$$

Note that a scalar product in Cartesian components, $\upsilon_i n_i$, can be replaced by the absolute notation $\boldsymbol{\upsilon} \cdot \boldsymbol{n}$. This relation is known in the literature as REYNOLDS' *general transport theorem* for material volumes. Such a complex form is only required if a singular surface passes through. In the case of a completely regular region the jump bracket in Eqs. (3.4.9 / 3.4.10) can be dropped and the following simpler relations hold:

$$\frac{d}{dt} \iiint_V \underset{V}{\psi} \, dV = \iiint_V \left[\frac{\partial \underset{V}{\psi}}{\partial t} + \nabla \cdot \left(\underset{V}{\psi} \boldsymbol{\upsilon}\right)\right] dV \equiv \iiint_V \left[\frac{\partial \underset{V}{\psi}}{\partial t} + \frac{\partial}{\partial x_i}\left(\underset{V}{\psi} \upsilon_i\right)\right] dV,$$

$$\frac{d}{dt} \iiint_V \underset{V}{\psi} \, dV = \iiint_V \frac{\partial \underset{V}{\psi}}{\partial t} dV + \oint\!\!\oint_{\partial V} \underset{V}{\psi} \boldsymbol{\upsilon} \cdot \boldsymbol{n} \, dA \equiv \iiint_V \frac{\partial \underset{V}{\psi}}{\partial t} dV + \oint\!\!\oint_{\partial V} \underset{V}{\psi} \upsilon_i n_i \, dA. \tag{3.4.12}$$

In this case the surface is closed, which is acknowledged by a circle in the double integral. This equation allows for a rather simple interpretation, which is often offered as a proof in elementary textbooks on continuum mechanics: The temporal change of an additive quantity has two sources. First, the local density of that quantity in a material point may change in time, which explains the volume integral. Second, the quantity may leave or enter the volume by convective flow. This corresponds to the surface integral. An influx corresponds to a negative value of $\boldsymbol{v} \cdot \boldsymbol{n}$ and a drain to a positive one. If the velocity and the normal are perpendicular to each other there is no net influx: nothing will leave nor enter the volume.

Joseph-Louis de LAGRANGE was born on January 25, 1736 in Torino (Italy) and died on April 10, 1813 in Paris. It is for that reason that both the Italians as well as the French consider him to be one of their countrymen (that is why the former call him "LAGRANGIA"). He coined the idea of a rational, analytical way of mechanics, based on mathematical principles alone, preferably without any illustrations. This line of reasoning is impressively demonstrated in his most famous work entitled (in French and not in Italian) *traité de mécanique analytique*.

However, we are not completely satisfied with the intuitive argument yet. Before we comment on how to treat the time derivative in front of the surface integral of Eq. (3.3.6) and discuss the corresponding transport theorem, we present a more formal proof of Eq. (3.4.1). For this purpose we need the concept of the *Lagrangian description* also known as *referential* or *material perception*. Recall that on the one hand the motion of a material point as shown in Fig. 3.4 can be described from the *Eulerian* (also known as *spatial*) *point of view*: At the time $t$ the body moves across a fixed grid in space, $\boldsymbol{x}$. According to LAGRANGE we may alternatively characterize the motion by moving along with the material point: Fig. 3.5. The material point itself—which can neither be created nor destroyed—is

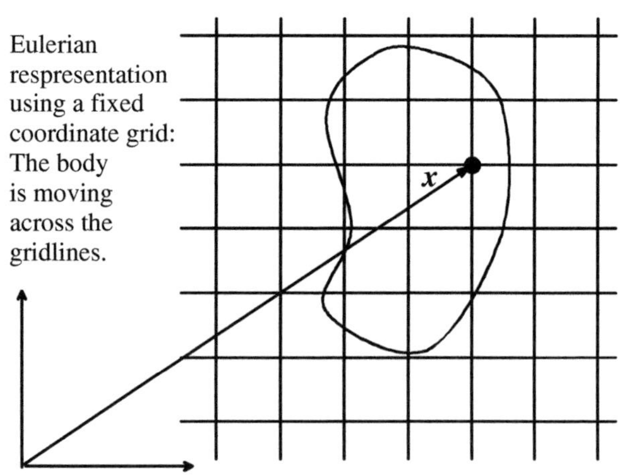


**Fig. 3.4** Eulerian description of motion by a fixed spatial grid

uniquely identified by its position vector, $\boldsymbol{X}$, in the so-called reference configuration (which can be stress free, say). We write for the current position of a material point (also known as *motion*)[2]:

$$\boldsymbol{x} = \tilde{\boldsymbol{x}}(\boldsymbol{X}, t). \tag{3.4.13}$$

Note that when we use the symbols $\boldsymbol{x}$ or $\boldsymbol{X}$ we have not specified the type of coordinate system yet. We now consider two neighboring points in their reference configuration expressed in Cartesian coordinates, $X_I$ and $X_I + \mathrm{d}X_I$, as well as in their current position, $\tilde{x}_i(X_K, t)$ and $\tilde{x}_i(X_K + \mathrm{d}X_K, t)$. Their distance in the current configuration is given by:

$$\begin{aligned} \mathrm{d}x_i &\equiv \tilde{x}_i(X_K + \mathrm{d}X_K, t) - \tilde{x}_i(X_K, t) \\ &= \tilde{x}_i(X_K, t) + \frac{\partial x_i}{\partial X_K}\mathrm{d}X_K - \tilde{x}_i(X_K, t) = F_{iK}\mathrm{d}X_K. \end{aligned} \tag{3.4.14}$$

Leonard EULER was born on April 15, 1707 in Basel. He died on September 18, 1783 in St. Petersburg. We owe him infinitely much as far as the concepts of analysis and of classical mechanics are concerned. Even though he is Swiss by birth, he does not stay in Basel and moves away soon: First to St. Petersburg (supported by his colleague Daniel BERNOULLI), then on to Berlin (thanks to FREDERICK THE GREAT), and then back to St. Petersburg (thanks to CATHERINE THE GREAT). EULER suffers a tragic fate: He loses his eyesight completely. Nevertheless he admirably continues to work in science, dictates his results to a secretary and leaves an *œvre* of gigantic proportions.

$F_{iK} = \partial\tilde{x}_i(X_L, t)/\partial X_K = \tilde{F}_{iK}(X_L, t)$ denotes the so-called *deformation gradient*. Its purpose is to connect current distances with distances in the reference configuration. Note that we have used capital Latin characters for some of the indices in the previous formulae. They run from 1 to 3 just like small letters. However, they are supposed to remind us of the reference configuration. Note that the deformation gradient possesses *both* types of indices since it connects the reference configuration with the current one. Although such a subtle distinction during the choice of indices is not imperative, it is helpful and provides an additional means of checking the correctness of tensor equations.

Moreover, the mathematical requirement $\det \boldsymbol{F} \neq 0$ can be interpreted as continuity of particle motion. In order to appreciate this notion, recall that the motion of a particle must be unique since particles cannot be created nor destroyed. Consequently, the following unique inversions of Eq. (3.4.13) are possible:

$$x_i \equiv \tilde{x}_i(X_K, t) \quad \Leftrightarrow \quad X_K = \hat{X}_K(x_i, t). \tag{3.4.15}$$

[2] In this book functions written in Eulerian notation are identified by a circumflex whereas functions in Lagrangian description are easily spotted by a tilde.

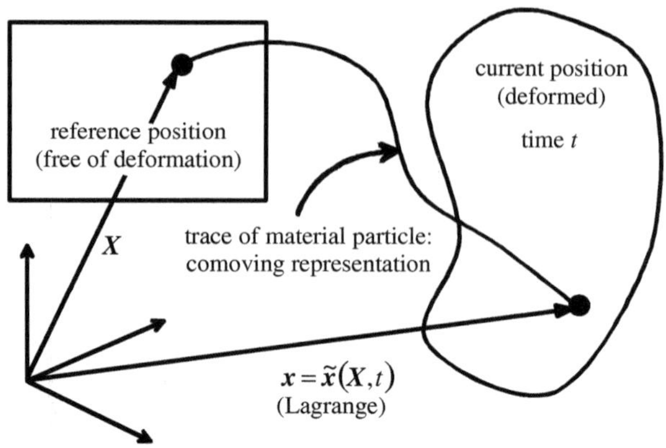


**Fig. 3.5** Lagrangian description of motion by comoving with the particle

Carl Gustav Jacob JACOBI was born on December 10, 1804 in Potsdam (Germany) and died on February 18, 1851 in Berlin. He was a child prodigy, at least in mathematics, ready to go to university at the tender age of 13. However, Berlin University would not accept students younger than 16 and so he did private studies on advanced mathematics for four more years. This allowed him to finish his dissertation in 1825 and his habilitation thesis the year after, both in Berlin. From 1826 to 1843 he taught in Königsberg (East Prussia). In 1844 he became a full member of the Prussian Academy of Science and, not surprisingly, had a nervous breakdown the very same year. For rehab he first went to Italy to finally retire in Berlin while enjoying his pension provided to him by the Prussian Crown. Foolishly he engaged himself in the revolution of 1848 and the payment of his monies was temporarily suspended.

In order to guarantee this property the function of motion $\tilde{x}_i(X_k, t)$ must be unique and continuously differentiable, i.e., the following determinant must not vanish (so-called *inverse function theorem*):

$$\left|\frac{\partial x_j}{\partial X_K}\right| \neq 0. \tag{3.4.16}$$

By comparison with the expression in Eq. (3.4.14), this corresponds to $\det F \neq 0$. Moreover, it is customary to write ($J$ stands for the Jacobian, i.e., the determinant of the functional derivatives):

$$J = \det \boldsymbol{F} \equiv \epsilon_{ijk} F_{i1} F_{j2} F_{k3} = \frac{1}{6} \epsilon_{ijk} \epsilon_{LMN} F_{iL} F_{jM} F_{kN}. \tag{3.4.17}$$

Here we have used the completely antisymmetric tensor of third order (a.k.a. the Levi-Civita symbol) in Cartesian components:

$$\epsilon_{ijk} = \begin{cases} +1, & \text{if} \quad i,j,k = 1,2,3 \text{ and cyclic permutations} \\ -1, & \text{if} \quad i,j,k = 2,1,3 \text{ and cyclic permutations} \\ 0, & \text{else.} \end{cases} \tag{3.4.18}$$

**Exercise 3.4.2:** ***A representation of the Jacobian***

Explain the validity of Eq. (3.4.17). For this purpose recall and prove the following auxiliary formula for the vector product:

$$\boldsymbol{a} \times \boldsymbol{b} \equiv \begin{vmatrix} \boldsymbol{e}_1 & \boldsymbol{e}_2 & \boldsymbol{e}_3 \\ a_1 & a_2 & a_3 \\ b_1 & b_2 & b_3 \end{vmatrix} = \boldsymbol{e}_i \epsilon_{ijk} a_j b_k. \tag{3.4.19}$$

Now identify $F_{i1}$, $F_{j2}$, $F_{k3}$ suitably and validate the form $\frac{1}{6}\epsilon_{ijk}\epsilon_{LMN}F_{iL}F_{jM}F_{kN}$.

**Exercise 3.4.3: The Levi-Civita** ***symbol in absolute tensor notation***

Recall the properties of an orthogonal Cartesian unit base and show that we may recover Eq. (3.4.18) by writing:

$$\epsilon_{ijk} \equiv \boldsymbol{e}_i \cdot \boldsymbol{e}_j \times \boldsymbol{e}_k. \tag{3.4.20}$$

Explain why it makes sense to define the absolute antisymmetric tensor of third order by:

$$\boldsymbol{\epsilon} = \epsilon_{ijk} \boldsymbol{e}_i \otimes \boldsymbol{e}_j \otimes \boldsymbol{e}_k. \tag{3.4.21}$$

Now the first assertion is that the following transformation rule holds for volume elements:

$$\mathrm{d}V = J\,\mathrm{d}V_0. \tag{3.4.22}$$

Here $\mathrm{d}V$ denotes the volume element in the current and $\mathrm{d}V_0$ the volume element in the reference configuration, respectively. For the proof we choose without loss of generality $\mathrm{d}V_0$ to be a rectangular volume. Consequently it can be described by a triple product of the following three vectors:

$$\mathrm{d}\overset{(1)}{X}_L = \left(\mathrm{d}\overset{(1)}{X}_1,\ 0,\ 0\right), \mathrm{d}\overset{(2)}{X}_M = \left(0,\ \mathrm{d}\overset{(2)}{X}_2,\ 0\right), \mathrm{d}\overset{(3)}{X}_N = \left(0,\ 0,\ \mathrm{d}\overset{(3)}{X}_3\right). \tag{3.4.23}$$

We check:

$$\begin{aligned} \mathrm{d}V_0 &= \mathrm{d}\overset{(1)}{\boldsymbol{X}} \cdot \left(\mathrm{d}\overset{(2)}{\boldsymbol{X}} \times \mathrm{d}\overset{(3)}{\boldsymbol{X}}\right) = \mathrm{d}\overset{(1)}{X}_L \epsilon_{LMN} \mathrm{d}\overset{(2)}{X}_M \mathrm{d}\overset{(3)}{X}_N \\ &= \epsilon_{123} \mathrm{d}\overset{(1)}{X}_1 \mathrm{d}\overset{(2)}{X}_2 \mathrm{d}\overset{(3)}{X}_3 = 1\,\mathrm{d}V_0 \end{aligned} \tag{3.4.24}$$

and prove Eq. (3.4.22):

$$\begin{aligned} \mathrm{d}V &= \mathrm{d}\overset{(1)}{\boldsymbol{x}} \cdot \left( \mathrm{d}\overset{(2)}{\boldsymbol{x}} \times \mathrm{d}\overset{(3)}{\boldsymbol{x}} \right) = \mathrm{d}\overset{(1)}{x}_i \, \epsilon_{ijk} \mathrm{d}\overset{(2)}{x}_j \, \mathrm{d}\overset{(3)}{x}_k \\ &= \epsilon_{ijk} \frac{\partial x_i}{\partial X_R} \frac{\partial x_j}{\partial X_S} \frac{\partial x_k}{\partial X_T} \mathrm{d}\overset{(1)}{X}_R \, \mathrm{d}\overset{(2)}{X}_S \, \mathrm{d}\overset{(3)}{X}_T \\ &= J \mathrm{d}V_0 \quad \Rightarrow \quad \mathrm{d}V = J \mathrm{d}V_0. \end{aligned} \tag{3.4.25}$$

Second, in context with the time derivative (indicated by a dot in what follows) of the deformation gradient, we will make use of the two following equations:

$$\frac{\partial \upsilon_i}{\partial x_j} = \dot{F}_{iK} \left( F^{-1} \right)_{Kj} \tag{3.4.26}$$

and:

$$\dot{J} = \frac{\partial \upsilon_i}{\partial x_i} J \quad \Rightarrow \quad (\ln J)^{\bullet} = \frac{\partial \upsilon_i}{\partial x_i}. \tag{3.4.27}$$

For the proof we first recall the basic definition of velocity in Lagrangian representation:

$$\upsilon_i = \tilde{\upsilon}_i(\boldsymbol{X}, t) = \left. \frac{\partial \tilde{x}_i(\boldsymbol{X}, t)}{\partial t} \right|_{\boldsymbol{X}} \equiv \dot{x}_i \tag{3.4.28}$$

and continue to argue as follows:

$$\begin{aligned} \frac{\partial \upsilon_i}{\partial x_j} &= \frac{\partial}{\partial x_j} \left( \frac{\partial x_i}{\partial t} \right) = \frac{\partial}{\partial X_K} \left( \frac{\partial x_i}{\partial t} \right) \frac{\partial X_K}{\partial x_j} \\ &= \frac{\partial}{\partial t} \left( \frac{\partial x_i}{\partial X_K} \right) \frac{\partial X_K}{\partial x_j} = \dot{F}_{iK} \left( F^{-1} \right)_{Kj}. \end{aligned} \tag{3.4.29}$$

We insert Eq. (3.4.26) into Eq. (3.4.17) after it has been differentiated with respect to time and expand:

$$\begin{aligned} \dot{J} &= \epsilon_{ijk} \left( \dot{F}_{i1} F_{j2} F_{k3} + F_{i1} \dot{F}_{j2} F_{k3} + F_{i1} F_{j2} \dot{F}_{k3} \right) \\ &= \epsilon_{ijk} \left( \frac{\partial \upsilon_i}{\partial x_r} F_{r1} F_{j2} F_{k3} + \frac{\partial \upsilon_j}{\partial x_r} F_{i1} F_{r2} F_{k3} + \frac{\partial \upsilon_k}{\partial x_r} F_{i1} F_{j2} F_{r3} \right) \\ &= \cdots (6 \text{ terms}). \end{aligned} \tag{3.4.30}$$

Now we expand Eq. $(3.4.27)_1$ and confirm in comparison with Eq. (3.4.30) the identity:

$$\dot{J} = \frac{\partial \upsilon_r}{\partial x_r} \epsilon_{ijk} F_{i1} F_{j2} F_{k3} = \cdots (6 \text{ terms}). \tag{3.4.31}$$

**Exercise 3.4.4:** ***Time derivative of the Jacobian (direct proof)***

Go once more through the proof of Eqs. (3.4.26 / 3.4.27) and discuss each step in the rather casually written chains of Eqs. (3.4.28–3.4.30) by meticulously stating which of the arguments is kept constant in each case.

Equations (3.4.22 / 3.4.27) will now be used to prove Eq. (3.4.1). We assume that the volume density is given in material coordinates, i.e., $\underset{V}{\psi} = \underset{V}{\tilde{\psi}}(X_K, t)$ and find:

$$\begin{aligned}\frac{\mathrm{d}}{\mathrm{d}t}\iiint\limits_{V^+\cup V^-} \underset{V}{\psi}\,\mathrm{d}V &= \frac{\mathrm{d}}{\mathrm{d}t}\iiint\limits_{V_0^+\cup V_0^-} \underset{V}{\psi}\,J\,\mathrm{d}V_0 = \iiint\limits_{V_0^+\cup V_0^-}\left[\frac{\partial \underset{V}{\psi}}{\partial t}J + \underset{V}{\psi}\,\dot{J}\right]\mathrm{d}V_0 \\ &= \iiint\limits_{V_0^+\cup V_0^-}\left[\frac{\partial \underset{V}{\psi}}{\partial t} + \underset{V}{\psi}\frac{\partial \upsilon_i}{\partial x_i}\right]J\,\mathrm{d}V_0 = \iiint\limits_{V^+\cup V^-}\left[\frac{\partial \underset{V}{\psi}}{\partial t} + \underset{V}{\psi}\frac{\partial \upsilon_i}{\partial x_i}\right]\mathrm{d}V.\end{aligned} \tag{3.4.32}$$

In a last step we convert the first term in the integrand from Lagrangian to Eulerian description:

$$\begin{aligned}\left(\frac{\partial \underset{V}{\psi}}{\partial t}\right)_{\text{Lagrange}} &\equiv \left.\frac{\partial \underset{V}{\tilde{\psi}}(\boldsymbol{X},t)}{\partial t}\right|_{\boldsymbol{X}} = \left.\frac{\partial \underset{V}{\tilde{\psi}}\big(\hat{\boldsymbol{X}}(\boldsymbol{x},t),t\big)}{\partial t}\right|_{\boldsymbol{X}} = \left.\frac{\partial \underset{V}{\hat{\psi}}(\tilde{\boldsymbol{x}}(\boldsymbol{X},t),t)}{\partial t}\right|_{\boldsymbol{X}} \\ &= \left.\frac{\partial \underset{V}{\hat{\psi}}(\boldsymbol{x},t)}{\partial x_i}\right|_t \left.\frac{\partial \tilde{x}_i(\boldsymbol{X},t)}{\partial t}\right|_{\boldsymbol{X}} + \left.\frac{\partial \underset{V}{\hat{\psi}}(\boldsymbol{x},t)}{\partial t}\right|_{\boldsymbol{x}} \\ &= \left.\frac{\partial \underset{V}{\hat{\psi}}(\boldsymbol{x},t)}{\partial t}\right|_{\boldsymbol{x}} + \left.\frac{\partial \underset{V}{\hat{\psi}}(\boldsymbol{x},t)}{\partial x_i}\right|_t \tilde{\upsilon}_i(\boldsymbol{X},t) \\ &= \left.\frac{\partial \underset{V}{\hat{\psi}}(\boldsymbol{x},t)}{\partial t}\right|_{\boldsymbol{x}} + \left.\frac{\partial \underset{V}{\hat{\psi}}(\boldsymbol{x},t)}{\partial x_i}\right|_t \tilde{\upsilon}_i\big(\hat{\boldsymbol{X}}(\boldsymbol{x},t),t\big) \\ &= \left.\frac{\partial \underset{V}{\hat{\psi}}(\boldsymbol{x},t)}{\partial t}\right|_{\boldsymbol{x}} + \left.\frac{\partial \underset{V}{\hat{\psi}}(\boldsymbol{x},t)}{\partial x_i}\right|_t \hat{\upsilon}_i(\tilde{\boldsymbol{x}}(\boldsymbol{X},t),t) \\ &= \left.\frac{\partial \underset{V}{\hat{\psi}}(\boldsymbol{x},t)}{\partial t}\right|_{\mathbf{x}} + \left.\frac{\partial \underset{V}{\hat{\psi}}(\boldsymbol{x},t)}{\partial x_i}\right|_t \hat{\upsilon}_i(\boldsymbol{x},t) \equiv \left(\frac{\partial \underset{V}{\psi}}{\partial t} + \frac{\partial \underset{V}{\psi}}{\partial x_i}\upsilon_i\right)_{\text{Euler}}.\end{aligned} \tag{3.4.33}$$

Note that before the first and after the second identity sign the arguments of the functions were not explicitly shown as is common practice in continuum theory, at least for non-ambiguous cases. We finally combine the result with the second part of the argument of the integral in Eq. (3.4.32):

$$\begin{aligned}\left(\frac{\partial \underset{V}{\psi}}{\partial t}+\underset{V}{\psi}\frac{\partial \upsilon_i}{\partial x_i}\right)_{\text{Lagrange}} &\equiv \left.\frac{\partial \underset{V}{\tilde{\psi}}(\boldsymbol{X},t)}{\partial t}\right|_{\boldsymbol{X}} + \underset{V}{\tilde{\psi}}(\boldsymbol{X},t)\left.\frac{\partial \tilde{\upsilon}_i(\boldsymbol{X}(\boldsymbol{x},t),t)}{\partial x_i}\right|_t \\ &= \left.\frac{\partial \underset{V}{\hat{\psi}}(\boldsymbol{x},t)}{\partial t}\right|_{\boldsymbol{x}} + \left.\frac{\partial \underset{V}{\hat{\psi}}(\boldsymbol{x},t)}{\partial x_i}\right|_t \hat{\upsilon}_i(\boldsymbol{x},t) + \underset{V}{\hat{\psi}}(\boldsymbol{x},t)\left.\frac{\partial \hat{\upsilon}_i(\boldsymbol{x},t)}{\partial x_i}\right|_t \\ &= \left.\frac{\partial \underset{V}{\hat{\psi}}(\boldsymbol{x},t)}{\partial t}\right|_{\boldsymbol{x}} + \left.\frac{\partial \left[\underset{V}{\hat{\psi}}(\boldsymbol{x},t)\hat{\upsilon}_i(\boldsymbol{x},t)\right]}{\partial x_i}\right|_t \\ &\equiv \left(\frac{\partial \underset{V}{\psi}}{\partial t}+\frac{\partial}{\partial x_i}\left(\underset{V}{\psi}\,\upsilon_i\right)\right)_{\text{Euler}} = \frac{\partial \underset{V}{\psi}}{\partial t}+\frac{\partial}{\partial x_i}\left(\underset{V}{\psi}\,\upsilon_i\right),\end{aligned} \tag{3.4.34}$$

where the expressions after the last identity sign were written in Eulerian manner. This concludes the proof of the transport theorem for volumes.

## 3.5 Transport Theorem for Surface Densities

We now turn to the problem of how to rewrite the time derivative of the surface integral from Eq. (3.3.6). As for the case of volume integrals we first state and discuss the result:

$$\frac{\mathrm{d}}{\mathrm{d}t}\iint_A \underset{A}{\psi}\,\mathrm{d}S = \iint_A \left[\frac{\partial \underset{A}{\psi}}{\partial t} + \underset{A}{\psi}\left(\underset{A}{\upsilon}{}^{\Delta}_{;\Delta} - 2K_{\mathrm{m}}\,\underset{A}{\upsilon}{}_{\perp}\right)\right]\mathrm{d}S. \tag{3.5.1}$$

A certain similarity to the transport theorem for volumes—see Eq. (3.4.1)—cannot be denied: In the first term the integrand is differentiated with respect to time. Two expressions follow. They contain velocities or, more specifically, the velocity of a point on the (curved) singular surface. In this context recall the results of Sect. 2.7 on differential geometry and note that this velocity can be obtained from the expression for the motion of the point on the surface in Lagrangian form, analogously to Eq. (3.4.26), as follows:

$$x_i = \tilde{x}_i(Z^1,Z^2,t) \equiv \tilde{x}_i(Z^\Gamma,t) \quad \Rightarrow \quad \underset{A}{\upsilon}{}_i = \frac{\partial \tilde{x}_i(Z^\Gamma,t)}{\partial t}. \tag{3.5.2}$$

Greek indices run from 1 to 2. The capital letters $Z^{\Gamma}$ denote the two surface coordinates of the material point *in the reference configuration*,[3] i.e., the terminology was chosen analogously to the symbol $\boldsymbol{X}$ from Sect. 3.4. The local velocity $\underset{A}{\boldsymbol{\upsilon}}$ of the singular surface is now decomposed w.r.t. the two tangential vectors, $\boldsymbol{\tau}_1$ and $\boldsymbol{\tau}_2$, and the normal vector, $\boldsymbol{e}$, of the corresponding point on the surface:

$$\underset{A}{\boldsymbol{\upsilon}} = \underset{A}{\upsilon}^{\Delta}\boldsymbol{\tau}_{\Delta} + \underset{A}{\upsilon}_{\perp}\boldsymbol{e} \text{ or } \underset{A}{\upsilon}_{i} = \underset{A}{\upsilon}^{\Delta}\tau^{i}_{\Delta} + \underset{A}{\upsilon}_{\perp} e_i. \tag{3.5.3}$$

Recall (cf., Sect. 2.7) that the components $\tau^{i}_{\Delta}$ and $e_i$ of all three vectors refer to a (global) Cartesian coordinate system. If we include time among the variables we find similarly to Eq. (2.7.3) that:

$$\tau^{i}_{\Delta} = \frac{\partial \tilde{x}_i(Z^{\Gamma}, t)}{\partial Z^{\Delta}}. \tag{3.5.4}$$

Elwin Bruno CHRISTOFFEL was born on November 10, 1829 in Montjoie (Montschau) near Aachen (Germany) and died on March 15, 1900 in Straßburg. He first attended a Jesuit college in Cologne and went for further studies to the University of Berlin. He completed his Ph.D. in 1856 with a thesis on the "motion of electricity" in homogeneous bodies. Then he returned to Montjoie where he lived for three years in seclusion from academia. In 1859 he became a private docent at Berlin University and in 1862 he moved to the ETH in Zurich where he became the successor to the famous mathematician DEDEKIND. One of his main functions in this position was setting up the Mathematics Department. He returns to Berlin to the Gewerbeakademie for a short while and, finally, becomes a professor at the University of Straßburg in 1872. In 1894 he reaches the retirement age and dies a few years later.

Moreover, the so-called *covariant derivative* of the tangential components of the velocity was introduced:

$$\underset{A}{\upsilon}^{\Delta}_{;\Gamma} = \frac{\partial \underset{A}{\upsilon}^{\Delta}}{\partial Z^{\Gamma}} + \Gamma^{\Delta}_{\Gamma\Sigma}\underset{A}{\upsilon}^{\Sigma}, \quad \Gamma^{\Delta}_{\Gamma\Sigma} = \frac{\partial Z^{\Delta}}{\partial x_k}\frac{\partial^2 x_k}{\partial Z^{\Gamma}\partial Z^{\Sigma}}. \tag{3.5.5}$$

$\Gamma^{\Delta}_{\Gamma\Sigma}$ denotes the so-called CHRISTOFFEL symbol*s*. Both concepts anticipate results from Chap. 4, where we will discuss in great detail how to differentiate vectors and tensors in skew curvilinear coordinates, albeit for three dimensions. At this point it may suffice to say that the CHRISTOFFEL symbols vanish for Cartesian coordinate transformations. In other words, for a moving singular plane the covariant derivative of the velocity will reduce to a partial one. This in turn, is a

[3] This is why we have used capital Greek characters for the indices.

direct analogue of the partial spatial derivative of Eq. (3.4.1). In this case the mean curvature, $K_m$, defined in Eq. (2.7.12) vanishes.

In order to prove Eq. (3.5.1) a few auxiliary equations are in order. A (directed) surface element results from a vector product between two non-collinear vectors $d\overset{(1)}{\boldsymbol{x}}$ and $d\overset{(2)}{\boldsymbol{x}}$. We use a Cartesian base, observe the chain rule in context with Eq. (3.5.2) as well as the definition for the tangent vectors of Eq. (3.5.4) and obtain:

$$\begin{aligned} dS_i &= \epsilon_{ijk} d\overset{(1)}{x}_j \, d\overset{(2)}{x}_k = \epsilon_{ijk} \frac{\partial x_j}{\partial Z^\Gamma} \frac{\partial x_k}{\partial Z^\Delta} d\overset{(1)}{Z}{}^\Gamma \, d\overset{(2)}{Z}{}^\Delta \\ &= \epsilon_{ijk} \tau^j_\Gamma \, \tau^k_\Delta \, d\overset{(1)}{Z}{}^\Gamma \, d\overset{(2)}{Z}{}^\Delta = dS \, e_i. \end{aligned} \tag{3.5.6}$$

Obviously $d\overset{(1)}{\boldsymbol{x}}$ and $d\overset{(2)}{\boldsymbol{x}}$ are tangential to the surface. Therefore their vector product points in the direction of the unit normal $\boldsymbol{e}$. This was explicitly acknowledged after the last equal sign so that we now possess a relation that links the current surface element to the time-independent coordinates of the reference configuration. This will be quite beneficial during differentiation. We obtain:

$$dS = \epsilon_{ijk} e_i \tau^j_\Gamma \tau^k_\Delta d\overset{(1)}{Z}{}^\Gamma \, d\overset{(2)}{Z}{}^\Delta. \tag{3.5.7}$$

In order to make calculations even easier we align the mesh of the reference configuration and the line elements $d\overset{(1)}{\boldsymbol{Z}}$ and $d\overset{(2)}{\boldsymbol{Z}}$ as follows:

$$d\overset{(1)}{Z}{}^\Gamma = \left(d\overset{(1)}{Z}{}^1, \, 0\right), \; d\overset{(2)}{Z}{}^\Delta = \left(0, \; d\overset{(2)}{Z}{}^2\right). \tag{3.5.8}$$

Thus Eq. (3.5.7) simplifies and reads:

$$dS = \epsilon_{ijk} e_i \tau^j_1 \tau^k_2 d\overset{(1)}{Z}{}^1 \, d\overset{(2)}{Z}{}^2. \tag{3.5.9}$$

The second auxiliary equation concerns the time derivative of the tangent vector. Based on its definition shown in Eq. (3.5.4) we find:

$$\dot{\tau}^j_\Gamma = \frac{\partial}{\partial t} \frac{\partial x_j}{\partial Z^\Gamma} = \frac{\partial \underset{A}{\upsilon}{}_j}{\partial Z^\Gamma}. \tag{3.5.10}$$

If we now observe (3.5.3) this turns into:

$$\dot{\tau}^j_\Gamma = \frac{\partial \underset{A}{\upsilon}{}^\Delta}{\partial Z^\Gamma} \tau^j_\Delta + \underset{A}{\upsilon}{}^\Delta \frac{\partial^2 x_j}{\partial Z^\Gamma \partial Z^\Delta} + \frac{\partial \underset{A}{\upsilon}{}_\perp}{\partial Z^\Gamma} e_j + \underset{A}{\upsilon}{}_\perp \frac{\partial e_j}{\partial Z^\Gamma}. \tag{3.5.11}$$

Since the length of the normal vector, $\boldsymbol{e}$, is equal to one, we conclude that its derivative with respect to $Z^\Gamma$ must be in the tangential plane:

$$e_j e_j = 1 \quad \Rightarrow \quad e_j \frac{\partial e_j}{\partial Z^\Gamma} = 0 \quad \Rightarrow \quad \frac{\partial e_j}{\partial Z^\Gamma} = \Lambda^\Delta_\Gamma \tau^j_\Delta. \tag{3.5.12}$$

The coefficients $\Lambda_\Gamma^\Delta$ can be related to the surface metric—cf., Eqs. (2.7.7)—and derivatives of the tangent vectors, which are orthogonal to the normal:

$$e_j\tau_\Delta^j = 0 \quad \Rightarrow \quad \frac{\partial e_j}{\partial Z^\Gamma}\tau_\Delta^j = -e_j\frac{\partial \tau_\Delta^j}{\partial Z^\Gamma} \quad \Rightarrow \quad \Lambda_\Gamma^\Delta = -g^{\Delta\Sigma}e_j\frac{\partial \tau_\Sigma^j}{\partial Z^\Gamma}. \tag{3.5.13}$$

If we finally observe the definition of the curvature tensor, see Eq. (2.7.8), we conclude that the derivative in last term of Eq. (3.5.11) can be written as follows:

$$\frac{\partial e_j}{\partial Z^\Gamma} = -g^{\Delta\Sigma}e_k\frac{\partial \tau_\Sigma^k}{\partial Z^\Gamma}\tau_\Delta^j = -g^{\Delta\Sigma}e_k\frac{\partial^2 x_k}{\partial Z^\Gamma \partial Z^\Sigma}\tau_\Delta^j = -g^{\Delta\Sigma}b_{\Gamma\Sigma}\tau_\Delta^j. \tag{3.5.14}$$

This result and application of the chain rule allows us to rewrite Eq. (3.5.11):

$$\begin{aligned}\dot{\tau}_\Gamma^j &= \left(\frac{\partial \underset{A}{\upsilon}^\Delta}{\partial Z^\Gamma} + \frac{\partial Z^\Delta}{\partial x_k}\frac{\partial^2 x_k}{\partial Z^\Gamma \partial Z^\Sigma}\underset{A}{\upsilon}^\Sigma\right)\tau_\Delta^j + \frac{\partial \underset{A}{\upsilon}_\perp}{\partial Z^\Gamma}e_j - g^{\Delta\Sigma}b_{\Gamma\Sigma}\tau_\Delta^j\underset{A}{\upsilon}_\perp \\ &= \underset{A}{\upsilon}^\Delta_{;\Gamma}\tau_\Delta^j + \frac{\partial \underset{A}{\upsilon}_\perp}{\partial Z^\Gamma}e_j - g^{\Delta\Sigma}b_{\Gamma\Sigma}\tau_\Delta^j\underset{A}{\upsilon}_\perp .\end{aligned} \tag{3.5.15}$$

We are now in a position to evaluate the second time derivative of Eq. (3.3.6)[4]:

$$\begin{aligned}\frac{\mathrm{d}}{\mathrm{d}t}\iint_{A(t)}\underset{A}{\tilde{\psi}}\,(Z^i,t)\,\mathrm{d}S &= \frac{\mathrm{d}}{\mathrm{d}t}\iint_{A_0}\underset{A}{\tilde{\psi}}\,(Z^i,t)\,\epsilon_{ijk}e_i\tau_1^j\tau_2^k\mathrm{d}\overset{(1)}{Z}{}^1\,\mathrm{d}\overset{(2)}{Z}{}^2 \\ &= \iint_{A(t)}\frac{\partial \underset{A}{\psi}}{\partial t}\,\mathrm{d}S + \iint_{A_0}\underset{A}{\psi}\,\epsilon_{ijk}e_i\left(\dot{\tau}_1^j\tau_2^k + \tau_1^j\dot{\tau}_2^k\right)\,\mathrm{d}\overset{(1)}{Z}{}^1\,\mathrm{d}\overset{(2)}{Z}{}^2 \\ &= \iint_{A(t)}\frac{\partial \underset{A}{\psi}}{\partial t}\,\mathrm{d}S + \iint_{A_0}\underset{A}{\psi}\,\epsilon_{ijk}\left(\underset{A}{\upsilon}^\Delta_{;\Delta} - g^{\Delta\Sigma}b_{\Gamma\Sigma}\underset{A}{\upsilon}_\perp\right)\tau_1^j\tau_2^k\,\mathrm{d}\overset{(1)}{Z}{}^1\,\mathrm{d}\overset{(2)}{Z}{}^2.\end{aligned} \tag{3.5.16}$$

In order to arrive at this result use has been made of the fact that whenever symmetric expressions, such as $e_ie_j$ or $\tau_\Gamma^j\tau_\Delta^k$, are multiplied by the antisymmetric tensor $\epsilon_{ijk}$ the result is simply zero.[5] If we now compare the second term in the second integral with the definition (2.7.12) of the mean curvature we finally arrive at Eq. (3.5.1).

---

[4] $A_0$ refers to the area of the singular surface in the reference configuration.

[5] For a proof one should simply expand the products.

## 3.6 Combining Balances and Transport Theorems

We now insert the equations for the transport theorems of volumes and surfaces from Sects. 3.4 and 3.5 into Eq. (3.3.6) and separate quantities that refer to the subvolumes $V^{\pm}$ and their surfaces $A^{\pm}$ as well as to the singular surface $A$:

$$\iiint\limits_{V^+\cup V^-} \frac{\partial \underset{V}{\psi}}{\partial t}\,\mathrm{d}V + \iint\limits_{A^+\cup A^-} \left[\underset{V}{\psi}\,\boldsymbol{\upsilon} + \underset{A}{\boldsymbol{\varphi}}\right]\cdot \boldsymbol{n}\,\mathrm{d}A - \iiint\limits_{V^+\cup V^-} \left(\underset{V}{p} + \underset{V}{s}\right)\mathrm{d}V$$

$$= -\iint\limits_{A} \left(\frac{\partial \underset{A}{\psi}}{\partial t} + \underset{A}{\psi}\left(\underset{A}{\upsilon}{}^{\Delta}_{;\Delta} - 2K_{\mathrm{m}}\underset{A}{\upsilon}_{\perp}\right) + \underset{L}{\varphi}{}^{\Delta}_{;\Delta} - [[\psi_V \boldsymbol{\upsilon}_A]]\cdot \boldsymbol{e} - \left(\underset{A}{p} + \underset{A}{s}\right)\right)\mathrm{d}A. \tag{3.6.1}$$

In the process the following integral theorem was used in which another covariant derivative occurs:

$$\oint\limits_{L} \underset{L}{\boldsymbol{\varphi}}\cdot \boldsymbol{\nu}\,\mathrm{d}l = \iint\limits_{S} \underset{L}{\varphi}{}^{\Delta}_{;\Delta}\,\mathrm{d}A \text{ since } \underset{L}{\varphi}_i = \underset{L}{\varphi}{}^{\Delta}\tau^{i}_{\Delta} + \underset{L}{\varphi}_{\perp} e_i \text{ and } \nu_i = \nu^{\Sigma}\tau^{i}_{\Sigma}. \tag{3.6.2}$$

We postpone its proof and evaluation until the concept of a covariant derivative has been established in detail. Eq. (3.6.1) represents the most general balance for the situation depicted in Fig. 3.1. Of course, this relation may also be applied to a (material) point and its volume d$V$. This way we obtain local balances for fields. From a mathematical point of view they are easier to treat, namely in terms of partial differential equations and jump conditions at boundaries. However, we have to distinguish between material points within the regular volume and on the singular surface.

## 3.7 General Balances in Regular and Singular Points

If we apply Eq. (3.6.1) to a small volume $V \to \mathrm{d}V$ corresponding to a regular material point, in other words a point completely located within $V^{\pm}$, all contributions from the singular surface on the right hand side will vanish. The left hand side can be rewritten by using the "regular" Gauss' theorem of Eq. (3.4.2):

$$\iiint\limits_{V} \left[\frac{\partial \underset{V}{\psi}}{\partial t} + \frac{\partial}{\partial x_j}\left(\underset{V}{\psi}\,\upsilon_j + \underset{A}{\phi}_j\right) - \left(\underset{V}{p} + \underset{V}{s}\right)\right]\mathrm{d}V = 0. \tag{3.7.1}$$

In the limit of an infinitely small volume we conclude that the integrand must vanish:

$$\frac{\partial \underset{V}{\psi}}{\partial t}+\frac{\partial}{\partial x_j}\left(\underset{V}{\psi}\, \upsilon_j+\underset{A}{\phi}_j\right)=\underset{V}{p}+\underset{V}{s}\,. \tag{3.7.2}$$

This is the general local balance in regular points of a body. If we recall GAUSS' theorem in its invariant form (3.4.8) we may write alternatively:

$$\frac{\partial \underset{V}{\psi}}{\partial t}+\nabla\cdot\left(\underset{V}{\psi}\ \boldsymbol{\upsilon}+\underset{A}{\boldsymbol{\phi}}\right)=\underset{V}{p}+\underset{V}{s}\,. \tag{3.7.3}$$

This is an aesthetically pleasing equation, but not much more. In contrast to the form shown in Eq. (3.7.2) it is not very useful during calculations, especially if a specific skew curvilinear coordinate system is concerned. We shall learn in Chap. 4 how to specify the del operator and the covariant derivative in arbitrary coordinates. For the time being we content ourselves with the explicit Cartesian way of writing shown in Eq. (3.7.2).

In order to derive the corresponding equation for the case of a singular point we consider the limit process shown in Fig. 3.6: A point on the singular surface is enclosed from both sides by a pillbox. We apply the general balance (3.6.1) to this very volume and, in a first step, shrink its height to zero. Then all the volume related contributions in the equation will vanish. In a second step the surfaces $A^{\pm}$ and $A$ will be contracted into a point so that the normals $\boldsymbol{n}$ and $\boldsymbol{e}$ are collinear. From the remaining surface contributions we obtain the general local balance in singular points which, for obvious reasons, is also know in the trade as the *jump condition*:

$$\frac{\partial \underset{A}{\psi}}{\partial t}+\underset{A}{\psi}\left(\underset{A}{\upsilon}{}^{\Delta}_{;\Delta}-2K_{\mathrm{m}}\underset{A}{\upsilon}_{\perp}\right)+\underset{L}{\varphi}{}^{\Delta}_{;\Delta}-\left(\underset{A}{p}+\underset{A}{s}\right)=-\boldsymbol{e}\cdot\left[\!\left[\psi_V\left(\boldsymbol{\upsilon}-\underset{A}{\upsilon}_{\perp}\boldsymbol{e}\right)+\boldsymbol{\varphi}_A\right]\!\right]. \tag{3.7.4}$$

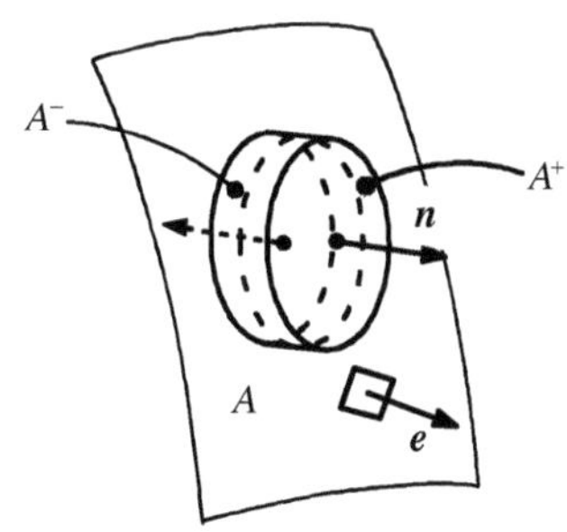


**Fig. 3.6** The pillbox argument

## 3.8 Local Balances of Mass and Momentum in Regular Points

In Sect. 3.2 the global balances for mass and momentum have already been introduced, at least for the case of a material volume without a singular surface. We now use Eqs. (3.2.2 / 3.2.10) in combination with the transport theorem $(3.4.12)_1$ and obtain:

$$\iiint\limits_{V(t)} \left[\frac{\partial \rho}{\partial t} + \frac{\partial}{\partial x_j}(\rho \upsilon_j)\right] \mathrm{d}V = 0 \Leftrightarrow \iiint\limits_{V(t)} \left[\frac{\partial \rho}{\partial t} + \nabla \cdot (\rho\, \boldsymbol{\upsilon})\right] \mathrm{d}V = 0 \tag{3.8.1}$$

and:

$$\begin{gathered}\iiint\limits_{V(t)} \left[\frac{\partial \rho \upsilon_i}{\partial t} + \frac{\partial}{\partial x_j}(\rho \upsilon_i \upsilon_j - \sigma_{ji}) - \rho f_i\right] \mathrm{d}V = 0 \quad \Leftrightarrow \\ \iiint\limits_{V(t)} \left[\frac{\partial \rho\, \boldsymbol{\upsilon}}{\partial t} + \nabla \cdot (\rho\, \boldsymbol{\upsilon} \otimes \boldsymbol{\upsilon} - \boldsymbol{\sigma}) - \rho \boldsymbol{f}\right] \mathrm{d}V = 0.\end{gathered} \tag{3.8.2}$$

If we observe the general balance structure shown in Eqs. (3.4.1 / 3.4.32) the balances of mass and momentum read in regular points:

$$\frac{\partial \rho}{\partial t} + \frac{\partial}{\partial x_j}(\rho \upsilon_j) = 0 \quad \Leftrightarrow \quad \frac{\partial \rho}{\partial t} + \nabla \cdot (\rho\, \boldsymbol{\upsilon}) = 0 \tag{3.8.3}$$

and:

$$\frac{\partial \rho \upsilon_i}{\partial t} + \frac{\partial}{\partial x_j}(\rho \upsilon_i \upsilon_j - \sigma_{ji}) = \rho f_i \Leftrightarrow \frac{\partial \rho\, \boldsymbol{\upsilon}}{\partial t} + \nabla \cdot (\rho\, \boldsymbol{\upsilon} \otimes \boldsymbol{\upsilon} - \boldsymbol{\sigma}) = \rho \boldsymbol{f}. \tag{3.8.4}$$

Equation (3.8.3) is also known as *continuity equation* in the literature. We shall now present a few alternative forms of these equations. Application of the product rule to $(3.8.3)_1$ yields:

$$\frac{\partial \rho}{\partial t} + \frac{\partial \rho}{\partial x_j} \upsilon_j + \rho \frac{\partial \upsilon_j}{\partial x_j} = 0. \tag{3.8.5}$$

If we switch between Eulerian and Lagrangian variables as outlined in Sect. 3.4 the first two terms can be combined as follows:

$$\frac{\mathrm{d}\rho}{\mathrm{d}t} = \frac{\mathrm{d}\hat{\rho}(\tilde{\boldsymbol{x}}(\boldsymbol{X},t),t)}{\mathrm{d}t} = \left.\frac{\partial \hat{\rho}(\boldsymbol{x},t)}{\partial t}\right|_{\boldsymbol{x}} + \left.\frac{\partial \hat{\rho}(\boldsymbol{x},t)}{\partial x_j}\right|_t \left.\frac{\partial \tilde{x}_j(\boldsymbol{X},t)}{\partial t}\right|_{\boldsymbol{X}}. \tag{3.8.6}$$

Thus if Eq. (3.4.26) is observed Eq. (3.8.5) can be rewritten as:

$$\frac{\mathrm{d}\rho}{\mathrm{d}t} + \rho \frac{\partial \upsilon_j}{\partial x_j} = 0 \quad \Leftrightarrow \quad \frac{\mathrm{d}\rho}{\mathrm{d}t} + \rho \nabla \cdot \boldsymbol{\upsilon} = 0. \tag{3.8.7}$$

This type of differentiation is known as a *material time derivative* in continuum theory. Note that material time derivatives can be defined for arbitrary fields. The mass density of Eq. (3.8.6) is just an example. Frequently material time derivatives are also denoted by a dot:

$$\frac{\mathrm{d}(\cdot)}{\mathrm{d}t} \equiv (\cdot)^{\boldsymbol{\cdot}}. \tag{3.8.8}$$

In fact, we have already tacitly used this form in context with Eq. (3.4.25). Mathematically speaking the sequence of relations shown in Eq. (3.8.6) is nothing else but forming a total differential w.r.t. the variables $\boldsymbol{x}$ and $t$. Without any reference to Eulerian or Lagrangian coordinates we could have written instead:

$$\mathrm{d}\rho(\boldsymbol{x},t) = \left.\frac{\partial\rho(\boldsymbol{x},t)}{\partial t}\right|_{\boldsymbol{x}} \mathrm{d}t + \left.\frac{\partial\rho(\boldsymbol{x},t)}{\partial x_j}\right|_{t} \mathrm{d}x_j. \tag{3.8.9}$$

"Division" by $\mathrm{d}t$ yields:

$$\frac{\mathrm{d}\rho}{\mathrm{d}t} = \frac{\partial\rho}{\partial t} + \frac{\partial\rho}{\partial x_j}\upsilon_j, \tag{3.8.10}$$

since velocity is nothing else but a change of position with time. However, note that Eq. (3.8.6) adds another flavor and, maybe, allows for a more subtle understanding of motion of matter both from the view of the comoving as well as of an external observer. Furthermore note that in the case of a Lagrangian representation we have:

$$\frac{\mathrm{d}\rho}{\mathrm{d}t} = \frac{\mathrm{d}\tilde{\rho}(\boldsymbol{X},t)}{\mathrm{d}t} \equiv \left.\frac{\partial\tilde{\rho}(\boldsymbol{X},t)}{\partial t}\right|_{\boldsymbol{X}}, \tag{3.8.11}$$

i.e., total (or material) and partial time derivative are identical.

**Exercise 3.8.1:** ***Variations of the local mass balance***

Recall Eqs. (3.4.25 / 3.4.27) and show by means of Eq. (3.8.7) that:

$$\rho = \frac{\rho_0}{J}, \tag{3.8.12}$$

where $\rho_0$ denotes the mass density in the reference configuration. Note that one of the objectives of Sect. 3.1 has now formally been reached: If the velocity $\boldsymbol{\upsilon}$, i.e., $\boldsymbol{F}$ and, consequently, $J$ are known, Eq. (3.8.12) allows us to predict the current mass density in all points of the body and for all times from the original density. Find a suitable definition of an incompressible material and show that the balance of mass assumes the following simple form:

$$\frac{\partial\upsilon_j}{\partial x_j} = 0 \quad \Leftrightarrow \quad \nabla\cdot\boldsymbol{\upsilon} = 0. \tag{3.8.13}$$

Interpret this equation in terms of a source or drain (also compare Chap. 4). What is the value of $J$ in that case?

Application of the product rule to Eq. $(3.8.4)_1$ while observing Eq. (3.8.3) yields:

$$\rho\frac{\mathrm{d}\upsilon_i}{\mathrm{d}t} = \frac{\partial\sigma_{ji}}{\partial x_j} + \rho f_i \quad \Leftrightarrow \quad \rho\frac{\mathrm{d}\boldsymbol{\upsilon}}{\mathrm{d}t} = \nabla\cdot\boldsymbol{\sigma} + \rho\boldsymbol{f}. \tag{3.8.14}$$

This equation is the epitome of NEWTON's law of motion, which in high school physics is frequently summarized by the slogan "force equals mass times acceleration."[6] Note that $\mathrm{d}\boldsymbol{\upsilon}/\mathrm{d}t$ is nothing else but the acceleration, $\boldsymbol{a}$, and the forces are decomposed into surface and volumetric parts, $(\nabla\cdot\boldsymbol{\sigma})$ and $(\rho\boldsymbol{f})$, respectively.

**Exercise 3.8.2:** ***Variations of the local balance of momentum***

Derive Eq. (3.8.14) and observe meticulously the difference between Eulerian and Lagrangian notation as during the sequence of transformations shown in Eq. (3.8.6) for the mass density.

Now that we have introduced the material time derivative, a reflection on the global balance equation in context with REYNOLDS' transport theorem is in order: It is sometimes required to consider a balance for a specific density $\psi$ instead of the volume density $\underset{V}{\psi}$. Clearly $\underset{V}{\psi} = \rho\psi$ holds and so we rewrite (3.3.7) for the case of material volumes without a singular surface:

$$\frac{\mathrm{d}}{\mathrm{d}t}\iiint_V \rho\psi\,\mathrm{d}V = -\oint\!\!\!\oint_{\partial V} \underset{A}{\boldsymbol{\phi}}\cdot\boldsymbol{n}\,\mathrm{d}A + \iiint_V \left(\underset{V}{p} + \underset{V}{s}\right)\mathrm{d}V. \tag{3.8.15}$$

Since $\rho\,\mathrm{d}V = \mathrm{d}M$ and the total mass $M$ of the volume is time independent (in contrast to $V(t)$) we may circumvent REYNOLDS' transport theorem by a simple substitution:

$$\frac{\mathrm{d}}{\mathrm{d}t}\iiint_V \rho\psi\,\mathrm{d}V = \frac{\mathrm{d}}{\mathrm{d}t}\int_M \psi\,\mathrm{d}M = \int_M \frac{\mathrm{d}\psi}{\mathrm{d}t}\,\mathrm{d}M = \int_M \rho\dot{\psi}\,\mathrm{d}V. \tag{3.8.16}$$

By combining these two equations and after applying GAUSS' theorem we immediately arrive at the following alternative form for a general balance in regular points:

$$\rho\dot{\psi} = -\nabla\cdot\underset{A}{\boldsymbol{\phi}} + \underset{V}{p} + \underset{V}{s}. \tag{3.8.17}$$

6 From the viewpoint of balance laws it would make more sense to say "mass times acceleration equals force" and to distinguish strictly between cause and effect as NEWTON did (see Chap. 8).

**Exercise 3.8.3: Circumventing Reynolds' transport theorem**

Insert the definition $\underset{V}{\psi} = \rho\psi$ into (3.7.3), observe the mass balance (3.8.3)$_1$ as well as the definition of the material time derivative shown in Eqs. (3.8.6 / 3.8.8) to arrive at the same result as shown in Eq. (3.8.17).

**Exercise 3.8.4:** ***Center of mass***

Define the center of mass of a material volume without a singular surface by:

$$\boldsymbol{x}^{\mathrm{c}} = \frac{1}{M} \iiint\limits_{V} \rho\, \boldsymbol{x}\, \mathrm{d}V. \tag{3.8.18}$$

Use Reynolds' transport theorem (3.4.1) or the alternative shown in Eq. (3.8.16) in context with the definition of material time derivatives of Eq. (3.8.6 / 3.8.8) to prove that:

$$M\ddot{\boldsymbol{x}}^{c} = \boldsymbol{T} + \boldsymbol{F} \tag{3.8.19}$$

For this purpose also recall Eq. (3.2.4). This relation is known as the *principle of motion of the mass center*: A deformable body moves translatorily as if it were a point mass under the combined action of forces $\boldsymbol{T}$ and $\boldsymbol{F}$. As we shall see it forms the first set of Euler's laws of mechanics for extended bodies.

## 3.9 Local Balances of Energy in Regular Points

The balance of momentum shown in Eq. (3.8.14) is the origin of two other mechanical concepts: kinetic energy and moment of momentum. In this section we shall look into kinetic and other forms of energy relevant to continuum physics. For this purpose we transform the vector equation (3.8.14)$_1$ by scalar multiplication with $\upsilon_i$ into a scalar relation:

$$\begin{aligned} \rho\frac{\mathrm{d}(\frac{1}{2}\upsilon_i\upsilon_i)}{\mathrm{d}t} &= \frac{\partial(\sigma_{ji}\upsilon_i)}{\partial x_j} - \sigma_{ji}\frac{\partial \upsilon_i}{\partial x_j} + \rho f_i \upsilon_i \quad \Leftrightarrow \\ \rho\frac{\mathrm{d}(\frac{1}{2}\upsilon^2)}{\mathrm{d}t} &= \nabla\cdot(\boldsymbol{\sigma}\cdot\boldsymbol{\upsilon}) - \boldsymbol{\sigma} : \nabla\boldsymbol{\upsilon} + \rho\boldsymbol{f}\cdot\boldsymbol{\upsilon}. \end{aligned} \tag{3.9.1}$$

In the latter formula we have introduced the so-called *double scalar product* ":" for a neighboring index. It is defined in a very specific way by double summation (see the indices $i$ and $j$ in the previous equation where $i$ is the neighboring index).[7] The result is now integrated w.r.t. the (regular) material volume $V(t)$ and GAUSS' theorem is observed:

$$\iiint\limits_{V(t)} \rho \frac{\mathrm{d}(\frac{1}{2}\upsilon_i\upsilon_i)}{\mathrm{d}t}\,\mathrm{d}V = \oiint\limits_{\partial V} \sigma_{ji}\upsilon_i n_j\,\mathrm{d}A - \iiint\limits_{V(t)} \sigma_{ji}\frac{\partial \upsilon_i}{\partial x_j}\,\mathrm{d}V + \iiint\limits_{V(t)} \rho f_i \upsilon_i\,\mathrm{d}V. \tag{3.9.2}$$

The integral on the left hand side is transformed by using the substitution already mentioned in context with Eq. (3.8.16): Recall that the mass of a material body does not change [see Eq. (3.2.2)], in other words it is time-independent in contrast to its volume. Also recall that $\mathrm{d}M = \rho\,\mathrm{d}V$. In context with Eq. (3.8.16) this leads to:

$$\begin{aligned}\frac{\mathrm{d}}{\mathrm{d}t}\iiint\limits_{V(t)} \frac{1}{2}\rho\upsilon_i\upsilon_i\,\mathrm{d}V &= \oiint\limits_{\partial V} \sigma_{ji}\upsilon_i n_j\,\mathrm{d}A - \iiint\limits_{V(t)} \sigma_{ji}\frac{\partial \upsilon_i}{\partial x_j}\,\mathrm{d}V \\ &+ \iiint\limits_{V(t)} \rho f_i\upsilon_i\,\mathrm{d}V \quad \Leftrightarrow \frac{\mathrm{d}}{\mathrm{d}t}\iiint\limits_{V(t)} \frac{\rho}{2}\upsilon^2\,\mathrm{d}V \\ &= \oiint\limits_{\partial V} \boldsymbol{n}\cdot\boldsymbol{\sigma}\cdot\boldsymbol{\upsilon}\,\mathrm{d}A - \iiint\limits_{V(t)} \boldsymbol{\sigma} : \nabla\boldsymbol{\upsilon}\,\mathrm{d}V + \iiint\limits_{V(t)} \rho\boldsymbol{f}\cdot\boldsymbol{\upsilon}\,\mathrm{d}V.\end{aligned} \tag{3.9.3}$$

Thus we have found another *balance equation,* namely one for the *kinetic energy*, whose volume density is given by $\frac{\rho}{2}\upsilon^2$.[8] This equation is known as the *work-energy equation* in the mechanics community. It states that the (temporal) change of kinetic energy of a system is equal to the sum of the power provided by the forces applied to the body either on its surface or to its volume (the first and the last term after the equal sign) minus the power loss due to non-conservative forces, i.e., friction (the second term). If we make use of the jargon established in context with balance equations and the remarks of Sect. 3.2 we may say alternatively that the temporal change of kinetic energy is given by the sum of the non-convective flow of kinetic energy across the surface and a corresponding volumetric supply, namely [according to CAUCHY's theorem of Eq. (3.2.6)]:

$$\oiint\limits_{\partial V} \boldsymbol{n}\cdot\boldsymbol{\sigma}\cdot\boldsymbol{\upsilon}\,\mathrm{d}A + \iiint\limits_{V(t)} \rho\boldsymbol{f}\cdot\boldsymbol{\upsilon}\,\mathrm{d}V \equiv \oiint\limits_{\partial V} \boldsymbol{t}\cdot\boldsymbol{\upsilon}\,\mathrm{d}A + \iiint\limits_{V(t)} \rho\boldsymbol{f}\cdot\boldsymbol{\upsilon}\,\mathrm{d}V. \tag{3.9.4}$$

---

[7] Frequently the stress tensor is symmetric and then we may as well write for the production term $\sigma_{ij}\partial\upsilon_i/\partial x_j \equiv \boldsymbol{\sigma}\cdot\cdot\,\nabla\boldsymbol{\upsilon}$ where the double scalar product for non-neighboring indices has been used, which will be introduced in Eq. (3.10.1).

[8] It is easily verified that the specific kinetic energy (energy per unit mass) is given by $\upsilon^2/2$.

Moreover, we must conclude that there is a (negative) production of kinetic energy, namely the second term on the right hand side of Eq. (3.9.3). Therefore it is justified to argue that the kinetic energy is *not* a conserved quantity, which (by definition) must not contain a production term in their balance. On first glance this is a surprising result, because physicists love to talk about the conservation of energy and exactly this is violated here. However, this is only seemingly the case. Indeed, kinetic energy is *not* conserved. However, the sum of kinetic *and* internal energy *is*. The latter (sometimes associated with the vague term *heat*) is a form of kinetic energy beyond the continuum scale, on which it "cannot be seen." It is a quantity defined on the atomic scale and manifests itself in the erratic movement of particles. In other words this ominous quantity is a measure of *temperature*. We write:

$$\begin{aligned}\frac{\mathrm{d}}{\mathrm{d}t}\iiint\limits_{V}\rho\left(u+\tfrac{1}{2}\upsilon_i\upsilon_i\right)\mathrm{d}V &= -\oiint\limits_{\partial V} n_j\left(q_j-\upsilon_i\sigma_{ji}\right)\mathrm{d}A+\iiint\limits_{V}\rho(f_i\upsilon_i+r)\,\mathrm{d}V \Leftrightarrow \\ \frac{\mathrm{d}}{\mathrm{d}t}\iiint\limits_{V}\rho\left(u+\tfrac{1}{2}\upsilon^2\right)\mathrm{d}V &= -\oiint\limits_{\partial V}\boldsymbol{n}\cdot(\boldsymbol{q}-\boldsymbol{\sigma}\cdot\boldsymbol{\upsilon})\,\mathrm{d}A+\iiint\limits_{V}\rho\,(\boldsymbol{f}\cdot\boldsymbol{\upsilon}+r)\,\mathrm{d}V.\end{aligned} \tag{3.9.5}$$

The symbol $u$ is used for the specific internal energy, i.e., internal energy per unit mass. $\boldsymbol{q}$ denotes the *heat flux vector*. The negative sign in the balance is pure convention: Internal energy (or temperature) increases if heat enters the body. This in combination with the scalar product between the heat flux vector and the outer normal motivates the use of a minus sign.[9] Moreover, surface tractions as well as volume forces power the body, and energy can also be supplied by radiation ($r$ denotes the specific radiation). All of them are supplies and not production terms, because we can (in principle) control them by shielding: *The total energy is conserved.*

Finally there is a balance just for the internal energy which, for historic reasons, has a certain "autonomy." It is also known as the *First Law of Thermodynamics*. It results by subtracting the balance of kinetic energy (3.9.3) from Eq. (3.9.5):

$$\begin{aligned}\frac{\mathrm{d}}{\mathrm{d}t}\iiint\limits_{V}\rho\,u\,\mathrm{d}V &= -\oiint\limits_{\partial V} q_j n_j\,\mathrm{d}A+\iiint\limits_{V(t)}\sigma_{ji}\frac{\partial\upsilon_i}{\partial x_j}\,\mathrm{d}V+\iiint\limits_{V}\rho\,r\,\mathrm{d}V \Leftrightarrow \\ \frac{\mathrm{d}}{\mathrm{d}t}\iiint\limits_{V(t)}\rho\,u\,\mathrm{d}V &= -\oiint\limits_{\partial V}\boldsymbol{q}\cdot\boldsymbol{n}\,\mathrm{d}A+\iiint\limits_{V(t)}\boldsymbol{\sigma}:\nabla\boldsymbol{\upsilon}\,\mathrm{d}V+\iiint\limits_{V(t)}\rho\,r\,\mathrm{d}V,\end{aligned} \tag{3.9.6}$$

or, alternatively, in local form:

[9] Note that this convention was not applied to the traction vector, $\boldsymbol{t}$, probably for historic reasons.

$$\rho \frac{\mathrm{d}u}{\mathrm{d}t} = -\frac{\partial q_j}{\partial x_j} + \sigma_{ji} \frac{\partial \upsilon_i}{\partial x_j} + \rho\, r \quad \Leftrightarrow \quad \rho \frac{\mathrm{d}u}{\mathrm{d}t} = -\nabla \cdot \boldsymbol{q} + \boldsymbol{\sigma} : \nabla \boldsymbol{\upsilon} + \rho\, r. \tag{3.9.7}$$

Just like kinetic energy internal energy is not conserved either. Except for the sign the corresponding production is equal to the production of kinetic energy.

In context with Eqs. (3.9.5–3.9.7) a *caveat* is in order: They hold in this (relatively simple) form only if the matter in the material volume does not possess any intrinsic moment of momentum, a.k.a. *spin*. We proceed to discuss this subtle point further below. At the end of this section it is only fair to remember that the development of the concept of various energy forms as well as the idea of energy conservation cost mankind several centuries of contemplation, intellectual struggle, as well as mutual personal animosity. The first ideas of momentum *versus* kinetic energy, or *vis viva* go back to LEIBNIZ. Then the industrial revolution took over, and it became important to think about energy generation as well as energy conversion. In this context mankind expected several scientists to do their duty, James Prescott JOULE, Robert Wilhelm MAYER, and Hermann von HELMHOLTZ, to mention just a few of them.

Julius Robert (von) MAYER was born on November 25, 1814 in Heilbronn (Germany) and he also died there on March 20, 1878. He was a German physician and physicist, known for his pioneering work in the establishment of the mechanical equivalent of heat. MAYER studied medicine at the University of Tübingen. In 1842 he published a paper in the journal Annales de Chimie, in which he gave a value for the mechanical equivalent of heat. His figure was based on the rise of temperature in paper pulp that was stirred by a horse-powered mechanism. MAYER was also the first to state the principle of conservation of energy, most notably for biological phenomena as well as for physical systems. He was fascinated by the concepts of heat and the conversion of thermal and mechanical energy. Being a physician by trade, he measured temperatures whenever he could in order to find evidence for his archaic ideas about energy and heat: During a trip to Java he has ample opportunity to take the temperatures of his patients. He states his findings (in a nowadays politically extremely incorrect language) as follows: "for a negro, lazy and idly laying in the cabana 37°; the same, however, sitting idly in the sun 40.20°; the same, however, working in the sun 39.75°." From this fascinating observation he concludes that heat is converted to mechanical work. However, his contemporaries, in particular JOULE, thought nothing of him and his findings. So he was at best ignored if not belittled by the scientific community. This did not really add to his psychological well-being: He attempts to commit suicide in 1850 after the sudden death of two of his children. In those days society was not too sensitive or even patient with the mentally ill, and so he is sent to a lunatic asylum right away. He returns to Heilbronn three years later, now truly broken. However, in his later years he is finally given credit for his work. In 1867 the local king awards him a medal, the Ritterkreuz 1st Class, which came with the privilege of a nobility title. From now on MAYER could call himself VON MAYER. In fact, this is an enormous advantage for the bearer of such an ordinary German name, as the author of this book can tell by personal experience.

Hermann von Helmholtz was born on August 31, 1821 in Potsdam (Germany) and died on September 8, 1894 in Berlin. He graduated as a physician from the Medical Institute in Berlin in 1843 and was then assigned to a military regiment at Potsdam, but spent all his spare time doing scientific research instead. In 1851 he became professor of anatomy and physiology in Bonn, then in 1871, professor of physics in Berlin. His most important work was in mathematical physics and acoustics where his major study of 1863 dealt with musical theory and the perception of sound. In mathematical appendices he advocated the use of Fourier series. In 1842 he published his crucial paper on the First Law.

## 3.10 Local Balances of Angular Momentum, Moment of Momentum, and Spin in Regular Points

Scalar multiplication of the balance of momentum by the velocity $\boldsymbol{\upsilon}$ led to the balance of kinetic energy. We now use the vector product and multiply Eq. (3.8.14) by $\mathbf{x}\times$ or (after renaming the indices $i$ and $j$ in the equation by $k$ and $l$, respectively) with $\epsilon_{ijk}x_j$ and find after a few manipulations:

$$\begin{aligned} \rho\frac{\mathrm{d}\left(\epsilon_{ijk}x_j\upsilon_k\right)}{\mathrm{d}t} &= \frac{\partial\left(\epsilon_{ijk}x_j\sigma_{lk}\right)}{\partial x_l} - \epsilon_{ijk}\sigma_{jk} + \epsilon_{ijk}x_j\rho f_k \quad \Leftrightarrow \\ \rho\frac{\mathrm{d}(\boldsymbol{x}\times\boldsymbol{\upsilon})}{\mathrm{d}t} &= \nabla\cdot(\boldsymbol{\sigma}\cdot\boldsymbol{\epsilon}\cdot\boldsymbol{x}) - \boldsymbol{\epsilon}\cdot\cdot\,\boldsymbol{\sigma} + \boldsymbol{x}\times\rho\boldsymbol{f}. \end{aligned} \tag{3.10.1}$$

Note that in the latter equation the double scalar product for non-neighboring indices has been used: $\boldsymbol{e}_i\epsilon_{ijk}\sigma_{jk} \equiv \boldsymbol{\epsilon}\cdot\cdot\,\boldsymbol{\sigma}$. Moreover, note that $\boldsymbol{e}_i\epsilon_{ijk}x_j\sigma_{lk} \equiv \boldsymbol{e}_i\sigma_{lk}\epsilon_{kij}x_j \equiv \boldsymbol{\sigma}\cdot\boldsymbol{\epsilon}\cdot\boldsymbol{x}$. The results are now integrated over a (regular) material volume, Gauss' theorem is applied, and the volume is temporarily substituted by mass. This yields:

$$\begin{aligned} \frac{\mathrm{d}}{\mathrm{d}t}\iiint\limits_{V(t)} \epsilon_{ijk}x_j\rho\upsilon_k\,\mathrm{d}V &= \oint\!\!\!\oint \epsilon_{ijk}x_j\sigma_{lk}n_l\,\mathrm{d}A - \iiint\limits_{V(t)} \epsilon_{ijk}\sigma_{jk}\,\mathrm{d}V + \iiint\limits_{V(t)} \epsilon_{ijk}\upsilon_j\rho f_k\,\mathrm{d}V \quad \Leftrightarrow \\ \frac{\mathrm{d}}{\mathrm{d}t}\iiint\limits_{V(t)} \boldsymbol{x}\times\rho\,\boldsymbol{\upsilon}\,\mathrm{d}V &= \oint\!\!\!\oint \boldsymbol{n}\cdot\boldsymbol{\sigma}\cdot\boldsymbol{\epsilon}\cdot\boldsymbol{x}\,\mathrm{d}A - \iiint\limits_{V(t)} \boldsymbol{\epsilon}:\boldsymbol{\sigma}\,\mathrm{d}V + \iiint\limits_{V(t)} \boldsymbol{x}\times\rho\boldsymbol{f}\,\mathrm{d}V. \end{aligned} \tag{3.10.2}$$

In engineering mechanics the vector products $\boldsymbol{n}\cdot\boldsymbol{\sigma}\cdot\boldsymbol{\epsilon}\cdot\boldsymbol{x} \equiv \boldsymbol{x}\times\boldsymbol{n}\cdot\boldsymbol{\sigma} \equiv \boldsymbol{x}\times\boldsymbol{t}$ and $\boldsymbol{x}\times\rho\boldsymbol{f}$ on the right hand side are known as *moments of forces* or just *moments*, for short. More specifically the first expression represents the moment of tractions and the second one the moment of volume force density. Therefore the vector product $\boldsymbol{x}\times\rho\,\boldsymbol{\upsilon}$ on the left side of the equation must consistently be referred to as *moment* (density) *of linear momentum*. It is sometimes—imprecisely—also called *angular momentum*. It obviously obeys a balance equation and, as known from engineering mechanics, its temporal change is dictated by the moments exerted by the forces. However, we also see that it is *not* a conserved quantity: The balance contains a production term, which vanishes only if the stress tensor is symmetric so that $\epsilon_{ijk}\sigma_{jk} = 0$. Indeed, the stress tensor is symmetric for most engineering materials and applications. In fact, we have tacitly assumed symmetry when we studied MOHR's circle in Exercises 2.4.6 and 2.6.3. However, there are materials which have an "intrinsic moment of momentum," invisible in terms of force couples applied at certain distances, at least on the continuum scale. This additional internal degree of freedom is known as *spin*. It is also a vector denoted by the symbol $\boldsymbol{s}$. Liquid crystals may serve as an example of materials with spin. The situation is similar to the case of energy, where it is possible to convert macroscopically visible kinetic energy into microscopic motion, i.e., a change of temperature or internal energy: Spin can be converted into macroscopically visible angular momentum or vice versa. Now total energy, i.e., the sum of internal energy and (translational as well as intrinsic rotational) kinetic energy, is conserved and so is *total angular momentum*, which is the sum of moment of momentum and spin. We write:

$$\begin{aligned}
\frac{\mathrm{d}}{\mathrm{d}t}\iiint\limits_{V(t)} \rho\big(s_i + \epsilon_{ijk}x_j\upsilon_k\big)\,\mathrm{d}V &= \oint\!\!\!\!\oint\limits_{\partial V} \big(m_{li} + \epsilon_{ijk}x_j\sigma_{lk}\big)n_l\,\mathrm{d}A \\
&\quad + \iiint\limits_{V(t)} \rho\big(\epsilon_{ijk}\upsilon_j f_k + l_i\big)\,\mathrm{d}V \quad \Leftrightarrow \\
\frac{\mathrm{d}}{\mathrm{d}t}\iiint\limits_{V(t)} \rho(\boldsymbol{s} + \boldsymbol{x}\times\boldsymbol{\upsilon})\,\mathrm{d}V &= \oint\!\!\!\!\oint\limits_{\partial V} \boldsymbol{n}\cdot(\boldsymbol{m} + \boldsymbol{\sigma}\cdot\boldsymbol{\epsilon}\cdot\boldsymbol{x})\,\mathrm{d}A \\
&\quad + \iiint\limits_{V(t)} \rho(\boldsymbol{x}\times\boldsymbol{f} + \boldsymbol{l})\,\mathrm{d}V.
\end{aligned} \tag{3.10.3}$$

The vector $\boldsymbol{s}$ denotes the *specific spin*, $\boldsymbol{m}$ is the so-called *surface couple-stresstensor*, and $\boldsymbol{l}$ refers to the vector of *specificspinsupply*, a.k.a. *specific body couple density*. The local form of the balance of total angular momentum in regular points reads:

$$\begin{aligned}
\rho\frac{\mathrm{d}\big(s_i + \epsilon_{ijk}x_j\upsilon_k\big)}{\mathrm{d}t} &= \frac{\partial\big(m_{li} + \epsilon_{ijk}x_j\sigma_{lk}\big)}{\partial x_l} + \rho\big(\epsilon_{ijk}\upsilon_j f_k + l_i\big) \quad \Leftrightarrow \\
\rho\frac{\mathrm{d}(\boldsymbol{s} + \boldsymbol{x}\times\boldsymbol{\upsilon})}{\mathrm{d}t} &= \nabla\cdot(\boldsymbol{m} + \boldsymbol{\sigma}\cdot\boldsymbol{\epsilon}\cdot\boldsymbol{x}) + \rho(\boldsymbol{x}\times\boldsymbol{f} + \boldsymbol{l}).
\end{aligned} \tag{3.10.4}$$

This settles a red-herring discussion frequently started in the mechanics community[10]: Is the *principle of angular momentum* independent of NEWTON's law of motion or not? The answer is very clear and simple: Yes *and* no! No, if we believe

[10] The argument was started by two eminent mechanics professors with a strong disposition and admiration for mathematics and clearly geared toward libeling physicists as numbskulls: see the paper (in German) and books by Truesdell [18, 19] as well as the book (also in German) by Szabó [16]. Until today many mechanics professors join the clamor of the Boeotians in a sycophant manner even without being able to give an explanation of what the problem really is.

that properties of matter can sufficiently be explained by the concept of mass points. Then position and momentum vectors in combination with NEWTON's law of motion is all that is needed to show that the change of moment of momentum is dictated by moments of forces. The answer is yes, however, if we believe that certain materials possess an intrinsic angular momentum (called spin) and, indeed, there is evidence (e.g., liquid crystals) that such a notion is not pure science fiction but very useful, indeed. Then the theorem on the temporal change of angular momentum needs to be postulated and tested by experiments just like NEWTON's law. Moreover, it should be pointed out that even on a subatomic scale—which is as close as we could possibly get to a mass point—there seems to be an intrinsic spin present.[11] It is also interesting to note that the great EULER [1] might have anticipated all of this. After much thinking (see the interesting historical discussion on the origin of the moment of momentum by Truesdell [18]) he insisted that the balance of moment of momentum is independent of the balance of linear momentum. His arguments start with a statement on body motion by linear momentum: "§. 27. Quod si nunc simili modo omnes vires, quibus corpus hoc tempore sollicitatur etiam secundum istas ternas directiones resoluantur, atque ex omnibus coniunctis pro directionibus IA, IB, IC vires oriantur P, Q et R, per principia motus necesse est, ut istae vires aequantur summis omnium virium acceleratricium, quae ex omnibus corporis elementis $d$ M iunctim sumtis nascuntur … impetrabimus tres aequationes sequentes

$$\int d\,\mathrm{M}\left(\frac{d\,d\,x}{d\,t^2}\right) = i\,\mathrm{P}$$
$$\int d\,\mathrm{M}\left(\frac{d\,d\,y}{d\,t^2}\right) = i\,\mathrm{Q}$$
$$\int d\,\mathrm{M}\left(\frac{d\,d\,z}{d\,t^2}\right) = i\,\mathrm{R}\text{''}.$$

where $i$ denotes a factor converting mass into weight. In fact these three formulae are nothing else but Eq. (3.2.10) if we only remind ourselves that the notion of a stress tensor was introduced fifty years later by CAUCHY. But this is just the first set of EULER's laws of mechanics and, indeed, he emphasizes that moments have to be considered as well: "§. 28. Cum igitur elemento $d$ M, quod in puncto $z$ concipimus, primo applicata fit vis $= d\,\mathrm{M}\left(\frac{d\,d\,x}{d\,t^2}\right)$" secundum directionem IA agens, ex ea nullum nascitur momentum pro hoc axe; … unde pro axe IA summa omnium momentorum erit

$$+\int z\,d\,\mathrm{M}\left(\frac{d\,d\,y}{d\,t^2}\right) - \int y\,d\,\mathrm{M}\left(\frac{d\,d\,z}{d\,t^2}\right) = i\,\mathrm{S}\text{''}.$$

---

[11] Scattering experiments make some particle physicists believe that the electron is a true point. However, it does have a (quantized) spin of $\pm\frac{1}{2}\hbar$, $\hbar = 1.055 \times 10^{-34}$Js being the normalized PLANCK constant (note the units of moment of momentum), which could easily be interpreted in terms of moment of momentum if the electron were only a rotating distributed mass.

The other two components are deduced similarly and we conclude (up to the sign) that this second set of EULER's laws of mechanics agrees indeed with $(3.10.2)_2$ if we do *not* explicitly specify the moments in terms of forces just like EULER did. He only has to say the following about them in §. 27.: "… quamobrem designemus ista momenta, quae ex omnibus viribus sollicitantibus pro ternis axibus IA, IB, IC nascuntur, litteris S, T, V, ita ut his quantitatibus per *i* multiplicatis summae omnium momentorum elementarium, quas singulae vires acceleratrices suppeditant aequari debeant." This concludes his argument and, to a certain degree, he makes it sounds like a conclusion from the balance of linear momentum if it were not for two things: First, he does not link the moments to the applied forces in an explicit manner and, second, he puts all of his *six* equations independently side-by-side: "§. 29. Hac igitur ratione sex nacti sumus aequationes, quas hic coniunctim conspectui exponamus

$$\begin{array}{ll} \text{I.}\ \int d\,\mathrm{M}\left(\frac{ddx}{dt^2}\right) = i\,\mathrm{P} & \text{IV.}\ \int z\,d\,\mathrm{M}\left(\frac{ddy}{dt^2}\right) - \int y\,d\,\mathrm{M}\left(\frac{ddz}{dt^2}\right) = i\,\mathrm{S} \\ \text{II.}\ \int d\,\mathrm{M}\left(\frac{ddy}{dt^2}\right) = i\,\mathrm{Q} & \text{V.}\ \int x\,d\,\mathrm{M}\left(\frac{ddz}{dt^2}\right) - \int z\,d\,\mathrm{M}\left(\frac{ddx}{dt^2}\right) = i\,\mathrm{T} \\ \text{III.}\ \int d\,\mathrm{M}\left(\frac{ddz}{dt^2}\right) = i\,\mathrm{R} & \text{VI.}\ \int y\,d\,\mathrm{M}\left(\frac{ddx}{dt^2}\right) - \int z\,d\,\mathrm{M}\left(\frac{ddy}{dt^2}\right) = i\,\mathrm{U}\text{''.} \end{array}$$

It is also interesting to note that the typo in the last equation[12] is not mentioned in the pertinent literature. Rather it was tacitly corrected in Szabó [16], p. 30.

After these philosophical remarks we now subtract the balance for the moment of momentum from the balance for the total angular momentum and obtain the balance of spin for a material volume:

$$\begin{aligned} \frac{\mathrm{d}}{\mathrm{d}t}\iiint\limits_{V(t)} \rho s_i\,\mathrm{d}V &= \oint\!\!\!\oint\limits_{\partial V} m_{li}n_l\,\mathrm{d}A + \iiint\limits_{V(t)} \epsilon_{ijk}\sigma_{jk}\,\mathrm{d}V + \iiint\limits_{V(t)} \rho l_i\,\mathrm{d}V \quad \Leftrightarrow \\ \frac{\mathrm{d}}{\mathrm{d}t}\iiint\limits_{V(t)} \rho \boldsymbol{s}\,\mathrm{d}V &= \oint\!\!\!\oint\limits_{\partial V} \boldsymbol{n}\cdot\boldsymbol{m}\,\mathrm{d}A + \iiint\limits_{V(t)} \boldsymbol{\epsilon}\cdots\boldsymbol{\sigma}\,\mathrm{d}V + \iiint\limits_{V(t)} \rho\boldsymbol{l}\,\mathrm{d}V. \end{aligned} \tag{3.10.5}$$

and in local form:

$$\rho\frac{\mathrm{d}s_i}{\mathrm{d}t} = \frac{\partial m_{li}}{\partial x_l} + \epsilon_{ijk}\sigma_{jk} + \rho l_i \quad \Leftrightarrow \quad \rho\frac{\mathrm{d}\boldsymbol{s}}{\mathrm{d}t} = \nabla\cdot\boldsymbol{m} + \boldsymbol{\epsilon}\cdot\boldsymbol{\sigma} + \rho\boldsymbol{l}. \tag{3.10.6}$$

[12] The typo actually appears twice in §. 28. *and* §. 29. of EULER's work so that we may suspect that it was incorrectly written down in his personal notes.

**Exercise 3.10.1: *Balances of angular momentum***

Explain in detail the various steps involved in the derivation of Eqs. (3.10.1–3.10.6). In particular, explain the double scalar product used in context with the production term, i.e., $\epsilon_{ijk}\sigma_{jk}$ versus $\boldsymbol{\epsilon} \cdot\cdot\, \boldsymbol{\sigma}$. Recall in this context the definition of the double scalar product for neighboring indices introduced in Eq. (3.9.1) and explain the difference.

**Table 3.1** Entries for balance equations in regular points

| $\psi$ | $\psi_v$ | $-\varphi_A$ | $p_V$ | $s_V$ |
|---|---|---|---|---|
| Mass | $\rho$ | $0$ | $0$ | $0$ |
| (linear) momentum | $\rho\,\boldsymbol{\upsilon}$ | $\boldsymbol{\sigma}$ | $0$ | $\rho\boldsymbol{f}$ |
| Kinetic energy | $\frac{\rho}{2}\upsilon^2$ | $\boldsymbol{\sigma}\cdot\boldsymbol{\upsilon}$ | $-\boldsymbol{\sigma}:\nabla\boldsymbol{\upsilon}$ | $\rho\boldsymbol{f}\cdot\boldsymbol{\upsilon}$ |
| Internal energy | $\rho\,u$ | $-\boldsymbol{q}$ | $\boldsymbol{\sigma}:\nabla\boldsymbol{\upsilon}$ | $\rho\,r$ |
| Total energy | $\rho\,(u+\frac{1}{2}\upsilon^2)$ | $-\boldsymbol{q}+\boldsymbol{\sigma}\cdot\boldsymbol{\upsilon}$ | $0$ | $\rho\,(\boldsymbol{f}\cdot\boldsymbol{\upsilon}+r)$ |
| Moment of momentum | $\boldsymbol{x}\times\rho\,\boldsymbol{\upsilon}$ | $\boldsymbol{\sigma}\cdot\boldsymbol{\epsilon}\cdot\boldsymbol{x}\equiv\epsilon_{ijk}x_j\sigma_{lk}\boldsymbol{e}_i\boldsymbol{e}_l$ | $-\boldsymbol{\epsilon}\cdot\cdot\,\boldsymbol{\sigma}\equiv-\epsilon_{ijk}\sigma_{jk}\boldsymbol{e}_i$ | $\boldsymbol{x}\times\rho\boldsymbol{f}$ |
| Spin | $\rho\,\boldsymbol{s}$ | $\boldsymbol{m}$ | $\boldsymbol{\epsilon}\cdot\cdot\,\boldsymbol{\sigma}\equiv\epsilon_{ijk}\sigma_{jk}\boldsymbol{e}_i$ | $\rho\,\boldsymbol{l}$ |
| Angular momentum | $\rho(\boldsymbol{s}+\boldsymbol{x}\times\boldsymbol{\upsilon})$ | $\boldsymbol{m}+\boldsymbol{\sigma}\cdot\boldsymbol{\epsilon}\cdot\boldsymbol{x}$ | $0$ | $\rho\,(\boldsymbol{x}\times\boldsymbol{f}+\boldsymbol{l})$ |

## 3.11 Summary of Local Balances in Regular Points

Recall the general form of a balance equation in regular points shown in Eq. (3.7.2). Table 3.1 makes it easy to reconstruct all of the balances shown so far.

The table needs two more lines of entry, namely for the balances of entropy as well as of electric charge. We will get back to that in Chaps. 12 and 13. It should also be noted that the entries for the various types of energy are valid for bodies *without* intrinsic moment of momentum, i.e., *spin*. We will reconsider them in Chap. 8 after angular velocity has been introduced as a kinematic quantity.

## 3.12 Summary of Local Balances in Singular Points

In contrast to Sect. 3.10 we start with a table which, if used in context with Eq. (3.7.4), directly leads to the balances for mass, momentum, energy, and angular momentum in singular points.

Note that, a singular surface is basically a mathematical model for a transition zone in a volume showing a very steep gradient. It is sometimes necessary to assign intrinsic properties to this structure. Therefore most of the entries in

Table 3.2 are in perfect analogy to the entries in the previous table showing volume properties.[13] Examples are a mass density, $\underset{A}{\rho}$, as well as related mechanical properties of the singular surface, such as momentum, kinetic energy, or moment of momentum. It can also have its own internal energy ("temperature") or spin. Such properties may become important when modeling soap bubbles or (more engineering-like) rubber membranes. However, the transition between the wall of a pressure vessel and the surrounding gas or fluid is very steep and not associated with any mass.

A few other surface properties are less intuitive and, therefore, deserve a comment. For example, as we shall see soon, the jump condition for the momentum dictates certain requirements regarding the continuity of the stress tensor. In Chap. 9 we will discuss which components are affected when we speak about boundary and interface conditions. Moreover, note that $\underset{L}{\boldsymbol{\sigma}}$ is known as the tensor of *surface tensions*. Recall that the production terms for kinetic and internal energy can be derived by scalar multiplication of the balance of momentum and suitable rearrangement. The situation is analogous to the procedure outlined in context with Eq. (3.9.1) for the volumetric kinetic energy. The production terms for spin and moment of momentum become clear after vector multiplication of the balance of momentum similarly as in context with Eq. (3.10.1).

As a specific example of how to use Table 3.2 we consider the balance of mass. The various entries lead to:

$$\frac{\partial \underset{A}{\rho}}{\partial t} + \underset{A}{\rho}\left(\underset{A}{\upsilon}{}^{\alpha}_{;\alpha} - 2K_{\mathrm{m}}\, \underset{A}{\upsilon}{}_{\perp}\right) = -[\![\, \rho(\boldsymbol{\upsilon} - \upsilon_{A\,\perp}\boldsymbol{e})]\!] \cdot \boldsymbol{e}. \tag{3.12.1}$$

If the singular surface has no mass of its own, in other words if $\underset{A}{\rho} = 0$, this simplifies to:

$$[\![\rho(\boldsymbol{\upsilon} - \upsilon_{A\,\perp}\boldsymbol{e})]\!] \cdot \boldsymbol{e} = 0. \tag{3.12.2}$$

If the singular surface does not move, in other words if $\underset{A}{\upsilon}{}_{\perp} = 0$, we obtain:

$$[\![\rho\boldsymbol{\upsilon}]\!] \cdot \boldsymbol{e} = 0 \quad \Leftrightarrow \quad (\rho\boldsymbol{\upsilon})^{+}\cdot\boldsymbol{e} = (\rho\boldsymbol{\upsilon})^{-}\cdot\boldsymbol{e}. \tag{3.12.3}$$

Intuitively speaking, this means that matter entering from one side has to leave the singular surface on the other: Mass cannot simply disappear. This is another possible version and interpretation of the *equation of continuity*. If the singular surface moves with the surrounding matter the expression in brackets shown in Eq. (3.9.2) vanishes and an identity results. Mass is neither entering nor leaving.

[13] We have used index notation in Table 3.2 since it makes it easier to distinguish operations referring to the volume and to the surface, respectively.

**Table 3.2** Entries for balance equations in singular points

| $\psi$ | $\underset{A}{\psi}$ | $-\underset{L}{\varphi}^{\Delta}$ | $\underset{A}{p}$ | $\underset{A}{s}$ |
|---|---|---|---|---|
| Mass | $\underset{A}{\rho}$ | $0$ | $0$ | $0$ |
| (linear) momentum | $\underset{A}{\rho}\underset{A}{v}_i$ | $\underset{L}{\sigma}^{\Delta}_{i}$ | $0$ | $\underset{A}{\rho}\underset{A}{f}_i$ |
| Kinetic energy | $\frac{\rho_A}{2}\underset{A}{\boldsymbol{v}}^2$ | $\underset{L}{\sigma}^{\Delta}_{i}\underset{A}{v}_i$ | $\underset{L}{\sigma}^{\Delta}_{i}\frac{\partial \underset{A}{v}_i}{\partial Z^{\Delta}}\left[\frac{1}{2}(\underset{A}{\boldsymbol{v}}-\boldsymbol{v})^2\rho(\underset{A}{v}\perp e_j)-\sigma_{ji}(\underset{A}{v}_i-v_i)\right]e_j$ | $\underset{A}{\rho}\underset{A}{f}_i\underset{A}{v}_i$ |
| Internal energy | $\underset{A}{\rho}\underset{A}{u}$ | $-\underset{L}{q}^{\Delta}$ | $\underset{L}{\sigma}^{\Delta}_{i}\frac{\partial \underset{A}{v}_i}{\partial Z^{\Delta}}+\left[\frac{1}{2}(\underset{A}{\boldsymbol{v}}-\boldsymbol{v})^2\rho(\underset{A}{v}\perp e_j)-\sigma_{ji}(\underset{A}{v}_i-v_i)\right]e_j$ | $\underset{A}{\rho}\underset{A}{r}$ |
| Total energy | $\underset{A}{\rho}\left(\underset{A}{u}+\frac{1}{2}\underset{A}{\boldsymbol{v}}^2\right)$ | $-\underset{L}{q}^{\Delta}+\underset{L}{\sigma}^{\Delta}_{i}\cdot\underset{A}{v}_i$ | $0$ | $\underset{A}{\rho}\left(\underset{A}{f}_i\underset{A}{v}_i+\underset{A}{r}\right)$ |
| Moment of momentum | $\underset{A}{\rho}\,\epsilon_{ijk}x_j\underset{A}{v}_k$ | $\epsilon_{ikl}x_k\underset{L}{\sigma}^{\Delta}_{l}$ | $-\epsilon_{ijk}\underset{L}{\sigma}^{\Delta}_{j}\underset{L}{\tau}^{k}_{\Delta}$ | $\underset{A}{\rho}\,\epsilon_{ijk}x_j\underset{A}{f}_k$ |
| Spin | $\underset{A}{\rho}\underset{A}{s}_i$ | $\underset{L}{m}^{\Delta}_{i}$ | $\epsilon_{ijk}\underset{L}{\sigma}^{\Delta}_{j}\underset{L}{\tau}^{k}_{\Delta}$ | $\underset{A}{\rho}\underset{A}{l}_i$ |
| Angular momentum | $\underset{A}{\rho}(\underset{A}{s}_i+\epsilon_{ijk}x_j\underset{A}{v}_k)$ | $\underset{L}{m}^{\Delta}_{i}+\epsilon_{ikl}x_k\underset{L}{\sigma}^{\Delta}_{l}$ | $0$ | $\underset{A}{\rho}\left(\epsilon_{ijk}x_j\underset{A}{f}_k+\underset{A}{l}_i\right)$ |

**Exercise 3.12.1:** ***Jump condition of momentum for the static case***

Specialize the balance of momentum in singular points to *static* conditions and *neglect gravitation*. Show that under these circumstances we have:

$$\boldsymbol{e} \cdot [\![\boldsymbol{\sigma}]\!] = \mathbf{0} \;\Leftrightarrow\; e_j [\![\sigma_{ji}]\!] = 0. \tag{3.12.4}$$

Specialize this equation now to the case of isotropic compression, for which ($p$ is known as the *pressure*):

$$\boldsymbol{\sigma} = -p\mathbf{1} \;\Leftrightarrow\; \sigma_{ij} = -p\delta_{ij}. \tag{3.12.5}$$

and show that at a very thin wall the inner and the outer pressure must be equal:

$$p^+ = p^-. \tag{3.12.6}$$

## 3.13 Would You Like to Know More?

The dominating topic of this chapter—the physics and mathematics of balance equations—is a neverending story for the enthusiastic continuum scientist. Consequently, we could only give a first impression. The simplified proof of REYNOLDS' transport theorem and its relation to LEIBNIZ' theorem on the differentiation of parametric integrals can be found in many variations, for example in Irgens [4], Sect. 8.2, Liu [5], Sect. 2.1, Müller [10],[14] Sect. 1.2.2 , Müller and Ferber (see Footnote 14) [11], Sect. 4.1.3, Müller and Müller [12], Sect. 1.2.1.

The full proof is presented in Grewe (see Footnote 14) [2], Sect. 2.1.1, Haupt [3], Sect. 3.5, Müller (see Footnote 14) [8], Chap. 1, Sect. 1.2, Chap. 2, Sect. 2.1, Müller [9], Sect. 3.1. The latter reference also presents many details to the balances in regular and singular points. The source of all wisdom, however, is certainly the famous handbook article by Truesdell and Toupin [17], Sections 79 ff and 172 ff.

The question how balances for open systems can be obtained is discussed in Schade (see Footnote 14) [15], Sect. 1.2.4 and, for the case of continuous fields, in detail in Müller and Muschik (see Footnote 14) [13], and Muschik and Müller (see Footnote 14) [14].

Jump conditions in index notation were derived in great detail by Moeckel [7]. In fact his paper was the basis for the discussion on general balances as outlined in Müller [9]. More recently this issue has seen a revival, e.g., in the paper by McBride et al. [6], which favors the abstract notation.

[14] In German.

# References


1. Euler L (1775) Nova methodus motum corporum rigidorum determinandi. Novi Commentarii Academiae Petropolitanae, pp 208–238
2. Greve R (2003) Kontinuumsmechanik: Ein Grundkurs für Ingenieure und Physiker. Springer, Berlin
3. Haupt P (2002) Continuum mechanics and theory of materials, 2nd edn. Springer, Berlin
4. Irgens F (2008) Continuum mechanics. Springer, Berlin
5. Liu I-S (2010) Continuum mechanics. Springer, Berlin
6. McBride AT, Javili A, Steinmann P, Bargmann S (2011) Geometrically nonlinear continuum thermomechanics with surface energies coupled to diffusion. J Mech Phys Solids 59:2116–2133
7. Moeckel GP (1974) Thermodynamics of an interface. ARMA 57:255–280
8. Müller I (1973) Thermodynamik Die Grundlagen der Materialtheorie. Bertelsmann Universitätsverlag, Düsseldorf
9. Müller I (1985) Thermodynamics. Pitman Advanced Publishing Program, Boston
10. Müller I (1994) Grundzüge der Thermodynamik mit historischen Anmerkungen, 1st edn. Springer, Berlin
11. Müller WH, Ferber F (2008) Technische Mechanik für Ingenieure, 4. aktualisierte Auflage, Carl Hanser, München
12. Müller I, Müller WH (2009) Fundamentals of thermodynamics and applications. Springer, Berlin
13. Müller WH, Muschik W (1983) Bilanzgleichungen offener mehrkomponentiger Systeme I: Massen- und Impulsbilanzen. J Non-Equilib Thermodyn 8:29–46
14. Muschik W, Müller WH (1983) Bilanzgleichungen offener mehrkomponentiger Systeme II: Energie und Entropiebilanz. J Non-Equilib Thermodyn 8:47–66
15. Schade H (1970) Kontinuumstheorie strömender Medien. Springer, Berlin
16. Szabó I (1977) Geschichte der mechanischen Prinzipien. Birkhäuser, Basel
17. Truesdell C, Toupin R (1960) The classical field theories. In: Flügge S (ed) Encyclopedia of physics, vol III/1, Principles of classical mechanics and field theory. Springer, Berlin, Göttingen, Heidelberg
18. Truesdell C (1968) Whence the law of moment of momentum. In: Essays in the history of mechanics. Springer, Berlin
19. Truesdell C (1969) Rational thermodynamics. McGraw-Hill, New York

# Chapter 4
# Spatial Derivatives of Fields

**Abstract** The balances of continuum theory as well as the constitutive equations contain spatial derivatives of scalar vector and of tensor fields. In this chapter we will learn how to express these in arbitrary curvilinear coordinates. This means that we will now truly switch from tensor algebra to tensor analysis. In index notation spatial derivatives will be handled by means of the covariant derivative and CHRISTOFFEL symbols. In absolute notation we will introduce the so-called del operator. Both will be specified for technically relevant coordinate systems, namely cylindrical, spherical, and elliptical coordinates. The formalism will be applied to balance and constitutive equations in subsequent chapters.

*In the fall of 1972 President Nixon announced that the rate of increase of inflation was decreasing. This was the first time a sitting president used the third derivative to advance his case for reelection.*

Hugo ROSSI, Professor Emeritus of Mathematics, University of Utah

## 4.1 Spatial Derivatives of Scalar Fields

The general balance equations (3.7.3/3.7.4) in combination with Tables 3.1 and 3.2 clearly show that it is necessary to investigate how to calculate spatial derivatives of scalar fields like mass density, vector fields, such as velocity and, finally, tensor fields like stress in arbitrary coordinate systems.

We start with the gradient of an arbitrary scalar field $\underset{(x)}{f} = \underset{(x)}{\tilde{f}}(x_i) = \underset{(x)}{\tilde{f}}(x_i(z^j)) = \underset{(x)}{\hat{f}}(z^j)$[1] w.r.t. a Cartesian coordinate system. By means of the chain rule we may write:

[1] The tildes and circumflexes do not refer to the Lagrangian and Eulerian ways of descriptions from Sect. 3.4. They are merely a hint that different functions were used depending upon the choice of coordinates.

W. H. Müller, *An Expedition to Continuum Theory*, Solid Mechanics and Its Applications 210, DOI: 10.1007/978-94-007-7799-6_4,

$$\frac{\partial \underset{(x)}{\tilde{f}}(x_i)}{\partial x_i} = \frac{\partial z^k}{\partial x_i} \frac{\partial \underset{(x)}{\hat{f}}(z^j)}{\partial z^k} \quad \text{or} \quad \frac{\partial \underset{(x)}{f}}{\partial x_i} = \frac{\partial z^k}{\partial x_i} \frac{\partial \underset{(x)}{f}}{\partial z^k} \text{ for short.} \tag{4.1.1}$$

Scalar functions are independent of the coordinate frame, and therefore:

$$\underset{(x)}{\hat{f}}(z^j) = \underset{(z)}{\hat{f}}(z^j) \quad \text{or } \underset{(x)}{f} = \underset{(z)}{f} \,. \tag{4.1.2}$$

Equation (4.1.1) can be inverted:

$$\frac{\partial \underset{(z)}{f}}{\partial z^k} = \frac{\partial x_i}{\partial z^k} \frac{\partial \underset{(x)}{f}}{\partial x_i} \,. \tag{4.1.3}$$

According to Eq. (2.4.1) we may therefore say that the quantity $\partial \underset{(z)}{f} / \partial z^k$ transforms like the components of a covariant vector field. This quantity is called the *gradient of the scalar field f.*[2]

## 4.2 Spatial Derivatives of Vector Fields

In what follows we will—for simplicity—no longer distinguish between the various functions of a field by tildes and circumflexes. In this spirit we consider now the vector field $\underset{(x)}{A_i}$ in Cartesian representation $(x)$ and its spatial derivative:

$$\frac{\partial \underset{(x)}{A_i}}{\partial x_j} = \frac{\partial z^k}{\partial x_j} \frac{\partial \underset{(x)}{A_i}}{\partial z^k} = \frac{\partial z^k}{\partial x_j} \frac{\partial}{\partial z^k} \left( \frac{\partial x_i}{\partial z^l} \underset{(z)}{A^l} \right) = \frac{\partial z^k}{\partial x_j} \frac{\partial x_i}{\partial z^l} \left( \frac{\partial \underset{(z)}{A^l}}{\partial z^k} + \frac{\partial z^l}{\partial x_s} \frac{\partial^2 x_s}{\partial z^k \partial z^n} \underset{(z)}{A^n} \right) . \tag{4.2.1}$$

In this equation use was made of the chain rule, the product rule, as well as Eq. (2.4.9). On this basis we may write:

$$\frac{\partial \underset{(z)}{A^l}}{\partial z^k} + \frac{\partial z^l}{\partial x_s} \frac{\partial^2 x_s}{\partial z^k \partial z^n} \underset{(z)}{A^n} = \frac{\partial z^l}{\partial x_i} \frac{\partial x_j}{\partial z^k} \frac{\partial \underset{(x)}{A_i}}{\partial x_j} \,. \tag{4.2.2}$$

If we compare this result with Eq. (2.4.15) we must conclude that the quantity on the left hand side, i.e.:

---

[2] Note that the symbol *f* for the scalar field does not contain any information about the coordinate system that was used. It is to be understood in an absolute sense just like the symbol ***A*** denotes an absolute vector. The contravariant components of the gradient in Eq. (4.1.3) can be obtained by multiplication with the metric $g^{lk}$.

$$\underset{(z)}{A}{}^{l}{}_{;k} = \frac{\partial \underset{(z)}{A}{}^{l}}{\partial z^k} + \Gamma^{l}_{kn} \underset{(z)}{A}{}^{n} , \quad \Gamma^{l}_{kn} = \frac{\partial z^l}{\partial x_s} \frac{\partial^2 x_s}{\partial z^k \partial z^n} \tag{4.2.3}$$

form the components of a mixed tensor field, namely the *gradient of the vector* $A$. The quantity $\underset{(z)}{A}{}^{l}{}_{;k}$ defined by Eq. (4.2.3) is also known as the *covariant derivative* of the contravariant components of the vector $\boldsymbol{A}$. The quantities $\Gamma^{l}_{kn}$ are known as *Christoffel symbols*. Note that they are *symmetric* w.r.t. the *lower indices*.

**Exercise 4.2.1: *Christoffel symbols expressed by the metric***

Proof the following alternative formula for calculating Christoffel symbols:

$$\Gamma^{l}_{kn} = \frac{1}{2} g^{lm} \left( \frac{\partial g_{mk}}{\partial z^n} + \frac{\partial g_{mn}}{\partial z^k} - \frac{\partial g_{kn}}{\partial z^m} \right). \tag{4.2.4}$$

**Exercise 4.2.2: *Christoffel symbols for cylindrical coordinates***

Show by using Eq. (2.2.13) for the metric and the alternative representation (4.2.4) for the Christoffel symbols that the only non-vanishing ones in cylindrical coordinates are:

$$\Gamma^{r}_{\vartheta\vartheta} = -r, \quad \Gamma^{\vartheta}_{r\vartheta} = \Gamma^{\vartheta}_{\vartheta r} = \frac{1}{r}. \tag{4.2.5}$$

**Exercise 4.2.3: *Christoffel symbols for spherical coordinates***

Show by using Eq. (2.2.16) for the metric and the alternative representation (4.2.4) for the Christoffel symbol that the only non-vanishing ones in spherical coordinates are:

$$\begin{aligned} &\Gamma^{\vartheta}_{\vartheta r} = \Gamma^{\vartheta}_{r\vartheta} = \Gamma^{\varphi}_{\varphi r} = \Gamma^{\varphi}_{r\varphi} = \frac{1}{r}, \quad \Gamma^{r}_{\vartheta\vartheta} = -r, \\ &\Gamma^{r}_{\varphi\varphi} = -r \sin^2 \vartheta \, , \; \Gamma^{\varphi}_{\varphi\vartheta} = \Gamma^{\varphi}_{\vartheta\varphi} = \cot \vartheta, \quad \Gamma^{\vartheta}_{\varphi\varphi} = - \sin \vartheta \cos \vartheta. \end{aligned} \tag{4.2.6}$$

**Exercise 4.2.4: *Christoffel symbols for elliptical coordinates***

Recall the results for the metric in elliptic coordinates from Exercise 2.3.2:

$$g_{ij} = \frac{c^2}{2}\begin{bmatrix} \cosh(2z^1) - \cos(2z^2) & 0 \\ 0 & \cosh(2z^1) - \cos(2z^2) \end{bmatrix}. \tag{4.2.7}$$

Use them to show that the corresponding non-vanishing Christoffel symbols are given by:

$$\begin{aligned} \Gamma^1_{11} &= -\Gamma^1_{22} = -\Gamma^2_{12} = \Gamma^2_{21} = \frac{\sinh(2z^1)}{\cosh(2z^1) - \cos(2z^2)}, \\ \Gamma^1_{12} &= \Gamma^1_{21} = -\Gamma^2_{11} = \Gamma^2_{22} = \frac{\sin(2z^2)}{\cosh(2z^1) - \cos(2z^2)}. \end{aligned} \tag{4.2.8}$$

Equation (4.2.3) defines the *covariant derivative of the contravariant components* $\underset{(z)}{A^l}$. But how does the covariant derivative of the *covariant* components of the vector field $\boldsymbol{A}$ read? In order to find out we write:

$$\begin{aligned} \frac{\partial \underset{(x)}{A_i}}{\partial x_j} &= \frac{\partial z^k}{\partial x_j}\frac{\partial}{\partial z^k}\left(\frac{\partial z^l}{\partial x_i}\underset{(z)}{A_l}\right) = \frac{\partial z^k}{\partial x_j}\frac{\partial z^l}{\partial x_i}\frac{\partial \underset{(z)}{A_l}}{\partial z^k} + \frac{\partial z^k}{\partial x_j}\frac{\partial^2 z^l}{\partial z^k \partial x_i}\underset{(z)}{A_l} \\ &= \frac{\partial z^k}{\partial x_j}\frac{\partial z^l}{\partial x_i}\left(\frac{\partial \underset{(z)}{A_l}}{\partial z^k} - \Gamma^t_{kl}\underset{(z)}{A_l}\right). \end{aligned} \tag{4.2.9}$$

**Exercise 4.2.5: *Covariant derivative of covariant vector components***

Verify the last step in the chain of Eq. (4.2.9). To this end make use of the definition (4.2.3) for the Christoffel symbols and the following auxiliary condition, which should be verified as well:

$$\frac{\partial^2 z^n}{\partial z^k \partial x_i}\frac{\partial x_i}{\partial z^t} + \frac{\partial z^n}{\partial x_i}\frac{\partial^2 x_i}{\partial z^k \partial z^t} = 0. \tag{4.2.10}$$

We conclude that:

$$\frac{\partial \underset{(z)}{A_l}}{\partial z^k} - \Gamma^t_{kl} \underset{(z)}{A_t} = \frac{\partial x_i}{\partial z^l}\frac{\partial x_j}{\partial z^k}\frac{\partial \underset{(x)}{A_i}}{\partial x_j}. \tag{4.2.11}$$

If we compare this with Eq. (2.4.15) we must conclude that the expression

$$\underset{(z)}{A_{l;k}} = \frac{\partial \underset{(z)}{A_l}}{\partial z^k} - \Gamma^t_{kl} \underset{(z)}{A_t} \tag{4.2.12}$$

transforms like the components of a covariant tensor field. This is just another representation of the gradient of $\boldsymbol{A}$. It is also known as the *covariant derivative* $\underset{(z)}{A_{l;k}}$ *of the covariant vector components* $\underset{(z)}{A_l}$.

In context with spatial derivatives of vector fields it should finally be pointed out that due to [*cp.*, Eqs. (2.4.1) and (2.4.15)]

$$\underset{(z)}{A^k} = \frac{\partial z^k}{\partial x_l}\underset{(x)}{A_l}, \quad \underset{(z)}{A^i}_{;k} = \frac{\partial z^i}{\partial x_n}\frac{\partial x_s}{\partial z^k}\frac{\partial \underset{(x)}{A_n}}{\partial x_s} \tag{4.2.13}$$

the combination

$$\underset{(z)}{A^k}\,\underset{(z)}{A^i}_{;k} = \frac{\partial z^i}{\partial x_n}\underset{(x)}{A_s}\frac{\partial \underset{(x)}{A_n}}{\partial x_s} \tag{4.2.14}$$

represents the components of a contravariant vector. It will reappear when rewriting the balance of momentum for arbitrary coordinates.

**Exercise 4.2.6:** ***The divergence of a vector field***

Use Eq. (4.2.2) and the definition for the covariant derivative shown in Eq. (4.2.3) and show that:

$$\underset{(z)}{A^l}_{;l} = \frac{\partial \underset{(x)}{A_i}}{\partial x_i}. \tag{4.2.15}$$

Interpret this equation in terms of its transformation character and conclude that $\underset{(z)}{A^l}_{;l}$ must be a scalar. It is known as the *divergence* of the vector field $\boldsymbol{A}$ and it is customary to write in absolute notation:

$$\underset{(z)}{A^l}_{;l} = \nabla\cdot\boldsymbol{A} \quad \text{or} \quad \underset{(z)}{A^l}_{;l} = \mathrm{div}\boldsymbol{A}. \tag{4.2.16}$$

**Exercise 4.2.7:** ***The LAPLACE operator***

Use the definition for the covariant derivative shown in Eq. (4.2.3) and show that the divergence of the gradient of a scalar field $\underset{(z)}{A}{}^{l}{}_{;l}$ in arbitrary coordinates, $z$, can be written as:

$$\Delta f = \left( g^{nm} \frac{\partial f}{\partial z^m} \right)_{;n} = \frac{\partial}{\partial z^n} \left( g^{nm} \frac{\partial f}{\partial z^m} \right) + \Gamma^n_{nk} g^{km} \frac{\partial f}{\partial z^m} . \tag{4.2.17}$$

Specialize this equation to the case of Cartesian coordinates, $x$, and conclude that it is the generalization of the LAPLACE operator to arbitrary coordinates, $z$. Use the metric and the CHRISTOFFEL symbols from Eqs. (2.2.13), (2.2.16), (4.2.5), (4.2.6), and show that the LAPLACE operator in cylindrical and spherical coordinates reads:

$$\Delta f = \frac{1}{r} \frac{\partial}{\partial r} \left( r \frac{\partial f}{\partial r} \right) + \frac{1}{r^2} \frac{\partial^2 f}{\partial \vartheta^2} + \frac{\partial^2 f}{\partial z^2} \tag{4.2.18}$$

and:

$$\Delta f = \frac{1}{r} \frac{\partial^2}{\partial r^2} (rf) + \frac{1}{r^2 \sin^2 \vartheta} \frac{\partial^2 f}{\partial \varphi^2} + \frac{1}{r^2 \sin \vartheta} \frac{\partial}{\partial \vartheta} \left( \sin \vartheta \frac{\partial f}{\partial \vartheta} \right) . \tag{4.2.19}$$

Pierre Simon de LAPLACE was born on March 28, 1749 in Beaumont-en-Auge (Normandy) and died on March 5, 1827 in Paris. Even though he is of noble birth he survives the French revolution completely unharmed. In particular LAPLACE is known for his work on celestial mechanics. This is most impressively demonstrated by his *œuvre* in five volumes entitled *Traité de Mécanique Céleste*. Here he provided us with a hypothesis on the formation of our solar system by gravitationally induced contraction out of a nebula. He also featured an estimate about the mass of a star from which no light can escape—without knowing about black holes.

## 4.3 Invariant Notation of Spatial Derivatives of Scalar Fields

Recall the result from Sect. 4.1 according to which we may write for the gradient of an arbitrary scalar field $\underset{(x)}{f} = \underset{(z)}{f} \equiv f$ w.r.t. a Cartesian and an arbitrary skew curvilinear coordinate system:

$$\frac{\partial f}{\partial z^k} = \frac{\partial x_i}{\partial z^k}\frac{\partial f}{\partial x_i}. \tag{4.3.1}$$

It was already noted that the gradient of a scalar transforms like a *covariant* vector field and we therefore define the corresponding absolute vector $\nabla f$[3] by means of the contravariant base, $\boldsymbol{g}^k$:

$$\nabla f = \frac{\partial f}{\partial z^k}\boldsymbol{g}^k, \tag{4.3.2}$$

where the *del* or *Nabla operator* has been defined as [also see Exercise 3.4.1, Eq. (3.4.5)]:

$$\nabla(\bullet) = \frac{\partial(\bullet)}{\partial z^k}\boldsymbol{g}^k. \tag{4.3.3}$$

In particular we find in a Cartesian base that:

$$\nabla(\bullet) = \frac{\partial(\bullet)}{\partial x_i}\boldsymbol{e}_i. \tag{4.3.4}$$

All of this is consistent with the remaining equations: If Eq. (2.3.4) is inserted into Eq. (4.3.2) application of the chain rule yields:

$$\nabla f = \frac{\partial f}{\partial z^k}\frac{\partial z^k}{\partial x_i}\boldsymbol{e}_i = \frac{\partial f}{\partial x_i}\boldsymbol{e}_i \Rightarrow \nabla(\bullet) = \frac{\partial(\bullet)}{\partial x_i}\boldsymbol{e}_i. \tag{4.3.5}$$

The del operator, which allowed to form the gradient of a scalar, can now be used to obtain the LAPLACE operator when applied to a scalar. In Exercise 4.2.7 it was shown that the LAPLACE operator can be identified as the divergence of a contravariant vector field. In Cartesian coordinates we may write:

$$\begin{aligned}\nabla\cdot(\nabla f) &= \boldsymbol{e}_i\cdot\frac{\partial}{\partial x_i}\left(\frac{\partial f}{\partial x_j}\boldsymbol{e}_j\right) = \frac{\partial^2 f}{\partial x_i\partial x_j}\boldsymbol{e}_i\cdot\boldsymbol{e}_j \\ &= \frac{\partial^2 f}{\partial x_i\partial x_j}\delta_{ij} = \frac{\partial^2 f}{\partial x_i\partial x_i}.\end{aligned} \tag{4.3.6}$$

[3] In absolute notation it is also customary to write grad $f$ instead of $\nabla f$.

Note that the differentiation of the base vectors $\boldsymbol{e}_i$ yields zero. Consequently we obtain the well known result in Cartesian coordinates for three dimensions:

$$\Delta(\bullet) = \frac{\partial^2(\bullet)}{\partial x_1^2} + \frac{\partial^2(\bullet)}{\partial x_2^2} + \frac{\partial^2(\bullet)}{\partial x_3^2}. \tag{4.3.7}$$

Thus we may write[4]:

$$\Delta(\bullet) = \nabla \cdot (\nabla(\bullet)). \tag{4.3.8}$$

This relation can now be used in context with Eq. (4.3.2) to obtain the following representation of the LAPLACE operator in skew curvilinear coordinates, which is equivalent to Eq. (4.2.17):

$$\begin{aligned} \Delta(\bullet) = \nabla \cdot (\nabla(\bullet)) &= \left[\frac{\partial}{\partial z^k}\left(\frac{\partial(\bullet)}{\partial z^l}\boldsymbol{g}^l\right)\right] \cdot \boldsymbol{g}^k = \frac{\partial^2(\bullet)}{\partial z^k \partial z^l}\boldsymbol{g}^l \cdot \boldsymbol{g}^k + \frac{\partial \boldsymbol{g}^l}{\partial z^k}\frac{\partial(\bullet)}{\partial z^l} \cdot \boldsymbol{g}^k \\ &= \frac{\partial^2(\bullet)}{\partial z^k \partial z^l}g^{lk} + \frac{\partial \boldsymbol{g}^l}{\partial z^k} \cdot \boldsymbol{g}^k \frac{\partial(\bullet)}{\partial z^l}. \end{aligned} \tag{4.3.9}$$

The following exercise makes explicit use of this relation.

**Exercise 4.3.1:** ***The LAPLACE operator in cylindrical coordinates revisited***

Recall the transformation rules between Cartesian and cylindrical coordinates and show in a first step that:

$$\boldsymbol{g}^r = \cos\vartheta\, \boldsymbol{e}_1 + \sin\vartheta\, \boldsymbol{e}_2,\ \boldsymbol{g}^\vartheta = -\frac{\sin\vartheta}{r}\boldsymbol{e}_1 + \frac{\cos\vartheta}{r}\boldsymbol{e}_2,\ \boldsymbol{g}^z = \boldsymbol{e}_3. \tag{4.3.10}$$

Now use these relations and show in context with Eq. (4.3.9) that the LAPLACE operator in cylindrical coordinates is given by the following formula known already from Eq. (4.2.17):

$$\Delta(\bullet) = \frac{\partial^2(\bullet)}{\partial r^2} + \frac{1}{r}\frac{\partial(\bullet)}{\partial r} + \frac{1}{r^2}\frac{\partial^2 f}{\partial \vartheta^2} + \frac{\partial^2 f}{\partial z^2}. \tag{4.3.11}$$

[4] or div grad $(\bullet)$.

**Exercise 4.3.2:** ***The Laplace operator in absolute and in index notation***

Show that Eqs. (4.2.17) and (4.3.9) are identical. To this end use the definitions (2.5.4) for the dual base as well as the identity (4.2.10) which is useful when rewriting Christoffel symbols.

## 4.4 Spatial Derivatives of Tensors

We now turn to the spatial derivatives of a second order tensor $\underset{(x)}{B}_{ij}$:

$$\frac{\partial \underset{(x)}{B}_{ij}}{\partial x_k} = \frac{\partial z^l}{\partial x_k}\frac{\partial}{\partial z^l}\left(\frac{\partial x_i}{\partial z^n}\frac{\partial x_j}{\partial z^p}\underset{(z)}{B}^{np}\right) = \frac{\partial z^l}{\partial x_k}\frac{\partial x_i}{\partial z^n}\frac{\partial x_j}{\partial z^p}\left(\frac{\partial \underset{(z)}{B}^{np}}{\partial z^l} + \Gamma^n_{lr}\underset{(z)}{B}^{rp} + \Gamma^p_{lr}\underset{(z)}{B}^{nr}\right). \quad (4.4.1)$$

The product rule and the definition $(4.2.3)_2$ for the Christoffel symbols were used during the various transformations. It follows that:

$$\frac{\partial \underset{(z)}{B}^{np}}{\partial z^l} + \Gamma^n_{lr}\underset{(z)}{B}^{rp} + \Gamma^p_{lr}\underset{(z)}{B}^{nr} = \frac{\partial x_k}{\partial z^l}\frac{\partial z^n}{\partial x_i}\frac{\partial z^p}{\partial x_j}\frac{\partial \underset{(x)}{B}_{ij}}{\partial x_k}. \quad (4.4.2)$$

We conclude in perfect analogy to the transformation rule for tensors of second order, see Eq. (2.4.15), that the combination

$$\underset{(z)}{B}^{np}{}_{;l} = \frac{\partial \underset{(z)}{B}^{np}}{\partial z^l} + \Gamma^n_{lr}\underset{(z)}{B}^{rp} + \Gamma^p_{lr}\underset{(z)}{B}^{nr} \quad (4.4.3)$$

represents the components of a mixed tensor of third order, namely the gradient of the second order tensors $\boldsymbol{B}$. We may also say that this equation represents the covariant derivative of the contravariant components of the second order tensor $\boldsymbol{B}$. Note that it is also customary in mathematics to denote the partial derivatives in such equations by a comma:

$$\underset{(z)}{B}^{np}{}_{;l} = \underset{(z)}{B}^{np}{}_{,l} + \Gamma^n_{lr}\underset{(z)}{B}^{rp} + \Gamma^p_{lr}\underset{(z)}{B}^{nr}. \quad (4.4.4)$$

This way of writing shows very nicely, which corrections are required to turn a partial derivative into a covariant one or, in other words, how to create an *invariant tensor expression*. The corrections are given by the Christoffel symbols which are suitably combined with both indices of the second order tensor. The "corrections" vanish for "straight" Cartesian coordinate systems. The equation also implies a *hierarchy of derivatives*. According to the results of the previous sections we may write for scalars (i.e., zero order tensors), vectors (i.e., first order tensors), and tensors that are higher than second order:

$$\begin{aligned}
\underset{(z)}{f}{}_{;l} &= \underset{(z)}{f}{}_{,l}, \\
\underset{(z)}{A}{}^{n}{}_{;l} &= \underset{(z)}{A}{}^{n}{}_{,l} + \Gamma^{n}_{lr}\underset{(z)}{A}{}^{r}, \qquad (4.4.5)\\
\underset{(z)}{C}{}^{npq}{}_{;l} &= \underset{(z)}{C}{}^{npq}{}_{,l} + \Gamma^{n}_{lr}\underset{(z)}{C}{}^{rpq} + \Gamma^{p}_{lr}\underset{(z)}{C}{}^{nrq} + \Gamma^{q}_{lr}\underset{(z)}{C}{}^{npr}, \ etc.
\end{aligned}$$

Each (contravariant) index gets "its own" CHRISTOFFEL symbol, so to speak. However, there is room for variations, especially in the case of higher order tensors, which we do not necessarily have to represent in fully contravariant form. For example note that the covariant derivative of the covariant components $\underset{(z)}{B}{}_{ij}$, can be obtained as follows:

$$\frac{\partial \underset{(x)}{B}{}_{ij}}{\partial x_k} = \frac{\partial z^l}{\partial x_k}\frac{\partial}{\partial z^l}\left(\frac{\partial z^n}{\partial x_i}\frac{\partial z^p}{\partial x_j}\underset{(z)}{B}{}_{np}\right) = \frac{\partial z^l}{\partial x_k}\frac{\partial z^n}{\partial x_i}\frac{\partial z^p}{\partial x_j}\left(\frac{\partial \underset{(z)}{B}{}_{np}}{\partial z^l} - \Gamma^{r}_{nl}\underset{(z)}{B}{}_{rp} - \Gamma^{r}_{lp}\underset{(z)}{B}{}_{nr}\right). \qquad (4.4.6)$$

This is a consequence of the product rule and the definition $(4.2.3)_2$ for the CHRISTOFFEL symbols, just like Eq. (4.4.1). We conclude that:

$$\frac{\partial \underset{(z)}{B}{}_{np}}{\partial z^l} - \Gamma^{r}_{nl}\underset{(z)}{B}{}_{rp} - \Gamma^{r}_{pl}\underset{(z)}{B}{}_{nr} = \frac{\partial x_k}{\partial z^l}\frac{\partial x_i}{\partial z^n}\frac{\partial x_j}{\partial z^p}\frac{\partial \underset{(x)}{B}{}_{ij}}{\partial x_k}. \qquad (4.4.7)$$

In full analogy to Eq. (2.4.17) we realize that the combination

$$\underset{(z)}{B}{}_{np;l} = \frac{\partial \underset{(z)}{B}{}_{np}}{\partial z^l} - \Gamma^{r}_{pl}\underset{(z)}{B}{}_{nr} - \Gamma^{r}_{nl}\underset{(z)}{B}{}_{pr} \qquad (4.4.8)$$

represents the components of a covariant tensor of third order, which is just another incarnation of the gradient of $\boldsymbol{B}$. Alternatively speaking, this expression represents the covariant derivative of the covariant components $\underset{(z)}{B}{}_{ij}$.

**Exercise 4.4.1:** ***The covariant derivative of the metric tensor***

Use Eqs. (4.4.3) and (4.4.8) as well as the alternative form (4.2.4) for the CHRISTOFFEL symbols and show that:

$$g^{np}{}_{;l} = 0, \quad g_{np;l} = 0. \qquad (4.4.9)$$

Finally note that the following formulae hold for the covariant derivative of a mixed tensor:

$$\underset{(z)}{B}{}^{n}{}_{p;l} = \frac{\partial \underset{(z)}{B}{}^{n}{}_{p}}{\partial z^{l}} + \Gamma^{n}_{lr} \underset{(z)}{B}{}^{r}{}_{p} - \Gamma^{r}_{pl} \underset{(z)}{B}{}^{n}{}_{r},$$
$$\underset{(z)}{B}{}_{p}{}^{n}{}_{;l} = \frac{\partial \underset{(z)}{B}{}_{p}{}^{n}}{\partial z^{l}} - \Gamma^{r}_{pl} \underset{(z)}{B}{}_{r}{}^{n} + \Gamma^{n}_{lr} \underset{(z)}{B}{}_{p}{}^{r} \tag{4.4.10}$$

As before each index gets "its own" Christoffel symbol. These "correction terms" come with a plus sign for contravariant indices and a minus sign in case of covariant ones.

Moreover, note that Eqs. (4.4.2) and (4.4.3) imply that $\underset{(z)}{B}{}^{nl}{}_{;l}$ are the components of a contravariant vector field, namely the *divergence* of $\boldsymbol{B}$:

$$\underset{(z)}{B}{}^{nl}{}_{;l} = \frac{\partial \underset{(z)}{B}{}^{nl}}{\partial z^{l}} + \Gamma^{n}_{lr} \underset{(z)}{B}{}^{rl} + \Gamma^{l}_{lr} \underset{(z)}{B}{}^{nr} = \frac{\partial x_k}{\partial z^l} \frac{\partial z^n}{\partial x_i} \frac{\partial z^l}{\partial x_j} \frac{\partial \underset{(x)}{B}{}_{ij}}{\partial x_k}$$
$$= \delta_{kj} \frac{\partial z^n}{\partial x_i} \frac{\partial \underset{(x)}{B}{}_{ij}}{\partial x_k} = \frac{\partial z^n}{\partial x_i} \frac{\partial \underset{(x)}{B}{}_{ij}}{\partial x_j}. \tag{4.4.11}$$

We will encounter this expression again in context with the spatial derivative of the stress tensor in the balance of momentum.

**Exercise 4.4.2:** ***The covariant derivative for mixed tensors of second order***

Provide a proof of Eq. (4.4.10), first, in analogy to the sequence of transformation steps shown in Eqs. (4.4.1/4.4.2) and, second, quasi in an indirect manner, by application of the rules for raising and lowering indices by means of the metric (see Sect. 2.4) when applied to Eqs. (4.4.3/4.4.8) and by observing (4.4.9).

**Exercise 4.4.3:** ***The covariant derivative for tensors of arbitrary order***

Derive a formula for the covariant derivative of a mixed tensor of arbitrary order. Apply the formula to the stiffness tensor [see Eq. (2.5.14)] and show that, for example:

$$\underset{(z)}{C}{}^{ijkl}{}_{;m} = \underset{(z)}{C}{}^{ijkl}{}_{,m} + \Gamma^{i}_{mr} \underset{(z)}{C}{}^{rjkl} + \Gamma^{j}_{mr} \underset{(z)}{C}{}^{irkl} + \Gamma^{k}_{mr} \underset{(z)}{C}{}^{ijrl} + \Gamma^{l}_{mr} \underset{(z)}{C}{}^{ijkr},$$
$$\underset{(z)}{C}{}^{ij}{}_{k}{}^{l}{}_{;m} = \underset{(z)}{C}{}^{ij}{}_{k}{}^{l}{}_{,m} + \Gamma^{i}_{mr} \underset{(z)}{C}{}^{rj}{}_{k}{}^{l} + \Gamma^{j}_{mr} \underset{(z)}{C}{}^{ir}{}_{k}{}^{l} - \Gamma^{r}_{km} \underset{(z)}{C}{}^{ij}{}_{r}{}^{l} + \Gamma^{l}_{mr} \underset{(z)}{C}{}^{ij}{}_{k}{}^{r}. \tag{4.4.12}$$

**Exercise 4.4.4: *Product rule and SCHWARZ' theorem for covariant derivatives***

Does a product rule hold for covariant derivatives, e.g.:

$$\left(\underset{(z)}{A}^i \underset{(z)}{B}^j\right)_{;k} = \underset{(z)}{A}^i{}_{;k} \underset{(z)}{B}^j + \underset{(z)}{A}^i \underset{(z)}{B}^j{}_{;k} \tag{4.4.13}$$

or:

$$\left(\underset{(z)}{\upsilon}^i \underset{(z)}{\sigma}^j{}_k\right)_{;l} = \underset{(z)}{\upsilon}^i{}_{;l} \underset{(z)}{\sigma}^j{}_k + \underset{(z)}{\upsilon}^i \underset{(z)}{\sigma}^j{}_{k;l}. \tag{4.4.14}$$

Can the sequence of covariant derivatives be interchanged ? In other words: Does SCHWARZ' theorem for mixed (partial) derivatives hold also for covariant derivatives:

$$\underset{(z)}{A}^i{}_{;jk} = \underset{(z)}{A}^i{}_{;kj} \ ? \tag{4.4.15}$$

If not, does the following hold:

$$\underset{(z)}{A}^i{}_{;ij} = \underset{(z)}{A}^i{}_{;ji} \ ? \tag{4.4.16}$$

## 4.5 Integral Theorems Revisited

Recall the proof of GAUSS' theorem from Exercise 3.4.1. It was based on Cartesian coordinates and finally led to Eq. (3.4.2). Then this equation was rewritten in absolute tensor form, Eq. (3.4.6), by means of the del operator. However, we never clarified the details and explained how to transform GAUSS' theorem in co-/contravariant notation (say) and, to begin with, we also never explained how the volume element $dV$ or the surface element $dA$ are evaluated in skew curvilinear coordinates. The solution is as follows. First, the volume element:

$$dV = \sqrt{\det(g_{ij})}\, d\underset{(z)}{V}, \quad d\underset{(z)}{V} = dz^1 dz^2 dz^3. \tag{4.5.1}$$

The proof is related to the arguments presented in context with Eq. (3.4.22). The volume element in curvilinear coordinates is given by a triple product[5] of three line segments:

$$dV = d\overset{(1)}{\boldsymbol{x}} \cdot \left(d\overset{(2)}{\boldsymbol{x}} \times d\overset{(3)}{\boldsymbol{x}}\right) \equiv \epsilon_{ijk} \frac{\partial x_i}{\partial z^l} \frac{\partial x_j}{\partial z^m} \frac{\partial x_k}{\partial z^n} d\overset{(1)}{z}{}^l d\overset{(2)}{z}{}^m d\overset{(3)}{z}{}^n \tag{4.5.2}$$

[5] The issue how the volume element transforms during reflections is not addressed in this section. See Sect. 8.4 for more details.

whose orientation is chosen along the skew curvilinear coordinate lines:

$$\mathrm{d}\overset{(1)}{z}{}^{l} = \left(\mathrm{d}z^1,\ 0,\ 0\right),\ \mathrm{d}\overset{(2)}{z}{}^{m} = \left(0,\ \mathrm{d}z^2,\ 0\right),\ \mathrm{d}\overset{(3)}{z}{}^{n} = \left(0,\ 0,\ \mathrm{d}z^3\right) \tag{4.5.3}$$

Mutual insertion yields:

$$\begin{aligned}\mathrm{d}V &= \epsilon_{ijk}\frac{\partial x_i}{\partial z^1}\frac{\partial x_j}{\partial z^2}\frac{\partial x_k}{\partial z^3}\mathrm{d}z^1\mathrm{d}z^2\mathrm{d}z^3 \equiv \det\left(\frac{\partial x_k}{\partial z^i}\right)\mathrm{d}\underset{(z)}{V} \\ &= \sqrt{\det\left(\frac{\partial x_k \partial x_k}{\partial z^i\, \partial z^j}\right)}\mathrm{d}\underset{(z)}{V} = \sqrt{\det(g_{ij})}\mathrm{d}\underset{(z)}{V}.\end{aligned} \tag{4.5.4}$$

It is sometimes required to write the relation (4.5.1) in terms of the corresponding Jacobi determinant. The following nomenclature has been established for this purpose:

$$J = \det\left(\frac{\partial x_k}{\partial z^i}\right) \equiv \frac{\partial(x_1, x_2, x_3)}{\partial(z^1, z^2, z^3)} \Rightarrow \mathrm{d}V = \frac{\partial(x_1, x_2, x_3)}{\partial(z^1, z^2, z^3)}\mathrm{d}z^1\mathrm{d}z^2\mathrm{d}z^3. \tag{4.5.5}$$

The notation is also quite useful for rewriting surface elements in skew curvilinear coordinates as we shall see now. We use the vector product of line elements:

$$\mathrm{d}\underset{(x)}{A}_i = \left(\mathrm{d}\overset{(1)}{\boldsymbol{x}} \times \mathrm{d}\overset{(2)}{\boldsymbol{x}}\right)_i = \epsilon_{ijk}\frac{\partial x_j}{\partial z^m}\frac{\partial x_k}{\partial z^n}\mathrm{d}\overset{(1)}{z}{}^m\mathrm{d}\overset{(2)}{z}{}^n. \tag{4.5.6}$$

whose orientation is chosen along the skew curvilinear surface coordinate lines:

$$\mathrm{d}\overset{(1)}{Z}{}^l = \left(\mathrm{d}z^1,\ 0\right),\quad \mathrm{d}\overset{(2)}{Z}{}^m = \left(0,\ \mathrm{d}z^2\right). \tag{4.5.7}$$

This leads to:

$$\mathrm{d}\underset{(x)}{A}_i = \left(\mathrm{d}\overset{(1)}{\boldsymbol{x}} \times \mathrm{d}\overset{(2)}{\boldsymbol{x}}\right)_i = \epsilon_{ijk}\frac{\partial x_j}{\partial z^1}\frac{\partial x_k}{\partial z^2}\mathrm{d}z^1\mathrm{d}z^2. \tag{4.5.8}$$

By expansion it follows that the three components of the surface element vector in Cartesian coordinates can be expressed in terms of the skew curvilinear coordinates by three Jacobi determinants:

$$\begin{aligned}\mathrm{d}\underset{(x)}{A}_1 &= \left(\frac{\partial x_2}{\partial z^1}\frac{\partial x_3}{\partial z^2} - \frac{\partial x_3}{\partial z^1}\frac{\partial x_2}{\partial z^2}\right)\mathrm{d}z^1\mathrm{d}z^2 \equiv \frac{\partial(x_2, x_3)}{\partial(z^1, z^2)}\mathrm{d}z^1\mathrm{d}z^2, \\ \mathrm{d}\underset{(x)}{A}_2 &= \left(\frac{\partial x_3}{\partial z^1}\frac{\partial x_1}{\partial z^2} - \frac{\partial x_1}{\partial z^1}\frac{\partial x_3}{\partial z^2}\right)\mathrm{d}z^1\mathrm{d}z^2 \equiv \frac{\partial(x_3, x_1)}{\partial(z^1, z^2)}\mathrm{d}z^1\mathrm{d}z^2, \\ \mathrm{d}\underset{(x)}{A}_3 &= \left(\frac{\partial x_1}{\partial z^1}\frac{\partial x_2}{\partial z^2} - \frac{\partial x_2}{\partial z^1}\frac{\partial x_1}{\partial z^2}\right)\mathrm{d}z^1\mathrm{d}z^2 \equiv \frac{\partial(x_1, x_2)}{\partial(z^1, z^2)}\mathrm{d}z^1\mathrm{d}z^2.\end{aligned} \tag{4.5.9}$$

We now consider the surface element itself and find by using Eqs. (4.5.8) and (8.3.35):

$$\begin{aligned} \mathrm{d}A &= \sqrt{\epsilon_{ijk}\frac{\partial x_j}{\partial z^1}\frac{\partial x_k}{\partial z^2}\epsilon_{irs}\frac{\partial x_r}{\partial z^1}\frac{\partial x_s}{\partial z^2}\mathrm{d}z^1\mathrm{d}z^2} \\ &= \mathrm{d}A = \sqrt{\frac{\partial x_r}{\partial z^1}\frac{\partial x_r}{\partial z^1}\frac{\partial x_s}{\partial z^2}\frac{\partial x_s}{\partial z^2} - \frac{\partial x_s}{\partial z^1}\frac{\partial x_s}{\partial z^2}\frac{\partial x_r}{\partial z^1}\frac{\partial x_r}{\partial z^2}\mathrm{d}z^1\mathrm{d}z^2} = \sqrt{\det(g_{\alpha\beta})}\,\mathrm{d}\underset{(z)}{A} \end{aligned} \tag{4.5.10}$$

where:

$$\mathrm{d}\underset{(z)}{A} = \mathrm{d}z^1\mathrm{d}z^2. \tag{4.5.11}$$

These two equations are the direct analogues to the expressions for volume elements shown in Eqs. (4.5.1). We conclude that:

$$\det(g_{\alpha\beta}) = \frac{\partial(x_2,x_3)}{\partial(z^1,z^2)}^2 + \frac{\partial(x_3,x_1)}{\partial(z^1,z^2)}^2 + \frac{\partial(x_1,x_2)}{\partial(z^1,z^2)}^2. \tag{4.5.12}$$

**Exercise 4.5.1:** ***Volume and surface elements in cylindrical and spherical coordinates***

Use the results from Eqs. (2.2.13/2.2.16) and derive the following expressions for the volume elements in cylindrical and spherical coordinates, respectively:

$$\mathrm{d}V = r\mathrm{d}r\ \mathrm{d}\vartheta\ \mathrm{d}z, \quad \mathrm{d}V = r^2\sin\vartheta\ \mathrm{d}r\ \mathrm{d}\vartheta\ \mathrm{d}\varphi. \tag{4.5.13}$$

Specialize now Eq. (2.2.13) to plane polar coordinates, recall the equations for the surface metric (2.7.14/2.7.17), and derive the following expression for the corresponding surface elements:

$$\mathrm{d}A = r\mathrm{d}r\ \mathrm{d}\vartheta, \quad \mathrm{d}A = r^2\sin\vartheta\ \mathrm{d}\vartheta\ \mathrm{d}\varphi, \tag{4.5.14}$$

and for the surface vectors by using Eq. (4.5.9) in both cases:

$$\begin{aligned} &\mathrm{d}\underset{(x)}{A_1} = r\ \cos\varphi\mathrm{d}\varphi\mathrm{d}z,\ \mathrm{d}\underset{(x)}{A_2} = r\ \sin\varphi\mathrm{d}\varphi\mathrm{d}z,\ \mathrm{d}\underset{(x)}{A_3} = 0, \\ &\mathrm{d}\underset{(x)}{A_1} = r^2\sin^2\vartheta\ \cos\varphi\mathrm{d}\vartheta\mathrm{d}\varphi, \\ &\mathrm{d}\underset{(x)}{A_2} = r^2\sin^2\vartheta\ \sin\varphi\mathrm{d}\vartheta\mathrm{d}\varphi, \\ &\mathrm{d}\underset{(x)}{A_3} = r^2\sin\vartheta\ \cos\vartheta\mathrm{d}\vartheta\mathrm{d}\varphi. \end{aligned} \tag{4.5.15}$$

Use these results to reconfirm the equations for the normal vectors shown in Eqs. (2.7.15/2.7.18). Finally prove Eq. (4.5.12).

We now assume that the symbol $g$ in Eq. (3.4.3) stands for a scalar $f = \underset{(x)}{f} = \underset{(z)}{f}$, and observe Eqs. (4.1.3) and $(4.4.5)_1$:

$$\iiint\limits_{V^+ \cup V^-} \frac{\partial z^k}{\partial x_i} \underset{(z)}{f}_{;k} \, \mathrm{d}V = \iint\limits_{A^+ \cup A^-} \underset{(z)}{f} \frac{\partial z^k}{\partial x_i} \underset{(z)}{n}_k \mathrm{d}A - \iint\limits_{A} \left[\!\left[ \underset{(z)}{f} \right]\!\right] \frac{\partial z^k}{\partial x_i} \underset{(z)}{e}_k \, \mathrm{d}A \tag{4.5.16}$$

Note that we cannot "cancel" the expression $\frac{\partial z^k}{\partial x_i}$, which is common to all three terms but depends on position and, hence, is part of the integration process. Eventually it will make the integration more cumbersome than the corresponding one in Cartesian coordinates. If now $\underset{(z)}{g}$ stands for a vector, which we can express in contravariant form $\underset{(z)}{A^j}$ (say), we have to observe Eqs. (4.2.2/4.2.3) and obtain:

$$\begin{aligned} \iiint\limits_{V^+ \cup V^-} \frac{\partial x_j}{\partial z^r} \frac{\partial z^s}{\partial x_i} \underset{(z)}{A^r}_{;s} \, \mathrm{d}V = & \iint\limits_{A^+ \cup A^-} \frac{\partial x_j}{\partial z^r} \frac{\partial z^s}{\partial x_i} \underset{(z)}{A^r} \underset{(z)}{n}_s \mathrm{d}A \\ & - \iint\limits_{A} \frac{\partial x_j}{\partial z^r} \frac{\partial z^s}{\partial x_i} [\![ \underset{(z)}{A^r} ]\!] \underset{(z)}{e}_s \, \mathrm{d}A. \end{aligned} \tag{4.5.17}$$

As in the previous case we have a common expression, $\frac{\partial x_j}{\partial z^r} \frac{\partial z^s}{\partial x_i}$, that cannot be cancelled. However, if we consider the special case of the divergence of a vector by putting $j = i$ the following interesting expression evolves:

$$\iiint\limits_{V^+ \cup V^-} \underset{(z)}{A^r}_{;r} \, \mathrm{d}V = \iint\limits_{A^+ \cup A^-} \underset{(z)}{A^r} \underset{(z)}{n}_r \mathrm{d}A - \iint\limits_{A} \left[\!\left[ \underset{(z)}{A^r} \right]\!\right] \underset{(z)}{e}_r \, \mathrm{d}A. \tag{4.5.18}$$

Under these circumstances GAUSS' theorem preserves the original form in Cartesian coordinates shown in Eq. (3.4.7). This is why the equation is known under a special name and people speak of the *divergence theorem*.

ARCHIMEDES was born around 287 in Syracuse, Sicily, and died there around 212 B.C. Rumor has it that he found the law of buoyancy when his king, HIERO of Syracuse, asked him whether his new crown was pure gold, as commissioned, or contained an addition of cheap copper or silver. ARCHIMEDES measured the volume of the crown by dipping it into water and watching the rise of the water level and measuring the buoyancy. He compared that to the case when a piece of gold of the same weight was immersed. They say that the successful solution of the task made him running around in the streets, stark naked and shouting, out of joy, Eureka!, which is Greek for I've got it! ARCHIMEDES is considered the greatest engineering scientist of ancient times. He made some progress with the rectification of the circle and the determination of the number $\pi$. He was also expelled from the Greek Academy of Science when he examined the volume of geometrical objects by experiment instead mathematically by pure thought. When Syracuse was conquered by the Romans he was slain by soldiers on the beach while hunched over some circles drawn in the sand.

We now turn to the pending proof of Eq. (3.6.2). Mutual insertion of the equations yields:

$$\oint_L \underset{L}{\boldsymbol{\phi}} \cdot \boldsymbol{v}\, \mathrm{d}l = \oint_L \left( \underset{L}{\phi}^{\Delta} \tau^i_{\Delta} + \underset{L}{\phi}_{\perp} e_i \right) v^{\Sigma} \tau^i_{\Sigma}\, \mathrm{d}l = \oint_L \underset{L}{\phi}^{\Delta} g_{\Delta\Sigma} v^{\Sigma} \mathrm{d}l = \oint_L \underset{L}{\phi}^{\Delta} v_{\Delta} \mathrm{d}l \tag{4.5.19}$$

if we only observe Eq. (2.7.7). The last step in this equation is the basis for a 2D analogue to Eq. (4.5.16) if the singular part is ignored. We simply identify $\underset{(z)}{A}^r \to \underset{L}{\phi}^{\Delta}$, $\underset{(z)}{n}_r \to v_{\Delta}$, $A \to L$, $V \to S$ and write without any hesitation:

$$\oint_L \underset{L}{\phi}^{\Delta} v_{\Delta}\, \mathrm{d}l = \iint_S \underset{L}{\phi}^{\Delta}{}_{;\Delta}\, \mathrm{d}A. \tag{4.5.20}$$

**Exercise 4.5.2:** ***A masochist's view on Archimedes' law of buoyancy***

Consider a spherical bathyscaphe of radius $R$ fully submersed in an (incompressible) fluid of density $\rho_0$ with its center being at depth $t$. From high school physics it is known that that the pressure $p$ on its hull varies linearly with depth like

$$p(R, \vartheta, \varphi) = p_0 + \rho_0 g(t + x_3), \tag{4.5.21}$$

($x_3$ is pointing "downward") where $p_0$ is the constant pressure at the fluid surface, $g$ denotes standard gravity acceleration, and $x_3 = R\cos(\vartheta)$ is an arbitrary position on the hull. Use the relation $\sigma_{ij} = -p\delta_{ij}$ and show with the results from Exercise 4.5.1 that the total force on the surface of the bathyscaphe is given by:

$$\oiint_A \sigma_{ji} \mathrm{d}\underset{(x)}{A}_j = -\left(0, 0, \frac{4\pi}{3} R^3 \rho_0 g\right). \tag{4.5.22}$$

Why does this confirm Archimedes' law of buoyancy? Now consider a very large spherical balloon in the (isothermal) atmosphere. We shall see in Chap. 10 that the pressure and the density of the surrounding air vary like:

$$p(R, \vartheta, \varphi) = p_0 \exp\left[-\frac{g\rho_0}{p_0} x_3\right], \quad \rho(R, \vartheta, \varphi) = \rho_0 \exp\left[-\frac{g\rho_0}{p_0} x_3\right], \tag{4.5.23}$$

where $p_0$ and $\rho_0$ denote the pressure and the density on the ground. Use results from Exercise 4.5.1 and show that the mass of the displaced air is given by:

$$M_{\mathrm{a}} = \iiint\limits_V \rho(R, \vartheta, \varphi)\, \mathrm{d}V$$
$$= \frac{4\pi\rho_0 \exp\left(-\frac{g\rho_0 h}{p_0}\right)}{\left(\frac{g\rho_0}{p_0}\right)^3}\left[\frac{g\rho_0 R}{p_0}\cosh\left(\frac{g\rho_0 R}{p_0}\right) - \sinh\left(\frac{g\rho_0 R}{p_0}\right)\right], \tag{4.5.24}$$

where $h$ denotes the height of the center of the balloon. Calculate now the total force acting on the balloon surface and prove the surprising result ($x_3$ is pointing "upward"):

$$\oint\!\!\!\oint\limits_A \sigma_{ji} \mathrm{d}\underset{(x)}{A_j} = -\oint\!\!\!\oint\limits_A p(R, \vartheta, \varphi)\, \mathrm{d}\underset{(x)}{A_i} = (0, 0, M_{\mathrm{a}} g). \tag{4.5.25}$$

How does this agree with ARCHIMEDES' law?

## 4.6 Would You Like to Know More?

Further information on tensor analysis in index-related as well as in absolute form can be found, for example, in Sect. 4.4 (in German) of Schade and Neemann [9], in Chap. 2 of Itskov [7], Liu [8], Appendix A.2, as well as in Bertram [1], Sects. 1.3 and 1.4 (with many additional mathematical concepts).

Of course the "classics" knew about the covariant derivative as well: Eisenhart [3], Chap. 2, Sects. 20–22, Ericksen [4], Section 18, Green and Zerna [6], Sect. 1.12, and Flügge [5], Chap. 12.

Finally note that after a study of the basic concepts of tensor analysis the literature on General Relativity becomes also accessible (e.g., [2]. Then the summation runs over four indices, since time is included as a fourth coordinate.

## References

1. Bertram A (2008) Elasticity and plasticity of large deformations, 2nd edn. Springer, Berlin
2. Einstein A (1983) Über die spezielle und die allgemeine Relativitätstheorie. Wissenschaftliche Taschenbücher 59. 21. Auflage. Vieweg, Braunschweig
3. Eisenhart LP (1947) An introduction to differential geometry with use of the tensor calculus. Princeton University Press, Princeton
4. Ericksen JL (1960) Appendix. Tensor fields. In: Flügge S (ed) Encyclopedia of physics, volume 3/1 principles of classical mechanics and field theory. Springer, Berlin
5. Flügge W (1972) Tensor analysis and continuum mechanics. Springer, New York, Berlin

6. Green AE, Zerna W (1968) Theoretical elasticity, 2nd edn. Dover Publications, Inc, New York
7. Itskov M (2007) Tensor algebra and tensor analysis for engineers with applications to continuum mechanics. Springer, Berlin, New York
8. Liu I-S (2010) Continuum mechanics. Springer, Berlin, New York
9. Schade H, Neemann N (2009) Tensoranalysis, 3. überarbeitete Auflage. de Gruyter. Berlin, New York

# Chapter 5
# Balance Equations in Skew Curvilinear Coordinate Systems

**Abstract** We now return to the balances of Chap. 3 and rewrite them for arbitrary coordinate systems. We start with the balances in regular points and, in this context, with the simplest one, namely the balance of mass, which is a *scalar* equation. We then move on to the more complex ones for momentum, energy, as well as total angular momentum, and specify them for cylindrical and spherical coordinates. As before we follow both ways and present the balances in index form as well as symbolically. The chapter ends with a discussion of the jump conditions and of global balances in arbitrary coordinates.

*There are two sides of the balance sheet—the left side and the right side.*
*On the left side, nothing is right, and on the right side, nothing is left !*

Answer by UBS to the journalist Dirk Maxeiner after the resignation of Ingrid Matthäus-Maier, CEO of KfW Bank

## 5.1 The Balance of Mass in Regular Points in Arbitrary Coordinate Systems

Mass density is a scalar quantity. Therefore we write in analogy to Eq. (4.1.2):

$$\underset{(x)}{\rho} = \underset{(z)}{\rho} \, . \tag{5.1.1}$$

For its gradient we find according to Eq. (4.1.1):

$$\frac{\partial \underset{(x)}{\rho}}{\partial x_i} = \frac{\partial z^k}{\partial x_i} \frac{\partial \underset{(z)}{\rho}}{\partial z^k} , \tag{5.1.2}$$

and with the result (4.2.15) of Exercise 4.2.6 for the divergence of the velocity,



W. H. Müller, *An Expedition to Continuum Theory*, Solid Mechanics and Its Applications 210, DOI: 10.1007/978-94-007-7799-6_5,

$$\underset{(z)}{\upsilon}^{j}{}_{;l} = \frac{\partial \underset{(x)}{\upsilon_i}}{\partial x_i} \tag{5.1.3}$$

we finally obtain for the *balance of mass in arbitrary coordinates* from Eqs. (3.8.3/3.8.5) in combination with Table 3.1:

$$\frac{\partial \underset{(x)}{\rho}}{\partial t} + \underset{(x)}{\upsilon}_j \frac{\partial \underset{(x)}{\rho}}{\partial x_j} + \underset{(x)}{\rho} \frac{\partial \underset{(x)}{\upsilon}_j}{\partial x_j} = 0 \quad \Rightarrow \quad \frac{\partial \underset{(z)}{\rho}}{\partial t} + \underset{(z)}{\upsilon}^k \frac{\partial \underset{(z)}{\rho}}{\partial z^k} + \underset{(z)}{\rho}\, \underset{(z)}{\upsilon}^k{}_{;k} = 0 \tag{5.1.4}$$

## 5.2 The Balance of Mass (Regular Form) in Cylindrical Coordinates

By using the definition of cylindrical coordinates (2.2.11), Eq. (2.2.13) for the metric, and the definition (2.6.3) for the physical components of a vector we find that:

$$\upsilon^r = \dot{r} = \upsilon_{\langle r \rangle}\ , \quad \upsilon^{\vartheta} = \dot{\vartheta} = \frac{1}{r}\upsilon_{\langle \vartheta \rangle}\ , \quad \upsilon^z = \dot{z} = \upsilon_{\langle z \rangle}. \tag{5.2.1}$$

The definition (4.2.3) for the covariant derivative of contravariant vector components in combination with the CHRISTOFFEL symbols for cylindrical coordinates shown in Eq. (4.2.5) yields:

$$\underset{(z)}{\upsilon}^k{}_{;k} = \frac{\partial \dot{r}}{\partial r} + \frac{\partial \dot{\vartheta}}{\partial \vartheta} + \frac{\partial \dot{z}}{\partial z} + \frac{1}{r}\dot{r}. \tag{5.2.2}$$

Thus Eq. (5.1.4) allows us to write for the *balance of mass in cylindrical coordinates*:

$$\frac{\partial \rho}{\partial t} + \frac{\partial}{\partial r}(\rho \dot{r}) + \frac{\partial}{\partial \vartheta}(\rho \dot{\vartheta}) + \frac{\partial}{\partial z}(\rho \dot{z}) + \frac{\rho}{r}\dot{r} = 0. \tag{5.2.3}$$

We may also insert into this relation the physical components for velocity from Eq. (5.2.1). This is customarily done in many textbooks on continuum mechanics (e.g., Segel [9], Appendix 3.1, or also Chandrasekhar [1], Chap. 9):

$$\frac{\partial \rho}{\partial t} + \frac{\partial}{\partial r}(\rho \upsilon_{\langle r \rangle}) + \frac{1}{r}\frac{\partial}{\partial \vartheta}(\rho \upsilon_{\langle \vartheta \rangle}) + \frac{\partial}{\partial z}(\rho \upsilon_{\langle z \rangle}) + \frac{\rho}{r}\upsilon_{\langle r \rangle} = 0. \tag{5.2.4}$$

It is interesting to note that this fairly cumbersome looking equation can be rewritten into a somewhat more convenient form if the material time derivative is used. Note that the total derivative in cylindrical coordinates reads:

$$\begin{aligned}\frac{\mathrm{d}\rho(r,\vartheta,z)}{\mathrm{d}t} &= \frac{\partial\rho}{\partial t}+\frac{\partial\rho}{\partial r}\frac{\mathrm{d}r}{\mathrm{d}t}+\frac{\partial\rho}{\partial\vartheta}\frac{\mathrm{d}\vartheta}{\mathrm{d}t}+\frac{\partial\rho}{\partial z}\frac{\mathrm{d}z}{\mathrm{d}t}\\ &= \frac{\partial\rho}{\partial t}+\frac{\partial\rho}{\partial r}\upsilon_{\langle r\rangle}+\frac{1}{r}\frac{\partial\rho}{\partial\vartheta}\upsilon_{\langle\vartheta\rangle}+\frac{\partial\rho}{\partial z}\upsilon_{\langle z\rangle}\end{aligned} \tag{5.2.5}$$

if we only observe Eq. (5.2.1). Thus Eq. (5.2.4) turns into:

$$\frac{\mathrm{d}\rho}{\mathrm{d}t} = -\rho\left(\frac{\partial\upsilon_{\langle r\rangle}}{\partial r}+\frac{1}{r}\frac{\partial\upsilon_{\langle\vartheta\rangle}}{\partial\vartheta}+\frac{\partial\upsilon_{\langle z\rangle}}{\partial z}+\frac{\upsilon_{\langle r\rangle}}{r}\right). \tag{5.2.6}$$

The right hand sight of this equation is nothing else but the divergence of the velocity written in cylindrical coordinates and it is this very quantity which dictates how the density of a material particle changes in time.

**Exercise 5.2.1:** ***The balance of mass (regular form) in spherical coordinates***

Proceed as in the case for cylindrical coordinates and use Eqs. (2.2.16), (4.2.3), and (4.2.6) to show that for spherical coordinates the following relations hold:

$$\upsilon^r = \dot{r} = \upsilon_{\langle r\rangle}\ ,\quad \upsilon^\vartheta = \dot{\vartheta} = \frac{1}{r}\upsilon_{\langle\vartheta\rangle}\ ,\quad \upsilon^\varphi = \dot{\varphi} = \frac{1}{r\sin\vartheta}\upsilon_{\langle\varphi\rangle}, \tag{5.2.7}$$

$$\underset{(z)}{\upsilon}{}^k_{;k} = \frac{\partial\dot{r}}{\partial r}+\frac{\partial\dot{\vartheta}}{\partial\vartheta}+\frac{\partial\dot{\varphi}}{\partial\varphi}+\frac{2}{r}\dot{r}+\dot{\vartheta}\cot\vartheta \tag{5.2.8}$$

and finally:

$$\frac{\partial\rho}{\partial t}+\frac{\partial}{\partial r}(\rho\dot{r})+\frac{\partial}{\partial\vartheta}(\rho\dot{\vartheta})+\frac{\partial}{\partial\varphi}(\rho\dot{\varphi})+\rho\left[\frac{2}{r}\dot{r}+\dot{\vartheta}\cot\vartheta\right]=0. \tag{5.2.9}$$

Rewrite the last equation by using physical components for the velocity in spherical coordinates and show that:

$$\begin{aligned}&\frac{\partial\rho}{\partial t}+\frac{\partial}{\partial r}(\rho\upsilon_{\langle r\rangle})+\frac{1}{r}\frac{\partial}{\partial\vartheta}(\rho\upsilon_{\langle\vartheta\rangle})+\frac{1}{r\sin\vartheta}\frac{\partial}{\partial\varphi}(\rho\upsilon_{\langle\varphi\rangle})\\ &\quad+\frac{\rho}{r}\left[2\upsilon_{\langle r\rangle}+\upsilon_{\langle\vartheta\rangle}\cot\vartheta\right]=0\end{aligned} \tag{5.2.10}$$

Finally show that the material time derivative of the mass density in spherical coordinates reads:

$$\frac{\mathrm{d}\rho(r,\vartheta,\varphi)}{\mathrm{d}t} = \frac{\partial\rho}{\partial t}+\frac{\partial\rho}{\partial r}\upsilon_{\langle r\rangle}+\frac{\partial\rho}{\partial\vartheta}\frac{\upsilon_{\langle\vartheta\rangle}}{r}+\frac{\partial\rho}{\partial\varphi}\frac{\upsilon_{\langle\varphi\rangle}}{r\sin\vartheta}. \tag{5.2.11}$$

## 5.3 Balance of Momentum in Regular Points in Arbitrary Coordinate Systems

In order to rewrite the balance of momentum in regular points for an arbitrary coordinate system $z$ we start with Eq. (3.8.14) from which the balance of mass has already been eliminated. Moreover, note the following useful relations, which follow from Eqs. (2.4.1), (4.2.13), and (4.4.5):

$$\begin{gathered}\underset{(x)}{\upsilon}_i = \frac{\partial x_i}{\partial z^k}\underset{(z)}{\upsilon}^k,\ \underset{(x)}{f}_i = \frac{\partial x_i}{\partial z^k}\underset{(z)}{f}^k,\\ \underset{(x)}{\upsilon}_i \frac{\partial \underset{(x)}{\upsilon}_i}{\partial x_j} = \frac{\partial x_i}{\partial z^k}\underset{(z)}{\upsilon}^j \underset{(z)}{\upsilon}^k{}_{;j},\ \frac{\partial \underset{(x)}{\sigma}_{ji}}{\partial x_j} = \frac{\partial x_i}{\partial z^k}\underset{(z)}{\sigma}^{lk}{}_{;l},\end{gathered} \tag{5.3.1}$$

observe Eq. (5.1.1) for the mass density and write:

$$\underset{(z)}{\rho}\left(\frac{\partial \underset{(z)}{\upsilon}^i}{\partial t} + \underset{(z)}{\upsilon}^j \underset{(z)}{\upsilon}^i{}_{;j}\right) = \underset{(z)}{\sigma}^{ji}{}_{;j} + \underset{(z)}{\rho}\,\underset{(z)}{f}^i \tag{5.3.2}$$

This is the *balance of momentum in arbitrary coordinates z*.

---

**Exercise 5.3.1:** ***Alternative forms of the balance of momentum (regular) in arbitrary coordinates***

Derive an alternative form to Eq. (5.3.2) in covariant coordinates. Is it also possible to write:

$$\begin{gathered}\frac{\partial \underset{(z)}{\rho}}{\partial z_j}\underset{(z)}{\upsilon}^j \underset{(z)}{\upsilon}^i + \underset{(z)}{\rho}\left(\underset{(z)}{\upsilon}^j \underset{(z)}{\upsilon}^i\right)_{;j} = \left(\underset{(z)}{\rho}\,\underset{(z)}{\upsilon}^j \underset{(z)}{\upsilon}_{;j}\right)_{;j}\\ \Rightarrow \frac{\partial\left(\underset{(z)}{\rho}\,\underset{(z)}{\upsilon}^i\right)}{\partial t} + \left(\underset{(z)}{\rho}\,\underset{(z)}{\upsilon}^j \underset{(z)}{\upsilon}^i\right)_{;j} = \underset{(z)}{\sigma}^{ji}{}_{;j} + \underset{(z)}{\rho}\,\underset{(z)}{f}^i?\end{gathered} \tag{5.3.3}$$

Recall in this context the hierarchy of derivatives explained in Sect. 4.4: What is the covariant derivative of a scalar? Is it allowed to pull the density under a covariant derivative as shown above? What are the pros and cons of this alternative balance of momentum?

---

**Exercise 5.3.2:** ***A note regarding the material time derivative***

Recall Sect. 3.8 where the material time derivative was introduced first for the scalar mass density in Eq. (3.8.6) and then in context with the velocity vector in Eq. (3.8.14). Show that we may write in arbitrary coordinates:

$$\frac{\mathrm{d}(\cdot)}{\mathrm{d}t} = \frac{\partial(\cdot)}{\partial t} + (\cdot)_{;i} \underset{(z)}{\upsilon}^{i}\,. \tag{5.3.4}$$

Argue that Eqs. $(5.1.4)_2$ and (5.3.2) may be rewritten as follows:

$$\frac{\mathrm{d}\underset{(z)}{\rho}}{\mathrm{d}t} + \underset{(z)}{\rho}\,\underset{(z)}{\upsilon}^{i}_{;i} = 0 \tag{5.3.5}$$

and:

$$\underset{(z)}{\rho}\,\frac{\mathrm{d}\underset{(z)}{\upsilon^{i}}}{\mathrm{d}t} = \underset{(z)}{\sigma}^{ji}_{;j} + \underset{(z)}{\rho}\,\underset{(z)}{f}^{i}\,. \tag{5.3.6}$$

## 5.4 The Balance of Momentum (Regular) in Cylindrical and Spherical Coordinates

Note that whenever expressions involving the stress tensor appear in the next sections we shall *not* assume that it is necessarily symmetric. The presented expressions can, of course, be simplified if this assumption is made and non-polar media are considered. In order to transform the balance of momentum into cylindrical coordinates we start from the general formula, Eq. (5.3.2), and obtain by using the equations for the Christoffel symbols (4.2.5) and the relations for the velocity from Eq. (5.2.1) for the $r$-component:

$$\begin{aligned} &\rho\left(\frac{\partial \dot{r}}{\partial t} + \dot{r}\frac{\partial \dot{r}}{\partial r} + \dot{\vartheta}\frac{\partial \dot{r}}{\partial \vartheta} + \dot{z}\frac{\partial \dot{r}}{\partial z} - r\dot{\vartheta}^{2}\right) + \\ &\quad -\frac{\partial \sigma^{rr}}{\partial r} - \frac{\partial \sigma^{\vartheta r}}{\partial \vartheta} - \frac{\partial \sigma^{zr}}{\partial z} + r\sigma^{\vartheta\vartheta} - \frac{1}{r}\sigma^{rr} = \rho f^{r}, \end{aligned} \tag{5.4.1}$$

for the $\vartheta$-component:

$$\begin{aligned}&\rho\left(\frac{\partial\dot{\vartheta}}{\partial t}+\dot{r}\frac{\partial\dot{\vartheta}}{\partial r}+\dot{\vartheta}\frac{\partial\dot{\vartheta}}{\partial\vartheta}+\dot{z}\frac{\partial\dot{\vartheta}}{\partial z}+\frac{2}{r}\dot{r}\dot{\vartheta}\right)+\\&-\frac{\partial\sigma^{r\vartheta}}{\partial r}-\frac{\partial\sigma^{\vartheta\vartheta}}{\partial\vartheta}-\frac{\partial\sigma^{z\vartheta}}{\partial z}-\frac{1}{r}\left(2\sigma^{r\vartheta}+\sigma^{\vartheta r}\right)=\rho f^{\vartheta},\end{aligned}\tag{5.4.2}$$

and for the $z$-component:

$$\begin{aligned}&\rho\left(\frac{\partial\dot{z}}{\partial t}+\dot{r}\frac{\partial\dot{z}}{\partial r}+\dot{\vartheta}\frac{\partial\dot{z}}{\partial\vartheta}+\dot{z}\frac{\partial\dot{z}}{\partial z}\right)+\\&-\frac{\partial\sigma^{rz}}{\partial r}-\frac{\partial\sigma^{\vartheta z}}{\partial\vartheta}-\frac{\partial\sigma^{zz}}{\partial z}-\frac{1}{r}\sigma^{rz}=\rho f^{z}.\end{aligned}\tag{5.4.3}$$

**Exercise 5.4.1:** ***The balance of moment of momentum (regular) in physical cylindrical coordinates***

Recall that quantities like $\sigma^{\vartheta\vartheta}$ do not have the proper physical units, which can be corrected by switching to physical components, e.g., $\sigma_{\langle r\vartheta\rangle}=\sqrt{g_{rr}}\sqrt{g_{\vartheta\vartheta}}\sigma^{r\vartheta}$ Rewrite the Eqs. (5.4.1–5.4.3) into their corresponding "physical" form and show that:

$$\begin{aligned}&\rho\left(\frac{\partial\upsilon_{\langle r\rangle}}{\partial t}+\upsilon_{\langle r\rangle}\frac{\partial\upsilon_{\langle r\rangle}}{\partial r}+\frac{\upsilon_{\langle\vartheta\rangle}}{r}\frac{\partial\upsilon_{\langle r\rangle}}{\partial\vartheta}+\upsilon_{\langle z\rangle}\frac{\partial\upsilon_{\langle r\rangle}}{\partial z}-\frac{\upsilon_{\langle\vartheta\rangle}^2}{r}\right)\\&=\frac{\partial\sigma_{\langle rr\rangle}}{\partial r}+\frac{1}{r}\frac{\partial\sigma_{\langle\vartheta r\rangle}}{\partial\vartheta}+\frac{\partial\sigma_{\langle zr\rangle}}{\partial z}+\frac{\sigma_{\langle rr\rangle}-\sigma_{\langle\vartheta\vartheta\rangle}}{r}+\rho f_{\langle r\rangle}\ ,\\&\rho\left(\frac{\partial\upsilon_{\langle\vartheta\rangle}}{\partial t}+\upsilon_{\langle r\rangle}\frac{\partial\upsilon_{\langle\vartheta\rangle}}{\partial r}+\frac{\upsilon_{\langle\vartheta\rangle}}{r}\frac{\partial\upsilon_{\langle\vartheta\rangle}}{\partial\vartheta}+\upsilon_{\langle z\rangle}\frac{\partial\upsilon_{\langle\vartheta\rangle}}{\partial z}+\frac{\upsilon_{\langle r\rangle}\upsilon_{\langle\vartheta\rangle}}{r}\right)\\&=\frac{\partial\sigma_{\langle r\vartheta\rangle}}{\partial r}+\frac{1}{r}\frac{\partial\sigma_{\langle\vartheta\vartheta\rangle}}{\partial\vartheta}+\frac{\partial\sigma_{\langle z\vartheta\rangle}}{\partial z}+\frac{1}{r}\left(\sigma_{\langle r\vartheta\rangle}+\sigma_{\langle\vartheta r\rangle}\right)+\rho f_{\langle\vartheta\rangle}\ ,\\&\rho\left(\frac{\partial\upsilon_{\langle z\rangle}}{\partial t}+\upsilon_{\langle r\rangle}\frac{\partial\upsilon_{\langle z\rangle}}{\partial r}+\frac{\upsilon_{\langle\vartheta\rangle}}{r}\frac{\partial\upsilon_{\langle z\rangle}}{\partial\vartheta}+\upsilon_{\langle z\rangle}\frac{\partial\upsilon_{\langle z\rangle}}{\partial z}\right)\\&=\frac{\partial\sigma_{\langle rz\rangle}}{\partial r}+\frac{1}{r}\frac{\partial\sigma_{\langle\vartheta z\rangle}}{\partial\vartheta}+\frac{\partial\sigma_{\langle zz\rangle}}{\partial z}+\frac{1}{r}\sigma_{\langle rz\rangle}+\rho f_{\langle z\rangle}\ ,\end{aligned}\tag{5.4.4}$$

Show that it is possible to combine the three left hand sides of the previous equations in terms of a material time derivative:

$$\rho\frac{\mathrm{d}\upsilon}{\mathrm{d}t}=\rho\frac{\mathrm{d}}{\mathrm{d}t}\left(\upsilon_{\langle r\rangle}\boldsymbol{e}_r+\upsilon_{\langle\vartheta\rangle}\boldsymbol{e}_\vartheta+\upsilon_{\langle z\rangle}\boldsymbol{e}_z\right).\tag{5.4.5}$$

For this purpose prove the following relations first:

$$\frac{\mathrm{d}\boldsymbol{e}_r}{\mathrm{d}t} = \frac{\upsilon_{\langle\vartheta\rangle}}{r}\boldsymbol{e}_\vartheta, \quad \frac{\mathrm{d}\boldsymbol{e}_\vartheta}{\mathrm{d}t} = -\frac{\upsilon_{\langle\vartheta\rangle}}{r}\boldsymbol{e}_r, \quad \frac{\mathrm{d}\boldsymbol{e}_z}{\mathrm{d}t} = 0, \tag{5.4.6}$$

How does this compare to Eqs. (5.3.4) and (5.3.6)?

**Exercise 5.4.2:** ***The balance of moment of momentum (regular) in physical spherical coordinates***

Follow the arguments for cylindrical coordinates of this section and show analogously that the balance of momentum in physical spherical coordinates reads:

$$\rho\left(\frac{\partial\upsilon_{\langle r\rangle}}{\partial t} + \upsilon_{\langle r\rangle}\frac{\partial\upsilon_{\langle r\rangle}}{\partial r} + \frac{\upsilon_{\langle\vartheta\rangle}}{r}\frac{\partial\upsilon_{\langle r\rangle}}{\partial\vartheta} + \frac{\upsilon_{\langle\varphi\rangle}}{r\sin\vartheta}\frac{\partial\upsilon_{\langle r\rangle}}{\partial\varphi} - \frac{1}{r}\left[\upsilon^2_{\langle\vartheta\rangle} + \upsilon^2_{\langle\varphi\rangle}\right]\right)$$
$$= \frac{\partial\sigma_{\langle rr\rangle}}{\partial r} + \frac{1}{r}\frac{\partial\sigma_{\langle\vartheta r\rangle}}{\partial\vartheta} + \frac{1}{r\sin\vartheta}\frac{\partial\sigma_{\langle\varphi r\rangle}}{\partial\varphi}$$
$$+\frac{1}{r}\left[2\sigma_{\langle rr\rangle} - \sigma_{\langle\vartheta\vartheta\rangle} - \sigma_{\langle\varphi\varphi\rangle} + \sigma_{\langle\vartheta r\rangle}\cot\vartheta\right] + \rho f_{\langle r\rangle},$$
$$\rho\left(\frac{\partial\upsilon_{\langle\vartheta\rangle}}{\partial t} + \upsilon_{\langle r\rangle}\frac{\partial\upsilon_{\langle\vartheta\rangle}}{\partial r} + \frac{\upsilon_{\langle\vartheta\rangle}}{r}\frac{\partial\upsilon_{\langle\vartheta\rangle}}{\partial\vartheta} + \frac{\upsilon_{\langle\varphi\rangle}}{r\sin\vartheta}\frac{\partial\upsilon_{\langle\vartheta\rangle}}{\partial\varphi} + \frac{\upsilon_{\langle r\rangle}\upsilon_{\langle\vartheta\rangle}}{r} - \frac{\upsilon^2_{\langle\varphi\rangle}\cot\vartheta}{r}\right)$$
$$= \frac{\partial\sigma_{\langle r\vartheta\rangle}}{\partial r} + \frac{1}{r}\frac{\partial\sigma_{\langle\vartheta\vartheta\rangle}}{\partial\vartheta} + \frac{1}{r\sin\vartheta}\frac{\partial\sigma_{\langle\varphi\vartheta\rangle}}{\partial\varphi}$$
$$+\frac{1}{r}\left[2\sigma_{\langle r\vartheta\rangle} + \sigma_{\langle\vartheta r\rangle} + \left(\sigma_{\langle\vartheta\vartheta\rangle} - \sigma_{\langle\varphi\varphi\rangle}\right)\cot\vartheta\right] + \rho f_{\langle\vartheta\rangle}$$
$$\rho\left(\frac{\partial\upsilon_{\langle\varphi\rangle}}{\partial t} + \upsilon_{\langle r\rangle}\frac{\partial\upsilon_{\langle\varphi\rangle}}{\partial r} + \frac{\upsilon_{\langle\vartheta\rangle}}{r}\frac{\partial\upsilon_{\langle\varphi\rangle}}{\partial\vartheta} + \frac{\upsilon_{\langle\varphi\rangle}}{r\sin\vartheta}\frac{\partial\upsilon_{\langle\varphi\rangle}}{\partial\varphi} + \frac{\upsilon_{\langle\varphi\rangle}\left[\upsilon_{\langle r\rangle} + \upsilon_{\langle\vartheta\rangle}\cot\vartheta\right]}{r}\right)$$
$$= \frac{\partial\sigma_{\langle r\varphi\rangle}}{\partial r} + \frac{1}{r}\frac{\partial\sigma_{\langle\vartheta\varphi\rangle}}{\partial\vartheta} + \frac{1}{r\sin\vartheta}\frac{\partial\sigma_{\langle\varphi\varphi\rangle}}{\partial\varphi}$$
$$+\frac{1}{r}\left[2\sigma_{\langle r\varphi\rangle} + \sigma_{\langle\varphi r\rangle} + \cot\vartheta\left(\sigma_{\langle\vartheta\varphi\rangle} + \sigma_{\langle\varphi\vartheta\rangle}\right)\right] + \rho f_{\langle\varphi\rangle} \tag{5.4.7}$$

Show that it is possible to combine the three left hand sides of the previous equations in terms of a material time derivative in absolute notation:

$$\rho\frac{\mathrm{d}\boldsymbol{\upsilon}}{\mathrm{d}t} = \rho\frac{\mathrm{d}}{\mathrm{d}t}\left(\upsilon_{\langle r\rangle}\boldsymbol{e}_r + \upsilon_{\langle\vartheta\rangle}\boldsymbol{e}_\vartheta + \upsilon_{\langle\varphi\rangle}\boldsymbol{e}_\varphi\right), \tag{5.4.8}$$

For this purpose prove the following relations first:

$$\frac{\mathrm{d}\boldsymbol{e}_r}{\mathrm{d}t} = \frac{\upsilon_{\langle\vartheta\rangle}}{r}\boldsymbol{e}_\vartheta + \frac{\upsilon_{\langle\varphi\rangle}}{r}\boldsymbol{e}_\varphi, \quad \frac{\mathrm{d}\boldsymbol{e}_\vartheta}{\mathrm{d}t} = -\frac{\upsilon_{\langle\vartheta\rangle}}{r}\boldsymbol{e}_r + \frac{\upsilon_{\langle\varphi\rangle}}{r}\cot\vartheta\,\boldsymbol{e}_\varphi,$$
$$\frac{\mathrm{d}\boldsymbol{e}_\varphi}{\mathrm{d}t} = -\frac{\upsilon_{\langle\varphi\rangle}}{r}\boldsymbol{e}_r - \frac{\upsilon_{\langle\varphi\rangle}}{r}\cot\vartheta\,\boldsymbol{e}_\vartheta. \tag{5.4.9}$$

How does this compare to Eqs. (5.3.4) and (5.3.6)?

## 5.5 Balance of Momentum in Statics

Statics problems occur very frequently in continuum theory. Two examples of engineering relevance have already been mentioned in Chap. 1: First, the sphere that was pressed into a bushing and, second, the fiber-matrix compound under thermal stress. For such problems the balance equations of mass and momentum simplify considerably since they do no longer contain time derivatives and velocities.

Thus Eq. (5.1.4) for the mass is identically satisfied and Eq. (5.3.2) for the momentum becomes:

$$\underset{(z)}{\sigma}{}^{ji}{}_{;j} = -\underset{(z)}{\rho}\,\underset{(z)}{f}{}^{i}. \tag{5.5.1}$$

This equation is also known as the *balance of momentum in statics*. In many cases even the volume forces can be neglected. Then the equation simplifies even more:

$$\underset{(z)}{\sigma}{}^{ji}{}_{;j} = 0. \tag{5.5.2}$$

In other words: The *divergence* of the stress tensor vanishes [cp., the remarks in context with Eq. (4.4.11)]. This is one of the relations that has already been mentioned in Chap. 1.

## 5.6 Balance of Momentum (Regular Form) of Statics in Cylindrical and Spherical Coordinates

If velocities and time derivatives are neglected in Eqs. (5.4.1–5.4.3) or (5.4.4) the static balance of momentum in (physical) cylindrical coordinates can immediately be read off. However, as an example of how to use absolute tensor notation we shall derive these equations differently. First we note that the del operator in cylindrical coordinates reads:

$$\nabla(\bullet) = \frac{\partial(\bullet)}{\partial r}\boldsymbol{e}_r + \frac{1}{r}\frac{\partial(\bullet)}{\partial \vartheta}\boldsymbol{e}_\vartheta + \frac{\partial(\bullet)}{\partial z}\boldsymbol{e}_z. \tag{5.6.1}$$

This is easy to prove either by using Eq. (4.3.3) in combination with Eqs. (2.3.8) and (2.5.8) or by combining Eqs. (4.3.4) and (4.3.10). This operator must now by applied to the (symmetric) stress tensor, which we also decompose w.r.t. the unit base $\boldsymbol{e}_r$, $\boldsymbol{e}_\vartheta$, $\boldsymbol{e}_z$. Clearly this requires us to use physical coordinates for the tensor components since the unit base vectors have no units. We write:

$$\sigma_{\langle ij\rangle}\boldsymbol{e}_i \otimes \boldsymbol{e}_j \equiv \sigma_{\langle rr\rangle}\boldsymbol{e}_r \otimes \boldsymbol{e}_r + \sigma_{\langle r\vartheta\rangle}\boldsymbol{e}_r \otimes \boldsymbol{e}_\vartheta + \cdots + \sigma_{\langle zz\rangle}\boldsymbol{e}_z \otimes \boldsymbol{e}_z, \tag{5.6.2}$$

and must now evaluate $\nabla \cdot \boldsymbol{\sigma}$. However, by doing so, we have to perform the differentiations very carefully. Note that not only the tensor components need to be differentiated with respect to $r$, $\vartheta$, and $\varphi$ but also some of the base vectors. In fact from Eq. (2.3.8) we must conclude that:

$$\begin{aligned}
&\frac{\partial \boldsymbol{e}_r}{\partial r} = \boldsymbol{0},\ \frac{\partial \boldsymbol{e}_r}{\partial \vartheta} = \boldsymbol{e}_\vartheta,\ \frac{\partial \boldsymbol{e}_r}{\partial z} = \boldsymbol{0},\\
&\frac{\partial \boldsymbol{e}_\vartheta}{\partial r} = \boldsymbol{0},\ \frac{\partial \boldsymbol{e}_\vartheta}{\partial \vartheta} = -\boldsymbol{e}_r,\ \frac{\partial \boldsymbol{e}_\vartheta}{\partial z} = \boldsymbol{0},\ \frac{\partial \boldsymbol{e}_z}{\partial r} = \boldsymbol{0},\ \frac{\partial \boldsymbol{e}_z}{\partial \vartheta} = \boldsymbol{0},\ \frac{\partial \boldsymbol{e}_z}{\partial z} = \boldsymbol{0}.
\end{aligned} \tag{5.6.3}$$

Thus we expect contributions from differentiations of two unit vectors. Hence by observing the product rule we finally obtain from Eq. (5.6.2):

$$\begin{aligned}
\nabla \cdot \boldsymbol{\sigma} = &\left(\frac{\partial \sigma_{\langle rr\rangle}}{\partial r} + \frac{1}{r}\frac{\partial \sigma_{\langle \vartheta r\rangle}}{\partial \vartheta} + \frac{\partial \sigma_{\langle zr\rangle}}{\partial z} + \frac{\sigma_{\langle rr\rangle} - \sigma_{\langle \vartheta\vartheta\rangle}}{r}\right)\boldsymbol{e}_r\\
&+ \left(\frac{\partial \sigma_{\langle r\vartheta\rangle}}{\partial r} + \frac{1}{r}\frac{\partial \sigma_{\langle \vartheta\vartheta\rangle}}{\partial \vartheta} + \frac{\partial \sigma_{\langle z\vartheta\rangle}}{\partial z} + \frac{1}{r}\left(\sigma_{\langle r\vartheta\rangle} + \sigma_{\langle \vartheta r\rangle}\right)\right)\boldsymbol{e}_\vartheta\ .\\
&+ \left(\frac{\partial \sigma_{\langle rz\rangle}}{\partial r} + \frac{1}{r}\frac{\partial \sigma_{\langle \vartheta z\rangle}}{\partial \vartheta} + \frac{\partial \sigma_{\langle zz\rangle}}{\partial z} + \frac{1}{r}\sigma_{\langle rz\rangle}\right)\boldsymbol{e}_z
\end{aligned} \tag{5.6.4}$$

**Exercise 5.6.1:** ***Divergence of the stress tensor in cylindrical coordinates***

Show in detail the correctness of Eqs. (5.6.1–5.6.4).

**Exercise 5.6.2:** ***Divergence of the stress tensor in spherical coordinates***

Show in complete analogy to the case of cylindrical coordinates that the del operator in spherical coordinates is given by:

$$\nabla(\bullet) = \frac{\partial(\bullet)}{\partial r}\boldsymbol{e}_r + \frac{1}{r}\frac{\partial(\bullet)}{\partial\vartheta}\boldsymbol{e}_\vartheta + \frac{1}{r\sin\vartheta}\frac{\partial(\bullet)}{\partial\varphi}\boldsymbol{e}_\varphi. \tag{5.6.5}$$

Use it and prove that the LAPLACE operator is given by:

$$\begin{aligned}\Delta(\bullet) &= \nabla\cdot(\nabla(\bullet))\\ &= \frac{\partial^2(\bullet)}{\partial r^2} + \frac{1}{r^2}\frac{\partial^2(\bullet)}{\partial\vartheta^2} + \frac{1}{r^2\sin^2\vartheta}\frac{\partial^2(\bullet)}{\partial\varphi^2} + \frac{2}{r}\frac{\partial(\bullet)}{\partial r} + \frac{\cot\vartheta}{r^2}\frac{\partial(\bullet)}{\partial\vartheta}.\end{aligned} \tag{5.6.6}$$

How does this compare to Eq. (4.2.19)? Now argue that the stress tensor in spherical coordinates can be decomposed as follows:

$$\sigma_{\langle ij\rangle}\boldsymbol{e}_i\otimes\boldsymbol{e}_j \equiv \sigma_{\langle rr\rangle}\boldsymbol{e}_r\otimes\boldsymbol{e}_r + \sigma_{\langle r\vartheta\rangle}\boldsymbol{e}_r\otimes\boldsymbol{e}_\vartheta + \cdots + \sigma_{\langle\varphi\varphi\rangle}\boldsymbol{e}_\varphi\otimes\boldsymbol{e}_\varphi. \tag{5.6.7}$$

Use the del operator to compute the divergence of that expression and show that:

$$\begin{aligned}\nabla\cdot\boldsymbol{\sigma} = &\left[\frac{\partial\sigma_{\langle rr\rangle}}{\partial r} + \frac{1}{r}\frac{\partial\sigma_{\langle\vartheta r\rangle}}{\partial\vartheta} + \frac{1}{r\sin\vartheta}\frac{\partial\sigma_{\langle\varphi r\rangle}}{\partial\varphi}\right.\\ &\left.+\frac{2\sigma_{\langle rr\rangle} - \sigma_{\langle\varphi\varphi\rangle} - \sigma_{\langle\vartheta\vartheta\rangle} + \sigma_{\langle\vartheta r\rangle}\cot\vartheta}{r}\right]\boldsymbol{e}_r + \left[\frac{\partial\sigma_{\langle r\vartheta\rangle}}{\partial r} + \frac{1}{r}\frac{\partial\sigma_{\langle\vartheta\vartheta\rangle}}{\partial\vartheta}\right.\\ &\left.+\frac{1}{r\sin\vartheta}\frac{\partial\sigma_{\langle\varphi\vartheta\rangle}}{\partial\varphi} + \frac{2\sigma_{\langle r\vartheta\rangle} + \sigma_{\langle\vartheta r\rangle} + \left(\sigma_{\langle\vartheta\vartheta\rangle} - \sigma_{\langle\varphi\varphi\rangle}\right)\cot\vartheta}{r}\right]\boldsymbol{e}_\vartheta\\ &+\left[\frac{\partial\sigma_{\langle r\varphi\rangle}}{\partial r} + \frac{1}{r}\frac{\partial\sigma_{\langle\vartheta\varphi\rangle}}{\partial\vartheta} + \frac{1}{r\sin\vartheta}\frac{\partial\sigma_{\langle\varphi\varphi\rangle}}{\partial\varphi}\right.\\ &\left.+\frac{2\sigma_{\langle r\varphi\rangle} + \sigma_{\langle\varphi r\rangle} + \cot\vartheta\left(\sigma_{\langle\vartheta\varphi\rangle} + \sigma_{\langle\varphi\vartheta\rangle}\right)}{r}\right]\boldsymbol{e}_\varphi.\end{aligned} \tag{5.6.8}$$

However, before that prove the following auxiliary formulae:

$$\frac{\partial\boldsymbol{e}_r}{\partial r} = \boldsymbol{0},\quad \frac{\partial\boldsymbol{e}_r}{\partial\vartheta} = \boldsymbol{e}_\vartheta,\quad \frac{\partial\boldsymbol{e}_r}{\partial\varphi} = \sin\vartheta\,\boldsymbol{e}_\varphi,\quad \frac{\partial\boldsymbol{e}_\vartheta}{\partial r} = \boldsymbol{0},\quad \frac{\partial\boldsymbol{e}_\vartheta}{\partial\vartheta} = -\boldsymbol{e}_r, \tag{5.6.9}$$

$$\frac{\partial\boldsymbol{e}_\vartheta}{\partial\varphi} = \cos\vartheta\,\boldsymbol{e}_\varphi,\quad \frac{\partial\boldsymbol{e}_\varphi}{\partial r} = \boldsymbol{0},\quad \frac{\partial\boldsymbol{e}_\varphi}{\partial\vartheta} = \boldsymbol{0},\quad \frac{\partial\boldsymbol{e}_\varphi}{\partial\varphi} = -[\sin\vartheta\,\boldsymbol{e}_r + \cos\vartheta\,\boldsymbol{e}_\vartheta].$$

## 5.7 Balances of Energy for Regular Points in Arbitrary Coordinate Systems

By inserting the corresponding entries in Table 3.1 in Eq. (3.7.2) we obtain the local regular balance for the total energy in a Cartesian coordinate system:

$$\begin{aligned}\frac{\partial}{\partial t}\left(\underset{(x)}{\rho}\left(\underset{(x)}{u}+\frac{1}{2}\underset{(x)}{\upsilon_i}\underset{(x)}{\upsilon_i}\right)\right)+\frac{\partial}{\partial x_j}\left(\underset{(x)}{\rho}\left(\underset{(x)}{u}+\frac{1}{2}\underset{(x)}{\upsilon_i}\underset{(x)}{\upsilon_i}\right)\underset{(x)}{\upsilon_j}\right)\\=\frac{\partial}{\partial x_j}\left(-\underset{(x)}{q_j}+\underset{(x)}{\sigma_{ji}}\underset{(x)}{\upsilon_i}\right)+\underset{(x)}{\rho}\left(\underset{(x)}{f_i}\underset{(x)}{\upsilon_i}+\underset{(x)}{r}\right).\end{aligned} \tag{5.7.1}$$

In order to rewrite it for the $z$-system it should be mentioned that the (invariant) scalar product $\underset{(x)}{\upsilon_i}\underset{(x)}{\upsilon_i}$ of the kinetic energy part can be written as:

$$\underset{(x)}{\upsilon_i}\underset{(x)}{\upsilon_i}=\frac{\partial z^k}{\partial x_i}\underset{(z)}{\upsilon_k}\frac{\partial x_i}{\partial z^l}\underset{(z)}{\upsilon^l}=\frac{\partial z^k}{\partial z^l}\underset{(z)}{\upsilon_k}\underset{(z)}{\upsilon^l}=\delta_l^k\underset{(z)}{\upsilon_k}\underset{(z)}{\upsilon^l}=\underset{(z)}{\upsilon_k}\underset{(z)}{\upsilon^k}. \tag{5.7.2}$$

Note that the rules of transformation for co- and contravariant vectors as well as the chain rule have been used. The result is not too surprising since a scalar product represents an invariant. Also note the following alternative formulae to Eq. (5.7.2), which are somewhat lengthier:

$$\begin{aligned}\underset{(x)}{\upsilon_i}\underset{(x)}{\upsilon_i}&=\frac{\partial z^k}{\partial x_i}\underset{(z)}{\upsilon_k}\frac{\partial z^l}{\partial x_i}\underset{(z)}{\upsilon_l}=g^{kl}\underset{(z)}{\upsilon_k}\underset{(z)}{\upsilon_l}\\\underset{(x)}{\upsilon_i}\underset{(x)}{\upsilon_i}&=\frac{\partial x_i}{\partial z^k}\underset{(z)}{\upsilon^k}\frac{\partial x_i}{\partial z^l}\underset{(z)}{\upsilon^l}=g_{kl}\underset{(z)}{\upsilon^k}\underset{(z)}{\upsilon^l}\end{aligned} \tag{5.7.3}$$

Consequently, the following expression is also a scalar:

$$\underset{(x)}{\rho}\left(\underset{(x)}{u}+\frac{1}{2}\underset{(x)}{\upsilon_i}\underset{(x)}{\upsilon_i}\right)=\underset{(z)}{\rho}\left(\underset{(z)}{u}+\frac{1}{2}\underset{(z)}{\upsilon_k}\underset{(z)}{\upsilon^k}\right), \tag{5.7.4}$$

because just like mass density the specific internal energy, $u$, is a scalar quantity $\underset{(x)}{u}=\underset{(z)}{u}$. The second expression on the left hand side of Eq. (5.7.1) corresponds to the divergence of a vectors multiplied by a scalar which, again, results in a vector quantity. Thus we obtain according to Exercise 4.2.6:

$$\frac{\partial}{\partial x_j}\left(\underset{(x)}{\rho}\left(\underset{(x)}{u}+\frac{1}{2}\underset{(x)}{\upsilon_i}\underset{(x)}{\upsilon_i}\right)\underset{(x)}{\upsilon_j}\right)=\left(\underset{(z)}{\rho}\left(\underset{(z)}{u}+\frac{1}{2}\underset{(z)}{\upsilon_k}\underset{(z)}{\upsilon^k}\right)\underset{(z)}{\upsilon^l}\right)_{;l} \tag{5.7.5}$$

In this equation we have interpreted the single velocity as a contravariant object. However, this is not the only possibility. We might have equally as well chosen a covariant notation for the velocity (note that because of Eq. (4.4.9) we have $g^{ij}{}_{;k} = 0$):

$$\frac{\partial}{\partial x_j}\left(\underset{(x)}{\rho}\left(\underset{(x)}{u}+\frac{1}{2}\underset{(x)}{\upsilon_i}\underset{(x)}{\upsilon_i}\right)\underset{(x)}{\upsilon_j}\right)=\left(\underset{(z)}{\rho}\left(\underset{(z)}{u}+\frac{1}{2}\underset{(z)}{\upsilon_k}\underset{(z)}{\upsilon^k}\right)\underset{(z)}{\upsilon^i}\right)_{;l} g^{li}. \tag{5.7.6}$$

The expressions on the right hand side of Eq. (5.7.1) are manipulated similarly. For the divergence of the heat flux vector it is possible to write:

$$\begin{aligned}\frac{\partial \underset{(x)}{q_j}}{\partial x_j} &= \underset{(z)}{q}{}^r{}_{;r} = \left(g^{rm}\underset{(z)}{q_m}\right)_{;r} = g^{rm}_{;r}\underset{(z)}{q_m} + g^{rm}\underset{(z)}{q_{m;r}} \\ &= 0 + g^{rm}\underset{(z)}{q_{m;r}} = g^{rm}\underset{(z)}{q_{m;r}}\end{aligned} \tag{5.7.7}$$

depending upon whether the heat flux vector is interpreted as a co- or contravariant object. Moreover, the scalar product between stress tensor and velocity results in a vector, which can be written the co- or contravariant way:

$$\begin{aligned}\underset{(x)}{\sigma_{ij}}\underset{(x)}{\upsilon_i} &= \frac{\partial z^r}{\partial x_j}\frac{\partial z^s}{\partial x_i}\underset{(z)}{\sigma}{}_{rs}\frac{\partial x_i}{\partial z^k}\underset{(z)}{\upsilon^k} = \frac{\partial z^r}{\partial x_j}\delta^s_k\underset{(z)}{\sigma}{}_{rs}\underset{(z)}{\upsilon^k} = \frac{\partial z^r}{\partial x_j}\underset{(z)}{\sigma}{}_{rs}\underset{(z)}{\upsilon^s} \\ &\Rightarrow \underset{(z)}{\sigma}{}_{rs}\underset{(z)}{\upsilon^s} = \frac{\partial x_j}{\partial z^r}\underset{(x)}{\sigma_{ji}}\underset{(x)}{\upsilon_i}, \\ \underset{(x)}{\sigma_{ji}}\underset{(x)}{\upsilon_i} &= \frac{\partial x_j}{\partial z^r}\frac{\partial x_i}{\partial z^s}\underset{(z)}{\sigma^{rs}}\frac{\partial z^k}{\partial x_i}\underset{(z)}{\upsilon}{}_k = \frac{\partial x_j}{\partial z^r}\delta^k_s\underset{(z)}{\sigma^{rs}}\underset{(z)}{\upsilon}{}_k = \frac{\partial x_j}{\partial z^r}\underset{(z)}{\sigma^{rs}}\underset{(z)}{\upsilon}{}_s \\ &\Rightarrow \underset{(z)}{\sigma^{rs}}\underset{(z)}{\upsilon_s} = \frac{\partial z^r}{\partial x_j}\underset{(x)}{\sigma_{ji}}\underset{(x)}{\upsilon_i},\end{aligned} \tag{5.7.8}$$

if we only apply the transformation rules for co- and contravariant vectors and tensors as well as the chain rule sufficiently often. Thus the second term on the right hand side of Eq. (5.7.1), which is the divergence of a vector, can either be written as the covariant derivative of a contravariant or a covariant vector field:

$$\begin{aligned}\frac{\partial}{\partial x_j}\left(-\underset{(x)}{q}{}_i + \underset{(x)}{\sigma}{}_{ji}\underset{(x)}{\upsilon}{}_i\right) &= \left(\underset{(z)}{q^r} + -\underset{(z)}{\sigma^{rs}}\underset{(z)}{\upsilon}{}_s\right)_{;r} \\ &= \left(-\underset{(z)}{q^r} + g^{rl}g^{sm}\underset{(z)}{\sigma_{lm}}\underset{(z)}{\upsilon_s}\right)_{;r} = g^{rl}\left(-\underset{(z)}{q_l} + \underset{(z)}{\sigma_{lm}}\underset{(z)}{\upsilon^m}\right)_{;r}.\end{aligned} \tag{5.7.9}$$

The body force term in Eq. (5.7.1)—another scalar product—can be treated analogously to the sequence of equations shown in (5.7.2/5.7.3). We omit the details and only note various equivalent ways of writing:

$$\underset{(x)}{\upsilon}_i \underset{(x)}{f}_i = \underset{(z)}{\upsilon}_k \underset{(z)}{f}^k = \underset{(z)}{\upsilon}^k \underset{(z)}{f}_k = g^{kl} \underset{(z)}{\upsilon}_k \underset{(z)}{f}_l = g_{kl} \underset{(z)}{\upsilon}^k \underset{(z)}{f}^l. \tag{5.7.10}$$

Finally, the radiation density is another scalar and we may write:

$$\underset{(x)}{\rho} \underset{(x)}{r} = \underset{(z)}{\rho} \underset{(z)}{r}. \tag{5.7.11}$$

We now combine Eqs. (5.7.4/5.7.5) and (5.7.9–5.7.11) and insert them in Eq. (5.7.1):

$$\begin{aligned} &\frac{\partial}{\partial t}\left(\underset{(z)}{\rho}\left(\underset{(z)}{u} + \frac{1}{2}\underset{(z)}{\upsilon}_k \underset{(z)}{\upsilon}^k\right)\right) + \left(\underset{(z)}{\rho}\left(\underset{(z)}{u} + \frac{1}{2}\underset{(z)}{\upsilon}_i \underset{(z)}{\upsilon}^i\right)\underset{(z)}{\upsilon}^j - \underset{(z)}{\sigma}^{ji}\underset{(z)}{\upsilon}_i + \underset{(z)}{q}^j\right)_{;j} \\ &= \underset{(z)}{\rho}\left(\underset{(z)}{\upsilon}_k \underset{(z)}{f}^k + \underset{(z)}{r}\right). \end{aligned} \tag{5.7.12}$$

**Exercise 5.7.1:** ***Rewriting the left hand side of the energy balance***

Recall the definition of the material time derivative first introduced in context with the mass density in Eq. (3.8.6) as well as the mass balance (3.8.3). Show that the left hand side of the energy balance (5.7.1) can alternatively be written as:

$$\begin{aligned} &\frac{\partial}{\partial t}\left(\underset{(x)}{\rho}\left(\underset{(x)}{u} + \frac{1}{2}\underset{(x)}{\upsilon}_i \underset{(x)}{\upsilon}_i\right)\right) + \frac{\partial}{\partial x_j}\left(\underset{(x)}{\rho}\left(\underset{(x)}{u} + \frac{1}{2}\underset{(x)}{\upsilon}_i \underset{(x)}{\upsilon}_i\right)\underset{(x)}{\upsilon}_j\right) \\ &\equiv \underset{(x)}{\rho}\frac{\mathrm{d}}{\mathrm{d}t}\left(\underset{(x)}{u} + \frac{1}{2}\underset{(x)}{\upsilon}_i \underset{(x)}{\upsilon}_i\right) \equiv \underset{(z)}{\rho}\frac{\mathrm{d}}{\mathrm{d}t}\left(\underset{(z)}{u} + \frac{1}{2}\underset{(z)}{\upsilon}_i \underset{(z)}{\upsilon}^i\right). \end{aligned} \tag{5.7.13}$$

Use results from Exercise 5.3.2 and show that:

$$\frac{\mathrm{d}}{\mathrm{d}t}\left(\underset{(z)}{u} + \frac{1}{2}\underset{(z)}{\upsilon}_i \underset{(z)}{\upsilon}^i\right) = \frac{\partial}{\partial t}\left(\underset{(z)}{u} + \frac{1}{2}\underset{(z)}{\upsilon}_i \underset{(z)}{\upsilon}^i\right) + \frac{\partial}{\partial z^j}\left(\underset{(z)}{u} + \frac{1}{2}\underset{(z)}{\upsilon}_i \underset{(z)}{\upsilon}^i\right)\underset{(z)}{\upsilon}^j. \tag{5.7.14}$$

Evaluate this expression in physical cylindrical as well as spherical coordinates or use Eqs. (5.2.5) and (5.2.9) as analogues to prove that:

$$\begin{aligned} \frac{\mathrm{d}e}{\mathrm{d}t} &= \frac{\partial e}{\partial t} + \frac{\partial e}{\partial r}\upsilon_{\langle r\rangle} + \frac{1}{r}\frac{\partial e}{\partial \vartheta}\upsilon_{\langle \vartheta\rangle} + \frac{\partial e}{\partial z}\upsilon_{\langle z\rangle}, \\ \frac{\mathrm{d}e}{\mathrm{d}t} &= \frac{\partial e}{\partial t} + \frac{\partial e}{\partial r}\upsilon_{\langle r\rangle} + \frac{\partial e}{\partial \vartheta}\frac{\upsilon_{\langle \vartheta\rangle}}{r} + \frac{\partial e}{\partial \varphi}\frac{\upsilon_{\langle \varphi\rangle}}{r\sin\vartheta}, \end{aligned} \tag{5.7.15}$$

with $e = u + \frac{1}{2}\left(\upsilon_{\langle r\rangle}^2 + \upsilon_{\langle \vartheta\rangle}^2 + \upsilon_{\langle z\rangle}^2\right)$ or $e = u + \frac{1}{2}\left(\upsilon_{\langle r\rangle}^2 + \upsilon_{\langle \vartheta\rangle}^2 + \upsilon_{\langle \varphi\rangle}^2\right)$, respectively.

**Exercise 5.7.2:** ***The balance of kinetic energy in regular points in arbitrary coordinates***

Show first that after scalar multiplication of the balance of momentum in regular points written in Cartesian coordinates (3.7.2) with the velocity components one may write:

$$\begin{gathered}\frac{\partial}{\partial t}\left(\frac{\underset{(x)}{\rho}}{2}\underset{(x)}{\upsilon}_i\,\underset{(x)}{\upsilon}_i\right)+\frac{\partial}{\partial x_j}\left(\frac{\underset{(x)}{\rho}}{2}\underset{(x)}{\upsilon}_i\,\underset{(x)}{\upsilon}_i\,\underset{(x)}{\upsilon}_j\right)\\ =\frac{\partial}{\partial x_j}\left(\underset{(x)}{\sigma}_{ji}\,\underset{(x)}{\upsilon}_i\right)+\underset{(x)}{\rho}\,\underset{(x)}{f}_i\,\underset{(x)}{\upsilon}_i-\underset{(x)}{\sigma}_{ji}\frac{\partial\underset{(x)}{\upsilon}_i}{\partial x_j}.\end{gathered}\tag{5.7.16}$$

Analyze this result by using the arguments presented at the beginning of Sect. 3.9: Interpret the first two terms on the right hand side as kinetic energy supplies and the third one as a production of kinetic energy. Show that the latter can be interpreted as a double scalar product, which can be transformed into the $z$-system as follows:

$$\underset{(x)}{\sigma}_{ji}\frac{\partial\underset{(x)}{\upsilon}_i}{\partial x_j}=\underset{(z)}{\sigma}^{ji}\,\underset{(z)}{\upsilon}_{i;j}\,.\tag{5.7.17}$$

Use further arguments from the previous section to show that the local balance of kinetic energy can be rewritten as:

$$\begin{gathered}\frac{\partial}{\partial t}\left(\frac{\underset{(z)}{\rho}}{2}\underset{(z)}{\upsilon}_i\,\underset{(z)}{\upsilon}^i\right)+\left(\frac{\underset{(z)}{\rho}}{2}\underset{(z)}{\upsilon}_i\,\underset{(z)}{\upsilon}^i\,\underset{(z)}{\upsilon}^j\right)_{;j}\\ =\left(\underset{(z)}{\sigma}^{ji}\,\underset{(z)}{\upsilon}_i\right)_{;j}+\underset{(z)}{\rho}\,\underset{(z)}{f}^k\,\underset{(z)}{\upsilon}_k-\underset{(z)}{\sigma}^{ji}\,\underset{(z)}{\upsilon}_{i;j}\,.\end{gathered}\tag{5.7.18}$$

In the same context comment on the following coordinate-free version of the kinetic energy balance:

$$\frac{\partial}{\partial t}\left(\frac{\rho}{2}\upsilon^2\right)+\nabla\cdot\left(\frac{\rho}{2}\upsilon^2\boldsymbol{\upsilon}\right)=\nabla\cdot(\boldsymbol{\sigma}\cdot\boldsymbol{\upsilon})+\rho\boldsymbol{f}\cdot\boldsymbol{\upsilon}-\boldsymbol{\sigma}:\nabla\boldsymbol{\upsilon}.\tag{5.7.19}$$

Interpret the use of the scalar and double scalar products as well as of the del operators. What are the pros and cons of this way of writing? Show that the terms containing the stress tensor can be written in physical cylindrical coordinates:

$$
\begin{aligned}
(\boldsymbol{\sigma}\cdot\boldsymbol{\upsilon}) = {} & \frac{\partial\sigma_{\langle rr\rangle}\upsilon_{\langle r\rangle}}{\partial r}+\frac{1}{r}\frac{\partial\sigma_{\langle\vartheta r\rangle}\upsilon_{\langle r\rangle}}{\partial\vartheta}+\frac{\partial\sigma_{\langle zr\rangle}\upsilon_{\langle r\rangle}}{\partial z}\\
&+\frac{\partial\sigma_{\langle r\vartheta\rangle}\upsilon_{\langle\vartheta\rangle}}{\partial r}+\frac{1}{r}\frac{\partial\sigma_{\langle\vartheta\vartheta\rangle}\upsilon_{\langle\vartheta\rangle}}{\partial\vartheta}+\frac{\partial\sigma_{\langle z\vartheta\rangle}\upsilon_{\langle\vartheta\rangle}}{\partial z}\\
&+\frac{\partial\sigma_{\langle rz\rangle}\upsilon_{\langle z\rangle}}{\partial r}+\frac{1}{r}\frac{\partial\sigma_{\langle\vartheta z\rangle}\upsilon_{\langle z\rangle}}{\partial\vartheta}+\frac{\partial\sigma_{\langle zz\rangle}\upsilon_{\langle z\rangle}}{\partial z}\\
&\frac{\sigma_{\langle rr\rangle}\upsilon_{\langle r\rangle}+\sigma_{\langle r\vartheta\rangle}\upsilon_{\langle\vartheta\rangle}+\sigma_{\langle rz\rangle}\upsilon_{\langle z\rangle}}{r},
\end{aligned} \tag{5.7.20}
$$

$$
\begin{aligned}
\boldsymbol{\sigma}:\nabla\boldsymbol{\upsilon} = {} & \sigma_{\langle rr\rangle}\frac{\partial\upsilon_{\langle r\rangle}}{\partial r}+\sigma_{\langle r\vartheta\rangle}\frac{\partial\upsilon_{\langle\vartheta\rangle}}{\partial r}+\sigma_{\langle rz\rangle}\frac{\partial\upsilon_{\langle z\rangle}}{\partial r}\\
&+\frac{1}{r}\left(\sigma_{\langle\vartheta r\rangle}\frac{\partial\upsilon_{\langle r\rangle}}{\partial\vartheta}+\sigma_{\langle\vartheta\vartheta\rangle}\frac{\partial\upsilon_{\langle\vartheta\rangle}}{\partial\vartheta}+\sigma_{\langle\vartheta z\rangle}\frac{\partial\upsilon_{\langle z\rangle}}{\partial\vartheta}\right)\\
&\sigma_{\langle zr\rangle}\frac{\partial\upsilon_{\langle r\rangle}}{\partial z}+\sigma_{\langle z\vartheta\rangle}\frac{\partial\upsilon_{\langle\vartheta\rangle}}{\partial z}+\sigma_{\langle zz\rangle}\frac{\partial\upsilon_{\langle z\rangle}}{\partial z}+\frac{\sigma_{\langle\vartheta\vartheta\rangle}\upsilon_{\langle r\rangle}-\sigma_{\langle\vartheta r\rangle}\upsilon_{\langle\vartheta\rangle}}{r},
\end{aligned}
$$

and in physical spherical coordinates:

$$
\begin{aligned}
\nabla\cdot(\boldsymbol{\sigma}\cdot\boldsymbol{\upsilon}) = {} & \frac{\partial\sigma_{\langle rr\rangle}\upsilon_{\langle r\rangle}}{\partial r}+\frac{1}{r}\frac{\partial\sigma_{\langle\vartheta r\rangle}\upsilon_{\langle r\rangle}}{\partial\vartheta}+\frac{1}{r\sin\vartheta}\frac{\partial\sigma_{\langle\varphi r\rangle}\upsilon_{\langle r\rangle}}{\partial\varphi}\\
&+\frac{\partial\sigma_{\langle r\vartheta\rangle}\upsilon_{\langle\vartheta\rangle}}{\partial r}+\frac{1}{r}\frac{\partial\sigma_{\langle\vartheta\vartheta\rangle}\upsilon_{\langle\vartheta\rangle}}{\partial\vartheta}+\frac{1}{r\sin\vartheta}\frac{\partial\sigma_{\langle\varphi\vartheta\rangle}\upsilon_{\langle\vartheta\rangle}}{\partial\varphi}\\
&+\frac{\partial\sigma_{\langle r\varphi\rangle}\upsilon_{\langle\varphi\rangle}}{\partial r}+\frac{1}{r}\frac{\partial\sigma_{\langle\vartheta\varphi\rangle}\upsilon_{\langle\varphi\rangle}}{\partial\vartheta}+\frac{1}{r\sin\vartheta}\frac{\partial\sigma_{\langle\varphi\varphi\rangle}\upsilon_{\langle\varphi\rangle}}{\partial\varphi}\\
&+\frac{2\left(\sigma_{\langle rr\rangle}\upsilon_{\langle r\rangle}+\sigma_{\langle r\vartheta\rangle}\upsilon_{\langle\vartheta\rangle}+\sigma_{\langle r\varphi\rangle}\upsilon_{\langle\varphi\rangle}\right)+\cot\vartheta\left(\sigma_{\langle\vartheta r\rangle}\upsilon_{\langle r\rangle}+\sigma_{\langle\vartheta\vartheta\rangle}\upsilon_{\langle\vartheta\rangle}+\sigma_{\langle\vartheta\varphi\rangle}\upsilon_{\langle\varphi\rangle}\right)}{r},
\end{aligned} \tag{5.7.21}
$$

$$
\begin{aligned}
\boldsymbol{\sigma}:\nabla\boldsymbol{\upsilon} = {} & \sigma_{\langle rr\rangle}\frac{\partial\upsilon_{\langle r\rangle}}{\partial r}+\sigma_{\langle r\vartheta\rangle}\frac{\partial\upsilon_{\langle\vartheta\rangle}}{\partial r}+\sigma_{\langle r\varphi\rangle}\frac{\partial\upsilon_{\langle\varphi\rangle}}{\partial r}\\
&+\frac{1}{r}\left(\sigma_{\langle\vartheta r\rangle}\frac{\partial\upsilon_{\langle r\rangle}}{\partial\vartheta}+\sigma_{\langle\vartheta\vartheta\rangle}\frac{\partial\upsilon_{\langle\vartheta\rangle}}{\partial\vartheta}+\sigma_{\langle\vartheta\varphi\rangle}\frac{\partial\upsilon_{\langle\varphi\rangle}}{\partial\vartheta}\right)\\
&+\frac{1}{r\sin\vartheta}\left(\sigma_{\langle\varphi r\rangle}\frac{\partial\upsilon_{\langle r\rangle}}{\partial\varphi}+\sigma_{\langle\varphi\vartheta\rangle}\frac{\partial\upsilon_{\langle\vartheta\rangle}}{\partial\varphi}+\sigma_{\langle\varphi\varphi\rangle}\frac{\partial\upsilon_{\langle\varphi\rangle}}{\partial\varphi}\right)\\
&+\frac{(\sigma_{\langle\vartheta\vartheta\rangle}+\sigma_{\langle\varphi\varphi\rangle})\upsilon_{\langle r\rangle}-(\sigma_{\langle\vartheta r\rangle}-\cot\vartheta\,\sigma_{\langle\varphi\varphi\rangle})\upsilon_{\langle\vartheta\rangle}-(\sigma_{\langle\varphi r\rangle}+\cot\vartheta\,\sigma_{\langle\varphi\vartheta\rangle})\upsilon_{\langle\varphi\rangle}}{r}.
\end{aligned}
$$

Use the method of covariant derivatives as well as the del operator technique in order to arrive at these results. Finally show that for the specific power of the volume forces in the two physical coordinate systems the following holds:

$$\boldsymbol{f} \cdot \boldsymbol{\upsilon} = f_{\langle r\rangle}\upsilon_{\langle r\rangle} + f_{\langle\vartheta\rangle}\upsilon_{\langle\vartheta\rangle} + f_{\langle z\rangle}\upsilon_{\langle z\rangle} \tag{5.7.22}$$

and

$$\boldsymbol{f} \cdot \boldsymbol{\upsilon} = f_{\langle r\rangle}\upsilon_{\langle r\rangle} + f_{\langle\vartheta\rangle}\upsilon_{\langle\vartheta\rangle} + f_{\langle\varphi\rangle}\upsilon_{\langle\varphi\rangle}. \tag{5.7.23}$$

**Exercise 5.7.3: *The balance of internal energy in regular points in arbitrary coordinates***

Subtract Eq. (5.7.16) from Eq. (5.7.12) and conclude that the balance of internal energy in regular points must read in an arbitrary coordinate system:

$$\frac{\partial}{\partial t}\left(\underset{(z)}{\rho}\,\underset{(z)}{u}\right) + \left(\underset{(z)}{\rho}\,\underset{(z)}{u}\,\underset{(z)}{\upsilon}{}^{j}\right)_{;j} = -\underset{(z)}{q}{}^{r}_{;r} + \underset{(z)}{\sigma}{}^{ji}\,\underset{(z)}{\upsilon}{}_{i;j} + \underset{(z)}{\rho}\,\underset{(z)}{r}\,. \tag{5.7.24}$$

Show that this can be rewritten as:

$$\underset{(z)}{\rho}\frac{\partial \underset{(z)}{u}}{\partial t} + \underset{(z)}{\rho}\,\underset{(z)}{\upsilon}{}^{j}\frac{\partial \underset{(z)}{u}}{\partial z^{j}} = -\underset{(z)}{q}{}^{r}_{;r} + \underset{(z)}{\sigma}{}^{ji}\,\underset{(z)}{\upsilon}{}_{i;j} + \underset{(z)}{\rho}\,\underset{(z)}{r}\,. \tag{5.7.25}$$

Use these results to discuss the pros and cons of the following coordinate independent form of the First Law of thermodynamics:

$$\rho\frac{\partial u}{\partial t} + \rho\,\boldsymbol{\upsilon}\cdot\nabla u = -\nabla\cdot\boldsymbol{q} + \boldsymbol{\sigma} : (\nabla\boldsymbol{\upsilon}) + \rho\, r. \tag{5.7.26}$$

Derive the following equations for the divergence of the heat flux in physical cylindrical as well as in physical spherical coordinates by using the del operator and, alternatively, the covariant derivative:

$$\nabla\cdot\boldsymbol{q} = \frac{\partial q_{\langle r\rangle}}{\partial r} + \frac{1}{r}\frac{\partial q_{\langle\vartheta\rangle}}{\partial\vartheta} + \frac{\partial q_{\langle z\rangle}}{\partial z} + \frac{q_{\langle r\rangle}}{r} \tag{5.7.27}$$

and:

$$\nabla\cdot\boldsymbol{q} = \frac{\partial q_{\langle r\rangle}}{\partial r} + \frac{1}{r}\frac{\partial q_{\langle\vartheta\rangle}}{\partial\vartheta} + \frac{1}{r\sin\vartheta}\frac{\partial q_{\langle\varphi\rangle}}{\partial\varphi} + \frac{2}{r}q_{\langle r\rangle} + \frac{\cot\vartheta}{r}q_{\langle\vartheta\rangle}. \tag{5.7.28}$$

## 5.8 Balances of Angular Momentum for Regular Points in Arbitrary Coordinate Systems

In the previous sections it was explained in great detail how to derive a balance equation for scalar and vector quantities in arbitrary coordinate systems, ($z$), once the corresponding balance has been established in a Cartesian system, ($x$). We therefore recall the balance of total angular momentum from Sect. 3.10:

$$\underset{(x)}{\rho}\frac{\mathrm{d}\left(\underset{(x)}{s}_i+\underset{(x)}{\epsilon}_{ijk}\underset{(x)}{x}_j\underset{(x)}{\upsilon}_k\right)}{\mathrm{d}t}=\frac{\partial\left(\underset{(x)}{m}_{li}+\underset{(x)}{\epsilon}_{ijk}\underset{(x)}{x}_j\underset{(x)}{\sigma}_{lk}\right)}{\partial x_l}+\underset{(x)}{\rho}\left(\underset{(x)}{\epsilon}_{ijk}\underset{(x)}{\upsilon}_j\underset{(x)}{f}_k+\underset{(x)}{l}_i\right) \tag{5.8.1}$$

and, by applying the same rules as before, i.e., partial derivatives turn into covariant ones, covariant and contravariant indices must be appropriately placed, and free indices must agree on both sides of a tensor equation, we arrive immediately at:

$$\underset{(z)}{\rho}\frac{\mathrm{d}\left(\underset{(z)}{s}^i+\underset{(z)}{\epsilon}^{ijk}\underset{(z)}{x}_j\underset{(z)}{\upsilon}_k\right)}{\mathrm{d}t}=\left(\underset{(z)}{m}^{li}+\underset{(z)}{\epsilon}^i{}_{jk}\underset{(z)}{x}^j\underset{(z)}{\sigma}^{lk}\right)_{;l}+\underset{(z)}{\rho}\left(\underset{(z)}{\epsilon}^i{}_{jk}\underset{(z)}{\upsilon}^j\underset{(z)}{f}^k+\underset{(z)}{l}^i\right), \tag{5.8.2}$$

which is just one way of writing this equation among many other co-/contravariant variations. Similarly the balance of moment of momentum and of spin read [cf., Eqs. (3.10.1) and (3.10.6)]:

$$\begin{aligned}\underset{(z)}{\rho}\frac{\mathrm{d}\left(\underset{(z)}{\epsilon}^{ijk}\underset{(z)}{x}_j\underset{(z)}{\upsilon}_k\right)}{\mathrm{d}t}&=\left(\underset{(z)}{\epsilon}^{ijk}\underset{(z)}{x}_j\underset{(z)}{\sigma}_{lk}\right)_{;l}-\underset{(z)}{\epsilon}^{ijk}\underset{(z)}{\sigma}_{jk}+\underset{(z)}{\epsilon}^{ijk}\underset{(z)}{x}_j\underset{(z)}{\rho}\underset{(z)}{f}_k,\\ \underset{(z)}{\rho}\frac{\mathrm{d}\underset{(z)}{s}^i}{\mathrm{d}t}&=\underset{(z)}{m}^{li}{}_{;l}+\underset{(z)}{\epsilon}^{ijk}\underset{(z)}{\sigma}_{jk}+\underset{(z)}{\rho}\underset{(z)}{l}^i.\end{aligned} \tag{5.8.3}$$

**Exercise 5.8.1:** ***The* Levi-Civita *symbol in cylindrical and spherical coordinates***

Recall Eq. (3.4.18) where the totally antisymmetric tensor of third order, or Levi-Civita symbol for short, had been defined in Cartesian coordinates. Use the transformation rules for tensors and show that in cylindrical and spherical coordinates we may write:

$$\epsilon^{ijk} = \begin{cases} +\frac{1}{r}, & \text{if } i,j,k = r,\vartheta,z \text{ and cyclic permutations} \\ -\frac{1}{r}, & \text{if } i,j,k = z,\vartheta,r \text{ and cyclic permutations} \\ 0, & \text{else} \end{cases} \tag{5.8.4}$$

and

$$\epsilon^{ijk} = \begin{cases} +\frac{1}{r^2 \sin\vartheta}, & \text{if } i,j,k = r,\vartheta,\varphi \text{ and cyclic permutations} \\ -\frac{1}{r^2 \sin\vartheta}, & \text{if } i,j,k = \vartheta,r,\varphi \text{ and cyclic permutations} \\ 0, & \text{else,} \end{cases} \tag{5.8.5}$$

respectively. Show that this can be rewritten as:

$$\epsilon_{\langle ijk\rangle} = \begin{cases} +1, & \text{if } i,j,k = r,\vartheta,z \text{ (or } \varphi) \text{ and cyclic permutations} \\ -1, & \text{if } i,j,k = z \text{ (or } \varphi),\vartheta,r \text{ and cyclic permutations} \\ 0, & \text{else.} \end{cases} \tag{5.8.6}$$

Write the last result in terms of scalar and vector products between the corresponding unit base vectors $\boldsymbol{e}_r, \boldsymbol{e}_\vartheta$, and $\boldsymbol{e}_{z/\varphi}$.

**Exercise 5.8.2:** ***Terms of the balance of angular momentum in cylindrical and spherical coordinates***

Use results from the previous exercise, Eqs. (5.8.1–5.8.3), or Eqs. $(3.10.1)_2$, $(3.10.4)_2$, and $(3.10.6)_2$ in absolute notation to express all occurring terms in cylindrical or spherical coordinates. In particular show that the production term reads:

$$\boldsymbol{\epsilon}\cdot\cdot\boldsymbol{\sigma} = \boldsymbol{e}_r(\sigma_{\langle\vartheta z\rangle} - \sigma_{\langle z\vartheta\rangle}) + \boldsymbol{e}_\vartheta(\sigma_{\langle zr\rangle} - \sigma_{\langle rz\rangle}) + \boldsymbol{e}_z(\sigma_{\langle r\vartheta\rangle} - \sigma_{\langle\vartheta r\rangle}) \tag{5.8.7}$$

and:

$$\boldsymbol{\epsilon}\cdot\cdot\boldsymbol{\sigma} = \boldsymbol{e}_r(\sigma_{\langle\vartheta\varphi\rangle} - \sigma_{\langle\varphi\vartheta\rangle}) + \boldsymbol{e}_\vartheta\big(\sigma_{\langle\varphi r\rangle} - \sigma_{\langle r\varphi\rangle}\big) + \boldsymbol{e}_\varphi(\sigma_{\langle r\vartheta\rangle} - \sigma_{\langle\vartheta r\rangle}). \tag{5.8.8}$$

Moreover, show that the divergence of the flux is given by:

$$\begin{aligned} \nabla\cdot(\boldsymbol{\sigma}\cdot\boldsymbol{\epsilon}\cdot\boldsymbol{x}) = &-\boldsymbol{e}_r\left[z\left(\frac{\partial\sigma_{\langle r\vartheta\rangle}}{\partial r} + \frac{1}{r}\frac{\partial\sigma_{\langle\vartheta\vartheta\rangle}}{\partial\vartheta} + \frac{\partial\sigma_{\langle z\vartheta\rangle}}{\partial z} + \frac{\sigma_{\langle r\vartheta\rangle} + \sigma_{\langle\vartheta r\rangle}}{r}\right)\right. \\ &\left. + \sigma_{\langle z\vartheta\rangle} - \sigma_{\langle\vartheta z\rangle}\right] + \boldsymbol{e}_\vartheta\left[-r\left(\frac{\partial\sigma_{\langle rz\rangle}}{\partial r} + \frac{1}{r}\frac{\partial\sigma_{\langle\vartheta z\rangle}}{\partial\vartheta} + \frac{\partial\sigma_{\langle zz\rangle}}{\partial z} + \frac{\sigma_{\langle rz\rangle}}{r}\right)\right. \\ &+ z\left(\frac{\partial\sigma_{\langle rr\rangle}}{\partial r} + \frac{1}{r}\frac{\partial\sigma_{\langle\vartheta r\rangle}}{\partial\vartheta} + \frac{\partial\sigma_{\langle zr\rangle}}{\partial z} + \frac{\sigma_{\langle rr\rangle} - \sigma_{\langle\vartheta\vartheta\rangle}}{r}\right) \\ &\left. + \sigma_{\langle zr\rangle} - \sigma_{\langle rz\rangle}\right] + \boldsymbol{e}_z\left[r\left(\frac{\partial\sigma_{\langle r\vartheta\rangle}}{\partial r} + \frac{1}{r}\frac{\partial\sigma_{\langle\vartheta\vartheta\rangle}}{\partial\vartheta} + \frac{\partial\sigma_{\langle z\vartheta\rangle}}{\partial z}\right) + 2\sigma_{\langle r\vartheta\rangle}\right] \end{aligned} \tag{5.8.9}$$

and

$$
\begin{aligned}
\nabla\cdot(\boldsymbol{\sigma}\cdot\boldsymbol{\epsilon}\cdot\boldsymbol{x}) &= \boldsymbol{e}_r(\sigma_{\langle\vartheta\varphi\rangle}-\sigma_{\langle\varphi\vartheta\rangle}) \\
&+\boldsymbol{e}_\vartheta\left[-r\left(\frac{\partial\sigma_{\langle r\varphi\rangle}}{\partial r}+\frac{1}{r}\frac{\partial\sigma_{\langle\vartheta\varphi\rangle}}{\partial\vartheta}+\frac{1}{r\sin\vartheta}\frac{\partial\sigma_{\langle\varphi\varphi\rangle}}{\partial\varphi}\right.\right. \\
&+\left.\left.\frac{+\cot\vartheta(\sigma_{\langle\varphi\vartheta\rangle}+\sigma_{\langle\vartheta\varphi\rangle})}{r}\right)-3\sigma_{\langle r\varphi\rangle}\right] \\
&+\boldsymbol{e}_\varphi\left[r\left(\frac{\partial\sigma_{\langle r\vartheta\rangle}}{\partial r}+\frac{1}{r}\frac{\partial\sigma_{\langle\vartheta\vartheta\rangle}}{\partial\vartheta}+\frac{1}{\sin\vartheta}\frac{\partial\sigma_{\langle\varphi\vartheta\rangle}}{\partial\varphi}\right.\right. \\
&+\left.\left.\frac{\cot\vartheta\,(2\sigma_{\langle\vartheta\vartheta\rangle}-\sigma_{\langle\varphi\varphi\rangle})}{r}\right)+3\sigma_{\langle r\vartheta\rangle}\right].
\end{aligned}
\tag{5.8.10}
$$

Prove that the $r$-component of the moment of momentum balance vanishes identically in spherical coordinates. Also verify that the balances of moment of momentum in cylindrical and spherical coordinates contain no further information than the balances of momentum shown in Eqs. (5.4.4) and (5.4.7). This also explains the way Eqs. (5.8.9) and (5.8.10) were written.

## 5.9 The Essential Balances in Singular Points for Arbitrary Coordinate Systems

In what follows we will only investigate how the fields of mass, velocity, etc. jump when passing through a singular surface with *no* intrinsic properties. This is mainly due to lack of space. For an extensive discussion the reader is referred to Müller [7], Sect. 3.2 or to the paper by Moeckel [6]. This means that the mass density per unit surface, $\underset{A}{\rho}$, in the general Eq. (3.10.1) is equal to zero as well as all the other surface related quantities on the left hand side. Thus we find in combination with Table 3.2 in Cartesian coordinates for the jump of mass density:

$$
\left[\!\left[\underset{(x)}{\rho}\left(\underset{(x)}{\upsilon}_i-\underset{A}{\upsilon}_\perp\underset{(x)}{e}_i\right)\right]\!\right]\underset{(x)}{e}_i=0,\ \underset{A}{\upsilon}_\perp=\underset{(x)}{\upsilon}_i\underset{(x)}{e}_i \tag{5.9.1}
$$

of momentum density:

$$
\left[\!\left[\underset{(x)}{\rho}\,\underset{(x)}{\upsilon_i}\left(\underset{(x)}{\upsilon_j}-\underset{A}{\upsilon}_\perp\underset{(x)}{e_j}\right)-\underset{(x)}{\sigma_{ji}}\right]\!\right]\underset{(x)}{e_j}=0, \tag{5.9.2}
$$

and, for the sake of brevity, only for the jump of the internal energy:

$$\left[\!\left[ \underset{(x)}{\rho}\,\underset{(x)}{u} \left( \underset{(x)}{v_i} - \underset{A}{v_\perp}\underset{(x)}{e_i} \right) + \underset{(x)}{q_i} \right]\!\right] \underset{(x)}{e_i} = 0 \tag{5.9.3}$$

The scalar, vectors, tensors, and scalar products appearing in these relations can now easily be transformed into an arbitrary skew curvilinear coordinate system according to the previously explained transformation rules. Consequently, possible co-/contravariant notations read for mass:

$$\left[\!\left[ \underset{(z)}{\rho} \left( \underset{(z)}{v}^i - \underset{A}{v_\perp}\underset{(z)}{e}^i \right) \right]\!\right] \underset{(z)}{e}_i = 0, \quad \underset{A}{v_\perp} = \underset{\underset{A}{(z)}}{v}^i \underset{(z)}{e}_i \tag{5.9.4}$$

for momentum:

$$\left[\!\left[ \underset{(z)}{\rho}\,\underset{(z)}{v}^i \left( \underset{(z)}{v}^j - \underset{A}{v_\perp}\underset{(z)}{e}^j \right) - \underset{(z)}{\sigma}^{ji} \right]\!\right] \underset{(z)}{e}_j = 0, \tag{5.9.5}$$

and, again for the sake of brevity, only for the internal energy:

$$\left[\!\left[ \underset{(z)}{\rho}\,\underset{(z)}{u} \left( \underset{(z)}{v}^i - \underset{A}{v_\perp}\underset{(z)}{e}^i \right) + \underset{(z)}{q}^i \right]\!\right] \underset{(z)}{e}_i = 0. \tag{5.9.6}$$

## 5.10 The Transport Theorem for Volume Integrals in Arbitrary Coordinate Systems

For rewriting the global balances of mass, momentum, and energy we shall first investigate GAUSS' theorem (3.4.7) in arbitrary coordinate systems:

$$\iiint\limits_V \frac{\partial \underset{(x)}{g}_i}{\partial x_i} \mathrm{d}V = \iint\limits_{A^+ \cup A^-} \underset{(x)}{g}_i \underset{(x)}{n}_i \,\mathrm{d}A - \iint\limits_A \left[\!\left[ \underset{(x)}{g}_i \right]\!\right] \underset{(x)}{e}_i \,\mathrm{d}A. \tag{5.10.1}$$

The divergence and the scalar products are easily converted:

$$\iiint\limits_V \underset{(z)}{g}^i_{;i} \,\mathrm{d}V = \iint\limits_{A^+ \cup A^-} \underset{(z)}{g}^i \underset{(z)}{n}_i \,\mathrm{d}A - \iint\limits_A \left[\!\left[ \underset{(z)}{g}^i \right]\!\right] \underset{(z)}{e}_i \,\mathrm{d}A. \tag{5.10.2}$$

This allows rewriting the transport theorem of Eq. (3.4.12) in the following form:

$$\frac{\mathrm{d}}{\mathrm{d}t}\iiint\limits_{V^+\cup V^-} \underset{(z)}{\psi_V}\,\mathrm{d}V = \iiint\limits_{V^+\cup V^-}\left[\frac{\partial \underset{(z)}{\psi_V}}{\partial t} + \left(\underset{(z)}{\psi_V}\,\underset{(z)}{\upsilon}^{\,i}\right)_{;i}\right]\mathrm{d}V$$

$$= \iiint\limits_{V^+\cup V^-}\frac{\partial \underset{(z)}{\psi_V}}{\partial t}\,\mathrm{d}V + \iint\limits_{A^{+-}} \underset{(z)}{\psi_V}\,\underset{(z)}{\upsilon}^{\,i}\,\underset{(z)}{n}_i\,\mathrm{d}A - \iint\limits_{A}\left[\!\left[\underset{(z)}{\psi_V}\,\underset{(z)}{\upsilon}^{\,i}\right]\!\right]\underset{(z)}{e}_i\,\mathrm{d}A \tag{5.10.3}$$

$$= \iiint\limits_{V^+\cup V^-}\frac{\partial \underset{(z)}{\psi_V}}{\partial t}\,\mathrm{d}V + \iint\limits_{A^{+-}} \underset{(z)}{\psi_V}\,\underset{(z)}{\upsilon}^{\,i}\,\underset{(z)}{n}_i\,\mathrm{d}A - \iint\limits_{A}\left[\!\left[\underset{(z)}{\psi_V}\right]\!\right]\underset{A}{\upsilon_\perp}\,\mathrm{d}A$$

## 5.11 Global Balances of Mass, Momentum, and Energy in Arbitrary Coordinate Systems

For convenience we restrict ourselves to material volumes without singular surfaces. In that case we may sum up, i.e., integrate the regular balances of mass, momentum, and energy from Eqs. (5.1.4), (5.3.2), (5.7.12) with respect to the volume. By observing GAUSS' rule and the transport theorem 5.10.2/5.10.3) we may then write:

$$\begin{aligned}
&\frac{\mathrm{d}}{\mathrm{d}t}\iiint\limits_{V}\underset{(z)}{\rho}\,\mathrm{d}V = 0,\\
&\frac{\mathrm{d}}{\mathrm{d}t}\iiint\limits_{V}\underset{(z)}{\rho}\,\underset{(z)}{\upsilon}^{\,i}\,\mathrm{d}V = \iint\limits_{\partial V}\underset{(z)}{\sigma}^{ji}\,\underset{(z)}{n}_j\,\mathrm{d}A + \iiint\limits_{V}\underset{(z)}{\rho}\,\underset{(z)}{f}^{\,i}\,\mathrm{d}V,\\
&\frac{\mathrm{d}}{\mathrm{d}t}\iiint\limits_{V}\underset{(z)}{\rho}\left(\underset{(z)}{u} + \frac{1}{2}\underset{(z)}{\upsilon}^{\,i}\,\underset{(z)}{\upsilon}_i\right)\mathrm{d}V\\
&= -\iint\limits_{\partial V}\left(\underset{(z)}{q}^{\,j} - \underset{(z)}{\sigma}^{ji}\,\underset{(z)}{\upsilon}_i\right)\underset{(z)}{n}_j\,\mathrm{d}A + \iiint\limits_{V^+\cup V^-}\underset{(z)}{\rho}\left(\underset{(z)}{r} + \underset{(z)}{f}^{\,i}\,\underset{(z)}{\upsilon}_i\right)\mathrm{d}V
\end{aligned} \tag{5.11.1}$$

If a singular surface with intrinsic properties passes through the material additional terms must be added to these equations. The left hand sides will then contain time derivatives of surface integrals and the right hand sides are complemented by line as well as surface integrals using entries from Table 3.2. However, note that not all indices in that table refer to 3D Cartesian coordinates and need to be transformed according to the rules that we have explained above. Greek indices already indicate curvilinear surface coordinates.

## 5.12 Would You Like to Know More?

Further information on the balances of mass, momentum, moment of momentum, and energy in invariant notation can, for example, be found in the books by Greve [2], Chap. 2, Haupt [3], Chaps. 2 and 3, as well as in the "bible of continuum theory" by Truesdell and Toupin [11], Sects. BIII, D1, E1. The entropy balance is mentioned in these monographs as well. We will explore it in Chap. 12 in more detail.

Further explanations regarding the balances of mass, momentum, and energy in specific, non-Cartesian coordinate systems can be found in Irgens [4], Chap. 13, the books by Chandrasekhar [1] and Segel [9], which have already been mentioned and the handbook article by Truesdell and Toupin, Sect. 112 where fluid mechanics problems are discussed.

In general textbooks on fluid mechanics and elasticity provide a most valuable source of *applications* of the balances of mass, momentum, etc., specifically in skew-curvilinear coordinates. For a start one may consult Landau and Lifschitz [5], Chap. 2 (application of the balance of momentum to cylindrical structures), Özişik [8], Chap. 3 and 4 (heat conduction problems in cylinders and spheres), and Sokolnikoff [10], Chap. 4, Sect. 48 (linear elasticity). While reading these books it will become obvious that it is *not* daily engineering practice to distinguish rigorously between balance and constitutive equations, which are then combined to obtain field equations so that everything stems from first principles. Of course, all of this is sometimes just a matter of taste and prejudice, since researchers in linear elasticity think of HOOKE's law as a paramount statement which is as least as fundamental as Newton's laws of motion. And equally researchers in the field of NAVIER–STOKES fluid mechanics, etc. have very similar selfish thoughts. However, we believe clarity and rationality goes beyond wishful thinking.

## References

1. Chandrasekhar S (1981) Hydrodynamic and hydromagnetic stability. Dover Publications Inc, New York
2. Greve R (2003) Kontinuumsmechanik—Ein Grundkurs für Ingenieure und Physiker. Springer, Berlin
3. Haupt P (2002) Continuum mechanics and theory of materials, 2nd edn. Springer, Berlin
4. Irgens F (2008) Continuum mechanics. Springer, Berlin
5. Landau LD, Lifschitz EM (1987) Fluid mechanics. Course of theoretical physics, vol 6, 2nd edn. Butterworth-Heinemann
6. Moeckel GP (1974) Thermodynamics of an interface. ARMA 57:255–280
7. Müller I (1985) Thermodynamics. Pitman Advanced Publishing Program, Boston
8. Özişik MN (1989) Boundary value problems of heat conduction. Dover Publications Inc, Mineola
9. Segel LA (1987) Mathematics applied to continuum mechanics. Dover Publications Inc, Mineola
10. Sokolnikoff IS (1956) Mathematical theory of elasticity. McGraw-Hill Book Company Inc, New York
11. Truesdell C, Toupin R (1960) The classical field theories. In: Flügge S (ed) Encyclopedia of physics. Volume III/1 Principles of classical mechanics and field theory. Springer, Berlin

# Chapter 6
# Constitutive Equations in Arbitrary Coordinate Systems

**Abstract** In this chapter we present a few simple but technically important constitutive equations. However, we will not derive them in a stringent manner in the sense of materials theory where constitutive equations follow as a consequence of superior principles. Rather we will argue in an informal, engineering manner, i.e., cast the constitutive equations into mathematical form after they have been motivated, and detail their form in arbitrary coordinate systems. In particular we will discuss Hooke's law for the anisotropic and for the isotropic linear solid, the NAVIER–STOKES law for viscous fluids, the thermal and the caloric equations of state for the ideal gas, the connection between the specific heats and the internal energy for simple solids according to DULONG-PETIT, and FOURIER's law for the heat flux vector.

*Living in a material world*
*And I am a material girl*
*You know that we are living in a material world*
*And I am a material girl...*

M. L. CICCONE a.k.a. MADONNA

## 6.1 Some Initial Remarks

It has already been mentioned in Chap. 3 that the main objective of a thermomechanical continuum theory consists of computing five fields, namely mass density, velocity, and temperature at all times and in all points of a body. To this end *constitutive equations* are required in addition to the balances of mass, momentum, and energy.

Indeed, the balance equations turn into *field equations of continuum thermomechanics* only if we clarify the dependence of the stress tensor, of internal energy, of the heat flux, of the specific volumetric force, and of the heat supply in

W. H. Müller, *An Expedition to Continuum Theory*, Solid Mechanics and Its Applications 210, DOI: 10.1007/978-94-007-7799-6_6,

terms of the aforementioned five fields or of their derivatives with respect to position and/or time or derivates of them. For example, it will turn out that in order to describe the behavior of solids it is much more advantageous to use the displacement (and related spatial and temporal derivatives) rather than the velocity field. The velocity is "related" to displacement, so to speak. It is nothing else but its time derivative, in other words a "derivate."

In what follows we will concentrate on constitutive equations for the stress tensor, the internal energy, and the heat flux. The specific volumetric force is assumed as "known," just like the heat supply. The former is essentially given by NEWTON's law of gravity, or (for the near field) equal to the constant gravitational acceleration. The latter is either given by the absorbed radiation per unit volume within the body (for example induced in a microwave) or by STEFAN-BOLTZMANN's $T^4$-law, i.e., it will also be considered as a phenomenological quantity in this book.

The equations for the stress tensor, the internal energy, and for the heat flux depend on the material in question. In fact it is the task of a (thermodynamic) constitutive or materials theory to establish restrictions on the possible (mathematical) form of these equations and to predict which dependencies on density, velocity, and temperature (and their derivatives) are permissible—at least in principle—as well as required in order to describe certain physical phenomena of fluid flow or deformation of solids.

A detailed presentation of constitutive theory and, in particular, the underlying, so-called "principles" is beyond the scope of this book. Instead we will argue mostly phenomenologically and state—without "proof"—some relevant constitutive equations for simple, rather idealized materials, namely the linear-elastic Hookean solid and the NAVIER–STOKES–FOURIER fluid, first in Cartesian and later in arbitrary coordinates. By doing so we follow an old tradition since these relations were originally not presented within a most general invariant notation or theoretical framework. Rather they were established empirically for simple, mostly one-dimensional configurations and only much later recast into a proper mathematical form.

We will use the rules of tensor analysis to convert the Cartesian equations to arbitrary coordinates. If they are inserted into the general balance equations, field equations are obtained, which will then be solved to obtain mass density, velocity, and temperature for various geometries. In fact we shall proceed as follows: The constitutive relations will be inserted into the local balance equations for *regular* points and the resulting partial differential equations for mass density, velocity, and temperature will be solved by taking initial and boundary conditions into account and by applying solution techniques for coupled partial differential equations. In fact, the boundary conditions follow from the local balance equations in *singular* form which must also be complemented by appropriate constitutive relations.

## 6.2 HOOKE's Law

The deformation of linear-elastic bodies is described by HOOKE's law which, in a Cartesian base, $x$, reads:

$$\underset{(x)}{\sigma}_{ij} = \underset{(x)}{C}_{ijkl}\left(\underset{(x)}{\varepsilon}_{kl} - \underset{(x)}{\varepsilon}^{*}_{kl}\right). \tag{6.2.1}$$

$\underset{(x)}{C}_{ijkl}$ denotes the so-called stiffness matrix, a tensor of forth order, which characterizes the resistance of a body against deformations of various type (tension/compression or shear) in the various directions of space. As we shall see in Chap. 13 it contains at most 21 independent elastic constants, which—in principle—can be obtained by deformation measurements performed on single crystals. Moreover, $\underset{(x)}{\varepsilon}_{ij}$ denotes the linear *strain tensor*, also w.r.t. the Cartesian base, $x$. Just like the (ordinary) stress tensor it is a *symmetric* quantity and related to gradients of the displacement, $\underset{(x)}{u}_i$, as follows:

$$\underset{(x)}{\varepsilon}_{ij} = \frac{1}{2}\left(\frac{\partial \underset{(x)}{u}_i}{\partial x_j} + \frac{\partial \underset{(x)}{u}_j}{\partial x_i}\right). \tag{6.2.2}$$

The *displacement*, $\underset{(x)}{u}_i$, is defined as the distance of the current position, $x_i$, of a material particle of the body from its reference position which, in Cartesian coordinates, is given by the vector $X_i$:

$$\underset{(x)}{u}_i = x_i - X_i. \tag{6.2.3}$$

Moreover, $\underset{(x)}{\varepsilon}^{*}_{ij}$ denotes the (symmetric) tensor of strains of inelastic origin, for example strains induced by thermal expansion:

$$\underset{(x)}{\varepsilon}^{*}_{ij} = \underset{(x)}{\alpha}_{ij}\,\Delta T. \tag{6.2.4}$$

$\Delta T = T - T_{\mathrm{R}}$ refers to the change of temperature ($T_{\mathrm{R}}$ is the temperature of the stress-free reference state and $T$ denotes the current temperature) and $\underset{(x)}{\alpha}_{ij}$ is a (symmetric) tensor comprising the coefficients of thermal expansion coefficient in the various directions of a crystal. These relations simplify considerably for isotropic materials. Since in that case direction does not matter the stiffness as well as the expansion matrix can be expressed by combinations of the unit tensor alone:

$$\underset{(x)}{C}_{ijkl} = \underset{(x)}{\lambda}\,\delta_{ij}\delta_{kl} + \underset{(x)}{\mu}\left(\delta_{ik}\delta_{jl} + \delta_{il}\delta_{jk}\right), \quad \underset{(x)}{\alpha}_{ij} = \underset{(x)}{\alpha}\,\delta_{ij}. \tag{6.2.5}$$

If this is inserted in Eq. (6.2.1) it follows that:

$$\underset{(x)}{\sigma}_{ij} = \underset{(x)}{\lambda}\, \underset{(x)}{\varepsilon}_{kk} \delta_{ij} + 2 \underset{(x)}{\mu}\, \underset{(x)}{\varepsilon}_{ij} - 3 \underset{(x)}{k}\, \underset{(x)}{\alpha} \Delta T \delta_{ij} \ , \ 3 \underset{(x)}{k} = 3 \underset{(x)}{\lambda} + 2 \underset{(x)}{\mu} \ . \tag{6.2.6}$$

The coefficient $3 \underset{(x)}{k}$ is also known as *compressibility*. The scalar quantities $\underset{(x)}{\lambda}$ and $\underset{(x)}{\mu}$, which may depend on temperature, are the so-called LAMÉ constants.

Siméon Denis POISSON was born on June 21, 1781 in the small town of Pithiviers (France) and died on April 25, 1840 in Paris. He was born into a poor family. Consequently, during his youth he had hardly any opportunity to acquire much more than elementary skills in reading and writing. However, his real talents finally emerged: He attempted to study mathematics and physics and passed the entry exam at the famous École Polytechnique in Paris in 1798 with highest honors. After that it did not take him very long to become one of the leading figures at the French Academy of his time.

**Exercise 6.2.1: *LAMÉ-NAVIER equations for an elementary pull test***

We consider a long and slender beam fixed on one end and subjected to a uniform tensile stress $\sigma$ on its opposite surface (both perpendicular to the $x_1$-axis). Otherwise the beam is free of any loading or clamping. Sketch the situation analogously to Fig. 6.1. Use the Cartesian form of the static balance of momentum in regular and in singular points [cf., Eqs. (5.5.2) and (3.12.4)], in other words NEWTON's law for points within the beam and on its surfaces, in combination with CAUCHY's theorem from Eq. (3.2.7) and show that in every material point the state of stress is given by:

$$\sigma_{ij} = \begin{bmatrix} \sigma & 0 & 0 \\ 0 & 0 & 0 \\ 0 & 0 & 0 \end{bmatrix} . \tag{6.2.7}$$

Combine this result with HOOKE's law (6.2.6) without thermal strains and show algebraically that:

$$\sigma = E\varepsilon \ , \ \varepsilon_{11} = \varepsilon \ , \ \varepsilon_{22} = \varepsilon_{33} = -\nu\varepsilon_{11} \tag{6.2.8}$$

with:

$$E = \frac{\mu(3\lambda + 2\mu)}{\lambda + \mu} \ , \ \nu = \frac{\lambda}{2(\lambda + \mu)} . \tag{6.2.9}$$

The last two relations relate *YOUNG's modulus*, *E*, and *POISSON's ratio*, $\nu$, with the two LAMÉ constants of elasticity, $\lambda$ and $\mu$.

Alternatively (and in preparation for Sect. 7.3) combine the static balance of momentum (by neglecting volumetric forces) with HOOKE's law (while neglecting thermal expansion) and the kinematic relations for small strains and obtain the *Lamé-Navier differential equations* for the displacements for the anisotropic as well as for the isotropic case:

$$\underset{(x)}{C}_{ijkl} \frac{\partial^2 \underset{(x)}{u}_k}{\partial x_i \partial x_l} = 0\,, \quad \left(\underset{(x)}{\lambda} + \underset{(x)}{\mu}\right) \frac{\partial^2 \underset{(x)}{u}_k}{\partial x_j \partial x_k} + \underset{(x)}{\mu} \frac{\partial^2 \underset{(x)}{u}_j}{\partial x_k \partial x_k} = 0. \tag{6.2.10}$$

Explain why in the present case it is reasonable to choose the following *ansatz* for the displacement:

$$\underset{(x)}{u}_1 = u_1(x_1), \quad \underset{(x)}{u}_2 = u_2(x_2), \quad \underset{(x)}{u}_3 = u_3(x_3). \tag{6.2.11}$$

Such a restriction of the possible form of the solution by an intelligent guess is also known as the *semiinverse method*. Insert the *ansatz* into the Lamé-Navier equations, solve the resulting differential equations for the displacements to rediscover the relations (6.2.8/6.2.9). Discuss the pros and cons of both approaches. How can the final result be used to define an experiment for measuring Young's modulus and Poisson's ratio?

Finally discuss the following line of arguments[1]: Consider the 1-D beam shown in Fig. 6.1. A constant displacement field $\underset{(x)}{u}_1 = u_1(x_1 = l, x_2, x_3) = u_0$, $\underset{(x)}{u}_2 = u_2(x_1 = l, x_2, x_3) = 0$, $\underset{(x)}{u}_3 = u_3(x_1 = l, x_2, x_3) = 0$ is applied to its surface at $x_1 = l$, which is oriented in $x_1$-direction. Note and discuss the difference to the previously considered beam. Use the semi-inverse *ansatz* $\partial/\partial x_2 = 0,\ \partial/\partial x_3 = 0$ for all fields and show by integrating the static balance of momentum (without volumetric forces) that:

$$\underset{(x)}{\sigma}_{11} = C_1(x_2, x_3), \quad \underset{(x)}{\sigma}_{12} = C_2(x_2, x_3), \quad \underset{(x)}{\sigma}_{13} = C_3(x_2, x_3). \tag{6.2.12}$$

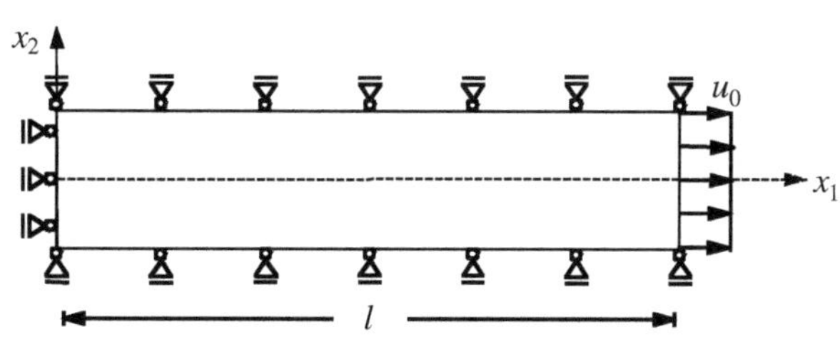


**Fig. 6.1** Semi-inverse *ansatz*

[1] Thanks for suggesting this problem are due to Dr. Wolf Weiss from the Weierstrass Institut in Berlin.

Now evaluate HOOKE's law from Eq. (6.2.6) without thermal strains with the *ansatz*, conclude that $C_1$, $C_2$, $C_3$ are true constants and show that:

$$\underset{(x)}{u}_1 = C_1 x_1, \quad \underset{(x)}{u}_2 = C_2 x_1, \quad \underset{(x)}{u}_3 = C_3 x_1. \tag{6.2.13}$$

Combine this result with the boundary conditions for the displacement to obtain:

$$\underset{(x)}{u}_1 = \frac{u_0}{l} x_1, \quad \underset{(x)}{u}_2 = 0, \quad \underset{(x)}{u}_3 = 0 \tag{6.2.14}$$

and:

$$\underset{(x)}{\sigma}_{11} = \left(\underset{(x)}{\lambda} + 2\underset{(x)}{\mu}\right)\frac{u_0}{l}, \quad \underset{(x)}{\sigma}_{22} = \underset{(x)}{\lambda}\frac{u_0}{l}, \quad \underset{(x)}{\sigma}_{33} = \underset{(x)}{\lambda}\frac{u_0}{l}. \tag{6.2.15}$$

Interpret these equations in terms of the forces required to guarantee clamped boundaries. Which value of POISSON's ratio is required for boundaries that are free of forces? What kind of material would that be? What happens in case of an incompressible material?

**Exercise 6.2.2: *LAMÉ-NAVIER equations for an elementary shear test***

Consider the solid block shown in Fig. 6.2 (left): One of its faces is loaded by a shear force $\tau$ per unit surface. The block deforms "slightly" into a parallelepiped. Use the static balances of momentum in regular and singular points and show that the state of stress within each point of the body is given by:

$$\sigma_{ij} = \begin{bmatrix} 0 & \tau & 0 \\ \tau & 0 & 0 \\ 0 & 0 & 0 \end{bmatrix}. \tag{6.2.16}$$

Ignore thermal strains and use this result in context with HOOKE's law shown in Eq. (6.2.6) to show that:

$$\tau = 2G\varepsilon_{12}, \quad G = \mu. \tag{6.2.17}$$

The last equation proves that the second LAMÉ constant $\mu$ is nothing else but the shear modulus $G$ known from elementary strength of materials theory. Now discuss the same problem by using the LAMÉ-NAVIER equations from Exercise 6.2.1. Explain why the following *ansatz* for the displacement is reasonable to describe the deformation shown in Fig. 6.2 (right):

**Fig. 6.2** The relation between shear modulus $G$ and Lamé's constant $\mu$

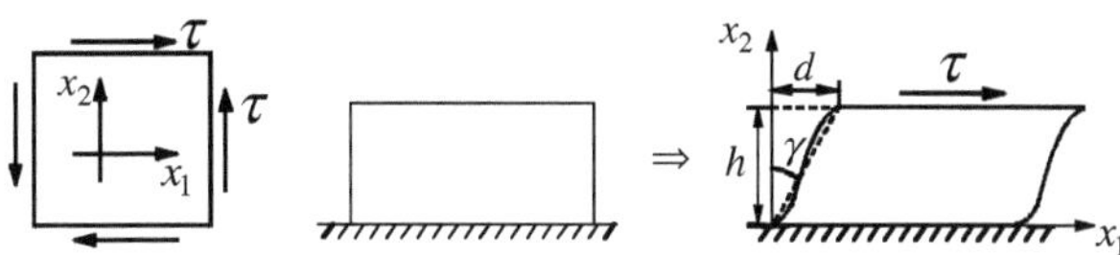


$$\underset{(x)}{u}_1 = u_1(x_2), \quad \underset{(x)}{u}_2 \approx 0\,, \quad \underset{(x)}{u}_3 = 0. \tag{6.2.18}$$

Exploit the Lamé-Navier equations with this *ansatz*, solve the resulting ordinary differential equation to prove that:

$$G = \mu \tag{6.2.19}$$

and:

$$\tau = G \tan \gamma \approx G\gamma, \tag{6.2.20}$$

where $\gamma$ denotes the *shear angle* indicated in Fig. 6.2 (right). Comment on the differences between this and the previous line of arguments. In particular explain why some of the flanks in Fig. 6.2 (left) are unloaded whereas in Fig. 6.2 (right) they are not, i.e., discuss as to whether the problem can easily be solved for a block of finite size in $x_1$ direction.

By observing the transformation rules (2.4.15) for tensors to Eq. (6.2.6) Hooke's law in covariant notation is obtained:

$$\underset{(z)}{\sigma}_{ij} = \underset{(z)}{\lambda}\, g^{kl} \underset{(z)}{\varepsilon}_{lk}\, g_{ij} + 2 \underset{(z)}{\mu} \underset{(z)}{\varepsilon}_{ij} - 3 \underset{(z)}{k} \underset{(z)}{\alpha} \Delta T g_{ij}. \tag{6.2.21}$$

The strain tensor in covariant notation can be found by applying Eqs. (4.2.11/4.2.12) and Eq. (6.2.2):

$$\underset{(z)}{\varepsilon}_{ij} = \frac{1}{2}\left( \underset{(z)}{u}_{i;j} + \underset{(z)}{u}_{j;i} \right). \tag{6.2.22}$$

**Exercise 6.2.3: *Hooke's law and the strain tensor in contravariant and other notations***

Repeat all the steps required to get from Eqs. (6.2.6) to (6.2.21), and from Eqs. (4.2.11/4.2.12) and (6.2.2) to Eq. (6.2.22). Also explain and verify the validity of the following alternative equations for the (trace of the) strain tensor:

$$
\begin{aligned}
&\underset{(z)}{\varepsilon}{}^{rs} = g^{ri} g^{sj} \underset{(z)}{\varepsilon}{}_{ij} \,, \quad \underset{(z)}{\varepsilon}{}^{r}{}_{j} = g^{ri} \underset{(z)}{\varepsilon}{}_{ij} \,, \quad \underset{(z)}{\varepsilon}{}_{i}{}^{s} = g^{sj} \underset{(z)}{\varepsilon}{}_{ij} \,, \\
&\varepsilon_{\langle ij\rangle} = \sqrt{g^{\underline{ii}}}\sqrt{g^{\underline{jj}}}\, \underset{(z)}{\varepsilon}{}_{ij} = \sqrt{g_{\underline{ii}}}\sqrt{g^{\underline{jj}}}\, \underset{(z)}{\varepsilon}{}^{i}{}_{j} = \sqrt{g_{\underline{ii}}}\sqrt{g^{\underline{jj}}}\, \underset{(z)}{\varepsilon}{}^{i}{}_{j} = \sqrt{g_{\underline{ii}}}\sqrt{g_{\underline{jj}}}\, \underset{(z)}{\varepsilon}{}^{ij} \,, \\
&\underset{(z)}{\varepsilon}{}_{i}{}^{i} = g^{ij} \underset{(z)}{\varepsilon}{}_{ij} = g_{jr} \underset{(z)}{\varepsilon}{}^{rj} \,, \quad \underset{(z)}{\varepsilon}{}^{j}{}_{j} = g^{ji} \underset{(z)}{\varepsilon}{}_{ij} = g_{jr} \underset{(z)}{\varepsilon}{}^{rj} \,, \\
&\underset{(z)}{\varepsilon}{}_{\langle ii\rangle} = g^{ji} \underset{(z)}{\varepsilon}{}_{ij} = g_{jr} \underset{(z)}{\varepsilon}{}^{rj} \,.
\end{aligned}
\tag{6.2.23}
$$

and for HOOKE's law:

$$
\begin{aligned}
\underset{(z)}{\sigma}{}^{ij} &= \underset{(z)}{\lambda}\, g^{kl} \underset{(z)}{\varepsilon}{}_{lk}\, g^{ij} + 2 \underset{(z)}{\mu} \underset{(z)}{\varepsilon}{}^{ij} - 3 \underset{(z)}{k} \underset{(z)}{\alpha}\, \Delta T g^{ij} \,, \\
\underset{(z)}{\sigma}{}_{i}{}^{j} &= \underset{(z)}{\lambda}\, g^{kl} \underset{(z)}{\varepsilon}{}_{lk}\, \delta_i{}^j + 2 \underset{(z)}{\mu} \underset{(z)}{\varepsilon}{}_{i}{}^{j} - 3 \underset{(z)}{k} \underset{(z)}{\alpha}\, \Delta T \delta_i{}^j \,, \\
\underset{(z)}{\sigma}{}^{i}{}_{j} &= \underset{(z)}{\lambda}\, g^{kl} \underset{(z)}{\varepsilon}{}_{lk}\, \delta^i{}_j + 2 \underset{(z)}{\mu} \underset{(z)}{\varepsilon}{}^{i}{}_{j} - 3 \underset{(z)}{k} \underset{(z)}{\alpha}\, \Delta T \delta^i{}_j \,.
\end{aligned}
\tag{6.2.24}
$$

Finally derive the following equation, which replaces Eq. (6.2.1) in arbitrary systems, and discuss other possible alternative representations:

$$
\underset{(z)}{\sigma}{}_{ij} = \underset{(z)}{C}{}_{ij}{}^{kl} \left( \underset{(z)}{\varepsilon}{}_{kl} - \underset{(z)}{\overset{*}{\varepsilon}}{}_{kl} \right) . \tag{6.2.25}
$$

**Exercise 6.2.4:** ***HOOKE's law and the strain tensor in cylindrical coordinates***

Use the metric tensor and the CHRISTOFFEL symbols in cylindrical coordinates from Eqs. (2.2.13) and (4.2.5) and evaluate the general Eqs. (6.2.21) and (6.2.22) to show that:

$$
\sigma_{\langle ij\rangle} = \lambda\, \varepsilon_{\langle ll\rangle} \delta_{\langle ij\rangle} + 2\mu\, \varepsilon_{\langle ij\rangle} \,, \qquad i,j \in (r, \vartheta, z) \tag{6.2.26}
$$

with:

$$
\begin{aligned}
\varepsilon_{\langle rr\rangle} &= \frac{\partial u_{\langle r\rangle}}{\partial r} \,, \quad \varepsilon_{\langle \vartheta\vartheta\rangle} = \frac{1}{r}\frac{\partial u_{\langle \vartheta\rangle}}{\partial \vartheta} + \frac{1}{r} u_{\langle r\rangle} \,, \\
\varepsilon_{\langle zz\rangle} &= \frac{\partial u_{\langle z\rangle}}{\partial z} \,, \quad \varepsilon_{\langle r\vartheta\rangle} = \frac{1}{2}\left( \frac{1}{r}\frac{\partial u_{\langle r\rangle}}{\partial \vartheta} + \frac{\partial u_{\langle \vartheta\rangle}}{\partial r} - \frac{1}{r} u_{\langle \vartheta\rangle} \right) , \\
\varepsilon_{\langle rz\rangle} &= \frac{1}{2}\left( \frac{\partial u_{\langle r\rangle}}{\partial z} + \frac{\partial u_{\langle z\rangle}}{\partial r} \right) , \quad \varepsilon_{\langle \vartheta z\rangle} = \frac{1}{2}\left( \frac{\partial u_{\langle \vartheta\rangle}}{\partial z} + \frac{1}{r}\frac{\partial u_{\langle z\rangle}}{\partial \vartheta} \right) .
\end{aligned}
\tag{6.2.27}
$$

**Exercise 6.2.5: *HOOKE's law and the strain tensor in spherical coordinates***

Use the metric tensor and the CHRISTOFFEL symbols in spherical coordinates from Eqs. (2.2.16) and (4.2.6), and evaluate the general Eqns. (6.2.21/ 6.2.22) to show that:

$$\sigma_{\langle ij\rangle} = \lambda\, \varepsilon_{\langle ll\rangle}\delta_{\langle ij\rangle} + 2\mu\, \varepsilon_{\langle ij\rangle}, \quad i,j \in (r, \varphi, \vartheta) \tag{6.2.28}$$

and:

$$\begin{aligned}
\varepsilon_{\langle rr\rangle} &= \frac{\partial u_{\langle r\rangle}}{\partial r}, \quad \varepsilon_{\langle\vartheta\vartheta\rangle} = \frac{1}{r}\frac{\partial u_{\langle\vartheta\rangle}}{\partial\vartheta} + \frac{1}{r}u_{\langle r\rangle},\\
\varepsilon_{\langle\varphi\varphi\rangle} &= \frac{1}{r\sin(\vartheta)}\frac{\partial u_{\langle\varphi\rangle}}{\partial\varphi} + \frac{1}{r}u_{\langle r\rangle} + \frac{\cot(\vartheta)}{r}u_{\langle\vartheta\rangle},\\
\varepsilon_{\langle r\varphi\rangle} &= \frac{1}{2}\left(\frac{1}{r\sin(\vartheta)}\frac{\partial u_{\langle r\rangle}}{\partial\varphi} - \frac{1}{r}u_{\langle\varphi\rangle} + \frac{\partial u_{\langle\varphi\rangle}}{\partial r}\right),\\
\varepsilon_{\langle r\vartheta\rangle} &= \frac{1}{2}\left(\frac{1}{r}\frac{\partial u_{\langle r\rangle}}{\partial\vartheta} - \frac{1}{r}u_{\langle\vartheta\rangle} + \frac{\partial u_{\langle\vartheta\rangle}}{\partial r}\right),\\
\varepsilon_{\langle\varphi\vartheta\rangle} &= \frac{1}{2}\left(\frac{1}{r}\frac{\partial u_{\langle\varphi\rangle}}{\partial\vartheta} - \frac{\cot(\vartheta)}{r}u_{\langle\varphi\rangle} + \frac{1}{r\sin(\vartheta)}\frac{\partial u_{\langle\vartheta\rangle}}{\partial\varphi}\right).
\end{aligned} \tag{6.2.29}$$

**Exercise 6.2.6: *Linear strain tensor and deformation gradient***

Recall the definitions of the deformation gradient, the linear strain tensor, and of the displacement shown in Eqs. (3.4.14), (6.2.2), and (6.2.3), respectively. Show by use of the chain rule that:

$$\left(\delta_{il} - \frac{\partial u_i}{\partial x_l}\right)F_{lk} = \delta_{ik}. \tag{6.2.30}$$

Now assume that the displacement gradients are small and argue why the last equation can be rewritten as follows:

$$F_{ij} = \delta_{ij} + \frac{\partial u_i}{\partial x_j} + \mathrm{O}\left((\nabla \boldsymbol{u})^2\right). \tag{6.2.31}$$

Also show that:

$$\varepsilon_{ij} \approx \frac{1}{2}\left(F_{ij} + F_{ji}\right) - \delta_{ij}. \tag{6.2.32}$$

Observe Eq. (3.4.17) and show that the Jacobian is given by:

$$J \approx 1 + \varepsilon_{kk}. \tag{6.2.33}$$

Recall the result from Exercise 3.8.1 that the current mass density can be calculated from the mass density of the reference state by means of the Jacobian. Show that for small deformations Eq. (3.8.12) can be rewritten as:

$$\rho = \rho_0(1 - \varepsilon_{kk}). \tag{6.2.34}$$

Use Eq. (6.2.8) and show that the mass density of a tensile bar can be obtained from:

$$\frac{\rho - \rho_0}{\rho_0} = -(1 - 2\nu)\frac{\Delta l}{l}. \tag{6.2.35}$$

By how many percent does the density of a bar made of steel decrease after an elongation of 5 %?

## 6.3 The Constitutive Equation of Navier–Stokes

Viscous fluids subjected to small or medium (shear) velocity gradients are frequently modeled by the constitutive equation of Navier–Stokes. It reads in a Cartesian base $x$:

$$\underset{(x)}{\sigma}_{ij} = -\underset{(x)}{p}\,\delta_{ij} + \underset{(x)}{\lambda}\frac{\partial \underset{(x)}{\upsilon}_k}{\partial x_k}\delta_{ij} + \underset{(x)}{\mu}\left(\frac{\partial \underset{(x)}{\upsilon}_i}{\partial x_j} + \frac{\partial \underset{(x)}{\upsilon}_j}{\partial x_i}\right). \tag{6.3.1}$$

The scalars $\underset{(x)}{\lambda}$ and $\underset{(x)}{\mu}$, which may depend on temperature and density, are known as viscosity coefficients. More precisely $\underset{(x)}{\mu}$ is known as the *shear viscosity*, and $\underset{(x)}{\lambda}$ is related to what is known as *bulk* or *volumetric* viscosity, respectively. We shall explore the terminology in more detail in Exercise 7.4.2. Note that in mathematical terms the Navier–Stokes relation bears a certain similarity to Hooke's law for isotropic solids [see Eq. (6.2.6) in context with the definition of strain shown in Eq. (6.2.2)]. This is not too surprising since fluids normally show isotropic behavior and in order to describe their deformation it seems adequate that velocity replaces displacements. Moreover, for a first approximation linear velocity gradients may suffice. Indeed, by "switching-off" the friction terms Eq. (6.3.1) yields:

$$\underset{(x)}{\sigma}_{ij} = -\underset{(x)}{p}\,\delta_{ij}. \tag{6.3.2}$$

The scalar constitutive term $\underset{(x)}{p}$ is known as the *pressure*, which—as one should suspect—acts equally in all directions. Mathematically speaking this is acknowledged by the presence of the "spherical tensor" $\delta_{ij}$, equally in Eq. (6.3.1) as well as in Eq. (6.3.2). In general, the pressure is also a function of the density and of the temperature. We will later provide a simple analytical constitutive equation for the pressure in an ideal gas and thus relate it to the primary fields. In context with friction-free flow the relation (6.3.2) is also known as the stress relation of an EULER fluid. In order to rewrite Eq. (6.3.1) in arbitrary coordinates we note the following transformation equations for all relevant quantities:

$$\begin{gathered} \underset{(x)}{\sigma}_{ij} = \frac{\partial z^k}{\partial x_i}\frac{\partial z^l}{\partial x_j}\underset{(z)}{\sigma}_{kl}, \quad \delta_{ij} = \frac{\partial z^k}{\partial x_i}\frac{\partial z^l}{\partial x_j} g_{kl}, \\ \frac{\partial \underset{(x)}{\upsilon}_i}{\partial x_j} = \frac{\partial z^k}{\partial x_i}\frac{\partial z^l}{\partial x_j}\underset{(z)}{\upsilon}_{k;l}, \quad \frac{\partial \underset{(x)}{\upsilon}_k}{\partial x_k} = \underset{(z)}{\upsilon}^k{}_{;k}, \\ \underset{(x)}{p} = \underset{(z)}{p}, \quad \underset{(x)}{\lambda} = \underset{(z)}{\lambda}, \quad \underset{(x)}{\mu} = \underset{(z)}{\mu}. \end{gathered} \tag{6.3.3}$$

If these relations are inserted in Eq. (6.3.1) the term $\frac{\partial z^k}{\partial x_i}\frac{\partial z^l}{\partial x_j}$ can be extracted as a common factor and the constitutive law of NAVIER–STOKES in covariant notation w.r.t. the free indices $k$ and $l$ in an arbitrary base $z$ results:

$$\underset{(z)}{\sigma}_{kl} = -\underset{(z)}{p}\, g_{kl} + \underset{(z)}{\lambda}\,\underset{(z)}{\upsilon}^r{}_{;r}\, g_{kl} + \underset{(z)}{\mu}\left(\underset{(z)}{\upsilon}_{k;l} + \underset{(z)}{\upsilon}_{l;k}\right). \tag{6.3.4}$$

**Exercise 6.3.1:** ***Alternative forms of the NAVIER–STOKES constitutive law***

Recall that it is possible to raise and lower indices by means of the metric tensor and that its covariant derivative vanishes: Exercise 4.4.1. Use that and prove the following alternative forms of Eq. (6.3.4):

$$\begin{aligned} \underset{(z)}{\sigma}^{kl} &= -\underset{(z)}{p}\, g^{kl} + \underset{(z)}{\lambda}\,\underset{(z)}{\upsilon}^r{}_{;r}\, g^{kl} + \underset{(z)}{\mu}\left(\underset{(z)}{\upsilon}^k{}_{;r}\, g^{rl} + \underset{(z)}{\upsilon}^l{}_{;r}\, g^{rk}\right), \\ \underset{(z)}{\sigma}^k{}_l &= -\underset{(z)}{p}\, \delta^k_l + \underset{(z)}{\lambda}\,\underset{(z)}{\upsilon}^r{}_{;r}\, \delta^k_l + \underset{(z)}{\mu}\left(\underset{(z)}{\upsilon}^k{}_{;l} + \underset{(z)}{\upsilon}_{l;r}\, g^{rk}\right), \\ \underset{(z)}{\sigma}_k{}^l &= -\underset{(z)}{p}\, \delta^l_k + \underset{(z)}{\lambda}\,\underset{(z)}{\upsilon}^r{}_{;r}\, \delta^l_k + \underset{(z)}{\mu}\left(\underset{(z)}{\upsilon}_{k;r}\, g^{rl} + \underset{(z)}{\upsilon}^l{}_{;k}\right). \end{aligned} \tag{6.3.5}$$

**Exercise 6.3.2:** ***The NAVIER–STOKES constitutive equation and the velocity gradient in physical cylindrical and spherical coordinates***

Use the results from Exercises 6.2.4/6.2.5 to prove the validity of the following forms of the NAVIER–STOKES constitutive equation:

$$\sigma_{\langle ij\rangle} = -p\delta_{\langle ij\rangle} + \lambda\, d_{\langle ll\rangle}\delta_{\langle ij\rangle} + 2\mu\, d_{\langle ij\rangle}, \tag{6.3.6}$$

where $i,j \in (r,\vartheta,z)$ for cylindrical coordinates and $i,j \in (r,\varphi,\vartheta)$ for spherical coordinates and the *symmetric part* of the *velocity gradient* is given by:

$$\underset{(z)}{d}_{kl} = \frac{1}{2}\left(\underset{(z)}{\upsilon}_{k;l} + \underset{(z)}{\upsilon}_{l;k}\right). \tag{6.3.7}$$

Moreover, by analogy, use the results from Exercises 6.2.4/6.2.5 to show that the symmetrical part of the velocity gradient in cylindrical and spherical coordinates is given by:

$$\begin{aligned}
d_{\langle rr\rangle} &= \frac{\partial \upsilon_{\langle r\rangle}}{\partial r},\ d_{\langle \vartheta\vartheta\rangle} = \frac{1}{r}\left(\frac{\partial \upsilon_{\langle \vartheta\rangle}}{\partial \vartheta} + \upsilon_{\langle r\rangle}\right),\ d_{\langle zz\rangle} = \frac{\partial \upsilon_{\langle z\rangle}}{\partial z},\\
d_{\langle r\vartheta\rangle} &= \frac{1}{2}\left(\frac{1}{r}\frac{\partial \upsilon_{\langle r\rangle}}{\partial \vartheta} + \frac{\partial \upsilon_{\langle \vartheta\rangle}}{\partial r} - \frac{1}{r}\upsilon_{\langle \vartheta\rangle}\right),\\
d_{\langle rz\rangle} &= \frac{1}{2}\left(\frac{\partial \upsilon_{\langle r\rangle}}{\partial z} + \frac{\partial \upsilon_{\langle z\rangle}}{\partial r}\right),\ d_{\langle \vartheta z\rangle} = \frac{1}{2}\left(\frac{\partial \upsilon_{\langle \vartheta\rangle}}{\partial z} + \frac{1}{r}\frac{\partial \upsilon_{\langle z\rangle}}{\partial \vartheta}\right),
\end{aligned} \tag{6.3.8}$$

and:

$$\begin{aligned}
d_{\langle rr\rangle} &= \frac{\partial \upsilon_{\langle r\rangle}}{\partial r},\ d_{\langle \vartheta\vartheta\rangle} = \frac{1}{r}\left(\frac{\partial \upsilon_{\langle \vartheta\rangle}}{\partial \vartheta} + \upsilon_{\langle r\rangle}\right),\\
d_{\langle \varphi\varphi\rangle} &= \frac{1}{r\sin(\vartheta)}\frac{\partial \upsilon_{\langle \varphi\rangle}}{\partial \varphi} + \frac{1}{r}\upsilon_{\langle r\rangle} + \frac{\cot(\vartheta)}{r}\upsilon_{\langle \vartheta\rangle},\\
d_{\langle r\varphi\rangle} &= \frac{1}{2}\left(\frac{1}{r\sin(\vartheta)}\frac{\partial \upsilon_{\langle r\rangle}}{\partial \varphi} - \frac{1}{r}\upsilon_{\langle \varphi\rangle} + \frac{\partial \upsilon_{\langle \varphi\rangle}}{\partial r}\right),\\
d_{\langle r\vartheta\rangle} &= \frac{1}{2}\left(\frac{1}{r}\left(\frac{\partial \upsilon_{\langle r\rangle}}{\partial \vartheta} - \upsilon_{\langle \vartheta\rangle}\right) + \frac{\partial \upsilon_{\langle \vartheta\rangle}}{\partial r}\right),\\
d_{\langle \varphi\vartheta\rangle} &= \frac{1}{2}\left(\frac{1}{r}\frac{\partial \upsilon_{\langle \varphi\rangle}}{\partial \vartheta} - \frac{\cot(\vartheta)}{r}\upsilon_{\langle \varphi\rangle} + \frac{1}{r\sin(\vartheta)}\frac{\partial \upsilon_{\langle \vartheta\rangle}}{\partial \varphi}\right).
\end{aligned} \tag{6.3.9}$$

Finally note that in both cases the physical components of the velocity can be expressed by the time derivatives of the corresponding coordinates $z^i$.

## 6.4 The Ideal Gas Law

In general the pressure in Eqs. (6.3.1/6.3.2) is not an analytical function of the primary fields for mass density $\rho$ and temperature $T$ (in units of Kelvin) and (eventually) their derivatives w.r.t. space and time. However, there is a "material" for which an analytic relationship holds, i.e., the ideal gas. We write (for all coordinate systems):

$$p = \rho \frac{R}{M} T. \tag{6.4.1}$$

$R = 8.314\ \mathrm{kJ/(kg\,K)}$ denotes the so-called *ideal gas constant* and $M$ is the so-called (dimensionless) molecular weight, which can be obtained from the periodic table of elements, e.g., $M = 12 + 2 \times 16 = 44$ for carbon dioxide, $CO_2$. A relationship between pressure, density, and temperature as in Eq. (6.4.1) is also known as *thermal equation of state*. On first glance Eq. (6.4.1) may not look like the equation we know from high school chemistry:

$$pV = \nu \tilde{R} T. \tag{6.4.2}$$

Lorenzo Romano Amedeo Carlo Bernadette AVOGADRO Conte di Quaregna e Cerreto was born on August 9, 1776 in Torino/Piemont and died there on July 9, 1856. He first studied law, however, in 1800 he switched to mathematics and physics. In 1809 he became a teacher for natural philosophy at the Liceo Vercelli in Torino. This is also the place where he develops his hypotheses on gases consisting of atoms and molecules. In 1820 he becomes a professor for mathematical physics at the University of Torino. He is also politically active and takes part in the rebellion against the king of Sardinia. As a consequence he loses his professorship in 1823 or, officially, retires to find more time for his scientific work. However, in 1833 he is allowed to return and work at the university until the end of his life.

In this equation $\nu$ stands for the number of moles of a substance in a vessel of volume $V$. If $\nu$ is multiplied by AVOGADRO's number $N_{\mathrm{Avo}} = 6.021 \times 10^{23}\,1/\mathrm{mol}$ the total number of molecules in the vessel is obtained. The total mass of the substance (in kg) can be obtained if the mass of one twelfth of the mass of a carbon atom is known. This mass is roughly $\mu = 1.66 \times 10^{-27}$ kg. Consequently we have:

$$m = \nu\, N_{\mathrm{Avo}} \mu\, M, \tag{6.4.3}$$

provided that the molecular weight in the periodic table of elements is based on multiples of the mass of a carbon atom. If the number of moles in Eq. (6.4.2) is now replaced we may write:

$$p = \frac{m}{V} \frac{\tilde{R}}{N_{\mathrm{Avo}} \mu\, M} T. \tag{6.4.4}$$

In the case of a homogenously filled vessel we may write:

$$\rho = \frac{m}{V} \tag{6.4.5}$$

and observe that:

$$N_{\text{Avo}}\mu = 10^{-3}\ \frac{\text{kg}}{\text{mol}}. \tag{6.4.6}$$

Hence Eq. (6.4.4) results in Eq. (6.4.2) if we only put:

$$\tilde{R} = R \cdot 10^{-3}\ \frac{\text{kg}}{\text{mol}} = 8.314\ \frac{\text{J}}{\text{K mol}}. \tag{6.4.7}$$

This is the ideal gas constant frequently used in school. Note that Eq. (6.4.1) is more general than Eq. (6.4.2). It holds locally or, in other words, it can also be applied to a vessel with a heterogeneous gas filling. Thus gradients of density and temperature may exist. Moreover, note that the term ideal gas equation does not refer in any way whatsoever to the phenomenon of friction. Friction is allowed and accounted for by a different term as shown in Eq. (6.3.1). Thus the ideal gas law may also be used in context with frictional flow of gases, at least as a first approximation, as an analytic constitutive relation for the pressure.

Ludwig BOLTZMANN was born on February 20, 1844 in Vienna and died on September 5, 1906 in Duino near Trieste. He studied physics at the University of Vienna where he became a scientific assistant in 1867. In 1869 he accepted a professorship for theoretical physics in Graz, followed by vocations to the University of Munich, Vienna, and Leipzig. Finally, in 1895 he became the successor to the chair of Josef STEFAN in Vienna. BOLTZMANN suffered from severe psychological problems. This is also reflected in a series of events following his vocation to a chair in theoretical physics in Berlin, which he would accept on one day and reject the day after—not only once, but repeatedly in several reiterations, simply because he felt unapt to cope with the duties of the new office. KAISER WILHELM finally put an end to all this and withdrew the offer. BOLTZMANN's most important contribution to science was probably the statistical interpretation of thermodynamics and, in particular, the statistical interpretation of entropy. We must realize that the notion of the atomistic nature of matter was revolutionary in BOLTZMANN's days. Consequently, BOLTZMANN had many famous adversaries, in particular those of the positivistic Viennese school, such as Wilhelm OSTWALD, but—surprisingly—also one of the fathers of quantum mechanics, namely Max PLANCK, who acted through his assistant ZERMELO. Rumor has it that the scientific discussions were so painful to BOLTZMANN that he finally committed suicide. But, maybe, he was just a thoroughbred Viennese, who lived his life according to Georg KREISLER's song: *Der Tod, das muss ein Wiener sein.*

Finally we will present a few alternative ways of writing Eqs. (6.4.1) and (6.4.2), which can frequently be found in thermodynamics textbooks. They are based on the definition of the number of moles as the ratio between the number of particles, $N$, and AVOGADRO's number:

$$\nu = \frac{N}{N_{\text{Avo}}} \tag{6.4.8}$$

and an atomistic gas constant, which is known as the BOLTZMANN constant:

$$k = \mu R = 1.38 \cdot 10^{-23} \, \frac{\text{J}}{\text{K}}. \tag{6.4.9}$$

Simple algebraic manipulations lead to:

$$pV = NkT, \; pV = \nu N_{\text{Avo}} kT. \tag{6.4.10}$$

## 6.5 The Internal Energy of Gases and Solids

This section deals with the constitutive equation for the scalar field of the internal energy. It is also known as the *caloric equation of state*, since it is related to the heat storage of a material as we shall see soon. We first turn to gases: In general the internal energy of a gas is a non-analytic function of (at least) two variables and it depends, just like the gas pressure, on the (local) mass density, $\rho$, (or on its inverse, the specific volume, $\upsilon$) and on the temperature, $T$:

$$u = \tilde{u}(\rho, T) \text{ or } u = \hat{u}(\upsilon, T). \tag{6.5.1}$$

Joseph Louis GAY-LUSSAC was born on December 6, 1778 in Saint-Léonard-de-Noblat and died in Paris on May, 9, 1850. He first attended the École Centrale des Travaux Publics and later the famous École Nationale des Ponts et Chaussées. In 1802 he became a tutor for chemistry at the École Polytechnique, where he also gave lectures. He is famous for his balloon rides during which he investigated the magnetic field of the Earth and the composition of air. In 1808 he was appointed to a professorship for practical chemistry at the École Polytechnique in Paris in unison with a professorship for physics and chemistry at the Sorbonne.

However, for an *ideal gas* the so-called GAY-LUSSAC-JOULE experiment shows that the internal energy is exclusively a function of temperature, $u = u(T)$ and, what is more, a *linear* function for which the following analytical relation holds:

$$u = \zeta \frac{R}{M} T + u_0. \tag{6.5.2}$$

James Prescott JOULE was born on December 24, 1818 in Salford near Manchester and died on October 11, 1889 in Sale, Greater Manchester. He was born into a family of brewers and continued to work in the family tradition together with one of his brothers. However, he also began to study mathematics and science in 1834, both as a hobby and for the benefit of the family enterprise. In 1837 he installed his own chemical lab which was also financially supported by various brewery organizations later. His greatest scientific achievement was the experimental determination of the so called mechanical heat equivalent at a time where thermodynamic notions like heat and internal energy just started to emerge. In his later years his health deteriorated and he was more and more troubled by financial disasters. The latter problem was solved, at least in part, when Queen VICTORIA awarded this famous son of the British Crown a pension in 1878.

$u_0$ denotes an integration constant. It vanishes during differentiation w.r.t. space or time if inserted into the balance equations. Thus its specific numerical value is irrelevant, at least for our purpose. Moreover, $\zeta$ is a number depending on the molecular type of ideal gas. It is equal to $\frac{3}{2}$, $\frac{5}{2}$, or 3, depending on whether the gas consists of one, two, or more atoms, respectively. As we shall see soon below as well as in Chap. 12, Eq. (6.5.2) is related to the internal heat storage of an ideal gas. Nevertheless, it turns out that it can also be used for the description of the heat storage of real gases, to a good approximation over a surprisingly large temperature range. For practical applications, such as the description of the combustion of a mix of air (oxygen) and fuel in turbines, Eq. (6.5.2) reaches its limits and needs to be "improved" by writing:

$$u = c_\upsilon(T)\, T + u_0. \tag{6.5.3}$$

The temperature dependent function $c_\upsilon(T)$ denotes the *specific heat at constant volume* (typically in $\frac{\text{kJ}}{\text{kgK}}$), which is available in tabular form as a function of temperature for many (elementary) gases. It may deviate from the constant value $\zeta\frac{R}{M}$ of the ideal gas depending on the temperature. Why $c_\upsilon(T)$ is called specific heat becomes obvious if we look at the local balance of the internal energy in regular points, or in other words at the First Law of thermodynamics, which can easily be obtained from Table 3.1 in combination with the general balance for regular points, Eq. (3.7.2), while neglecting radiation:

$$\frac{\partial \rho\, u}{\partial t} + \frac{\partial}{\partial x_j}\left(\rho\, u \upsilon_j + q_j\right) = \sigma_{ji}\frac{\partial \upsilon_i}{\partial x_j}. \tag{6.5.4}$$

If we observe the product rule, the balance of mass from Eq. (3.8.3) and the definition for the material time derivative, which has been introduced in Eq. (3.8.6) in context with the mass density, we obtain:

$$\rho\frac{\mathrm{d}u}{\mathrm{d}t} = -\frac{\partial q_j}{\partial x_j} + \sigma_{ji}\frac{\partial \upsilon_i}{\partial x_j}. \tag{6.5.5}$$

We now focus on "slow" processes meaning that squares of velocity gradients are ignored. Thus if we insert the constitutive equation (6.3.1) of NAVIER–STOKES in Eq. (6.5.5) it follows that:

$$\rho \frac{\mathrm{d}u}{\mathrm{d}t} = -\frac{\partial q_j}{\partial x_j} - p\frac{\partial \upsilon_i}{\partial x_i}. \tag{6.5.6}$$

The balance of mass (3.8.7) yields:

$$\frac{\partial \upsilon_i}{\partial x_i} = -\frac{1}{\rho}\frac{\mathrm{d}\rho}{\mathrm{d}t} \equiv +\frac{1}{\upsilon}\frac{\mathrm{d}\upsilon}{\mathrm{d}t}, \tag{6.5.7}$$

and this allows us to rewrite Eq. (6.5.6) as follows:

$$\frac{\mathrm{d}u}{\mathrm{d}t} = -\frac{1}{\rho}\frac{\partial q_j}{\partial x_j} - p\frac{\mathrm{d}\upsilon}{\mathrm{d}t}. \tag{6.5.8}$$

This is the *First Law for slow processes*. It turns into the well known high school formula

$$\mathrm{d}U = \delta Q - p\,\mathrm{d}V \tag{6.5.9}$$

if the latter is referred to unit of mass and unit of time. By comparison it becomes also clear why $\delta Q$ is not a total differential: It is essentially given by the divergence of the heat flux vector and, depending on the sign may be interpreted as a heat supply or a heat drain. However, it does *not* correspond to a differential quotient $\mathrm{d}Q/\mathrm{d}t$. It is customary to speak in context of Eq. (6.5.8) of "$p\mathrm{d}V$-thermodynamics." We will explore this special case a little further. For this purpose we observe Eq. $(6.5.1)_2$, form the total differential, and obtain:

$$\left.\frac{\partial u}{\partial \upsilon}\right|_T \frac{\mathrm{d}\upsilon}{\mathrm{d}t} + \left.\frac{\partial u}{\partial T}\right|_\upsilon \frac{\mathrm{d}T}{\mathrm{d}t} = -\frac{1}{\rho}\frac{\partial q_j}{\partial x_j} - p\frac{\mathrm{d}\upsilon}{\mathrm{d}t}. \tag{6.5.10}$$

If now heat is added to or subtracted from the material particle at a constant volume, we have $\mathrm{d}\upsilon \equiv 0$ and conclude that:

$$-\left.\frac{1}{\rho}\frac{\partial q_j}{\partial x_j}\right|_\upsilon \frac{\mathrm{d}t}{\mathrm{d}T} = \left.\frac{\partial u}{\partial T}\right|_\upsilon. \tag{6.5.11}$$

The left hand side describes the amount of heat energy per unit mass required in order to observe a change in temperature $\mathrm{d}T$. This explains why it is called the *specific heat at a constant volume*:

$$c_\upsilon = \left.\frac{\partial u}{\partial T}\right|_\upsilon. \tag{6.5.12}$$

For an ideal gas we find from Eq. (6.5.2):

$$c_\upsilon = \zeta\frac{R}{M} = \text{const.} \tag{6.5.13}$$

**Exercise 6.5.1:** ***Specific heat of gases at a constant pressure***

Repeat the line of arguments leading to the First Law for slow processes in Eq. (6.5.8). Use this relation and proof the following equation:

$$\frac{\mathrm{d}h}{\mathrm{d}t} = -\frac{1}{\rho}\frac{\partial q_j}{\partial x_j} + \upsilon\frac{\mathrm{d}p}{\mathrm{d}t}, \tag{6.5.14}$$

where the so-called specific free enthalpy has been introduced:

$$h = u + p\,\upsilon. \tag{6.5.15}$$

Now define analogously as in Eq. (6.5.12) a specific heat at a constant pressure:

$$c_p = \left.\frac{\partial h}{\partial T}\right|_p \tag{6.5.16}$$

and show that:

$$-\frac{1}{\rho}\left.\frac{\partial q_j}{\partial x_j}\right|_p \frac{\mathrm{d}t}{\mathrm{d}T} = \left.\frac{\partial h}{\partial T}\right|_p. \tag{6.5.17}$$

Recall the ideal gas relations (6.4.1) and (6.5.2). Use them to show that for ideal gases we have:

$$h = (\zeta+1)\frac{R}{M}T + u_0\ ,\ c_p = (\zeta+1)\frac{R}{M}\ ,\ c_p = c_\upsilon + \frac{R}{M}. \tag{6.5.18}$$

Obviously the specific heat of an (ideal) gas at a constant pressure is greater than the one at a constant volume. Standard textbooks on thermodynamics interpret this observation by saying that the piston which keeps a gas under a constant pressure will be lifted when adding heat. This is an additional amount of work, which is not required in the situation of a gas in a vessel of fixed size. In the latter case the supplied heat would directly and completely affect the internal energy alone. Obviously this line of arguments is rather intuitive and it is rather difficult to judge its range of validity. However, it should be mentioned in this context that by combining pdV thermodynamics with the notion of entropy and the Second Law it can be shown that, in general, the specific heat at a constant pressure is always greater than the specific heat at a constant volume.

The internal energy of ideal solids is governed by a relation similarly to Eq. (6.5.13). On the basis of statistical mechanics we may argue that the internal energy of a gas increases per "molecular degree of freedom" by $\frac{1}{2}\frac{R}{M}T$. If the gas is monatomic there are three *translational* degrees of freedom. This is why we have

$\zeta = \frac{3}{2}$ in this case. If the gas consists of bi-atomic molecules two *rotational* degrees of freedom must be added so that rotational motions perpendicular to the atomic bond are acknowledged. Consequently this leads to $\zeta = \frac{5}{2}$. Finally, if molecules of three or more atoms are involved three degrees of freedom for translational as well as rotational motion result so that $\zeta = 6/2 = 3$.

With a little imagination this kind of argument can easily be extended to the case of solids: The solid is envisioned as a three-dimensional system of mass points (the atoms) that are connected to each other by nonlinear springs (representing the bonding forces). Such a solid has three translational degrees of freedom and (because of the 3D spring arrangement) three degrees of freedom of potential energy as well. Each of them we assign $\frac{1}{2}\frac{R}{M}T$ and, consequently, the specific heat must be:

$$c = 3\,\frac{R}{M}. \tag{6.5.19}$$

This is DULONG-PETIT's rule. However, there is a catch in our line of arguments: In contrast to the case of a gas we did not specify what is kept constant when measuring the specific heat of a solid, its volume or the pressure acting on it, or …? We start over again and just like the gas the internal energy of a solid depends on two variables, one of which is temperature. Now recall Eq. (6.2.34): It shows that the mass density can be determined from the trace of the strain tensor (for small deformations). This is why it is useful to replace the mass density in the constitutive equation for the specific internal energy by the strain tensor (components) instead:

$$u = \tilde{u}(\boldsymbol{\varepsilon}, T). \tag{6.5.20}$$

Thus DULONG-PETIT's rule reads more precisely:

$$c_{\varepsilon} \overset{\text{Def.}}{=} \left.\frac{\partial u}{\partial T}\right|_{\varepsilon} = 3\,\frac{R}{M}. \tag{6.5.21}$$

That this is really the specific heat at a constant strain can be shown with statistical mechanics arguments by using simple atomistic models for solids. Frequently the range of validity of DULONG-PETIT's rule is shown in a high school experiment which will be discussed in-depth in the following exercise.

Eduard GRÜNEISEN was born on May 26, 1877 in Giebichenstein near Halle (Germany) and died on April 5, 1949 in Marburg (Germany). At the age of 17 he studied physics in Halle and Berlin and obtains his doctoral degree in 1900 under the scientific guidance of WARBURG and PLANCK. In 1911 he becomes a professor at the Physikalisch-Technische Reichsanstalt (today's Federal Institute for Materials Research and Testing), advances to departmental manager in 1919, and moves in 1927 to the University of Marburg, where he stays until the end of his life. GRÜNEISEN worked predominantly on equations of state for solids. How- ever, one of his tasks was also the examination of medical students in physics, which can be very disillusioning for a true physicist to say the least. According to an anecdote of the author's father he was very pleased to hear that at least one of the physicians-to-be knew what a differential quotient was and that velocity was defined by one.

**Exercise 6.5.2: *Specific heat of a solid***

Consider the lump of iron of mass $m_{Fe}$ with initial temperature $T_{Fe}$ as shown in Fig. 6.3. At time $t_s$ it is thrown into a water bucket (mass $m_{H_2O}$, initial temperature $T_{H_2O}$). Due to the (initial) temperature difference the two substances start to exchange heat over their surfaces so that the temperature of the iron sinks and the temperature increases (say) until at the time $t_e$ both are at the same temperature, $T_e$. Derive a formula to compute the temperature from the initial data.

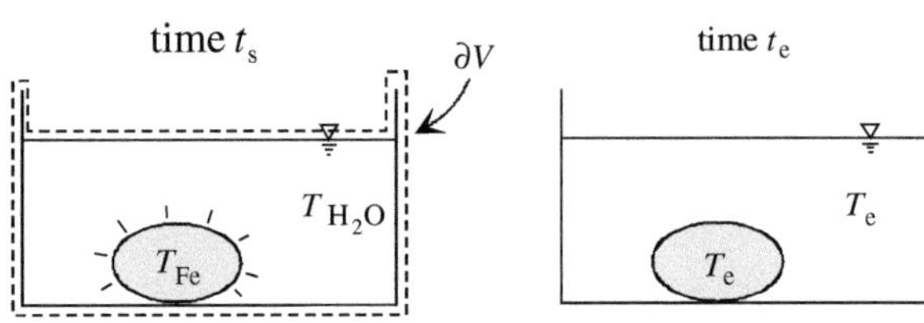


**Fig. 6.3** Heat exchange experiment for measuring the specific heat of solids

For this purpose start from the global balance for the total energy of Eq. (3.9.5) and apply it to the material volume *V*(*t*) with surface $\partial V$ consisting of the iron lump and of the water (see Fig. 6.3). Assume that the heat exchange is exclusively between the water and the iron: The container allows for no other heat exchange and is adiabatically sealed. Argue that under these circumstances we have:

$$\oint\!\!\!\oint_{\partial V} n_j q_j \, dA = 0 \,, \quad \oint\!\!\!\oint_{\partial V} n_j \upsilon_i \sigma_{ji} \, dA = -p_0 \frac{dV}{dt} \,, \tag{6.5.22}$$

where $p_0$ is the (constant) pressure of the surrounding air. Now integrate the energy balance w.r.t. time from the beginning to the end of the internal heat exchange. Which additional assumptions will then lead to:

$$U(t_e) + p_0 V(t_e) = U(t_s) + p_0 V(t_s)\ ? \tag{6.5.23}$$

Do the quantities in this relation add up to an enthalpy? Can this relation be obtained if we start from the global balance of the internal energy (in other words from the First Law) instead? What alternative surfaces to $\partial V$ are there in order to arrive at the same result? What are their pros and cons? Now observe the piecewise homogeneity of the material properties at the beginning and at the end of the heat exchange and explain why the following relation follows from Eq. (6.5.23):

$$\begin{aligned} & m_{Fe}[u_{Fe} + p_0 \upsilon](t_e) + m_{H_2O}[u_{H_2O} + p_0 \upsilon](t_e) \\ &= m_{Fe}[u_{Fe} + p_0 \upsilon](t_s) + m_{H_2O}[u_{H_2O} + p_0 \upsilon](t_s)\ ? \end{aligned} \tag{6.5.24}$$

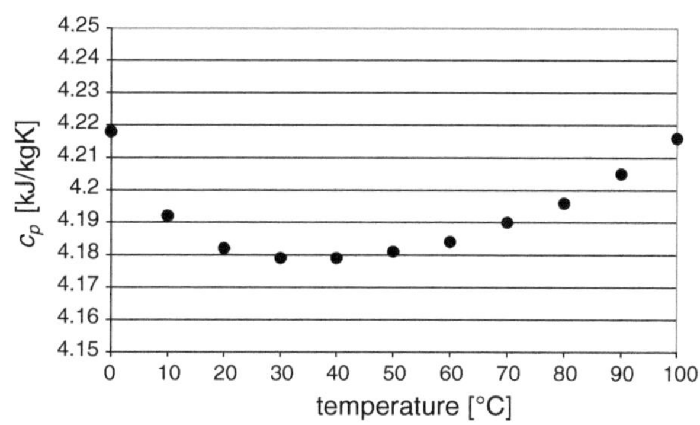


**Fig. 6.4** Specific heat at a constant pressure for water after http://www.wissenschaft-technik-ethik.de/wasser_eigenschaften.html

Obviously the experiment is performed at a constant external pressure $p_0$. If the specific heat of fluid water is measured at a constant pressure at various surrounding temperatures by adding a well defined amount of energy it turns out that it is nearly constant and about $c_{p,H_2O} = 4.18\ \mathrm{kJ/kg\ K}$ (see Fig. 6.4). The specific heat of iron (determined at a constant normal pressure) is also nearly constant within a huge temperature interval and of the order $c_{p,Fe} \approx (0.46 - 0.54)\mathrm{kJ/kg\ K}$ (depending on sort of iron, e.g., alloyed steels). Explain with Eq. (6.5.16) why we may write:

$$m_{Fe} c_{p,Fe}(T_e - T_{Fe}) + m_{H_2O} c_{p,H_2O}(T_e - T_{H_2O}) = 0 \tag{6.5.25}$$

or:

$$T_e = \frac{m_{Fe} c_{p,Fe} T_{Fe} + m_{H_2O} c_{p,H_2O} T_{H_2O}}{m_{Fe} c_{p,Fe} + m_{H_2O} c_{p,H_2O}}. \tag{6.5.26}$$

Moreover, use DULONG-PETIT's rule from Eq. (6.5.21) to show that $c_{\upsilon,Fe} \approx 0.45$ kJ/kg K. Note that for solids $c_p > c_\upsilon$ just like fluids. In fact,

GRÜNEISEN derived the following formula for the difference between the two specific heats of a solid:

$$c_p - c_v = T\alpha^2 k v. \tag{6.5.27}$$

$\alpha$ is the coefficient of thermal expansion, $k$ the modulus of compression (see Eq. (6.2.6) for both quantities) and $T$ is the absolute temperature. Use the results shown in Eq. (6.2.9) from Exercise 6.2.1 to show that:

$$k = \frac{E}{3(1-2\nu)} \tag{6.5.28}$$

and evaluate Eq. (6.5.25) with typical material data for iron at room temperature to show that $c_{p,\mathrm{Fe}} - c_{v,\mathrm{Fe}} \approx 10^{-3} \frac{\mathrm{kJ}}{\mathrm{kg\,K}}$. Comment on the difference from a practical point of view.

**Exercise 6.5.3:** ***The adiabatic relation for ideal gases***

Recall the balance for the internal energy for slow processes from Eq. (6.5.8). Which assumptions are required to arrive at the following so-called adiabatic relation for an ideal gas:

$$p v^{\kappa} = \mathrm{const.}_t \,, \quad \kappa = \frac{\zeta + 1}{\zeta} \; ? \tag{6.5.29}$$

Use results from Exercise 12.1.2 and explain why this relation is a.k.a. *isentropic* relation for an ideal gas. Also prove the following alternative forms:

$$v^{\kappa-1} T = \mathrm{const.}_t \,, \quad p T^{-\frac{\kappa}{\kappa-1}} = \mathrm{const.}_t \; ? \tag{6.5.30}$$

Are adiabatic relations constitutive equations like the ideal gas law? Is it thermodynamically correct to refer to them as adiabatic equations of state?

## 6.6 FOURIER's Law of Heat Conduction

It was FOURIER's great deed to realize that the heat flux is proportional and opposite in direction to the temperature gradient. He was one of the first to recast these verbal statements into mathematics. Following him we write in a Cartesian system:

$$\underset{(x)}{q}_i = -\underset{(x)}{\kappa} \frac{\partial \underset{(x)}{T}}{\partial x_i}. \tag{6.6.1}$$

The coefficient $\kappa$ is a (positive) scalar material parameter that, eventually, depends on the primary fields mass density and temperature. It is known as *thermal conductivity*. In order to rewrite this equation for an arbitrary $z$-system we multiply by $\partial x_i / \partial z^k$ and observe that:

$$\underset{(z)}{q}_k = \frac{\partial x_i}{\partial z^k} \underset{(x)}{q}_i , \quad \underset{(z)}{T} = \underset{(x)}{T} , \quad \underset{(z)}{\kappa} = \underset{(x)}{\kappa} . \tag{6.6.2}$$

By application of the chain rule we arrive at the following relation in covariant form:

$$\underset{(z)}{q}_k = - \underset{(z)}{\kappa} \frac{\partial \underset{(z)}{T}}{\partial z^k} . \tag{6.6.3}$$

Its contravariant equivalent can be obtained after multiplication with the metric tensor:

$$\underset{(z)}{q}^r = - \underset{(z)}{\kappa} g^{rk} \frac{\partial \underset{(z)}{T}}{\partial z^k} = - \underset{(z)}{\kappa} \frac{\partial z^r}{\partial x_j} \frac{\partial z^k}{\partial x_j} \frac{\partial \underset{(z)}{T}}{\partial z^k} = - \underset{(z)}{\kappa} \frac{\partial z^r}{\partial x_j} \frac{\partial \underset{(z)}{T}}{\partial x_j} . \tag{6.6.4}$$

**Exercise 6.6.1: *Direction of the heat flux and the temperature gradient***

Consider a wall of thickness $d$ whose left and right side are kept at temperature levels $T_1$ and $T_2$, respectively. Assume that $T_2 > T_1$. Determine the direction of the temperature gradient and of the heat flux vector. Use the expression for the total heat flux in the First Law in global form shown in Eq. (3.9.6) and confirm the rule that "heat flows from hot to cold." How could the analysis be used to obtain a numerical value for the heat conductivity?

## 6.7 Would You Like to Know More?

The mathematical setup of constitutive equations as well as establishing physically necessary restrictions to their potential form are currently subject of research of the so-called *theory of materials* in combination with *thermodynamically consistent principles*. More detailed information on this subject can be obtained from the monographs by [4–7]. The GAY-LUSSAC-JOULE experiment discussed in the present chapter is analyzed in more detail in [8], Sect. 2.3.4, [3], Sect. 3.8, as well as in [1], Sect. 2.1.3. "Curious notations" of the heat differential, as shown in Eq. (6.5.9) for the First Law, can be found (for example) in many internet wikis about the concept of specific heats. Reference [10] rightfully ridicules on page 3 of his book this desolate state of engineering thermodynamics. Further information on the statistical derivation of DULONG-PETIT's rule can be found in the book by Reif [9], Sect. 10.1, as well as in [2], §61.

## References


1. Baehr HD, Kabelac S (2009) Thermodynamik—Grundlagen und technische Anwendungen, 14th edn. Springer, New York
2. Becker R (1967) Theory of heat. Second edition revised by G. Leibfried. Springer, New York
3. Çengel YA, Boles MA (1998) Thermodynamics—an engineering approach, 6th edn. McGraw Hill, Boston
4. Greve R (2003) Kontinuumsmechanik—Ein Grundkurs für Ingenieure und Physiker. Springer, Berlin
5. Haupt P (2002) Continuum mechanics and theory of materials, 2nd edn. Springer, Berlin
6. Müller I (1973) Thermodynamik. Die Grundlagen der Materialtheorie. Bertelsmann Universitätsverlag, Düsseldorf
7. Müller I (1985) Thermodynamics. Pitman Advanced Publishing Program, Boston
8. Müller I (1994) Grundzüge der Thermodynamik mit historischen Anmerkungen, 1st edn. Springer, Berlin
9. Reif F (1965) Fundamentals of statistical and thermal physics. McGraw-Hill, Boston
10. Truesdell C (1969) Rational thermodynamics. McGraw-Hill, New York

# Chapter 7
# A First Glance on Field Equations

**Abstract** We now combine balance and constitutive equations and obtain field equations for fluids and solids, all in Cartesian coordinates. They are used to pose and to solve initial-boundary value problems for simple geometries and to reach the primary goal of continuum theory, namely the determination of the five fields for mass density, velocity, and temperature in each point of a material body and at all times.

*The iron fist of the real, inside the velvet glove of airy mathematics.*

Gregory Benford in Timescape on (Einstein's) field equations

## 7.1 A Preliminary Remark

In order to solve thermo-mechanical continuum problems we proceed as follows: The *local* balances of mass, momentum, and energy in regular points are combined with suitable constitutive equations so that five partial, coupled differential equations for the five primary fields of interest are obtained, i.e., mass density, velocity (or displacement), and temperature. These are then complemented by sufficiently many boundary and initial conditions. Moreover, at sharp boundaries in the body (e.g., regions of different material parameters), we must satisfy certain *interface conditions*. These result from the *local balances in singular points*, which, if need be, must be complemented by suitable constitutive relations. In total this suite of equations is known as a *well-posed system of field equations*.

As we shall prove shortly by an example a specific solution of the system of field equations requires, in general, numerical methods. Analytical or (more precisely) closed form solutions (e.g., in terms of series) are only possible for highly symmetrical problems in combination with relatively simple constitutive relations, for example Hooke's law and the Navier–Stokes–Fourier case.

W. H. Müller, *An Expedition to Continuum Theory*, Solid Mechanics and Its Applications 210, DOI: 10.1007/978-94-007-7799-6_7,

In the following sections we present the corresponding partial differential equations for mass, velocity (or displacement), and temperature that form the main body of the system of field equations. The analytical solution to simple boundary and initial value problems are subject of Chaps. 9, 10, 11.

However, sometimes a detailed numerical solution of a complicated initial/boundary value problem can be avoided, in particular, if the details of the evolution between a (homogeneous, static) initial and final state are *not* of interest. Initially and finally the system is in a state of *thermodynamic equilibrium*. Then, as we shall see, it is eventually easily possible to predict the end from the beginning even though in-between highly irreversible processes will take place. In the following section we shall detail the involved procedures by a non-trivial example. However, at this point we may already declare that such a solution is based on the exploitation of the *global* balances for mass, momentum, and (total) energy, where it is imperative to choose the balance volume and, in particular, its surface suitably so that all occurring terms can be evaluated easily and precisely. It should be mentioned that this so to speak "global method" has already been applied successfully in Exercise 6.5.2 without emphasizing it particularly.

## 7.2 Globally Stated Problems Involving Control Volumes

Consider the situation shown in Fig. 7.1: An adiabatically sealed cylinder (mass $m_c$) with an adiabatic piston of cross-sectional area $A$ contains an ideal gas of mass $m_g$ under the initial pressure $p_s$ (s = start). Initially this pressure is not completely counterbalanced by the weight of the piston, $m_p g$, in combination with an external pressure $p_0$: $p_s \neq m_p g/A + p_0$.

In other words, it is initially necessary to fix the piston at a starting position, $z_s$, so that it does not move. If the fixtures are detached the piston will start moving and turns the gas into turbulent motion. However, due to internal friction the motion will eventually come to a standstill.

Our objective is to determine the height $z_e$ (e=end) at which the piston will finally stop as well as the corresponding gas temperature $T_e$ and pressure $p_e$,

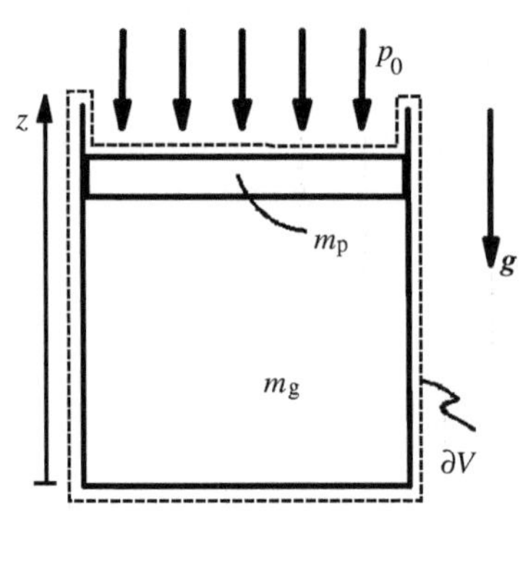


**Fig. 7.1** An adiabatically sealed cylinder with a heavy piston containing an ideal gas

respectively. The latter is very easy to compute since a purely mechanical problem is involved: In the end equilibrium of forces must prevail and thus:

$$-m_{\mathrm{p}}g - p_0 A + p_{\mathrm{e}} A = 0 \quad \Rightarrow \quad p_{\mathrm{e}} = p_0 + \frac{m_{\mathrm{p}}g}{A}. \tag{7.2.1}$$

The term "adiabatic" has been used several times in context with the problem statement. Within the scope of a beginner's course on thermodynamics one would be tempted to use the adiabatic relations from Exercise 6.5.3 in order to determine the remaining unknowns. Thus we start with Eq. (6.5.29), connect it with the (constant) mass $m_{\mathrm{g}}$ of the gas and conclude that:

$$\begin{aligned} &\upsilon_{\mathrm{s}} = \frac{V_{\mathrm{s}}}{m_{\mathrm{g}}} = \frac{A z_{\mathrm{s}}}{m_{\mathrm{g}}},\ \upsilon_{\mathrm{e}} = \frac{V_{\mathrm{e}}}{m_{\mathrm{g}}} = \frac{A z_{\mathrm{e}}}{m_{\mathrm{g}}} \quad \Rightarrow \quad z_{\mathrm{e}} = z_{\mathrm{s}} \left(\frac{p_{\mathrm{s}}}{p_{\mathrm{e}}}\right)^{\frac{1}{\kappa}} \\ &\Rightarrow \quad z_{\mathrm{e}} = z_{\mathrm{s}} \left(\frac{p_{\mathrm{s}}}{p_0 + \frac{m_{\mathrm{p}}g}{A}}\right)^{\frac{1}{\kappa}}. \end{aligned} \tag{7.2.2}$$

The final temperature follows from the ideal gas law, which can be applied to the initial and final states of equilibrium:

$$\begin{aligned} &p_{\mathrm{s/e}} z_{\mathrm{s/e}} A = m_{\mathrm{g}} \frac{R}{M} T_{\mathrm{s/e}} \quad \Rightarrow \quad T_{\mathrm{e}} = T_{\mathrm{s}} \frac{p_{\mathrm{e}}}{p_{\mathrm{s}}} \frac{z_{\mathrm{e}}}{z_{\mathrm{s}}} = T_{\mathrm{s}} \left(\frac{p_{\mathrm{e}}}{p_{\mathrm{s}}}\right)^{\frac{\kappa-1}{\kappa}} \\ &\Rightarrow \quad T_{\mathrm{e}} = T_{\mathrm{s}} \left(\frac{p_0 + \frac{m_{\mathrm{p}}g}{A}}{p_{\mathrm{s}}}\right)^{\frac{\kappa-1}{\kappa}}. \end{aligned} \tag{7.2.3}$$

Brook TAYLOR was born on August 18, 1685 in Edmonton and died on December 29, 1731 in London. He obtained his mathematical training at St. John's College in London and is known as an enthusiastic admirer of NEWTON. From 1712 onwards he published several papers in the Philosophical Transactions of the Royal Society, on the motion of projectiles and the shape of liquid surfaces. The famous TAYLOR expansion was established in 1715 and can be found in Proposition 7 of his paper *Methodus Incrementorum Directa et Inversa.*

Nevertheless the results shown in Eqs. (7.2.2/7.2.3) are dubious for several reasons. The main point of criticism is related to the fact that for large initial pressure differences—in other words for a very heavy piston—it cannot be avoided that turbulence in the gas will set in and that the resulting thermodynamic process is highly irreversible. However, under such circumstances the adiabatic relations of Exercise 6.5.3 do definitely *not* hold. Unfortunately they form the backbone in our previous line of arguments. On the other hand it is to be suspected that the equations represent the situation almost correctly for small pressure differences. This is why we expand the results in TAYLOR series using a smallness parameter $\Delta p$:

$$p_{\mathrm{s}} \approx p_0 + \frac{m_{\mathrm{p}}g}{A} + \Delta p. \tag{7.2.4}$$

If the series resulting from Eqs. (7.2.2/7.2.3) is truncated after the linear term we obtain:

$$z_{\mathrm{e}} \approx z_{\mathrm{s}}\left(1+\frac{1}{\kappa}\frac{\Delta p}{p_0+\frac{m_{\mathrm{p}}g}{A}}\right),\ T_{\mathrm{e}} \approx T_{\mathrm{s}}\left(1-\frac{\kappa-1}{\kappa}\frac{\Delta p}{p_0+\frac{m_{\mathrm{p}}g}{A}}\right). \tag{7.2.5}$$

These equations show very clearly what to expect intuitively: If initially an excess pressure is present, in other words $\Delta p > 0$, the final height $z_{\mathrm{e}}$ increases and the final temperature $T_{\mathrm{e}}$ decreases.[1] An analogous remark holds for the case $\Delta p < 0$.

Moreover, it is curious that in our solution the thermal properties of the piston and of the cylinder do not enter. Both have no heat capacity so to speak and this is hard to believe, in particular for the case of a heavy piston. Such shortcomings will now be removed.

For the ultimate answer to the problem we use the energy balance in its integral form (3.9.5) and apply it to the dashed *surface of the control volume* $\partial V(t)$ shown in Fig. 7.1. Since across the boundary $\partial V(t)$ heat cannot be transferred we must conclude that:

$$\oiint_{\partial V(t)} \boldsymbol{n} \cdot \boldsymbol{q}\, \mathrm{d}A = 0. \tag{7.2.6}$$

Note that the proper choice of a control volume was crucial in order to arrive at this simple result: If part of the envelope is positioned (for example) on the inner side of cylinder instead, Eq. (7.2.6) would not hold since the gas will exchange heat with the cylinder during the process. Moreover, there is no radiation supply in this problem:

$$\iiint_{V(t)} \rho r \mathrm{d}V = 0. \tag{7.2.7}$$

The power supply due to surface forces can be calculated as follows by acknowledging that $p_0 = \text{const.}$ on $\partial V(t)$:

$$\begin{aligned}\oiint_{\partial V(t)} \upsilon_i \sigma_{ji} n_j\, \mathrm{d}A &= -\oiint_{\partial V(t)} \upsilon_i p_0 \delta_{ji} n_j\, \mathrm{d}A = -p_0 \oiint_{\partial V(t)} \upsilon_i n_i\, \mathrm{d}A \\ &= -p_0 \frac{\mathrm{d}V}{\mathrm{d}t} \equiv -\frac{\mathrm{d}(p_0 A z)}{\mathrm{d}t}.\end{aligned} \tag{7.2.8}$$

The second to last step is most easily explained by using REYNOLDS' transport theorem (3.4.12) with $\underset{v}{\psi} = 1$. However, it can also be confirmed by direct evaluation. For this purpose the piston is considered as a rigid body with the normal

[1] The latter effect is best known from the air pump which heats up during fast compression (i.e., decrease of air volume).

$n_i = (0,\ 0,\ 1)$ so that each of its material points shows the same velocity $\upsilon_i = (0,\ 0,\ \mathrm{d}z/\mathrm{d}t)$. Thus:

$$\oiint\limits_{\partial V(t)} \upsilon_i n_i \,\mathrm{d}A = \oiint\limits_{\partial V(t)} \frac{\mathrm{d}z}{\mathrm{d}t} \,\mathrm{d}A = \frac{\mathrm{d}z}{\mathrm{d}t} \oiint\limits_{\partial V(t)} \mathrm{d}A = \frac{\mathrm{d}z}{\mathrm{d}t} A = \frac{\mathrm{d}(zA)}{\mathrm{d}t} = \frac{\mathrm{d}V}{\mathrm{d}t}. \tag{7.2.9}$$

In order to rewrite the power supply of the volumetric force in the balance of energy we note that gravitation is a (static) conservative force. In general conservative forces can be obtained from a spatial derivative of a scalar field, the potential $\varphi$, as follows:

$$f_i = -\frac{\partial \varphi}{\partial x_i}. \tag{7.2.10}$$

Consequently the power assigned to a (static) conservative specific force is given by:

$$\begin{aligned} \iiint\limits_{V(t)} \rho \upsilon_i f_i \,\mathrm{d}V &= -\iiint\limits_{V(t)} \rho \frac{\partial \varphi}{\partial x_i} \frac{\mathrm{d}x_i}{\mathrm{d}t} \,\mathrm{d}V = -\iiint\limits_{V(t)} \rho \frac{\mathrm{d}\varphi}{\mathrm{d}t} \,\mathrm{d}V \\ &= -\int\limits_M \frac{\mathrm{d}\varphi}{\mathrm{d}t} \,\mathrm{d}m = -\frac{\mathrm{d}}{\mathrm{d}t} \int\limits_M \varphi \,\mathrm{d}m = -\frac{\mathrm{d}}{\mathrm{d}t} \iiint\limits_{V(t)} \rho \varphi \,\mathrm{d}V, \end{aligned} \tag{7.2.11}$$

and, therefore, we obtain for the case of gravity near the surface of the Earth, $\varphi = gz$:

$$\begin{aligned} \iiint\limits_{V(t)} \rho \upsilon_i f_i \,\mathrm{d}V &= -\frac{\mathrm{d}}{\mathrm{d}t} \iiint\limits_{V_\mathrm{p}} \rho_\mathrm{p} g z \,\mathrm{d}V - \frac{\mathrm{d}}{\mathrm{d}t} \iiint\limits_{V_\mathrm{g}(t)} \rho_\mathrm{g} g z \,\mathrm{d}V \\ &= -g \left( \frac{\mathrm{d}}{\mathrm{d}t} \iiint\limits_{V_\mathrm{p}} \rho_\mathrm{p} z \,\mathrm{d}V + \frac{\mathrm{d}}{\mathrm{d}t} \iiint\limits_{V_\mathrm{g}(t)} \rho_\mathrm{g} z \,\mathrm{d}V \right) \\ &= -g \left( m_\mathrm{p} \frac{\mathrm{d}z_\mathrm{p}^\mathrm{c}}{\mathrm{d}t} + m_\mathrm{g} \frac{\mathrm{d}z_\mathrm{g}^\mathrm{c}}{\mathrm{d}t} \right) \equiv -\frac{\mathrm{d}}{\mathrm{d}t} \left( m_\mathrm{p} g z_\mathrm{p}^\mathrm{c} + m_\mathrm{g} g z_\mathrm{g}^\mathrm{p} \right). \end{aligned} \tag{7.2.12}$$

$V_\mathrm{p}$ and $V_\mathrm{g}(t)$ denote the (current) volumes of the piston and of the enclosed gas, respectively. Note that only the latter volume must be considered as time-dependent. The piston has already been idealized as a rigid body. Moreover, $z_\mathrm{p}^\mathrm{c}$ and $z_\mathrm{g}^\mathrm{c}$ denote the current positions of the center of gravity in vertical direction for the piston and for the gas, respectively. Note that the center of gravity is defined as follows:

$$z_\mathrm{p}^\mathrm{c} = \frac{\iiint_{V_\mathrm{p}} \rho_\mathrm{p} z \,\mathrm{d}V}{m_\mathrm{K}}, \quad z_\mathrm{g}^\mathrm{c} = \frac{\iiint_{V_\mathrm{g}(t)} \rho_\mathrm{g} z \,\mathrm{d}V}{m_\mathrm{g}}. \tag{7.2.13}$$

We now insert these results in the balance of energy (3.9.5) to find:

$$\frac{\mathrm{d}}{\mathrm{d}t}\left[\iiint\limits_{V(t)} \rho\left(u+\frac{1}{2}\upsilon^2\right)\mathrm{d}V + p_0 A z + m_\mathrm{p} g z_\mathrm{p}^\mathrm{c} + m_\mathrm{g} g z_\mathrm{g}^\mathrm{c}\right] = 0\,. \tag{7.2.14}$$

This equation can be integrated between the beginning and the end of the process, i.e., times $t_\mathrm{s}$ and $t_\mathrm{e}$, respectively:

$$\begin{aligned}&\iiint\limits_{V(t_\mathrm{s})} \rho\, u\,\mathrm{d}V + p_0 A z_\mathrm{i} + m_\mathrm{p} g z_\mathrm{p}^\mathrm{c}(t_i) + m_\mathrm{g} g z_\mathrm{g}^\mathrm{c}(t_\mathrm{i})\\ &= \iiint\limits_{V(t_\mathrm{e})} \rho\, u\,\mathrm{d}V + p_0 A z_\mathrm{e} + m_\mathrm{p} g z_\mathrm{p}^\mathrm{c}(t_\mathrm{e}) + m_\mathrm{g} g z_\mathrm{g}^\mathrm{c}(t_\mathrm{e}).\end{aligned} \tag{7.2.15}$$

The contributions from kinetic energy vanish because initially and at the end the system is at rest. For the difference of the internal energies we may write according to Eqs. (6.5.2/6.5.20/ 6.5.21):

$$\begin{aligned}\int\limits_{V(t_\mathrm{e})} \rho\, u\,\mathrm{d}V - \int\limits_{V(t_\mathrm{s})} \rho\, u\,\mathrm{d}V =& \int\limits_{V_\mathrm{p}\cup V_\mathrm{c}} \rho\, c_\varepsilon (T_\mathrm{e}-T_\mathrm{s})\,\mathrm{d}V\\ &+ \int\limits_{V_\mathrm{g}} \rho\,\zeta\frac{R}{M}(T_\mathrm{e}-T_\mathrm{s})\,\mathrm{d}V\\ &= \left[c_\varepsilon(m_\mathrm{p}+m_\mathrm{c}) + \zeta\frac{R}{M}m_\mathrm{g}\right](T_\mathrm{e}-T_\mathrm{s}),\end{aligned} \tag{7.2.16}$$

if we assume that the initial and final temperature fields are homogeneous and the piston and the cylinder are made of the same material, i.e., they possess the same specific heat $c_\varepsilon$. Note that the constants of Eq. (6.5.2) drop out during the subtraction since masses are conserved during the process. Moreover, the geometry requires that:

$$z_\mathrm{p}^\mathrm{c}(t_\mathrm{e}) - z_\mathrm{p}^\mathrm{c}(t_\mathrm{s}) = 2\left[z_\mathrm{g}^\mathrm{c}(t_\mathrm{e}) - z_\mathrm{g}^\mathrm{c}(t_\mathrm{s})\right] = z_\mathrm{e} - z_\mathrm{s}, \tag{7.2.17}$$

where $z_\mathrm{e}$ and $z_\mathrm{s}$ denote the bottom position of the piston initially and at the end, respectively. Thus Eq. (7.2.15) yields:

$$\begin{aligned}&\left[c_\varepsilon(m_\mathrm{p}+m_\mathrm{c}) + \zeta\frac{R}{M}m_\mathrm{g}\right](T_\mathrm{e}-T_\mathrm{s})\\ &= -\left[\left(m_\mathrm{p}+\frac{1}{2}m_\mathrm{g}\right)g + p_0 A\right](z_\mathrm{e}-z_\mathrm{s}).\end{aligned} \tag{7.2.18}$$

Furthermore recall that equilibrium of forces must be guaranteed in the end: Eq. (7.2.1). The initial and the final pressure in the gas, $p_\mathrm{s}$ and $p_\mathrm{e}$, obey the ideal gas law (6.4.1) applied to a homogeneous state:

$$p_s V_s = m_g \frac{R}{M} T_s, \quad p_e V_e = m_g \frac{R}{M} T_e. \tag{7.2.19}$$

Because $V_e = A z_e$ it follows that:

$$T_s = \frac{p_s A}{m_g \frac{R}{M}} z_s, \quad T_e = \frac{m_p g + p_0 A}{m_g \frac{R}{M}} z_e. \tag{7.2.20}$$

Decoupling of Eqs. (7.2.18/7.2.20) finally yields:

$$\begin{aligned} z_e &= \frac{\left(m_p + \frac{1}{2} m_g\right) g + p_0 A + p_s A \zeta \left(1 + \frac{m_p + m_c}{m_g} \frac{c_\varepsilon}{\zeta \frac{R}{M}}\right)}{\left(\zeta + \frac{m_p + m_c}{m_g} \frac{c_\varepsilon}{\frac{R}{M}}\right) \left(m_p g + p_0 A\right) + \left(m_p + \frac{1}{2} m_g\right) g + p_0 A} z_s, \\ T_e &= \frac{m_p g + p_0 A}{p_s A} \frac{\left(m_p + \frac{1}{2} m_g\right) g + p_0 A + p_s A \zeta \left(1 + \frac{m_p + m_c}{m_g} \frac{c_\varepsilon}{\zeta \frac{R}{M}}\right)}{\left(\zeta + \frac{m_p + m_c}{m_g} \frac{c_\varepsilon}{\frac{R}{M}}\right) \left(m_p g + p_0 A\right) + \left(m_p + \frac{1}{2} m_g\right) g + p_0 A} T_s. \end{aligned} \tag{7.2.21}$$

How are these results related to the quasistatic argument based on the adiabatic relation resulting in Eq. (7.2.5)? In order to find out we have to omit in the last two equations all the quantities that were irrelevant back then. These involve the (gravitational) mass $m_g$ of the gas and the specific heat $c_\varepsilon$ of the piston and of the cylinder. We then obtain:

$$\begin{aligned} z_e &= \frac{m_p g + p_0 A + p_s A \zeta}{(\zeta + 1)\left(m_p g + p_0 A\right)} z_s \approx \left(1 + \frac{\zeta}{\zeta + 1} \frac{\Delta p}{\left[\frac{m_p g}{A} + p_0\right]}\right) z_s, \\ T_e &= \frac{m_p g + p_0 A}{p_s A} \frac{m_p g + p_0 A + p_s A \zeta}{(\zeta + 1)\left(m_p g + p_0 A\right)} T_s \approx \left(1 - \frac{1}{\zeta + 1} \frac{\Delta p}{p_0 + \frac{m_p g}{A}}\right) T_s \end{aligned} \tag{7.2.22}$$

if we use the approximation shown in Eq. (7.2.4). Moreover, we have to acknowledge that:

$$\frac{\kappa - 1}{\kappa} = \frac{\frac{\zeta + 1}{\zeta} - 1}{\frac{\zeta + 1}{\zeta}} = \frac{1}{\zeta + 1}, \quad \frac{1}{\kappa} = \frac{\zeta}{\zeta + 1}. \tag{7.2.23}$$

In other words the approximate results (7.2.5) obtained by using the adiabatic relations are re-derived:

$$\Rightarrow \quad z_e \approx z_s \left(1 + \frac{1}{\kappa} \frac{\Delta p}{p_0 + \frac{m_p g}{A}}\right), \quad T_e \approx T_s \left(1 - \frac{\kappa - 1}{\kappa} \frac{\Delta p}{p_0 + \frac{m_p g}{A}}\right). \tag{7.2.24}$$

We now consider a few special cases and start with the exact relations (7.2.21). In the limit of an infinitely heavy piston, i.e., for $m_p \to \infty$ we find:

$$\lim_{m_p \to \infty} z_e = 0, \quad \lim_{m_p \to \infty} T_e = T_s + \frac{g}{c_\varepsilon} z_s. \tag{7.2.25}$$

In other words: An infinitely heavy piston compresses the gas completely to zero volume while the temperature attains a finite size value. Observe that this result is obtained only for a specific heat $c_\varepsilon \neq 0$. However, in Müller and Müller [1] the quantities of Eq. (7.2.21) were calculated while neglecting the change of the center of gravity of the gas for the case $c_\varepsilon = 0$. Under these circumstances we must write:

$$z_e = \frac{m_p g + p_0 A + p_s A \zeta}{(\zeta + 1)\left(m_p g + p_0 A\right)} z_s \equiv \frac{1}{\zeta + 1} z_s + \frac{\zeta}{\zeta + 1} \frac{m_g \frac{R}{M}}{m_p g + p_0 A} T_s,$$
$$T_e = \frac{1}{p_s A} \frac{m_p g + p_0 A + p_s A \zeta}{\zeta + 1} T_s \equiv \frac{\zeta}{\zeta + 1} T_s + \frac{1}{\zeta + 1} \frac{m_p g + p_0 A}{m_g \frac{R}{M}} z_s. \tag{7.2.26}$$

Interestingly we now find in the limit $m_p \to \infty$:

$$\lim_{m_p \to \infty} z_e = \frac{1}{\zeta + 1} z_s, \quad \lim_{m_K \to \infty} T_e \to \infty. \tag{7.2.27}$$

This means that even an infinitely heavy piston cannot compress the gas completely because the temperature of the gas becomes infinitely large since internal energy cannot be absorbed neither by the piston nor by the cylinder, both of which were assumed to have no heat storage capacity.

**Exercise 7.2.1:** ***Failure of the safety latch between two gas vessels***

Consider the situation shown in Fig. 7.2: The gas in the vessel on the right (start volume $V_2^s$, mass $m_2$, molecular weight $M_2$) is initially subjected to a much higher pressure $p_2^s$ than the gas in left vessel (volume $V_1^s$, mass $m_1$, molecular weight $M_1$, pressure $p_1^s$).

Both gases have the same temperature $T^s$ at the start. After failure of the safety latch the piston separating both vessels starts moving. Turbulent motion is induced in both gases until friction has turned all kinetic energy into heat and stationary, homogeneous conditions prevail again. Assume that the piston is permeable to heat whereas the chamber walls are not (i.e., they are adiabatic) so that the temperature at the end of the process is the same in both chambers: $T_1^e = T_2^e = T^e$. For simplicity both gases are assumed to be monatomic. In the beginning and at the end they can be described by the ideal gas law. Proceed analogously to the previous arguments in this section and show that:

$$T^e = T^s, \quad p_1^e = p_2^e = p^e = \frac{m_1 \frac{R}{M_1} + m_2 \frac{R}{M_2}}{V^{tot}} T^s,$$
$$V_1^e = V^{tot} \frac{\frac{m_1}{M_1}}{\frac{m_1}{M_1} + \frac{m_2}{M_2}}, \quad V_2^e = V^{tot} \frac{\frac{m_2}{M_2}}{\frac{m_1}{M_1} + \frac{m_2}{M_2}}, \quad V^{tot} = V_1^s + V_2^s. \tag{7.2.28}$$

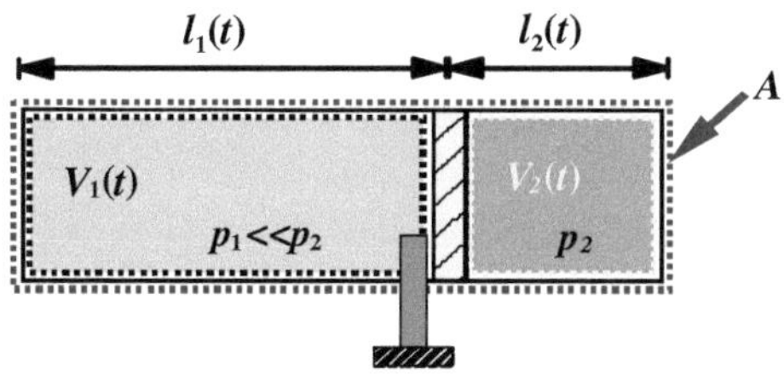


**Fig. 7.2** Two vessels filled with ideal gases

Can the problem be solved for the case of an adiabatic piston?

## 7.3 The Navier-Lamé Equations

We will first deal with the case of Cartesian coordinates $x$ and combine the local static balance of momentum (without volumetric forces) with Hooke's law, i.e., Eqs. (5.5.2) and (6.2.6):

$$\frac{\partial \underset{(x)}{\sigma}_{ji}}{\partial x_j} = 0, \quad \underset{(x)}{\sigma}_{ji} = \underset{(x)}{\lambda}\frac{\partial \underset{(x)}{u}_k}{\partial x_k}\delta_{ji} + \underset{(x)}{\mu}\left(\frac{\partial \underset{(x)}{u}_j}{\partial x_i} + \frac{\partial \underset{(x)}{u}_i}{\partial x_j}\right) - 3\underset{(x)}{k}\,\underset{(x)}{\alpha}\,\Delta T\delta_{ij}. \tag{7.3.1}$$

Moreover, we assume that the temperature does not depend on position and obtain:

$$\frac{\partial^2 \underset{(x)}{u}_j}{\partial x_i \partial x_j} + \frac{\underset{(x)}{\mu}}{\underset{(x)}{\lambda} + \underset{(x)}{\mu}}\frac{\partial^2 \underset{(x)}{u}_i}{\partial x_j \partial x_j} = 0\,, \quad i, j = 1,\ 2,\ 3. \tag{7.3.2}$$

These are *three coupled partial differential equations of second order* for the three unknown displacements $\underset{(x)}{u}_i$. They are already *field equations* in continuum mechanics terms because they only contain derivates of the primary field "*velocity*" and no further unknowns (Lamé's elastic constants are assumed to be known). These relations are also known as Navier-Lamé's equations after the names of their discoverers. In general they can only be solved numerically after suitable boundary conditions have been chosen. However, for a few simple cases it is possible to solve them analytically. Examples are one-dimensional states of stress, such as the beam under tension or the simple shear from Exercises 6.2.1 and 6.2.2. In this context it is advisable to use the so-called *semi-inverse method.* In other words we anticipate the solution and insert a simple but useful *ansatz* as shown in Eqs. (6.2.11) and (6.2.18) into the originally complex partial differential equations and prove that the solution of the resulting simplified differential equations does not suffer from internal contradictions. If such contradictions occur the suggested *ansatz* was too simple, i.e., simply "wrong" and a more complex *ansatz* must be tried instead: In the extreme it may become necessary to look for a purely numerical solution.

It is interesting to note that the first term in Eq. (7.3.2) represents the gradient of a scalar, i.e., the divergence of the displacement vector: $\frac{\partial^2 \underset{(x)}{u}_j}{\partial x_i \partial x_j} = \frac{\partial}{\partial x_i} \frac{\partial \underset{(x)}{u}_j}{\partial x_j}$. Indeed, we have already shown in Sect. 4.1 that such an expression transforms like the components of a covariant vector field. The same results if we start directly from the equations in the $z$-system corresponding to (7.3.2). Moreover, it is worth mentioning that the derivatives in the second term of Eq. (7.3.2) look like a LAPLACE operator applied to the single components of the displacement vector. We must investigate how this term can be transformed into an arbitrary coordinate system $z$ since, more precisely, the LAPLACE operator was not introduced as an operator acting on a vector but on a scalar instead (see Exercise 4.2.7). In order to clarify such issues we start from relations corresponding to Eq. (7.3.2) but in a $z$-system, as they result from Eqs. (5.5.2), (6.2.21/ 6.2.22):

$$\begin{aligned} &\underset{(z)}{\sigma}{}^{ji}{}_{;j} = 0, \\ &\underset{(z)}{\sigma}{}^{ji} = \underset{(z)}{\lambda}\, \underset{(z)}{u}{}^{k}{}_{;k}\, g^{ji} + \underset{(z)}{\mu} \left( g^{ik}\, \underset{(z)}{u}{}^{j}{}_{;k} + g^{jk}\, \underset{(z)}{u}{}^{i}{}_{;k} \right) - 3\, \underset{(z)}{k}\, \underset{(z)}{\alpha}\, \Delta T\, g^{ji}. \end{aligned} \tag{7.3.3}$$

Mutual insertion leads to the contravariant relation:

$$g^{ij} \underset{(z)}{u}{}^{k}{}_{;kj} + \frac{\underset{(z)}{\mu}}{\underset{(z)}{\lambda} + \underset{(z)}{\mu}} g^{jk}\, \underset{(z)}{u}{}^{i}{}_{;kj} = 0, \tag{7.3.4}$$

or to the corresponding covariant version:

$$\underset{(z)}{u}{}^{k}{}_{;ki} + \frac{\underset{(z)}{\mu}}{\underset{(z)}{\lambda} + \underset{(z)}{\mu}} g^{jk}\, \underset{(z)}{u}{}_{i;kj} = 0. \tag{7.3.5}$$

**Exercise 7.3.1: Co-** ***and contravariant forms of the*** **N*AVIER*-L*AMÉ* *equations***

Show the validity of Eqs. $(7.3.3)_2$ and (7.3.4/7.3.5). In particular, explain why we may write:

$$\underset{(z)}{u}{}^{k}{}_{;ki} = \frac{\partial \underset{(z)}{u}{}^{k}{}_{;k}}{\partial z^i}, \tag{7.3.6}$$

and that it makes sense to define the LAPLACE operator as follows:

$$\Delta \underset{(z)}{u}_i = \left( g^{jk} \underset{(z)}{u}_i \right)_{;kj} . \tag{7.3.7}$$

Explain how the first equation can also be derived from the following definition of the covariant derivative of a mixed tensor of second order:

$$\underset{(z)}{u}^k{}_{;ji} = \left( \underset{(z)}{u}^k{}_{;j} \right)_{;i} = \frac{\partial \underset{(z)}{u}^k{}_{;j}}{\partial z^i} + \Gamma^k_{ir} \underset{(z)}{u}^r{}_{;j} - \Gamma^r_{ji} \underset{(z)}{u}^k{}_{;r} . \tag{7.3.8}$$

For the case of orthogonal coordinate systems it is of course possible to rewrite the partial differential equations for the displacements in physical components and then start with a (numerical) solution. In this case it is necessary to apply the following relations to Eqs. (7.3.4/7.3.5):

$$\begin{aligned} g_{kj} &= \begin{bmatrix} g_{11} & 0 & 0 \\ 0 & g_{22} & 0 \\ 0 & 0 & g_{33} \end{bmatrix}, \; g^{jk} = \begin{bmatrix} 1/g_{11} & 0 & 0 \\ 0 & 1/g_{22} & 0 \\ 0 & 0 & 1/g_{33} \end{bmatrix}, \\ u_{\langle 1 \rangle} &= \sqrt{g_{11}} u^1 = \frac{1}{\sqrt{g_{11}}} u_1, \; u_{\langle 2 \rangle} = \sqrt{g_{22}} u^2 = \frac{1}{\sqrt{g_{22}}} u_2, \\ u_{\langle 3 \rangle} &= \sqrt{g_{33}} u^3 = \frac{1}{\sqrt{g_{33}}} u_3 . \end{aligned} \tag{7.3.9}$$

In order to become even more specific the coordinate transformations must be made explicit. This would result in NAVIER-LAMÉ equations in physical components analogous to Eqs. (7.3.4/7.3.5), for cylindrical or spherical coordinates (say). Indeed this is a possible but not an imperative way because in Sect. 5.6 and in Exercises 6.2.4 and 6.2.5 we have already detailed the form of the static balance of momentum as well as HOOKE's law for cylindrical as well as for spherical coordinates including a representation of the components of the strain tensor. In the following exercise a specific problem will be investigated that shows how these relations are evaluated by means of the semi-inverse method.

**Exercise 7.3.2:** ***A linear elastic beam of circular cross-section under torsion***

Consider a slender cylindrical beam of length $l$ of circular cross-section with outer radius $R$. The beam is clamped on one side. On the other side it is twisted so that all cross-sections perpendicular to the cylindrical axis remain planar and all material particles of the front cross-section are turned by the angle $\vartheta_0$ w.r.t. the origin of the beam axis (cf., Fig. 7.3).

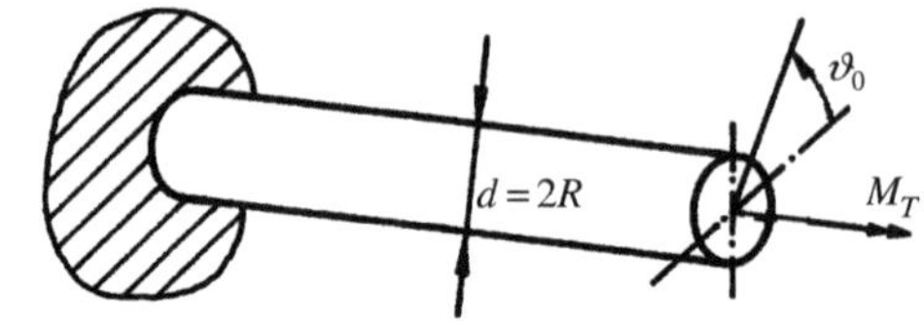


**Fig. 7.3** Circular beam subjected to torsion

Justify the following *ansatz* for the resulting displacements:

$$u^r = 0,\ u^\vartheta = f(z),\ u^z = 0. \tag{7.3.10}$$

Convert the *ansatz* into physical components and insert it in combination with HOOKE's law of Eqs. (6.2.26/ 6.2.27) in the static balance of momentum (5.6.4). Neglect volumetric forces. Show that all but one component of the momentum balance are identically satisfied. Verify by integration of the remaining ordinary differential equation and adjustment of the solution to the boundary condition that:

$$u_{\langle\vartheta\rangle} = r\frac{z}{l}\vartheta_0 \tag{7.3.11}$$

and that only the following strain and stress components are different from zero:

$$\varepsilon_{\langle\vartheta z\rangle} = \frac{r}{2l}\vartheta_0,\ \sigma_{\langle\vartheta z\rangle} = \mu\frac{r}{l}\vartheta_0. \tag{7.3.12}$$

## 7.4 The NAVIER-STOKES Equations

As in the previous section we first look at the case of Cartesian coordinates $x$. However, this time we make use of the complete balance of momentum shown in Eq. (3.8.14). We detail the material time derivative and combine it with the NAVIER-STOKES constitutive equation from Eq. (6.3.1):

$$\begin{gathered}
\underset{(x)}{\rho}\frac{\partial \underset{(x)}{\upsilon}_i}{\partial t} + \underset{(x)}{\rho}\,\underset{(x)}{\upsilon}_j\frac{\partial \underset{(x)}{\upsilon}_i}{\partial x_j} = \frac{\partial \underset{(x)}{\sigma}_{ji}}{\partial x_j} + \underset{(x)}{\rho}\,\underset{(x)}{f}_i, \\
\underset{(x)}{\sigma}_{ij} = -\underset{(x)}{p}\,\delta_{ij} + \underset{(x)}{\lambda}\frac{\partial \underset{(x)}{\upsilon}_k}{\partial x_k}\delta_{ij} + \underset{(x)}{\mu}\left(\frac{\partial \underset{(x)}{\upsilon}_i}{\partial x_j} + \frac{\partial \underset{(x)}{\upsilon}_j}{\partial x_i}\right).
\end{gathered} \tag{7.4.1}$$

This yields:

$$\underset{(x)}{\rho}\frac{\partial \underset{(x)}{\upsilon}_i}{\partial t} + \underset{(x)}{\rho}\,\underset{(x)}{\upsilon}_j\frac{\partial \underset{(x)}{\upsilon}_i}{\partial x_j} = -\frac{\partial \underset{(x)}{p}}{\partial x_i} + \left(\underset{(x)}{\lambda} + \underset{(x)}{\mu}\right)\frac{\partial^2 \underset{(x)}{\upsilon}_k}{\partial x_i \partial x_k} + \underset{(x)}{\mu}\frac{\partial^2 \underset{(x)}{\upsilon}_i}{\partial x_j \partial x_j} + \underset{(x)}{\rho}\,\underset{(x)}{f}_i . \tag{7.4.2}$$

In other words we obtain *three* coupled partial differential equations of second order that ultimately have the purpose to determine the three unknown components of velocity $\underset{(x)}{\upsilon}_i$. They also represent *field equations* in the spirit of continuum theory since they only contain the primary fields "mass density" and "velocity." There are no other unknowns: We assume bulk as well as shear viscosity to be known functions of mass density and of temperature. The same remark holds for the thermal equation of state $\underset{(x)}{p}$ and for the volumetric force density $\underset{(x)}{f}_i$. After the names of their discoverers these relations are known as the *non-stationary* NAVIER-STOKES equations for *compressible* fluids. In general they can only be solved numerically provided that suitable initial and boundary conditions are prescribed. However, for simple forms of flow analytical solutions can be obtained as we shall see below. The *semi-inverse method* proves to be very useful during their solution, just like for the NAVIER-LAMÉ equations. However, two restrictions are frequently made in the same context: Eqs. (7.4.2) are specialized to the *stationary* case:

$$\frac{\partial \underset{(x)}{\upsilon}_i}{\partial t} = 0 \tag{7.4.3}$$

and the fluid is idealized as *incompressible*. In order to realize the mathematical consequences we start with the mass balance from Eq. $(3.8.3)_1$:

$$\frac{\partial \underset{(x)}{\rho}}{\partial t} + \frac{\partial \underset{(x)}{\rho}\,\underset{(x)}{\upsilon}_j}{\partial x_j} = 0. \tag{7.4.4}$$

Incompressible means that the mass density is a constant that does not depend on space or time. Thus we conclude:

$$\frac{\partial \underset{(x)}{\upsilon}_j}{\partial x_j} = 0. \tag{7.4.5}$$

Consequently the divergence of the velocity vector vanishes and Eq. (7.4.2) simplifies:

$$\underset{(x)}{\rho}\,\underset{(x)}{\upsilon}_j\frac{\partial \underset{(x)}{\upsilon}_i}{\partial x_j} = -\frac{\partial \underset{(x)}{p}}{\partial x_i} + \underset{(x)}{\mu}\frac{\partial^2 \underset{(x)}{\upsilon}_i}{\partial x_j \partial x_j} + \underset{(x)}{\rho}\,\underset{(x)}{f}_i . \tag{7.4.6}$$

Note that the first term after the equal sign is the gradient of a scalar, namely of the pressure field. We have shown in Sect. 4.1 that gradients of scalars transform like the components of a covariant vector field. The second term in Eq. (7.4.2) can also be interpreted as the gradient of a scalar, i.e., the divergence of the velocity vector: $\frac{\partial^2 \underset{(x)}{\upsilon}_j}{\partial x_i \partial x_j} = \frac{\partial}{\partial x_i} \frac{\partial \underset{(x)}{\upsilon}_j}{\partial x_j}$ and the third term is essentially a LAPLACE operator acting on the single components of velocity. How it transforms has already been investigated in Exercise 7.3.1. All of this must be observed during transformation of Eq. (7.4.6) into the $z$-system. Also note that very similar arguments have been used in context with the NAVIER-LAMÉ equations.

**Exercise 7.4.1:** ***Parallel-plate flow***

Consider the stationary frictional flow of an incompressible NAVIER-STOKES-fluid in a channel between two very large, parallel plates, one at uniform speed, $V$, and the other one at rest: Fig. 7.4. Show that the speed in horizontal direction increases linearly with height:

$$\upsilon_1 = V \frac{x_2}{h}. \tag{7.4.7}$$

For this purpose motivate and make use of the following semi-inverse *ansatz* in Eq. (7.4.6):

$$\upsilon_1 = \upsilon_1(x_2),\ \upsilon_2 = 0,\ \upsilon_3 = 0. \tag{7.4.8}$$

Neglect gravitational forces and solve the resulting differential equation of second order in space. Finally adjust the two constants of integration to the boundary constraints in order to guarantee that the fluid sticks to the walls (no-slip condition).

Consider now the case that both plates are at rest and the fluid is moving stationary in horizontal direction driven by a constant pressure gradient $p'$. Show by using the same *ansatz* that now a parabolic velocity profile results:

$$\upsilon_1(x_2) = -\frac{p'}{8\mu} h^2 \left[1 - \left(\frac{2x_2}{h}\right)^2\right]. \tag{7.4.9}$$

Note that during integration of the balance of momentum it is advisable to position the coordinate system in the center between the two plates. How can a constant pressure gradient be realized in practice, especially for very long channels?

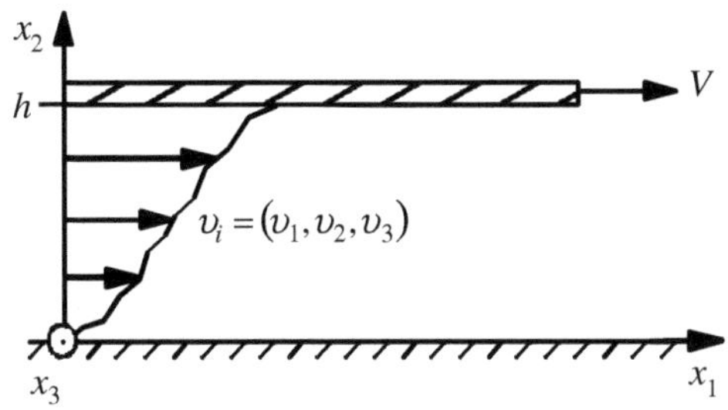


**Fig. 7.4** Sketch of the plan-parallel fluid flow between two large plates

**Exercise 7.4.2: *Alternative forms of the NAVIER-STOKES equations***

Start from the NAVIER-STOKES constitutive equation (6.3.1) and show that it can be rewritten as (Cartesian coordinates are used exclusively and the clumsy suffix $(x)$ has been omitted for convenience):

$$\sigma_{ij} = -p\delta_{ij} + \mu' \frac{\partial \upsilon_k}{\partial x_k} \delta_{ij} + \mu \left( \frac{\partial \upsilon_i}{\partial x_j} + \frac{\partial \upsilon_j}{\partial x_i} - \frac{2}{3} \frac{\partial \upsilon_k}{\partial x_k} \delta_{ij} \right). \qquad (7.4.10)$$

Recall that the divergence of velocity is related to volumetric changes (*cf.*, Sects. 3.4 and 3.8 for the details) and explain (in words) why the gradient terms behind $\mu' = \lambda + \frac{2}{3}\mu$ and $\mu$ are related to volumetric and to shear deformation, respectively. This is why $\mu'$ is known as *bulk* and $\mu$ as *shear viscosity*. Show that the NAVIER-STOKES equations (7.4.2) now take the following form:

$$\underset{(x)}{\rho} \frac{\partial \upsilon_i}{\partial t} + \rho \upsilon_j \frac{\partial \upsilon_i}{\partial x_j} = -\frac{\partial p}{\partial x_i} + \left( \mu' + \frac{1}{3}\mu \right) \frac{\partial^2 \upsilon_k}{\partial x_i \partial x_k} + \mu \frac{\partial^2 \upsilon_i}{\partial x_j \partial x_j} + \rho f_i. \qquad (7.4.11)$$

It was STOKES who suggested in 1845 (see [2] that the bulk viscosity vanishes, i.e., $\lambda = -\frac{2}{3}\mu$. Whilst this may be true for ideal monatomic gases it is certainly not the case for more complex gas molecules. Also note [3] that the combination $\mu' + \frac{1}{3}\mu$ is (sometimes) called bulk viscosity as well. Shear viscosity is typically measured by using a so-called cone-on-plate viscosimeter. Consult the literature as well as the internet and find out how this is done exactly. The bulk viscosity, however, is difficult to measure. Follow up on some references in Gad-el-Hak [2] and find out what is known about it so far.

Now define the mechanical pressure by:

$$p_{\mathrm{me}} = -\frac{1}{3}\sigma_{kk} \qquad (7.4.12)$$

and use Eq. (7.4.10) to show that:

$$p_{\mathrm{me}} = p - \mu' \frac{\partial v_k}{\partial x_k}. \tag{7.4.13}$$

Recall that according to Eqs. (6.4.1) and (6.5.2) the pressure and the internal energy density of an ideal gas are related by:

$$u - u_0 = \zeta \frac{p}{\rho}. \tag{7.4.14}$$

Use this result to comment on the following text passage from Gad-el-Hak [2]: "The mechanical pressure is a measure of the translational energy of the molecules. The thermodynamic pressure, on the other hand, is a measure of the total energy, which include(s) additional vibrational and rotational modes and, for liquids and dense gases, intermolecular attraction. For dilute monatomic gases, the translational energy is the only mode of molecular energy."

Other mechanics books use notations different from ours. For example we find in the textbook by Ziegler [4] the following form of the NAVIER-STOKES equations:

$$\rho\, \vec{a} = \vec{k} - \operatorname{grad} p + \eta\, \Delta \vec{v}. \tag{7.4.15}$$

Under which circumstances is this equivalent to Eqs. (7.4.2/7.4.11)? What assumptions are required? Ziegler (p. 198) calls the material parameter $\eta$ *dynamic or absolute viscosity* and quotes $\eta = 10^{-3}\,\mathrm{Ns/m^2}$ for water at 20 °C. Is that a new type of viscosity when compared to $\lambda$ and $\mu$? Is there also a *static viscosity*?

In context with viscous fluids the term NEWTONian fluid is also frequently used. In which way is a NEWTONian fluid "different" from a NAVIER-STOKES fluid?

We will now transform Eq. (7.4.2) into the $z$-system. To this end we start from the balance of momentum (5.3.2) in contravariant form:

$$\underset{(z)}{\rho} \left( \frac{\partial \underset{(z)}{v^i}}{\partial t} + \underset{(z)}{v^j}\, \underset{(z)}{v^i}{}_{;j} \right) = \underset{(z)}{\sigma^{ji}}{}_{;j} + \underset{(z)}{\rho}\, \underset{(z)}{f^i} \tag{7.4.16}$$

and insert the NAVIER-STOKES constitutive equation $(6.3.5)_1$ in contravariant form:

$$\underset{(z)}{\sigma^{ji}} = - \underset{(z)}{p}\, g^{ji} + \underset{(z)}{\lambda}\, \underset{(z)}{v^r}{}_{;r} g^{ji} + \underset{(z)}{\mu} \left( \underset{(z)}{v^j}{}_{;r} g^{ri} + \underset{(z)}{v^i}{}_{;r}\, g^{rj} \right). \tag{7.4.17}$$

Thus:

$$\underset{(z)}{\rho}\left(\frac{\partial \underset{(z)}{\upsilon^i}}{\partial t} + \underset{(z)}{\upsilon^j}\,\underset{(z)}{\upsilon^i}{}_{;j}\right) = -\frac{\partial \underset{(z)}{p}}{\partial z^j}\, g^{ji} + \underset{(z)}{\lambda}\,\underset{(z)}{\upsilon^r}{}_{;rj} g^{ji}$$
$$+ \underset{(z)}{\mu}\left(\underset{(z)}{\upsilon^j}{}_{;rj} g^{ri} + \underset{(z)}{\upsilon^i}{}_{;rj} g^{rj}\right) + \underset{(z)}{\rho}\,\underset{(z)}{f^i}\,. \tag{7.4.18}$$

---

**Exercise 7.4.3:** ***NAVIER-STOKES equations in contravariant notation***

Explain why in Eq. (7.4.18) $\frac{\partial \underset{(z)}{p}}{\partial z^j}$ was written instead of $\underset{(z)}{p}{}_{;j}$. Is it correct to write:

$$\underset{(z)}{\upsilon^r}{}_{;rj} = \frac{\partial \underset{(z)}{\upsilon^r}{}_{;r}}{\partial z^j}\,? \tag{7.4.19}$$

Finally convert the contravariant Eq. (7.4.18) into invariant notation similarly to Eq. (7.4.15).

---

## 7.5 The Semi-Inverse Method Applied to Dynamic Gas Flow

We now return to the problem of the dropping piston from Sect. 7.2.[2] However, this time we want to study the dynamic problem, i.e., the oscillatory motion of the piston which, due to friction, will slowly be damped down. The semi-inverse *ansatz* for this problem reads:

$$\upsilon_1 = 0,\ \upsilon_2 = 0,\ \upsilon_3 = \frac{\dot{z}(t)}{z(t)} x_3. \tag{7.5.1}$$

It is worth several comments. First, as weird as it may sound, it can be motivated from HUBBLE's law of astrophysics. According to that the universe expands such that the recessional velocity, $\upsilon$, of a far-away galaxy increases proportionally to its distance, $x$, from Earth:

$$\upsilon = H_0 x. \tag{7.5.2}$$

[2] Credit must be given to Prof. Ingo Müller (TU Berlin) and to Dr. Wolf Weiss (WIAS Berlin) who brought the arguments in this section to my attention.

Edwin Powell HUBBLE was born on November 20, 1889 in Marshfield Missouri and died on September 28, 1953 in San Marino, California. He was one of the eminent American scientists of the twentieth century known for his fundamental work on experimental astronomy which enlarged our local view of the universe to include the outermost galaxies and nebulae. His disciple Alan SANDAGE reports that HUBBLE believed that the redshift of the far-away galaxies was not necessarily based on their own movement (known as recession speed) but due to a "hitherto unrecognized principle of nature."

$H_0$ is HUBBLE's "constant." It is a (time-dependent) parameter that needs to be determined experimentally from redshift observations on receding distant galaxies. The receding velocity is limited by the speed of light, $c$, which defines the size $z(t)$ of the currently observable universe. Hence $H_0 = c/z(t)$. In our case the outer rim of the universe corresponds to the position $z(t)$ of the piston with its speed $c \to \dot{z}(t)$, so we end up with Eq. (7.5.1)$_3$.

Second, Eq. (7.5.1)$_3$ meets the constraints imposed to the motion of a fluid showing friction, which is supposed to stick at the boundaries of the vessel (cf., Exercise 7.4.1):

$$\upsilon_3(x_3 = 0,\ t) = 0,\ \upsilon_3(x_3 = z(t),\ t) = \dot{z}(t). \tag{7.5.3}$$

In order to take the cosmological analogy one step further, recall that in astronomy it is customary to believe that the universe is homogeneous in space and that it was much denser in the past. We also model the density, the pressure, and the temperature of the gas in the cylinder as homogeneous but time-dependent:

$$\rho = \rho(t) = \frac{m_g}{V(t)} = \frac{N_{\text{Avo}}\mu_{\text{H}}M\nu}{Az(t)},\ p(t) = \frac{\nu \widetilde{R}}{Az(t)}T(t), \tag{7.5.4}$$

where the ideal gas law of Sect. 6.4 in its various forms has been used. This, however, means that the mass balance is identically satisfied and gives us no further information as we shall convince ourselves now. The first term in Eq. (3.8.7)$_1$ yields:

$$\frac{\mathrm{d}\rho}{\mathrm{d}t} = \frac{\mathrm{d}}{\mathrm{d}t}\left(\frac{N_{\text{Avo}}\mu_{\text{H}}M\nu}{Az(t)}\right) = -\frac{N_{\text{Avo}}\mu_{\text{H}}M\nu}{Az^2(t)}\frac{\mathrm{d}z}{\mathrm{d}t}. \tag{7.5.5}$$

And if we observe the *ansatz* (7.5.1) the second one reads:

$$\rho\frac{\partial \upsilon_j}{\partial x_j} = \frac{N_{\text{Avo}}\mu_{\text{H}}M\nu}{Az(t)}\frac{\dot{z}(t)}{z(t)}, \tag{7.5.6}$$

which is just the negative of (7.5.5). We now turn to the balance of momentum in global form (3.2.10), apply it to the piston, which as assumed to behave like a rigid body, and obtain for the three terms:

$$\frac{\mathrm{d}}{\mathrm{d}t}\iiint\limits_{V_\mathrm{p}} \rho_\mathrm{p}\, \upsilon_3(x_3 = z(t), t)\,\mathrm{d}V = m_p \frac{\mathrm{d}^2 z}{\mathrm{d}t^2},$$
$$\oiint\limits_{\partial V_\mathrm{p}} n_j \sigma_{j3}\,\mathrm{d}A = \left(p - \frac{4}{3}\mu\frac{\dot z(t)}{z(t)}\right)A,\quad \iiint\limits_{V_\mathrm{p}} \rho_\mathrm{p} f_3\,\mathrm{d}V = -m_\mathrm{p} g. \tag{7.5.7}$$

In the second equation we neglected the pressure $p_0$ acting on the upper side of the piston. The traction on the bottom side of the piston, $\boldsymbol{n} = (0,0,-1)$, results from the viscous gas, which was modeled as a NAVIER-STOKES fluid with $\lambda = -\frac{2}{3}\mu$: Eqs. (7.4.10) and (7.5.1). If we finally replace $p$ by the ideal gas equation and take the homogeneity of the mass density into account, cf., Eq. (7.5.4), the equation of motion for the piston results:

$$m_p \frac{\mathrm{d}^2 z}{\mathrm{d}t^2} = \frac{\nu\,\widetilde{R}T(t)}{z(t)} - \frac{4}{3}\mu A \frac{\dot z(t)}{z(t)} - m_\mathrm{p} g. \tag{7.5.8}$$

This is the first of two ordinary differential equation for the two unknowns $z(t)$ and $T(t)$. Note that it resembles the equation for the harmonic oscillator but with a nonlinear damping term. However, the damping is still proportional to velocity as known from standard textbooks. The equation is obviously nonlinear and must be solved numerically. In this context it is useful to define dimensionless variables:

$$\bar t = \sqrt{\frac{\widetilde{R}T(0)}{m_p z^2(0)}}\, t,\ \bar T(\bar t) = \frac{T(t)}{T(0)},\ \bar z(\bar t) = \frac{z(t)}{z(0)}. \tag{7.5.9}$$

Thus Eq. (7.5.8) becomes:

$$\frac{\mathrm{d}^2 \bar z}{\mathrm{d}\bar t^2} = \nu \frac{\bar T(\bar t)}{\bar z(\bar t)} - C_1 \frac{1}{\bar z(\bar t)}\frac{\mathrm{d}\bar z}{\mathrm{d}\bar t} - C_2,\ C_1 = \frac{4}{3}\frac{\mu A}{\sqrt{m_\mathrm{p}\widetilde{R}T(0)}},$$
$$C_2 = \frac{m_\mathrm{p} g z(0)}{\widetilde{R}T(0)}. \tag{7.5.10}$$

We now turn to the First Law in global form, Eq. (3.9.6) and apply it exclusively to the gas volume $V_\mathrm{g} = A\, z(t)$. The left hand side reads:

$$\frac{\mathrm{d}}{\mathrm{d}t}\iiint\limits_{V_\mathrm{g}} \rho\, u\,\mathrm{d}V = \frac{\zeta \nu\,\widetilde{R}}{A z(t)}\frac{\mathrm{d}T}{\mathrm{d}t} \times A z(t) = \zeta \nu\,\widetilde{R}\frac{\mathrm{d}T}{\mathrm{d}t}. \tag{7.5.11}$$

The only non-vanishing term on the right side is the working term. If we insert our *ansatz* in Eq. (7.4.10) and ignore bulk viscosity we find:

$$\iiint\limits_{V_\mathrm{g}} \sigma_{ji}\frac{\partial \upsilon_i}{\partial x_j}\,\mathrm{d}V = -\frac{\nu\,\widetilde{R}T(t)}{z(t)}\frac{\mathrm{d}z}{\mathrm{d}t} + \frac{4}{3}\mu\frac{A}{z(t)}\left(\frac{\mathrm{d}z}{\mathrm{d}t}\right)^2. \tag{7.5.12}$$

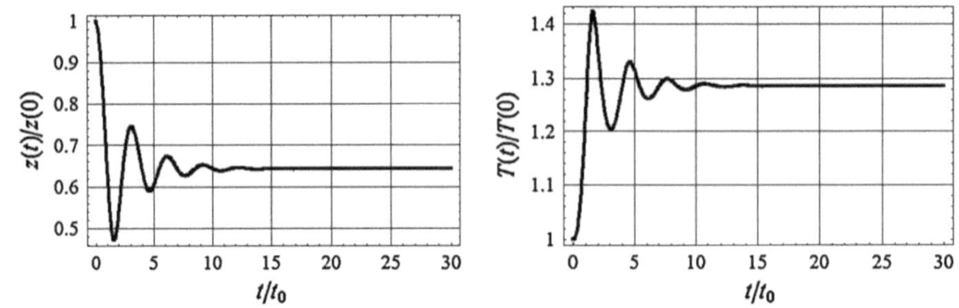


**Fig. 7.5** Falling of a piston in an adiabatically sealed cylinder (case of strong damping)

If we observe the definitions of dimensionless quantities in Eq. (7.5.9) we obtain the second ordinary differential equation, which is also non-linear:

$$\zeta \nu \frac{\mathrm{d}\overline{T}}{\mathrm{d}\overline{t}} = -\nu \frac{\overline{T}(\overline{t})}{\overline{z}(\overline{t})} \frac{\mathrm{d}\overline{z}}{\mathrm{d}\overline{t}} + C_1 \frac{1}{\overline{z}(\overline{t})} \left( \frac{\mathrm{d}\overline{z}}{\mathrm{d}\overline{t}} \right)^2 . \tag{7.5.13}$$

We choose $C_1 = 0.5$ and $C_2 = 2.0$, $\zeta = \frac{5}{2}$, $\nu = 1$, $\overline{z}(0) = 1$, $\dot{\overline{z}}(0) = 0$, $\overline{T}(0) = 1$ and solve the two differential equations numerically. The corresponding results are shown in Fig. 7.5: Due to the weight of the piston the ideal gas is adiabatically compressed. However, the piston overshoots the equilibrium height during first compression and is eventually pushed back. Oscillatory motion sets in and, due to damping by viscous friction, the piston finally stops at the equilibrium height. Corresponding to the movement of the piston the temperature is first increased, then decreased, etc. After a while equilibrium temperature is (nearly) reached. A comparison with Eq. (7.2.26) leads to the following formulae for the equilibrium values:

$$\begin{aligned} \overline{z}(\infty) \equiv \overline{z}_{\mathrm{e}} &= \frac{1}{\zeta + 1} + \frac{\zeta}{\zeta + 1} \frac{\nu}{C_2} \approx 0.64, \\ \overline{T}(\infty) \equiv \overline{T}_{\mathrm{e}} &= \frac{\zeta}{\zeta + 1} + \frac{1}{\zeta + 1} \frac{C_2}{\nu} \approx 1.29. \end{aligned} \tag{7.5.14}$$

Note that this result is consistent with the result for the stationary state following from Eq. $(7.5.10)_1$:

$$\overline{z}(\infty)\, C_2 = \nu\, \overline{T}(\infty). \tag{7.5.15}$$

However, if we now choose somewhat realistic values for the viscosity, $\mu = 1.8 \cdot 10^{-5}\,\mathrm{Pa\,s}$, the cross-section, $A = 10^{-2}\,\mathrm{m}^2$, the mass of the piston, $m_{\mathrm{p}} = 10^{-3}\,\mathrm{kg}$, the initial height, $z(0) = 0.5\,\mathrm{m}$, and the initial temperature, $T = 298\,\mathrm{K}$, it turns out that $C_1 = 1.5 \cdot 10^{-7}$, i.e., the damping effect of the viscosity is much smaller. Indeed, if we use this value for $C_1$ while $C_2 = 2.0$ the

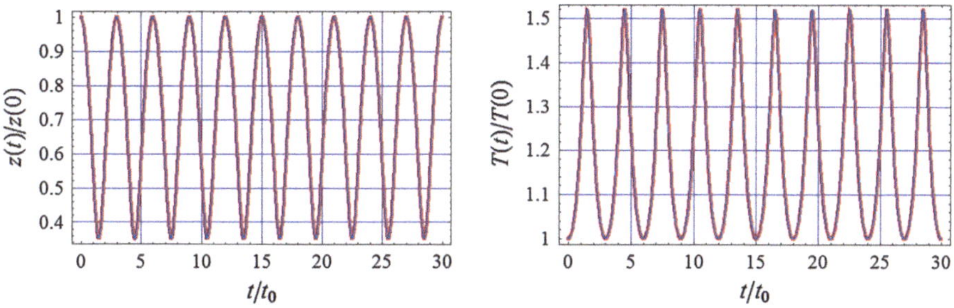


**Fig. 7.6** Falling of a piston in an adiabatically sealed cylinder (case of small damping in comparison to the solution with the adiabatic equation)

piston simply oscillates like a mass on a harmonic spring without friction: Fig. 7.6 (left).

This again can be compared with an analysis of the situation on the basis of the adiabatic relation shown in Eq. (6.5.29). In this case we find the following differential equation for the motion of the piston:

$$\frac{d^2\bar{z}}{d\bar{t}^2} = \nu\,\bar{z}^{-\kappa} - C_2, \quad \kappa = \frac{\zeta + 1}{\zeta}. \tag{7.5.16}$$

The temperature follows algebraically and not from a differential equation:

$$\overline{T}(\bar{t}) = \bar{z}^{1-\kappa}(\bar{t}). \tag{7.5.17}$$

Figure 7.6 shows that there is no difference when compared to the more complex analysis that takes frictional effects of the gas into account. This also explains why the adiabatic relation of $p\ dV$-thermodynamics is so successful even if applied to rapid processes as the combustion cycle in a car engine.

We now turn to the solution of the mysterious problem of the inner adiabatic wall. Reconsider the situation shown in Fig. 7.2. However, this time assume that the separating piston is adiabatic so that once equilibrium has been reached the temperatures in the two chambers are no longer the same. In his 1965 textbook Callen makes this problem look unsolvable. In fact he says on p. 321: "The problem of prediction of the equilibrium state in a composite system with an adiabatic internal wall is a uniquely delicate problem." In the 1985 edition, however, the problem has already been downgraded to a homework exercise, although to a mysterious one, and a solution is not provided. Callen says on p. 53: "The hypothetical problem of equilibrium in a closed composite system with an internal moveable adiabatic wall is a unique indeterminate problem. Physically, release of the piston would lead it to perpetual oscillation in the absence of viscous damping. With viscous damping the piston would eventually come to rest at such a position that the pressures on either side would be equal, but the temperatures in each subsystem would then depend on the relative viscosity in each subsystem. The solution of this problem depends on *dynamical* considerations."

> **Exercise 7.5.1:** ***Adiabatic relation versus irreversible thermodynamics***
>
> Show the validity of Eqs. (7.5.16/7.5.17). Note that in Fig. 7.1 (left) the piston initially moves downward. Find a set of suitable data so that the piston moves upward first.

All of this is very true, indeed. However, words help at most some philosophers but certainly not the engineers, especially if the "dynamical considerations" are neither revealed nor rationally explained. It is therefore not surprising that a vast amount of literature on this problem started to emerge immediately, and even today the discussion is not over yet. From the literature we only mention two papers: Crosignani et al. [5] and Gislason [6]. Their final formulae can directly be compared with ours even though their derivation is totally different and essentially based on kinetic theory.

For the solution of the dynamic problem we make exactly the same assumptions as in the previous example of the falling piston. Thus the *ansatz* for the velocities in both vessels (1) and (2) reads (*cf.*, Fig. 7.2 for the meaning of certain symbols):

$$\begin{aligned} \upsilon_1^{(1)} &= \frac{\dot{l}^{(1)}(t)}{l^{(1)}(t)} x_1^{(1)}, \ \upsilon_1^{(2)} = \frac{\dot{l}^{(2)}(t)}{l - l^{(1)}(t)} x_1^{(2)} \equiv -\frac{\dot{l}^{(1)}(t)}{l - l^{(1)}(t)} x_1^{(2)}, \\ \upsilon_{2/3}^{(1)/(2)} &= 0. \end{aligned} \tag{7.5.18}$$

Moreover, from Eq. (7.5.4) it follows that:

$$\rho^{(1)/(2)} = \frac{N_{\mathrm{Avo}} \mu_{\mathrm{H}} M^{(1)/(2)} \nu^{(1)/(2)}}{A l^{(1)/(2)}(t)}, \quad p^{(1)/(2)} = \frac{\nu^{(1)/(2)} \widetilde{R}}{A l^{(1)/(2)}(t)} T^{(1)/(2)}(t). \tag{7.5.19}$$

We conclude that the mass balances are identically satisfied. For the balance of momentum of the piston we find in perfect analogy to Eq. $(7.5.7)_2$:

$$m_p \frac{\mathrm{d}^2 l^{(1)}}{\mathrm{d}t^2} = \left( p^{(1)} - p^{(2)} - \frac{4}{3} \left[ \frac{\mu^{(1)}}{l^{(1)}} + \frac{\mu^{(2)}}{l - l^{(1)}} \right] \frac{\mathrm{d}l^{(1)}}{\mathrm{d}t} \right) A. \tag{7.5.20}$$

If we now apply the First Law separately to each chamber we obtain similarly to Eqs. (7.5.11/7.5.12):

$$\begin{aligned} \frac{\zeta^{(1)} \nu^{(1)} \widetilde{R}}{A l^{(1)}(t)} \frac{\mathrm{d}T^{(1)}}{\mathrm{d}t} &= -\frac{\nu^{(1)} \widetilde{R} T^{(1)}(t)}{A l^{(1)2}(t)} \frac{\mathrm{d}l^{(1)}}{\mathrm{d}t} + \frac{4}{3} \mu^{(1)} \frac{1}{l^{(1)2}(t)} \left( \frac{\mathrm{d}l^{(1)}}{\mathrm{d}t} \right)^2 \\ \frac{\zeta^{(2)} \nu^{(2)} \widetilde{R}}{A(l - l^{(1)}(t))} \frac{\mathrm{d}T^{(2)}}{\mathrm{d}t} &= \frac{\nu^{(2)} \widetilde{R} T^{(2)}(t)}{A(l - l^{(1)}(t))^2} \frac{\mathrm{d}l^{(1)}}{\mathrm{d}t} + \frac{4}{3} \mu^{(2)} \frac{1}{(l - l^{(1)}(t))^2} \left( \frac{\mathrm{d}l^{(1)}}{\mathrm{d}t} \right)^2. \end{aligned} \tag{7.5.21}$$

We now introduce the following dimensionless variables:

$$\bar{t} = \sqrt{\frac{\tilde{R}T^{(1)}(0)}{m_p l^2}}\, t,\ \overline{T}^{(1)}(\bar{t}) = \frac{T^{(1)}(t)}{T^{(1)}(0)},\ \overline{T}^{(2)}(\bar{t}) = \frac{T^{(2)}(t)}{T^{(1)}(0)},\ \bar{l}(\bar{t}) = \frac{l^{(1)}(t)}{l} \tag{7.5.22}$$

and obtain three coupled, non-linear, ordinary differential equations:

$$\begin{aligned}
\bar{l}(1-\bar{l})\frac{\mathrm{d}^2\bar{l}}{\mathrm{d}\bar{t}^2} &= \nu^{(1)}\overline{T}^{(1)}(1-\bar{l}) - \nu^{(2)}\overline{T}^{(2)}\bar{l} - C\left(1+\left[\frac{\mu^{(2)}}{\mu^{(1)}}-1\right]\bar{l}\right)\frac{\mathrm{d}\bar{l}}{\mathrm{d}\bar{t}},\\
\zeta^{(1)}\nu^{(1)}\bar{l}\frac{\mathrm{d}\overline{T}^{(1)}}{\mathrm{d}\bar{t}} &= -\nu^{(1)}\overline{T}^{(1)}\frac{\mathrm{d}\bar{l}}{\mathrm{d}t} + C\left(\frac{\mathrm{d}\bar{l}}{\mathrm{d}\bar{t}}\right)^2,\\
\zeta^{(2)}\nu^{(2)}(1-\bar{l})\frac{\mathrm{d}\overline{T}^{(2)}}{\mathrm{d}\bar{t}} &= \nu^{(2)}\overline{T}^{(2)}\frac{\mathrm{d}\bar{l}}{\mathrm{d}t} + C\frac{\mu^{(2)}}{\mu^{(1)}}\left(\frac{\mathrm{d}\bar{l}}{\mathrm{d}\bar{t}}\right)^2,\\
C &= \frac{4}{3}\frac{\mu^{(1)}A}{\sqrt{m_p\tilde{R}T^{(1)}(0)}}.
\end{aligned} \tag{7.5.23}$$

This corresponds exactly to Eqs. (6–8) from the paper by Crosignani et al. (1995) or Eqs. (8–10) from Gislason [6] with one exception: In our derivation the shear viscosities were constant whereas in the cited papers a square root dependence on temperature was assumed. This is basically due to the aforementioned strong linkage of their derivation to the kinetic theory of gases, which also predicts the square root dependence of viscosity on temperature.

If these equations are solved numerically the results are very similar to those shown in Fig. 7.5, provided the friction due to viscosity is strong enough. It should also be noted that if $\overline{T}^{(1)}$ goes down $\overline{T}^{(2)}$ goes up and *vice versa*: If the gas is compressed in one of the cylinders the gas in the other cylinder must expand. This in turn reminds us of an air pump, which gets warmer if compressed. Typically the latter behavior is described with the adiabatic equation. Hence similarly to Eq. (7.5.16) we obtain:

$$\frac{\mathrm{d}^2\bar{l}}{\mathrm{d}\bar{t}^2} = \frac{\nu^{(1)}}{\bar{l}(0)}\left(\frac{\bar{l}}{\bar{l}(0)}\right)^{-\kappa} - \frac{\nu^{(2)}}{1-\bar{l}(0)}\frac{\overline{T}^{(2)}(0)}{\overline{T}^{(1)}(0)}\left(\frac{1-\bar{l}}{1-\bar{l}(0)}\right)^{-\kappa}. \tag{7.5.24}$$

As in Eq. (7.5.17) the temperatures follow from algebraic equations:

$$\overline{T}^{(1)} = \left(\frac{\bar{l}}{\bar{l}(0)}\right)^{1-\kappa},\quad \overline{T}^{(2)} = \frac{\overline{T}^{(2)}(0)}{\overline{T}^{(1)}(0)}\left(\frac{1-\bar{l}}{1-\bar{l}(0)}\right)^{1-\kappa}. \tag{7.5.25}$$

And as in the previous case we observe at small shear viscosities an almost undamped periodic movement as shown in Fig. 7.6, and the differences between the numerical solution of Eq. (7.5.23) and the adiabatic one given by Eqs. (7.5.24) and (7.5.25) are hardly discernable.

Thus continuum theory can provide a happy ending to the otherwise never ending story of the internal adiabatic wall.

## 7.6 The Heat Conduction Equation

We start by deriving the heat conduction equation for a gas (at rest). To this end we start from the First Law in local regular form according to Eq. (6.5.5) in Cartesian coordinates:

$$\underset{(x)}{\rho}\frac{\mathrm{d}\underset{(x)}{u}_i}{\mathrm{d}t} = -\frac{\partial \underset{(x)}{q}_i}{\partial x_i} + \underset{(x)}{\sigma}_{ji}\frac{\partial \underset{(x)}{\upsilon}_i}{\partial x_j} + \underset{(x)}{\rho}\,\underset{(x)}{r}, \tag{7.6.1}$$

We now observe the constitutive equation for the internal energy of an ideal gas, Eq. (6.5.3), FOURIER's law (6.6.1), the constitutive equation for a frictionless fluid (6.3.2), and the balance of mass, Eq. (6.5.7):

$$\underset{(x)}{\rho}\, c_\upsilon \frac{\mathrm{d}\underset{(x)}{T}}{\mathrm{d}t} = \underset{(x)}{\kappa}\frac{\partial^2 \underset{(x)}{T}}{\partial x_i^2} - \underset{(x)}{\rho}\,\underset{(x)}{p}\frac{\mathrm{d}\underset{(x)}{\upsilon}}{\mathrm{d}t} + \underset{(x)}{\rho}\,\underset{(x)}{r}\,. \tag{7.6.2}$$

If we restrict ourselves to isochoric heat conduction processes the second term after the equals sign will drop out:

$$\underset{(x)}{\rho}\, c_\upsilon \frac{\mathrm{d}\underset{(x)}{T}}{\mathrm{d}t} = \underset{(x)}{\kappa}\frac{\partial^2 \underset{(x)}{T}}{\partial x_i^2} + \underset{(x)}{\rho}\,\underset{(x)}{r}\,. \tag{7.6.3}$$

However, for isobaric processes we find by using the ideal gas law (6.4.1) and the relation $(6.5.18)_3$ for the specific heat at a constant volume:

$$\underset{(x)}{\rho}\, c_p \frac{\mathrm{d}\underset{(x)}{T}}{\mathrm{d}t} = \underset{(x)}{\kappa}\frac{\partial^2 \underset{(x)}{T}}{\partial x_i^2} + \underset{(x)}{\rho}\,\underset{(x)}{r}\,. \tag{7.6.4}$$

Consequently, the differential equation takes the same form in both cases:

$$\underset{(x)}{\rho}\, c_{\upsilon/p}\left(\frac{\partial \underset{(x)}{T}}{\partial t} + \underset{(x)}{\upsilon}_i\frac{\partial \underset{(x)}{T}}{\partial x_i}\right) = \underset{(x)}{\kappa}\frac{\partial^2 \underset{(x)}{T}}{\partial x_i^2} + \underset{(x)}{\rho}\,\underset{(x)}{r}\,. \tag{7.6.5}$$

Frequently the gas is at rest so the second term on the left side vanishes. This is the classical equation of heat conduction, namely a *parabolic* partial differential equation for the temperature field. Parabolic equations lead to an artifact: They predict an infinitely fast propagation of disturbances. This means that a temperature peak applied to the center of a large beam (the initial condition) could immediately be felt everywhere in space independently of the beam size. This is a consequence of the constitutive laws, in particular of FOURIER's law. However, the effect is irrelevant for most engineering applications and the heat conduction Eq. (7.6.5) serves well as a field equation for temperature.

Transforming the heat conduction equation to other arbitrary coordinate systems is easy. Recall that material time derivatives of scalars (in the present case of temperature) can directly be transferred to arbitrary curvilinear coordinate systems. Moreover, the LAPLACE operator of the (scalar) temperature can directly be read off from Eq. (4.2.17). Hence:

$$\underset{(z)}{\rho}\, c_{p/v} \frac{\mathrm{d}\underset{(z)}{T}}{\mathrm{d}t} = \underset{(z)}{\kappa} \left( g^{nm} \frac{\partial \underset{(z)}{T}}{\partial z^m} \right)_{;n} + \underset{(z)}{\rho}\, \underset{(z)}{r} \,. \tag{7.6.6}$$

The derivation of a heat conduction equation for solids is more complex. We need to refer to the concept of entropy, which will be discussed more extensively in Chap. 12. In fact we assume that the internal energy, the specific entropy, as well as the stress tensor are functions of temperature and of the linear strain tensor as has previously been assumed in Eq. (6.5.20):

$$u = \tilde{u}(T, \boldsymbol{\varepsilon}), \; s = \tilde{s}(T, \boldsymbol{\varepsilon}), \; \boldsymbol{\sigma} = \tilde{\boldsymbol{\sigma}}(T, \boldsymbol{\varepsilon}). \tag{7.6.7}$$

While keeping this in mind we start from the First Law of Eq. (6.5.5) and replace the working term by the following expression which is particularly suitable for solids:

$$\sigma_{ji} \frac{\partial \upsilon_i}{\partial x_j} \approx \sigma_{ji} \frac{\mathrm{d}\varepsilon_{ij}}{\mathrm{d}t} \,. \tag{7.6.8}$$

The result is:

$$\rho \frac{\mathrm{d}u}{\mathrm{d}t} \approx \rho_0 \left( \left. \frac{\partial u}{\partial T} \right|_{\boldsymbol{\varepsilon}} \frac{\mathrm{d}T}{\mathrm{d}t} + \left. \frac{\partial u}{\partial \varepsilon_{ij}} \right|_T \frac{\mathrm{d}\varepsilon_{ij}}{\mathrm{d}t} \right) = -\frac{\partial q_j}{\partial x_j} + \sigma_{ji} \frac{\mathrm{d}\varepsilon_{ij}}{\mathrm{d}t} \,. \tag{7.6.9}$$

Various calculations on the side are required in order to rewrite the partial derivatives: We need the GIBBS' equation for solids under small deformations. In . Chap. 12 it will result as a consequence of the entropy inequality (a.k.a. the Second Law of thermodynamics). It connects certain quantities from Eq. (7.6.8) to the material time derivative of the specific entropy:

$$T \frac{\mathrm{d}s}{\mathrm{d}t} = \frac{\mathrm{d}u}{\mathrm{d}t} - \frac{1}{\rho_0} \sigma_{ji} \frac{\mathrm{d}\varepsilon_{ij}}{\mathrm{d}t} \,. \tag{7.6.10}$$

Josiah Willard GIBBS was born on February 11, 1839 in New Haven, Connecticut, where he also died on April 28, 1903. He was a theoretical physicist and is well known for his fundamental work on thermodynamics. It is fair to say that GIBBS was probably the first American physicist of worldwide reputation. During his early years he visited the strongholds of European physics to carry his newly acquired knowledge back home to the U.S. After that he decided to stay preferably in his home town. Rumor has it that he also hated to attend scientific congresses.

By taking into account the dependencies of Eq. (7.6.7) we can expand the time derivatives by means of the chain rule and relate the resulting factors to $\mathrm{d}T/\mathrm{d}t$ and $\mathrm{d}\varepsilon_{ij}/\mathrm{d}t$:

$$T\left.\frac{\partial s}{\partial T}\right|_{\varepsilon}=\left.\frac{\partial u}{\partial T}\right|_{\varepsilon},\quad T\left.\frac{\partial s}{\partial \varepsilon_{ij}}\right|_{T}=\left.\frac{\partial u}{\partial \varepsilon_{ij}}\right|_{T}-\frac{1}{\rho_0}\sigma_{ji}. \tag{7.6.11}$$

Mutual differentiation of both equations and exchanging the sequence of derivatives yields:

$$\left.\frac{\partial s}{\partial \varepsilon_{ij}}\right|_{T}=-\frac{1}{\rho_0}\left.\frac{\partial \sigma_{ji}}{\partial T}\right|_{\varepsilon} \tag{7.6.12}$$

If this is inserted into Eq. $(7.6.11)_2$ we obtain:

$$\left.\frac{\partial u}{\partial \varepsilon_{ij}}\right|_{T}=-\frac{T}{\rho_0}\left.\frac{\partial \sigma_{ji}}{\partial T}\right|_{\varepsilon}+\frac{1}{\rho_0}\sigma_{ji}. \tag{7.6.13}$$

This relates the second partial derivative in Eq. (7.6.9) to the stress tensor (which in thermodynamic terms corresponded to the thermal equation of state for the case of gases (*cf.*, Sect. 6.4). Moreover, the first partial derivative is already known: According to Eq. (Sect. 6.5.21) it is related to the specific heat of solids at a constant strain which, in principle, can be measured, see Exercise 6.5.2. And therefore:

$$\rho_0 c_\varepsilon \frac{\mathrm{d}T}{\mathrm{d}t}=-\frac{\partial q_j}{\partial x_j}+T\left.\frac{\partial \sigma_{ji}}{\partial T}\right|_{\varepsilon}\frac{\mathrm{d}\varepsilon_{ij}}{\mathrm{d}t}. \tag{7.6.14}$$

If we now apply FOURIER's law (6.6.1) the heat conduction equation for solids results in a form similar to Eq. (7.6.5):

$$\rho_0 c_\varepsilon\left(\frac{\partial T}{\partial t}+\upsilon_i\frac{\partial T}{\partial x_i}\right)=\kappa\frac{\partial^2 T}{\partial x_i^2}+T\left.\frac{\partial \sigma_{ji}}{\partial T}\right|_{\varepsilon}\frac{\mathrm{d}\varepsilon_{ij}}{\mathrm{d}t} \tag{7.6.15}$$

For a solid at rest we must choose $\upsilon_i = 0$. If we then insert HOOKE's law in the form (6.2.6) we find:

$$\rho_0 c_\varepsilon\frac{\partial T}{\partial t}=\kappa\frac{\partial^2 T}{\partial x_i^2}-3k\alpha T\frac{\mathrm{d}\varepsilon_{kk}}{\mathrm{d}t}. \tag{7.6.16}$$

**Exercise 7.6.1: *Heat conduction through a wall***

A wall of total surface $A$ with thickness $d$ separates a warm room of temperature $T_\mathrm{r}$ from frigid external air at a temperature $T_\mathrm{a}$. Show that the temperature profile within the wall is linear:

$$T(x_1) = (T_\mathrm{r} - T_\mathrm{a})\frac{x_1}{d} + T_\mathrm{a}. \tag{7.6.17}$$

To this end start from the First Law of thermodynamics and explain first why:

$$\frac{\partial \underset{(x)}{q}_i}{\partial x_i} = 0. \tag{7.6.18}$$

Then comment on the following semi-inverse *ansatz*:

$$\underset{(x)}{q}_i = \left( \underset{(x)}{q}_1(x_1) \ , \ 0 \ , \ 0 \right) \tag{7.6.19}$$

and use it in combination with FOURIER's law of heat conduction (6.6.1) to derive a differential equation for $T(x_1)$. Integrate and adjust the constants of integration by suitable boundary conditions. Calculate $\underset{(x)}{q}_1(x_1)$ and depict $\boldsymbol{q}$ in a schematic of the situation (direction?). Calculate the total heat loss of the room per unit time.

**Exercise 7.6.2: *Temperature distribution around a hot wire***

Consider a very long, straight, electric wire of circular cross-section in which JOULE heating is released due to dissipation of the electric current. The wire releases the heat flux $q_0$ at its surface $r = r_0$ to the surrounding air. In order to guarantee stationary conditions a metallic cylinder surrounding the wire axisymmetrically at $r = r_\mathrm{c}$ is used as a heat sink: The cylinder is kept at a constant temperature $T_\mathrm{c}$. Use the local stationary balance for the internal energy in cylindrical coordinates and solve it in combination with a suitable semi-inverse *ansatz* for the heat flux vector. Recall in this context the equations for the divergence of a vector field, (4.2.15) and (5.2.2). Rewrite FOURIER's law (6.6.3) in physical cylindrical coordinates and calculate the temperature distribution around the wire within $r_0 \leq r \leq r_\mathrm{c}$.

What happens to the solution if $r_\mathrm{c} \to \infty$? How can the result be interpreted ? How does the solution look if the surface temperature $T_0$ of the wire is prescribed instead of the heat flux ? How are both problems related to each other?

## 7.7 Would You Like to Know More?

The concept of a control volume is highly appreciated in fluid mechanics as well as in thermodynamics, but frequently without the continuum theoretical touch. Examples can be found in Sect. 1.3 and Chap. 4 of Çengel and Boles [7] or in Baehr and Kabelac [8] in Sects. 1.2.1, 2.3.2, 3.1.6, and 3.1.7. The continuum aspects are emphasized in various examples in Müller [9], namely in Sects. 1.3.2, 1.4.6–1.4.9, 1.5.4, 1.5.6, and 1.5.7, as well as in Müller and Müller [1] in Sects. 1.5.5–1.5.9. All of the three balances are considered: mass, momentum, and energy (in its variations for the total and for the internal energy). Application of the semi-inverse method during the solution of the field equations is also subject of the two latter textbooks.

A final remark is in order: One could be tempted and assume that the parallel plate flow from Exercise 7.4.1 occurs *isothermally* and the fluid has the same *constant* temperature everywhere. However, isothermal conditions are in contradiction with the First Law. This is not surprising since there is friction due to the shear movement between the fluid layers, which increases the temperature locally! Thus heat has to be removed from the surrounding plates and this results in a stationary, height dependent temperature distribution. The details can be found in Müller and Müller [1] in Sect. 12.1.2.

## References

1. Müller I, Müller WH (2009) Fundamentals of thermodynamics and applications. Springer, Berlin
2. Gad-el-Hak M (1995) Questions in fluid mechanics Stokes' hypothesis for a Newtonian isotropic fluid. J Fluids Eng 117:3–5
3. Dellar PJ (2001) Bulk and shear viscosities in lattice Boltzmann equations. Phys Rev E 64:031203 (11 pages)
4. Ziegler F (1995) Mechanics of solids and fluids, 2nd edn. Springer, New York
5. Crosignani B, Di Porto P, Segev M (1996) Approach to thermal equilibrium in a system with adiabatic constraints. Am J Phys 64:610–613
6. Gislason EA (2010) A close examination of the motion of an adiabatic piston. Am J Phys 78(10):995–1001
7. Çengel YA, Boles MA (1998) Thermodynamics: an engineering approach, 6th edn. McGraw Hill, Boston
8. Baehr HD, Kabelac S (2009) Thermodynamik: Grundlagen und technische Anwendungen, 14th edn. Springer, Dordrecht
9. Müller I (1994) Grundzüge der Thermodynamik mit historischen Anmerkungen, 1st edn. Springer, Berlin
10. Callen HB (1960) Thermodynamics: an introduction to the physical theories of equilibrium thermostatics and irreversible thermodynamics. Wiley, New York
11. Callen HB (1985) Thermodynamics and an introduction to thermostatistics, 2nd edn. Wiley, Singapore

# Chapter 8
# Observers and Frames of Reference in Classical Continuum Theory

**Abstract** This chapter is dedicated to a relatively abstract question: Do balance laws and constitutive equations keep their form if we switch from an observer at rest to an arbitrarily moving one? In mathematical terms this problem can be analyzed by using so-called Euclidean transformations and establish an almost philosophical principle according to which true laws of nature must keep their form, independently of the frame of reference.

*Here's looking at you, kid!*

Humphrey Bogart in Casablanca

## 8.1 Introduction

In Chap. 5 we have already extensively dealt with the question how to rewrite equations of balance in arbitrary, curvilinear coordinate systems. In conclusion we may say that the balances have the same form in all coordinate system. Additional coordinate dependent quantities do not occur during a change of coordinate system. In Eq. (3.7.3) we have already stated the generic most general form of a balance: It contains a local time derivative of the quantity to-be-balanced, a convective and a non-convective flux, a supply and a production term. Indeed, this structure stays the same, no matter which coordinate system is being used: If for practical reasons we wish to obtain the equations of balance in a non-Cartesian coordinate system, we transform each term, which is easily written down in Cartesian coordinates, from the Cartesian system onto a curvilinear one. This transition can be made explicit by using the co-/contravariant equations for mass, momentum, and energy from Chap. 5. Note again that during the transformation process no additional quantities of a coordinate dependent nature arise, which otherwise would need to be interpreted, for example, as an additional supply, which only occurs because of the choice of special coordinates.

W. H. Müller, *An Expedition to Continuum Theory*, Solid Mechanics and Its Applications 210, DOI: 10.1007/978-94-007-7799-6_8,

In fact, this is not too surprising since a "decent" law of nature must not depend on the choice of coordinate system. First, its form should be invariant against a change of coordinates and, second, a change of coordinates must not result in the creation of new fluxes, supplies, or productions that only exist in certain coordinate systems. Moreover, the same remark holds analogously for "decent" constitutive equations.

Note that so far all of our changes of coordinate systems have one thing in common: We transform the spatial positions, but in a *time independent* manner. We are going to change that in this chapter. The coordinate system will now be switched in a *time-dependent* manner. In other words, the new coordinate system is going to "move" w.r.t. the old one. In this context it is customary to refer to a *change of observers*. The old coordinate system is a.k.a. the reference frame and the new one as the moving frame. As we shall see this is going to bring in a whole new quality to our former arguments. In fact, it will become immediately clear that this is due to the time derivative in Eq. (3.7.3), which so far could be ignored completely!

## 8.2 Euclidean Transformations

Initially we want to assume naively that the balances for mass, momentum, and energy were initially verified experimentally by an observer at rest. Such a resting frame of reference is also known as a *Galilean inertial system*. We want to answer the question if and how the balance laws are affected if we change from the inertial system to an "arbitrarily" moving one. The most general change of observer within the framework of classical physics allows that, first, the origins of both systems separate at an arbitrary velocity. Second, it is possible that the coordinate axes of both systems turn against each other at an arbitrary speed or rather angular velocity.

Galileo GALILEI was born on February 15, 1564 in Pisa and he died on January 8, 1642 in Arcetri near Florence. He introduces and establishes a new way of thinking in science: In the end it is the experiment that decides if a (mathematical) theory is right or wrong. It is thus not too surprising that Galileo carries the nimbus of a revolutionary. We have all heard the story that he was one of the first scholars who preferred to write down his ideas in his mother tongue (Italian) instead of using the twisted Latin language, for the people, so to speak, like a Martin Luther of science. He is also famous for his manly appearance when dragged before the court of the inquisition. Wishful thinking or really the truth? Be it as it may, we all love heroes and drama, especially when something as sterile as science is concerned.

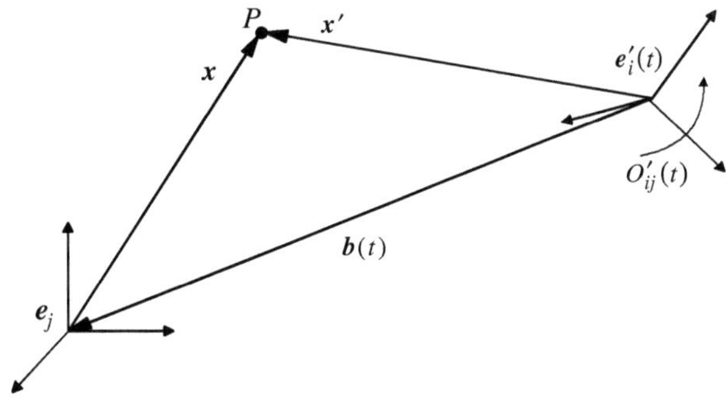


**Fig. 8.1** Schematic of a Euclidean transformation

Now these notions need to be converted into mathematical terms. The Cartesian unit base of the resting, hence time-independent inertial system is denoted by $\boldsymbol{e}_j$. It represents an observer who is doing measurements within the *inertial frame*. In contrast to that the moving, i.e., time dependent Cartesian unit base $\boldsymbol{e}'_i(t)$ represents the so-called *Euclidean observer*. In the simplest case both fixate a position $P$ in space *at the same time*. This point could be, for example, a material point. The moving coordinate frame can be (rigidly) attached to a moving material point but is does not have to be. From Fig. 8.1 we conclude:

$$\boldsymbol{x}' = \boldsymbol{x} + \boldsymbol{b} \Rightarrow x'_i(t)\boldsymbol{e}'_i(t) = x_j(t)\boldsymbol{e}_j + b'_i(t)\boldsymbol{e}'_i(t). \tag{8.2.1}$$

Note that, in general, the dashes at the vector components refer to the moving frame. In this spirit we should have written $x''_i(t)$ instead of $x'_i(t)$. However, since we will never represent $\boldsymbol{x}'$ in the inertial base $\boldsymbol{e}_j$ we avoid the clumsy looking double dash, which is customary in the pertinent literature. By scalar multiplication with $\boldsymbol{e}'_i(t)$ the following relation in Cartesian coordinates results:

$$x'_i = O'_{ij}(t)x_j + b'_i(t) \tag{8.2.2}$$

where the so-called *rotation matrix* has been introduced:

$$O'_{ij}(t) = \boldsymbol{e}'_i(t) \cdot \boldsymbol{e}_j \tag{8.2.3}$$

or rather:

$$x_j = O'_{ij}(t)\left(x'_i - b'_i(t)\right), \tag{8.2.4}$$

since the rotation matrix is *orthogonal*:

$$O'_{ik}O'_{jk} = \delta'_{ij} = O'_{ki}O'_{kj}. \tag{8.2.5}$$

These relations are visualized in Fig. 8.1. Note that the rotation matrix also gets a dash: Eq. (8.2.2) shows that it is responsible for converting the Cartesian components $x_j$ of the inertial frame into components of the moving system with the dashed base (summation w.r.t. the second index). However, in Eq. (8.2.4) it is used to transform components of the moving frame (see the two terms shown within the parentheses) to quantities of the inertial system (summation w.r.t. the first index). Alternatively to the component-wise representations of Eqs. (8.2.2) and (8.2.4) we may also write:

$$\boldsymbol{x}' = \boldsymbol{O}' \cdot \boldsymbol{x} + \boldsymbol{b}' \Leftrightarrow \boldsymbol{x} = \boldsymbol{O}'^{\mathrm{T}} \cdot (\boldsymbol{x}' - \boldsymbol{b}') \tag{8.2.6}$$

where:

$$\boldsymbol{O}' \cdot \boldsymbol{O}'^{T} = \mathbf{1}, \quad \boldsymbol{O}'^{T} \cdot \boldsymbol{O}' = \mathbf{1}. \tag{8.2.7}$$

On first glance Eq. (8.2.6) seems to be some other, maybe contradictory form of the vector relation $(8.2.1)_1$. However, this is only apparently so, as we shall show in Exercise 8.2.1. Some remarks are in order beforehand: In Eqs. (8.2.1/8.2.2) $\boldsymbol{x}'$ and $\boldsymbol{x}$ refer to position vectors, and $x_i'$ and $x_j$ are components of these position vectors in Cartesian coordinates and w.r.t. two different bases. In the spirit of the nomenclature introduced in Sect. 2.4 we should even write:

$$\underset{(x')}{x'}_i = \underset{(x',x)}{O'}_{ij}(t) \underset{(x)}{x}_j + \underset{(x')}{b'}_i(t). \tag{8.2.8}$$

However, this notation is undoubtedly very clumsy, maybe even redundant regarding the upper dash in the symbols $\underset{(x')}{x'}_i$, $\underset{(x',x)}{O'}_{ij}(t)$, and $\underset{(x')}{b'}_i(t)$. Moreover, we could have rewritten Eq. (8.2.1) as follows:

$$\boldsymbol{x} = \boldsymbol{x}' - \boldsymbol{b} \Rightarrow x_i(t)\boldsymbol{e}_i = x_j'(t)\boldsymbol{e}_j'(t) - b_i(t)\boldsymbol{e}_i. \tag{8.2.9}$$

Scalar multiplication with $\boldsymbol{e}_i$ then leads to the following result:

$$x_i = O_{ij}(t)x_i' - b_i(t) \tag{8.2.10}$$

where the following rotation matrix has been defined:

$$O_{ij}(t) = \boldsymbol{e}_i \cdot \boldsymbol{e}_j'(t). \tag{8.2.11}$$

By comparison with Eq. (8.2.3) we find that:

$$O_{ij} = O_{ji}', \tag{8.2.12}$$

i.e., because of Eq. (8.2.5) also:

$$O_{ik}O_{jk} = \delta_{ij} = O_{ki}O_{kj}. \tag{8.2.13}$$

Finally we conclude that in the spirit of Sect. 2.4 we should have written for Eq. (8.2.10) instead:

$$\underset{(x)}{x}_i(t) = \underset{(x',x)}{O}_{ij}(t)\underset{(x')}{x'}_j(t) - \underset{(x)}{b}_i(t). \tag{8.2.14}$$

Of course the upper dash on the symbol $\underset{(x')}{x'}_j$ is in a certain sense redundant.

**Exercise 8.2.1:** ***Alternative notions of the Euclidean transformation***

Start from the vector relation $(8.2.1)_1$. Evaluate it with the orthonormal bases $\boldsymbol{e}'_i$ and $\boldsymbol{e}_k$ in order to arrive at Eq. (8.2.2). To this end it is necessary to introduce $\boldsymbol{O}'$ as a scalar product between both bases, cf., the definition (8.2.3). Consequently $\boldsymbol{O}'$ is a quantity stemming from both bases with two indices.

Recall that any vector and, in particular, unit vectors $\boldsymbol{e}'_i$ can always be represented by using the other base $\boldsymbol{e}_j$ and vice versa (linear mapping): $\boldsymbol{e}'_i = A_{ij}\boldsymbol{e}_j$ and $\boldsymbol{e}_i = B_{ij}\boldsymbol{e}'_j$. In a first step show that $A_{ij} = O'_{ij}$ and $B_{ij} = O'_{ji}$ or $B_{ij} = O_{ij}$ and $A_{ij} = O_{ji}$. Then, in a second step, show the validity of Eqs. (8.2.5, 8.2.13) and (8.2.12).

Now we shall investigate representations in curvilinear coordinates analogously to Eqs. (8.2.1–8.2.5). Use results from Sects. 2.4 and 2.5 to comment on the validity of the following equations (note the co- and contravariant forms of notation):

$$\boldsymbol{x}' = \boldsymbol{x} + \boldsymbol{b} \Rightarrow$$

$$\underset{(z')}{x'^i}\,\underset{(z')}{\boldsymbol{g}'}_i(t) = \underset{(z)}{x^j}(t)\underset{(z)}{\boldsymbol{g}}_j + \underset{(z')}{b'^i}(t)\underset{(z')}{\boldsymbol{g}'}_i(t), \quad \underset{(z')}{x'}_i\,\underset{(z')}{\boldsymbol{g}'^i}(t) = \underset{(z)}{x}_j(t)\underset{(z)}{\boldsymbol{g}^j} + \underset{(z')}{b'}_i(t)\underset{(z')}{\boldsymbol{g}'^i}(t). \tag{8.2.15}$$

Use Eq. (2.5.8) and show that:

$$\underset{(z')}{x'}^i = \underset{(z',z)}{O'^i}_j(t)\,\underset{(z)}{x}^j(t) + \underset{(z')}{b'}^i(t), \quad \underset{(z')}{x'}_i = \underset{(z',z)}{O'}_i{}^j(t)\,\underset{(z)}{x}_j(t) + \underset{(z')}{b'}_i(t) \tag{8.2.16}$$

where the mixed co-/contravariant forms of the rotation matrix are defined as follows:

$$\underset{(z',z)}{O'^i}_j V = \underset{(z')}{\boldsymbol{g}'}^i(t)\cdot\underset{(z)}{\boldsymbol{g}}_j \quad \underset{(z',z)}{O'}_i{}^j = \underset{(z')}{\boldsymbol{g}'}_i(t)\cdot\underset{(z)}{\boldsymbol{g}}^j \tag{8.2.17}$$

Now derive the following relations:

$$\underset{(z')}{x'}_i = \underset{(z',z)}{O'}_{ij}\underset{(z)}{x}^j + \underset{(z')}{b'}_i \quad \underset{(z)}{x}^i = \underset{(z',z)}{O'}^{ji}\left(\underset{(z')}{x'}_j - \underset{(z')}{b'}_j\right) \tag{8.2.18}$$

where:

$$\underset{(z',z)}{O'}_{ij} = \underset{(z')}{\boldsymbol{g}'}_i(t)\cdot\underset{(z)}{\boldsymbol{g}}_j \quad \underset{(z',z)}{O'}^{ij} = \underset{(z')}{\boldsymbol{g}'}^i(t)\cdot\underset{(z)}{\boldsymbol{g}}^j \tag{8.2.19}$$

Which metric tensor must be used in order to convert $\underset{(z',z)}{O'}{}_{ij}$ into $\underset{(z',z)}{O'}{}^{i}{}_{j}$ or $\underset{(z',z)}{O'}{}_{i}{}^{j}$ (same question for $\underset{(z',z)}{O'}{}^{ij}$)? Discuss the pros and cons of all the different forms. Can the upper dash in the symbols $\underset{(z',z)}{O'}{}_{ij}$, $\underset{(z')}{b'}{}_{i}$, $\underset{(z')}{x'}{}_{i}$ be omitted? Show the validity of the rules of transformation such as:

$$\underset{(z',z)}{O'}{}_{ij} = \frac{\partial x'_k}{\partial z'^i}\frac{\partial x_l}{\partial z^j}\underset{(x',x)}{O'}{}_{kl}, \quad \underset{(x',x)}{O'}{}_{ij} \equiv O'_{ij}, \tag{8.2.20}$$

$\underset{(z')}{b'}{}_{i} = \frac{\partial x'_l}{\partial z'^i}\underset{(x')}{b'}{}_{l}$, $\underset{(x')}{b'}{}_{l} \equiv b'_l$, $\underset{(z)}{x}{}^{j} = \frac{\partial z^j}{\partial x_l}\underset{(x)}{x}{}_{l}$, $\underset{(x)}{x}{}_{l} \equiv x_l$, and so on.

Does this make $\boldsymbol{O}'$ a tensor of second order in the sense of Eq. (2.4.15)? And, finally, rewrite Eqs. (8.2.10–8.2.14) in curvilinear coordinates.

Now turn to the relations (8.2.6/8.2.7) in invariant notation. Why is there no dash on the distance vector $\boldsymbol{b}$? In the same context comment on the formula

$$\boldsymbol{e}'_i = \boldsymbol{O}' \cdot \boldsymbol{e}_j \tag{8.2.21}$$

and on the following statement translated from the book by Greve [8]: "The apparent contradiction between the equations (Note: He is referring to Eqs. $(8.2.1)_1$ and $(8.2.6)_1$) can be explained that on the one hand side the equations refer to different objects, namely to the vectors themselves as geometric objects, and on the other hand side to their representation in different coordinate systems in form of number triples."

In future calculations we will make use of the following useful relations that follow from Eqs. (8.2.2) and (8.2.4):

$$\frac{\partial x'_i}{\partial x_j} = O'_{ij}, \quad \frac{\partial x_i}{\partial x'_j} = O'_{ji}. \tag{8.2.22}$$

Note that Eq. (8.2.2) can also be applied to the case when switching from one inertial system to another. Such a change of observer is referred to as a GALILEAN *transformation* in classical continuum theory. It is characterized by a *time-independent* rotation matrix. Moreover, both origins of the bases move *uniformly* away from each other in a *straight* direction. We write:

$$x'_i = O'_{ij}x_j + V'_i\,t \quad \text{where} \quad O'_{ij}, V'_i = \text{const.}_t. \tag{8.2.23}$$

How to find out experimentally as to whether we are based in an inertial, a Galilean, or in a Euclidean frame, will be discussed in Sect. 8.5 when dealing with the behavior of the balance of momentum during Euclidean transformations. Finally note that the quantities referring to the Galilean system of this section can be eliminated so that relations for the positions as determined from *two* Euclidean observers can be derived. They look completely alike to Eqs. (8.2.2) or (8.2.10).

## 8.3 Objective Tensors and Kinematic Applications

We define a so-called *Euclidean tensor* of arbitrary order as follows: A quantity $\boldsymbol{A}$, which possesses w.r.t. the original Cartesian inertial system the components $A_{j_1 j_2 \ldots j_l}$ and w.r.t. the moving frame the Cartesian components $A'_{i_1 i_2 \ldots i_l}$ where

$$A'_{i_1 i_2 \ldots i_l} = (\det \boldsymbol{O}')^p O'_{i_1 j_1} O'_{i_2 j_2} \ldots O'_{i_l j_l} A_{j_1 j_2 \ldots j_l} \tag{8.3.1}$$

is called a *tensor* of $l^{\text{th}}$ order w.r.t. *Euclidean transformations*, and is a.k.a. an *objective tensor of* $l^{\text{th}}$ *order, if* $p = 0$.[1] If $p = 1$ it is customary to refer to it as *axial* tensor. Note that if Eq. $(8.2.22)_1$ is observed we may also write

$$A'_{i_1 i_2 \ldots i_l} = \left(\det \frac{\partial x'_i}{\partial x_j}\right)^p \frac{\partial x'_{i_1}}{\partial x_{j_1}} \frac{\partial x'_{i_2}}{\partial x_{j_2}} \cdots \frac{\partial x'_{i_l}}{\partial x_{j_l}} A_{j_1 j_2 \ldots j_l} \tag{8.3.2}$$

instead of Eq. (8.3.1).

**Exercise 8.3.1: *Alternative forms of the Euclidean tensor***

Discuss and show the validity of the following definitions of a Euclidean tensor of $l$th order by starting from Eq. (8.3.2) and with the results from Exercise 8.2.1:

$$\begin{aligned} &\boldsymbol{A}' = \boldsymbol{A} \Leftrightarrow A'_{i_1 i_2 \ldots i_l} = \boldsymbol{e}'_{i_1} \cdot \boldsymbol{e}_{j_1} \boldsymbol{e}'_{i_2} \cdot \boldsymbol{e}_{j_2} \ldots \boldsymbol{e}'_{i_l} \cdot \boldsymbol{e}_{j_l} A_{j_1 j_2 \ldots j_l} \Leftrightarrow \\ &\underset{(z')}{A}^{i_1 i_2 \ldots i_l} = \underset{(z')}{\boldsymbol{g}}^{\prime i_1} \cdot \underset{(z)}{\boldsymbol{g}}^{j_1} \underset{(z')}{\boldsymbol{g}}^{\prime i_2} \cdot \underset{(z)}{\boldsymbol{g}}^{j_2} \ldots \underset{(z')}{\boldsymbol{g}}^{\prime i_l} \cdot \underset{(z)}{\boldsymbol{g}}^{j_l} \underset{(z)}{A}_{j_1 j_2 \ldots j_l}. \end{aligned} \tag{8.3.3}$$

In this context also recall the following vector representations:

$$\begin{aligned} &\boldsymbol{A}' = \boldsymbol{e}'_{i_1} \boldsymbol{e}'_{i_2} \cdots \boldsymbol{e}'_{i_l} A'_{i_1 i_2 \cdots i_l}, \qquad \boldsymbol{A} = \boldsymbol{e}_{j_1} \boldsymbol{e}_{j_2} \cdots \boldsymbol{e}_{j_l} A_{j_1 j_2 \cdots j_l}, \\ &\boldsymbol{A}' = \underset{(z')}{\boldsymbol{g}'}_{i_1} \underset{(z')}{\boldsymbol{g}'}_{i_2} \cdots \underset{(z')}{\boldsymbol{g}'}_{i_l} \underset{(z')}{A}^{i_1 i_2 \cdots i_l}, \quad \boldsymbol{A} = \underset{(z)}{\boldsymbol{g}}^{j_1} \underset{(z)}{\boldsymbol{g}}^{j_2} \cdots \underset{(z)}{\boldsymbol{g}}^{j_l} \underset{(z)}{A}_{j_1 j_2 \cdots j_l}. \end{aligned} \tag{8.3.4}$$

[1] For the determinant of Eq. (8.3.1) we could simply write $\det(\boldsymbol{O}') = \pm 1$. This is a consequence of Eqs. (8.2.5/8.2.7), which follows from the fact that the inverse of a rotation matrix is given by its transposed. However, there are didactic reasons why we do not write it that way in Eq. (8.3.1): In Chap. 13 we will introduce so-called world tensors in complete analogy to Eq. (8.3.1/8.3.2). However, in contrast to the Euclidean transforms the corresponding world transforms are not normalized.

Discuss possible variations of these representations, in particular mixed co-/contravariant forms. Derive the following transformation rules (as well as other co-/contravariant derivates that are not explicitly mentioned here) by successive application of the chain rule:

$$A'_{i_1 i_2 \ldots i_l} = \frac{\partial x'_{i_1}}{\partial x_{j_1}} \frac{\partial x'_{i_2}}{\partial x_{j_2}} \cdots \frac{\partial x'_{i_l}}{\partial x_{j_l}} A_{j_1 j_2 \ldots j_l},$$
$$\underset{(z')}{A}{}'^{i_1 i_2 \ldots i_l} = \frac{\partial z'^{i_1}}{\partial x'_{i_1}} \frac{\partial z'^{i_2}}{\partial x'_{i_2}} \cdots \frac{\partial z'^{i_l}}{\partial x'_{i_l}} \underset{(x')}{A'}_{i_1 i_2 \ldots i_l}? \tag{8.3.5}$$

In what way are these relations different or in agreement when compared to Eqs. (2.4.15), say? Do relations of the following kind hold [recall in this context in particular Eqs. (8.2.2) and (8.2.16)]:

$$\underset{(x',x)}{O'}{}_{ij} = \frac{\partial x'_i}{\partial x_j}, \quad \underset{(z',z)}{O'}{}^i{}_j = \frac{\partial z'^i}{\partial z^j}, \quad , \underset{(z',z)}{O'}{}_i{}^j = \frac{\partial x'_k}{\partial z'^i} \frac{\partial z^j}{\partial x_l} \underset{(x',x)}{O'}{}_{kl},$$
$$\underset{(z')}{A}{}'^{i_1 i_2 \cdots i_l} = \frac{\partial z'^{i_1}}{\partial z^{j_1}} \frac{\partial z'^{i_2}}{\partial z^{j_2}} \cdots \frac{\partial z'^{i_l}}{\partial z^{j_l}} \underset{(z)}{A}{}^{j_1 j_2 \cdots j_l}? \tag{8.3.6}$$

As to whether a quantity with indices is objective or not, must be decided on an individual basis and should be motivated by physically based arguments. Geometric objects are comparatively easy to tackle. For example, the position vector is obviously not objective since it follows the rule for Euclidean transformations:

$$x'_i = O'_{ij} x_j + b'_i. \tag{8.3.7}$$

In contrast to that the distance vector $\boldsymbol{\delta}$ between two (material and simultaneous) points of space, $\overset{(1)}{P}$ and $\overset{(2)}{P}$, is objective as we can see easily:

$$\delta_i = \overset{(2)}{x}{}_i - \overset{(1)}{x}{}_i \Rightarrow \delta'_i = \overset{(2)}{x'}{}_i - \overset{(1)}{x'}{}_i = O'_{ij}\left(\overset{(2)}{x}{}_j - \overset{(1)}{x}{}_j\right) = O'_{ij}\delta_j \tag{8.3.8}$$

The transformation properties of the displacement of a material point follow in a rather complex manner. On the one hand side we have:

$$u_i = x_i - X_i \quad \text{and} \quad u'_i = x'_i - X'_i. \tag{8.3.9}$$

On the other hand side we have to observe Eq. (8.3.7) for the current positions. Analogously we have to write for the reference positions:

$$X'_i = O'_{ij}(t_0) X_j + b'_i(t_0), \tag{8.3.10}$$

where:

$$X'_i = x'_i(t_0), \quad X_j = x_j(t_0). \tag{8.3.11}$$

Hence we find that:

$$u'_i = O'_{ij}(t)u_j + \left[O'_{ij}(t) - O'_{ij}(t_0)\right]x_j(t_0) + b'_i(t) - b'_i(t_0). \tag{8.3.12}$$

Thus we conclude that the displacement vector is by no means objective. However, the gradient of the displacement is an objective quantity. For a proof we apply the chain rule to the last equation:

$$\frac{\partial u'_i}{\partial x'_j} = \frac{\partial u'_i}{\partial u_k}\frac{\partial u_k}{\partial x_l}\frac{\partial x_l}{\partial x'_j} = O'_{ik}O'_{jl}\frac{\partial u_k}{\partial x_l}. \tag{8.3.13}$$

It follows that the (linear) strain tensor is also objective:

$$\begin{aligned} \varepsilon'_{ij} &= \frac{1}{2}\left(\frac{\partial u'_i}{\partial x'_j} + \frac{\partial u'_j}{\partial x'_i}\right) = \frac{1}{2}\left(O'_{ik}O'_{jl}\frac{\partial u_k}{\partial x_l} + O'_{jk}O'_{il}\frac{\partial u_k}{\partial x_l}\right) \\ &= \frac{1}{2}O'_{ik}O'_{jl}\left(\frac{\partial u_k}{\partial x_l} + \frac{\partial u_l}{\partial x_k}\right) = O'_{ik}O'_{jl}\varepsilon_{kl}. \end{aligned} \tag{8.3.14}$$

This is a very important result because the strain tensor is one of the main "ingredients" in the constitutive equation for the stress tensor, in other words HOOKE's law as Eqs. (6.2.1) or (6.2.6) show. Just like balance equations good constitutive equations should be objective as well, in other words, they should be independent of the observer.

In order to examine the transformation properties of the velocity vector we note that:

$$\upsilon_i \equiv \dot{x}_i \Rightarrow \dot{x}'_i = \dot{O}'_{ij}x_j + O'_{ij}\dot{x}_j + \dot{b}'_i. \tag{8.3.15}$$

Moreover:

$$O'_{ik}O'_{jk} = \delta'_{ij} \Rightarrow \dot{O}'_{ik}O'_{jk} + O'_{ik}\dot{O}'_{jk} = 0. \tag{8.3.16}$$

We now define the *matrix of angular velocities* by:

$$\Omega'_{ij} = \dot{O}'_{ik}O'_{jk}. \tag{8.3.17}$$

If we multiply $\Omega'_{ij}$ by $O'_{jr}$ the following relation results:

$$\begin{aligned} &\Omega'_{ij}O'_{jr} = \dot{O}'_{ik}O'_{jk}O'_{jr} = \dot{O}'_{ik}\delta'_{kr} = \dot{O}'_{ir} \\ &\Rightarrow \upsilon'_i = \Omega'_{ir}O'_{rj}x_j + O'_{ij}\upsilon_j + \dot{b}'_i = O'_{ij}\upsilon_j + \Omega'_{ir}\left(x'_r - b'_r\right) + \dot{b}'_i. \end{aligned} \tag{8.3.18}$$

Consequently *velocity is not an objective vector*. This is not too surprising since the position and the displacement vector were not objective either. Suggestively speaking, velocity is nothing else but a momentary change of position, in other words a “small” displacement per unit time. In classical physics time is an objective scalar:

$$t' = t, \tag{8.3.19}$$

but a change of position is *not*.

Moreover, Eq. $(8.3.18)_2$ deserves another comment: If it were only for the first term after the equals sign, velocity would be an objective vector, too. $\upsilon_j$ denotes the components of the velocity of the point $P$ w.r.t. the Cartesian base of an inertial system. By means of the matrix $O'_{ij}$ these components are turned onto the coordinate system of the Euclidean observer, so to speak. Correspondingly, $O'_{rj}x_j \equiv x'_r - b'_r$ are the components of the position vector $\boldsymbol{x}$ in the Euclidean system. The quantity $\Omega'_{ir}$ is characteristic for a non-inertial system, and this is why we add a dash to the symbol. The matrix $\Omega'_{ir}$ vanishes for transformations between inertial systems. This becomes evident by looking at the Galilean transformation (8.2.23): Due to the time-independence of the rotation matrix the matrix of angular velocities $\Omega'_{ir}$ vanishes by its very definition for a Galilean observer: Eq. (8.3.17). The matrix of angular velocities is a 3 × 3 matrix. However, it is not completely populated and contains only three independent components because $\Omega'_{ir}$ is anti-symmetric, as can be seen from Eq. $(8.3.16)_2$:

$$\dot{O}'_{ik}O'_{jk} = -\dot{O}'_{jk}O'_{ik} \Rightarrow \Omega'_{ij} = -\Omega'_{ji}. \tag{8.3.20}$$

Since only three independent components exist, it becomes possible to define a *vector of angular velocity* $\boldsymbol{\omega}$ instead of a matrix. This is realized by means of the so-called *Levi-Civita symbol* $\boldsymbol{\epsilon}$, an axial tensor of third order, which will be introduced in the following exercise:

$$\Omega'_{ij} = \epsilon'_{ijk}\omega'_k \Leftrightarrow \omega'_i = \frac{1}{2}\epsilon'_{ijk}\Omega'_{jk} \quad \text{with} \quad \Omega'_{ij} = \begin{bmatrix} 0 & \omega'_3 & -\omega'_2 \\ -\omega'_3 & 0 & \omega'_1 \\ \omega'_2 & -\omega'_1 & 0 \end{bmatrix}. \tag{8.3.21}$$

Tullio Levi-Civita was born on March 29, 1873 in Padua and died on December 29, 1941 in Rome. He was a mathematician down to the very bottom of his heart. However, first he accepted a professorship for mechanics in Padua in 1898 and obtained a chair for mathematics in Rome much later in 1918. Being Jewish the fascists kicked him out of office in 1938. We owe him important contributions to tensor analysis. Rumor has it that he was the “inventor” of the covariant derivative. It is certain though, that he provided the mathematical framework which allowed Albert Einstein to create his theory of general relativity.

**Exercise 8.3.2:** ***The vector product revisited***

Consider the vector product between two vectors $\boldsymbol{a}$ and $\boldsymbol{b}$:

$$\boldsymbol{c} = \boldsymbol{a} \times \boldsymbol{b}, \tag{8.3.22}$$

and recall that the direction of the resulting vector $\boldsymbol{c}$ follows from the *right-hand-rule*: If we turn with the right hand $\boldsymbol{a}$ onto $\boldsymbol{b}$, the thumb will point in direction of $\boldsymbol{c}$. Thus, $\boldsymbol{c}$ is perpendicular to $\boldsymbol{a}$ and $\boldsymbol{b}$, and its length is given by:

$$|\boldsymbol{c}| = |\boldsymbol{a}| \times |\boldsymbol{b}| \cos(\alpha), \tag{8.3.23}$$

where $\alpha$ denotes the angle between $\boldsymbol{a}$ and $\boldsymbol{b}$. Show that we may write in Cartesian components:

$$\boldsymbol{c} = \begin{vmatrix} \boldsymbol{e}_1 & \boldsymbol{e}_2 & \boldsymbol{e}_3 \\ a_1 & a_2 & a_3 \\ b_1 & b_2 & b_3 \end{vmatrix} \equiv \epsilon_{ijk}\boldsymbol{e}_i a_j b_k \Rightarrow c_i = \epsilon_{ijk} a_j b_k \tag{8.3.24}$$

by using the LEVI-CEVITA *symbol* (a.k.a. totally antisymmetric tensor of third order):

$$\epsilon_{ijk} = \begin{cases} +1 & \\ -1 & \text{if} \\ 0 & \end{cases} \begin{matrix} i,j,k = 1,2,3 \text{ and cyclic permutations} \\ i,j,k = 2,1,3 \text{ and cyclic permutations} \\ \text{else.} \end{matrix} \tag{8.3.25}$$

Now consider a reflection about the (2,3) plane:

$$\boldsymbol{e}'_1 = -\boldsymbol{e}_1, \quad \boldsymbol{e}'_2 = \boldsymbol{e}_2, \quad \boldsymbol{e}'_3 = \boldsymbol{e}_3 \tag{8.3.26}$$

and calculate the corresponding transformation matrix $O'_{ij}$ according to the rule (8.2.3) as well as $\det(\boldsymbol{O}')$. In particular compute $\epsilon'_{123}$ and show that only if we require the LEVI-CEVITA symbol to behave like an axial tensor of third order, i.e.,

$$\epsilon'_{ijk} = \det(\boldsymbol{O}') O'_{ir} O'_{js} O'_{kt} \epsilon_{rst}, \tag{8.3.27}$$

it is guaranteed that

$$\epsilon'_{ijk} = \begin{cases} +1 & \\ -1 & \text{if} \\ 0 & \end{cases} \begin{matrix} i,j,k = 1,2,3 \text{ and cyclic permutations} \\ i,j,k = 2,1,3 \text{ and cyclic permutations} \\ \text{else} \end{matrix} \tag{8.3.28}$$

in the transformed system as well. Moreover, consider the following two vectors:

$$\boldsymbol{a} = a\,\boldsymbol{e}_1, \quad \boldsymbol{b} = b\,\boldsymbol{e}_2 \tag{8.3.29}$$

and show by using Eq. (8.3.24) that:

$$\boldsymbol{c} = ab\,\boldsymbol{e}_3. \tag{8.3.30}$$

Now use the transformation rules for true vectors:

$$a'_i = O'_{ij}a_j, \quad b'_i = O'_{ij}b_j, \quad c'_i = O'_{ij}c_j \tag{8.3.31}$$

and show that:

$$\boldsymbol{a}' = -a\boldsymbol{e}'_1, \quad \boldsymbol{b}' = b\boldsymbol{e}'_2, \quad \boldsymbol{c}' = ab\boldsymbol{e}'_3. \tag{8.3.32}$$

Multiply Eq. $(8.3.24)_2$ by $O'_{ri}$, apply the orthogonal conditions (8.2.5), and transform onto the dashed coordinate system by observing the relation (8.3.27) to prove that:

$$c'_r = \det(\boldsymbol{O}')^{-1}\epsilon'_{rst}a'_s b'_t. \tag{8.3.33}$$

Evaluate this relation with Eq. $(8.3.32)_{1,2}$ and show that in agreement with Eq. $(8.3.32)_3$:

$$\boldsymbol{c}' = +ab\boldsymbol{e}'_3. \tag{8.3.34}$$

Finally confirm with the results of this exercise the equations shown in (8.3.21). In particular confirm that the factor $1/2$ is correct. Use and prove in the same context the following auxiliary formula:

$$\epsilon_{ijk}\epsilon_{klm} = \delta_{il}\delta_{jm} - \delta_{im}\delta_{jl}. \tag{8.3.35}$$

By observing the anti-symmetry of $\boldsymbol{\epsilon}$ we obtain from Eq. (8.3.21):

$$\upsilon'_i = O'_{ij}\upsilon_j - \epsilon'_{irs}\omega'_r(x'_s - b'_s) + \dot{b}'_i = O'_{ij}\upsilon_j - \epsilon'_{irs}\omega'_r x'_s + \dot{b}'_i + \epsilon'_{irs}\omega'_r b'_s. \tag{8.3.36}$$

If we now wish to establish a vector relation for the velocity in analogy to the transition from Eqs. (8.2.1) to (8.2.2) we must write:

$$\boldsymbol{\upsilon}' = \boldsymbol{\upsilon} - \boldsymbol{\omega} \times \boldsymbol{x}' + \dot{\boldsymbol{b}}. \tag{8.3.37}$$

Note that the vector of angular velocity does not carry a dash just like the distance vector $\boldsymbol{b}$ in the vector relation $(8.2.1)_1$. And just like the matrix $\Omega'_{ij}$ the vector of angular velocity is a quantity that indicates the presence of a non-inertial system (which is what the dashed system in general represents) if it does not vanish. Like all vectors we can span the vector of angular velocity $\boldsymbol{\omega}$ w.r.t. the base $\boldsymbol{e}_i$ (with the components $\omega_i$) or in the base $\boldsymbol{e}'_i$ (with the components $\omega'_i$). The latter has already been done in Eq. (8.3.36). By its very definition, shown in Eq. (8.3.21), the vector of angular velocity is an axial quantity and transforms like:

$$\omega'_i = \det(\boldsymbol{O}')O'_{ir}\omega_r. \tag{8.3.38}$$

The correctness of the first two terms after the equal sign in Eq. (8.3.37) is easily confirmed in analogy to the quantity $O'_{ij}x_j$ from Eqs. (8.3.36) and (8.2.1/8.2.2), and by observing the results from Exercise 8.3.2. The last term in Eq. (8.3.36) becomes obvious if we recall that the base $\boldsymbol{e}'_i$ is time dependent:

$$\dot{\boldsymbol{b}} = \left(b'_i\boldsymbol{e}'_i\right)^{\cdot} = \dot{b}'_i\boldsymbol{e}'_i + b'_i\dot{\boldsymbol{e}}'_i. \tag{8.3.39}$$

Moreover, we may write:

$$\dot{\boldsymbol{e}}'_i = \dot{O}'_{ij}\boldsymbol{e}_j = \dot{O}'_{ij}O'_{kj}\boldsymbol{e}'_k = \Omega'_{ik}\boldsymbol{e}'_k = \epsilon'_{ikl}\omega'_l\boldsymbol{e}'_k \tag{8.3.40}$$

and also:

$$b'_i\dot{\boldsymbol{e}}'_i = \epsilon'_{ikl}\omega'_l b'_i\boldsymbol{e}'_k = \epsilon'_{ilk}\omega'_l b'_k\boldsymbol{e}'_i. \tag{8.3.41}$$

Thus:

$$\dot{\boldsymbol{b}}' = \left(\dot{b}'_i + \epsilon'_{ilk}\omega'_l b'_k\right)\boldsymbol{e}'_i. \tag{8.3.42}$$

Note that Eq. (8.3.37) is the transformation rule for the velocity typically found in elementary mechanics textbooks. In the same context strange looking formulae like

$$\frac{\mathrm{d}(\cdot)}{\mathrm{d}t} = \frac{\mathrm{d}'(\cdot)}{\mathrm{d}t} + \boldsymbol{\omega} \times (\cdot) \tag{8.3.43}$$

are often presented, where the dash at the differentiation symbol is supposed to indicate a time derivative in the moving system. Such definitions and considerations are not required here. Be that as it may, the final result is the same.

In contrast to the gradients of displacements, gradients of velocity are not objective. In order to show that explicitly we differentiate Eq. (8.3.36) w.r.t. the position:

$$\frac{\partial v'_i}{\partial x'_j} = O'_{ik}\frac{\partial v_k}{\partial x_l}\frac{\partial x_l}{\partial x'_j} - \epsilon'_{irs}\omega'_r\delta_{sj} = O'_{ik}O'_{jl}\frac{\partial v_k}{\partial x_l} + \epsilon'_{ijr}\omega'_r. \tag{8.3.44}$$

If we now calculate the *symmetric* velocity gradient according to Eq. (6.3.7) in Cartesian coordinates we obtain:

$$\begin{aligned} d'_{ij} &= \frac{1}{2}\left(\frac{\partial v'_i}{\partial x'_j} + \frac{\partial v'_j}{\partial x'_i}\right) = O'_{ik}O'_{jl}\frac{1}{2}\left(\frac{\partial v_k}{\partial x_l} + \frac{\partial v_l}{\partial x_k}\right) + \epsilon'_{ijr}\omega'_r + \epsilon'_{jir}\omega'_r \\ &\equiv O'_{ik}O'_{jl}d'_{kl}. \end{aligned} \tag{8.3.45}$$

This is a very important and satisfying result in context with the NAVIER–STOKES constitutive relation, which guarantees that it keeps its form in the moving frame. But what about acceleration? We start from Eq. $(8.3.18)_2$ by identifying:

$$
\begin{aligned}
a_i &\equiv \dot{\upsilon}_i \\
&\Rightarrow a'_i = O'_{ij}a_j + \dot{O}'_{ij}\upsilon_j + \dot{\Omega}'_{ir}(x'_r - b'_r) + \Omega'_{ir}\dot{O}'_{rj}x_j + \Omega'_{ir}\dot{O}'_{rj}\upsilon_j + \ddot{b}'_i \\
&\qquad = O'_{ij}a_j + 2\Omega'_{ir}(\upsilon'_r - \dot{b}'_r) - \Omega'_{ir}\Omega'_{rs}(x'_s - b'_s) + \dot{\Omega}'_{ir}(x'_r - b'_r) + \ddot{b}'_i\,.
\end{aligned}
\tag{8.3.46}
$$

Here we have used the relation:

$$
\dot{O}'_{ij}\upsilon_j = \Omega'_{ir}O'_{rj}\upsilon_j = \Omega'_{ir}\left(\upsilon'_r - \Omega'_{rs}\left[x'_s - b'_s\right] - \dot{b}'_r\right). \tag{8.3.47}
$$

It is customary to refer to

- $2\Omega'_{ir}\left(\upsilon'_r - \dot{b}'_r\right)$ as *Coriolis acceleration*,
- $-\Omega'_{ir}\Omega'_{rs}\left(x'_s - b'_s\right)$ as *centrifugal acceleration*,
- $\dot{\Omega}'_{ir}\left(x'_r - b'_r\right)$ as *Euler acceleration* and
- $\ddot{b}'_i$ as *relative acceleration*.

Gaspard-Gustave de Coriolis was born on May 21, 1792 in Paris where he also died on September 19, 1843. He also graduated from the famous École Polytechnique where he accepted a position as a tutor for analysis and mechanics in 1816 forced by the death of his father. At a very early stage he became interested in rotational motion and studied the conservation of the moment of momentum during impact or the motion of bodies on rotating surfaces. By doing so he finally "discovered" the "fictitious" force that was named after him. Coriolis also published several papers on business mathematics and he became a member of the Académie in 1836. Just like Ampère, Carnot, Cauchy, Coulomb, Dulong, Fourier, Foucault, Gay-Lussac, Lagrange, Lamé, Laplace, Legendre, Poisson who all appear in this book, he was one of the 72 great French scientists whose names are engraved on the Eiffel tower.

Thus we finally obtain:

$$
a_j = O'_{ij}\Big(a'_i - 2\Omega'_{ik}\left(\upsilon'_k - \dot{b}'_k\right) + \Omega'_{ir}\Omega'_{rs}\left(x'_s - b'_s\right) - \dot{\Omega}'_{ir}\left(x'_r - b'_r\right) - \ddot{b}'_i\Big). \tag{8.3.48}
$$

Consequently *acceleration is no objective vector* either. Finally note that in order to check the objectivity of a kinematic quantity a straightforward calculation will give the answer. Additional physics arguments are not required.

**Exercise 8.3.3:** ***Centrifugal and other accelerations in absolute notation***

Show analogously to the case of the velocity in Eqs. (8.3.36–8.3.41) that the transformation rule for the acceleration vector can be written as follows:

$$\boldsymbol{a}' = \boldsymbol{a} - 2\boldsymbol{\omega} \times \boldsymbol{v}' - \boldsymbol{\omega} \times (\boldsymbol{\omega} \times \boldsymbol{x}') - \dot{\boldsymbol{\omega}} \times \boldsymbol{x}' + \ddot{\boldsymbol{b}}\,. \tag{8.3.49}$$

In this context observe the following useful relations similar to Eq. (8.3.35):

$$\epsilon'_{ijk}\epsilon'_{klm} = \delta'_{il}\delta'_{jm} - \delta'_{im}\delta'_{jl}. \tag{8.3.50}$$

What are the names of the various acceleration terms in this equation (observe the signs)? Is there also a centripetal acceleration or how would it look like?

## 8.4 Balance of Mass and Euclidean Transformation

We now turn to the question how the balance of mass in its form $(3.8.3)_1$ behaves under Euclidean transformations. Judging by the results of Sect. 5.1 we expect that it will have the same form in the Euclidean system as in the inertial system and we may simply replace all quantities by dashed ones. However, this is just a guess and we should be cautious since the Euclidean transformation is time dependent and the balance of mass contains a time derivative that might cause trouble. In other words without further calculations we cannot be sure of a potential impact of the time derivative. Hence we argue as follows:

First of all we shall assume that the mass of a material particle is an objective scalar, at least within the framework of classical physics:

$$\mathrm{d}m' = \mathrm{d}m. \tag{8.4.1}$$

The same holds for the corresponding volume element:

$$\mathrm{d}V' = \mathrm{d}V, \tag{8.4.2}$$

if interpreted as the *absolute value of a scalar triple product* of three (infinitesimal) distances $\mathrm{d}\overset{(i)}{\boldsymbol{x}}$:

$$\mathrm{d}V = \left|\mathrm{d}\overset{(1)}{\boldsymbol{x}}\cdot\left(\mathrm{d}\overset{(2)}{\boldsymbol{x}}\times\mathrm{d}\overset{(3)}{\boldsymbol{x}}\right)\right|. \tag{8.4.3}$$

This makes sense since because of Eq. (8.3.8) the distance vector between two points is objective. Note that a volume element defined by a pure scalar triple product would be an axial scalar due to the vector product and this is why we take its absolute value. If we now observe the definition $\rho = \mathrm{d}m/\mathrm{d}V$ we conclude that mass density is an objective scalar:

$$\rho' = \rho. \tag{8.4.4}$$

Next we consider the *material time derivative* of an objective scalar, in particular of the mass density, $\rho$. We recall that by definition we may write in Lagrangian or in Eulerian representation (cf., Sect. 3.8):

$$\dot{\rho} \equiv \frac{\mathrm{d}\rho}{\mathrm{d}t} = \frac{\partial \rho(\boldsymbol{X},t)}{\partial t} = \frac{\partial \rho}{\partial t} + \upsilon_i \frac{\partial \rho}{\partial x_i}. \tag{8.4.5}$$

The Lagrangian representation is easily converted:

$$\frac{\partial \rho(\boldsymbol{X},t)}{\partial t} = \frac{\partial \rho'(\boldsymbol{X},t)}{\partial t} \Leftrightarrow \dot{\rho} = \dot{\rho}'. \tag{8.4.6}$$

Thus the material time derivative of an objective scalar is also objective. The proof is a little more cumbersome in Eulerian representation. In that case we find for the first part of the material time derivative:

$$\begin{aligned} \left.\frac{\partial \rho(\boldsymbol{x},t)}{\partial t}\right|_{\boldsymbol{x}} &= \left.\frac{\partial \rho'(\boldsymbol{x}',t)}{\partial x'_i}\right|_t \frac{\partial x'_i}{\partial t} + \left.\frac{\partial \rho'}{\partial t}\right|_{\boldsymbol{x}'} = \frac{\partial \rho'}{\partial x'_i}\left(\dot{O}'_{ij}x_j + \dot{b}'_i\right) + \frac{\partial \rho'}{\partial t} \\ &= \frac{\partial \rho'}{\partial x'_i}\left(\Omega'_{ik}\left(x'_k - b'_k\right) + \dot{b}'_i\right) + \frac{\partial \rho'}{\partial t}; \end{aligned} \tag{8.4.7}$$

and for the second:

$$\upsilon_k \frac{\partial \rho}{\partial x_k} = \upsilon_k \frac{\partial x'_l}{\partial x_k}\frac{\partial \rho}{\partial x'_l} = \upsilon_k O'_{lk}\frac{\partial \rho}{\partial x'_l} = \left(\upsilon'_l - \Omega'_{lk}\left(x'_k - b'_k\right) - \dot{b}'_l\right)\frac{\partial \rho'}{\partial x'_l}. \tag{8.4.8}$$

Therefore we have in total:

$$\begin{aligned} \frac{\partial \rho}{\partial t} + \upsilon_k \frac{\partial \rho}{\partial x_k} &= \frac{\partial \rho'}{\partial x'_i}\left(\Omega'_{ik}\left(x'_k - b'_k\right) + \dot{b}'_i\right) + \frac{\partial \rho'}{\partial t} \\ &\quad + \left(\upsilon'_l - \Omega'_{lk}\left(x'_k - b'_k\right) - \dot{b}'_l\right)\frac{\partial \rho'}{\partial x'_l} = \frac{\partial \rho'}{\partial t} + \upsilon'_l \frac{\partial \rho'}{\partial x'_l}. \end{aligned} \tag{8.4.9}$$

In this context Eq. $(8.3.18)_2$ and the fundamental equations for Euclidean transformations of Eqs. (8.2.2/8.2.4) were used. Moreover, note that in a Eulerian description the variables $x_k$ and $t$ are independent of each other. This was indirectly observed during the differentiations with respect to time.

Finally we examine the transformation properties of the divergence of the velocity under Euclidean transformations. From Eq. (8.3.36) it follows that:

$$\begin{aligned} \frac{\partial \upsilon'_i}{\partial x'_i} &= O'_{ik}\frac{\partial \upsilon_k}{\partial x_l}\frac{\partial x_l}{\partial x'_i} - \epsilon'_{irs}\omega'_r\delta_{si} = O'_{ik}O'_{il}\frac{\partial \upsilon_k}{\partial x_l} + \epsilon'_{iir}\omega'_r \\ &= \delta_{kl}\frac{\partial \upsilon_k}{\partial x_l} + 0 = \frac{\partial \upsilon_k}{\partial x_k}. \end{aligned} \tag{8.4.10}$$

By combination of Eqs. (8.4.9/8.4.10) we conclude that:

$$\frac{\partial \rho}{\partial t} + \frac{\partial \rho \upsilon_k}{\partial x_k} = \frac{\partial \rho'}{\partial t} + \frac{\partial \rho' \upsilon'_l}{\partial x'_l} \Leftrightarrow \dot{\rho} + \rho \frac{\partial \upsilon_k}{\partial x_k} = \dot{\rho}' + \rho' \frac{\partial \upsilon'_l}{\partial x'_l}. \tag{8.4.11}$$

Thus the form of the balance of mass is invariant under the most general changes of observers of classical physics. Moreover, during the transformation no system-dependent terms arise, and this is exactly how a true law of nature should behave.

## 8.5 The Balance of Momentum in a Moving Coordinate System: An Almost Philosophical Discourse

If the notion of a force is physically sound, forces should exist independently of an observer. In other words, forces should be *objective* vectors. Volumetric forces, i.e., forces per unit volume should also be objective according to our discussion in context with Eq. (8.4.3). Thus, we may write:

$$f'_i = O'_{ij} f_j. \tag{8.5.1}$$

The classification of surface related forces, in particular of the traction $\boldsymbol{t}$, i.e., a force per unit surface, is a little more subtle. A directed surface element $\mathrm{d}\boldsymbol{A}$ results from a vector product (!) of two (objective) distance vectors:

$$\mathrm{d}A_i = \epsilon_{ijk} \mathrm{d}x_j \mathrm{d}x_k. \tag{8.5.2}$$

We know from Exercise 8.3.2 that the totally antisymmetric tensor is an *axial* objective quantity. Consequently a surface element is of the same type:

$$\mathrm{d}A'_i = \det(\boldsymbol{O}') O'_{ij} \mathrm{d}A_j. \tag{8.5.3}$$

The same holds for the unit normal vector. It results from $\mathrm{d}\boldsymbol{A}$ by dividing it by the surface area $|\mathrm{d}\boldsymbol{A}|$. The latter is always positive and an objective scalar, just like the volume element, because the size of an area should not depend on the observer, at least not in classical physics [note that this property also follows from Eq. (8.5.3)]. Hence we find:

$$|\mathrm{d}A'_i| = |\mathrm{d}A_j| \Rightarrow n'_i = \det(\boldsymbol{O}') O'_{ij} n_j. \tag{8.5.4}$$

According to Eq. (3.2.7) the traction is a linear function of the unit normal. For this reason we require that it is an axial tensor of first order:

$$t'_i = \det(\boldsymbol{O}') O'_{ij} t_j, \tag{8.5.5}$$

so that the stress tensor becomes an objective tensor of second order:

$$\sigma'_{ij} = O'_{ir} O'_{js} \sigma_{rs}. \tag{8.5.6}$$

Thus in case of surface forces we shift our preference for objective (and not *axially* objective) force-related quantities from the traction vector to the stress tensor.

**Exercise 8.5.1:** ***The stress tensor as an objective quantity***

Start from the CAUCHY relation:

$$t_i = \sigma_{ji} n_j \tag{8.5.7}$$

and verify by using Eqs. (8.5.2/8.5.3) relation (8.5.6) according to which the CAUCHY stress tensor is an objective tensor of second order.

And therefore the divergence of the stress tensor transforms like:

$$\frac{\partial \sigma'_{ji}}{\partial x'_j} = \frac{\partial \sigma'_{ji}}{\partial x_k}\frac{\partial x_k}{\partial x'_j} = O'_{jr} O'_{is} \frac{\partial \sigma_{rs}}{\partial x_k} O'_{jk} = \delta'_{rk} O'_{is} \frac{\partial \sigma_{rs}}{\partial x_k} = O'_{is} \frac{\partial \sigma_{ks}}{\partial x_k}. \tag{8.5.8}$$

Finally recall the transformation rule for the acceleration according to Eq. (8.3.46) as well as the balance of momentum in the form $(3.8.14)_1$, from which the mass balance has already been extracted, and the validity of which has been established in an inertial system, so to speak. Hence the momentum balance assumes the following form in a Euclidean system:

$$\rho' \dot{\upsilon}'_i - \frac{\partial \sigma'_{ji}}{\partial x'_j} = \rho' f'_i + \rho' \left[ 2\Omega'_{ik}(\upsilon'_k - \dot{b}'_k) - \Omega'_{il}\Omega'_{lk}(x'_k - b'_k) + \dot{\Omega}'_{ik}(x'_k - b'_k) + \ddot{b}'_i \right]. \tag{8.5.9}$$

We must conclude that the form of the balance of momentum in a Euclidean system differs considerably from the one in an inertial system: In contrast to the transformation behavior of the mass balance all of a sudden system dependent terms appear and the form invariance is at stake. However, a simple trick comes to the rescue: The various accelerations are combined to form a new specific volumetric force:

$$\hat{f}'_i = f'_i + 2\Omega'_{ik}(\upsilon'_k - \dot{b}'_k) - \Omega'_{il}\Omega'_{lk}(x'_k - b'_k) + \dot{\Omega}'_{ik}(x'_k - b'_k) + \ddot{b}'_i. \tag{8.5.10}$$

Thus we may write:

$$\rho' \dot{\upsilon}'_i = \frac{\partial \sigma'_{ji}}{\partial x'_j} + \rho' \hat{f}'_i, \tag{8.5.11}$$

which corresponds exactly to the form of the balance of momentum in the inertial frame. To this extend we may talk of form invariance of the balance of momentum under the presence of system depending quantities. However, the whole story gets a mystical touch if we begin to refer to the various accelerations multiplied by the mass density as *fictitious forces* after moving them to the right hand side of the momentum balance. The adjective "fictitious" is an extremely unfortunate choice of a word. On the contrary! These "fictitious forces" are extremely real. We can convince ourselves immediately of their presence during a car accident. And what is more, in their effect they cannot be distinguished at all from the protagonist of the force density $f_i'$, which is gravity. In the latter case we got used to saying that its cause is the attraction between gravitational masses. But we neither explain why gravitational masses attract each other nor what a gravitational mass really is, at least within the framework of classical continuum physics. However, we can compare gravitational masses with each other by placing them on the bowls of a beam scale. In fact this phenomenological comparison works in all gravitational fields, on Earth, on the Moon, etc., and will lead to the same result everywhere.

**Exercise 8.5.2: *Balance of momentum in a moving system in absolute notation***

Observe Eq. (8.5.9) and the result from Exercise 8.3.3 and show that:

$$\rho\dot{\boldsymbol{v}}' = \nabla' \cdot \boldsymbol{\sigma} + \rho\left[\boldsymbol{f} - 2\boldsymbol{\omega} \times \boldsymbol{v}' - \boldsymbol{\omega} \times (\boldsymbol{\omega} \times \boldsymbol{x}') - \dot{\boldsymbol{\omega}} \times \boldsymbol{x}' + \ddot{\boldsymbol{b}}\right]. \qquad (8.5.12)$$

Why do the symbols $\rho$, $\boldsymbol{\sigma}$, and $\boldsymbol{f}$ carry no dashes? Note that the expressions $-2\rho\boldsymbol{\omega} \times \boldsymbol{v}'$, $-\rho\boldsymbol{\omega} \times (\boldsymbol{\omega} \times \boldsymbol{x}')$, $-\rho\dot{\boldsymbol{\omega}} \times \boldsymbol{x}'$, and $\rho'\ddot{\boldsymbol{b}}$ are known as Coriolis force, centrifugal force, Euler force, and force of relative translation, respectively. They are all inertia forces.

Instead of using the term fictitious forces one should rather talk of *inertia forces* (cf., Meriam (1978), p. 201), since this is exactly what these system dependent forces are: Masses seem to have a certain perseverance to remain in their original state of motion. In other words they are "inert" and in order to accelerate them or to change the course of their motion forces are required. This was Newton's great discovery. At the beginning of his famous book *Philosophiae Naturalis Principia Mathematica* (1726)[2] (Mathematical Principles of Natural Philosophy) he distinguishes in the *Definitiones* two kinds of forces. On the one hand side there is the *vis insita*, in other words a force proper to matter, which we call inertia force today (with the exception of the sign):

[2] The Latin citations stem from the third edition of Newton's book and can be found in the most carefully edited two volumes by Koyré et al. [9]. When compared to the first edition of the *Principia* (1687) we notice considerable differences in the wording, in terms of alterations as well as additional comments. This is an indication of Newton's lifelong struggle with his findings, and it also shows how his understanding grew steadily over time. Moreover, note, that all but one translation of the original Latin text stem from the book of Chandrasekhar (1995). The translation of the "hypotheses non fingo" passage is from Cohen and Whitman (1999).

*Definitio III. Materiae vis insita est potentia resistendi, qua corpus unumquodque, quantum in se est, perseverat in statu suo vel quiescendi vel movendi uniformiter in directum.* (Definition III. The vis insita, or innate force of matter, is a power of resisting, by which every body, as much as in it lies, continues in its present state, whether it be of rest, or of moving uniformly forwards in a right line.)

On the other hand side there is the *vis impressa* (in other words: an "applied" force, which comes from outside and which belongs on the right hand side of the momentum balance):

*Definitio IV. Vis impressa est actio in corpus exercita, ad mutandum ejus statum vel quiescendi vel movendi uniformiter in directum.* (Definition IV. An impressed force is an action exerted upon a body, in order to change its state, either of rest, or of uniform motion in a right line.)

It becomes particularly fascinating when NEWTON explains in his fifth definition the notion of a centripetal force for the first time, since his centripetal force is not what we mean by that term today, i.e., not $\rho\boldsymbol{\omega} \times (\boldsymbol{\omega} \times \boldsymbol{x}')$:

*Definitio V. Vis centripeta est, qua corpora versus punctum aliquod, tanquam ad centrum, undique trahuntur, impelluntur, vel utcunque tendunt.* (Definition V. A centripetal force is that by which bodies are drawn or impelled, or any way tend, towards a point as to a centre.)

He continues to explain and says: *Hujus generis est gravitas, qua corpora tendunt ad centrum terrae; vis magnetica, qua ferrum petit magnetem; .... Lapis, in funda circumactus, a circumagente manu abire conatur; & conatu suo fundam distendit, eoque fortius quo celerius revolvitur; &, quamprimum dimittitur, avolat. Vim conatui illi contrariam, qua funda lapidem in manum perpetuo retrahit & in orbe retinet, quoniam in manum ceu orbis centrum dirigitur, centripetam appello.* (Of this sort is gravity, by which bodies tend to the centre of the Earth; magnetism, by which iron tends to the loadstone; .... A stone, whirled about a sling, endeavours to recede from the hand that turns it; and by that endeavour, distends the sling, and that with so much the greater force, as it is revolved with the greater velocity, and by which the sling continually draws back the stone towards the hand, and retains it in its orbit, because it is directed to the hand as the centre to the orbit, I call the centripetal force). Thus NEWTON's concept of a centripetal force is not a kinematic one. To him a centrifugal force is linked to physical causes, which we summarize today in the term $\rho\boldsymbol{f}$ (for gravity and, in context with magnetism, the LORENTZ force) and/or $\nabla \cdot \boldsymbol{\sigma}$ (for the tension in the string of a sling to which a stone is attached).

Then NEWTON states his First Law:

*Lex I. Corpus omne perseverare in statu suo quiescendi vel movendi uniformiter in directum, nisi quatenus illud a viribus impressis cogitur statum illum mutare.* (Law I. Every body continues in its state of rest, or of uniform motion in a right line, unless it is compelled to change that state by forces impressed upon it.)

On first glance one might think that NEWTON's First Law is just a special case of his Second Law, if applied to a case where no forces are present:

*Lex II. Mutationem motus proportionalem esse vi motrici impressae, & fieri secundum lineam rectam qua vis illa imprimitur.* (Law II. The change of motion is proportional to the motive force impressed; and is made in the direction of the right line in which that force is impressed.)

However, in its final consequence NEWTON's First Law implies the existence of an inertial system, where the body is at rest or moves uniformly along a straight line due to lack of forces. Moreover, in his Second Law we recognize the simplest form of a balance: A quantity (namely "*motus*") changes in time ("*mutatio*") due to an applied propelling force ("*vi motrici impressae*"). The former is the effect and the latter the cause. Also note that NEWTON explains a few pages before that his "*motus*" is exactly that what we call momentum today:

*Definitio II. Quantitas motus est mensura ejusdem orta ex velocitate et quantitate materiae conjunctim.* (Definition II. The quantity of motion is a measure of the same, arising from the velocity and quantity of matter conjointly.)

Frequently the question is asked as to whether the force (in other words the right hand side of the Second Law) is defined by the left hand side, i.e., the change of momentum or, in other words, for a constant mass by a change of position, i.e., geometry. This is not so. In this context force is considered as a "primitive" quantity, and the Second Law serves by no means as an equation for its definition. Thus it is only fair to ask how a measurement standard for a force can be defined. The answer to this is an evasive one: In the end the standard must be based on a deeper understanding of the origin of force, for example on an atomic or subatomic level. This holds for the volumetric force density $\rho \boldsymbol{f}$, in particular gravity but also other inertia forces, as well as for surface force related terms like $\nabla \cdot \boldsymbol{\sigma}$. In the latter case we have constitutive relations for the stress tensor and the underlying materials theory (which is not subject of this book, at least not in detail) in mind. But again, a profound understanding of these concepts also takes place on various scales, on the macro-, the meso-, and on the micro-level.

When NEWTON spoke of forces the attraction of masses was definitely on his mind, which he so aptly described by his famous law of gravitation. Rather at the end of his *Principia* (in the *Scholium Generale* after *Propositio XLII*) NEWTON muses that so far he had not been able to find the cause for the existence of gravity: *Rationem vero harum gravitatis proprietatum ex phaenomenis nondum potui deducere, & hypotheses non fingo. Quicquid enim ex phaenomenis non deducitur, hypothesis vocanda est; & hypotheses seu metaphysicae, seu physicae, seu qualitatum occultarum, seu mechanicae, in philosophia experimentali locum non habent.* I have not as yet been able to discover the reason for these properties of gravity from phenomena, and I do not feign hypotheses. For whatever is not deduced from the phenomena must be called a hypothesis; and hypotheses, whether metaphysical or physical, or based on occult qualities, or mechanical, have no place in experimental philosophy. In this philosophy particular propositions are inferred from the phenomena, and afterwards rendered general by induction.) In the spirit of his laconic phrase "*hypotheses non fingo*" force is for us nothing else but a *primitive quantity*.

Besides CORIOLIS force and Co. it is customary in mechanics to call the (negative) temporal change of momentum an inertia force as well. This allows rewriting NEWTON's Second Law in the form that "the sum of all forces including inertial ones is equal to zero." Thus all of the computational techniques we have been introduced to in statics can usefully be transferred to dynamics. As convenient as this may be, it is conceptionally counterproductive: That way the important aspect that the additive quantity momentum obeys a balance law is completely lost. The dubious method is due to D'ALEMBERT [4], and I am inclined to say that it is probably more of a reaction of a member of the Académie Française to the dominant Anglo-Saxon way of mechanics than a deep insight.

Jean le Rond D'ALEMBERT was born on November 16, 1717 in Paris where he also died on October 29, 1783. He graduated from the Collège des Quatre Nations in 1735. First he worked as an attorney of law and as a physician. Finally, however, he switched to mathematics and mechanics and worked in Paris at the Academy of Science and also at the Académie Française. In 1775 he became the Secretary of this noble organization where one of his more pleasant tasks was the writing and setup of the obituaries of its members.

Let us now turn to NEWTON's Third Law according to which to each force (the "action") there belong a responding force (the "reaction"):

*Lex III. Actioni contrariam semper & aequalem esse reactionem: sive corporum duorum actiones in se mutuo semper esse aequales & in partes contrarias dirigi.* (Law III. To every action there is always opposed an equal reaction: or, the mutual actions of two bodies upon each other are always equal, and directed to contrary parts.)

Ernst Waldfried Josef Wenzel MACH was born on February 18, 1838 in Chirlitz-Turas, Moravia, today Brno, and he died on February 19, 1916 in Vaterstetten near Munich. As a true Austrian multi-talent MACH was physicist, philosopher, psychologist, and theoretician of the sciences. Above all his name is well known from the MACH number used in supersonic aviation. However, he is also one of the most influential representatives if not the founder of empirio-criticism. In psychology he prepared the way for the so-called *gestalt* psychology and *gestalt* theory.

Note that NEWTON does not say anything about how fast this reaction builds up, it is just "there:" The Earth is pulling the moon, and the moon strikes back, instantaneously. Moreover, in some mechanics textbooks (e.g., Hauger, Schnell, Gross, 1993) one can find the following interesting comment of D'ALEMBERT's inertia force: "This force is no force in the sense of NEWTON since it does not come with a reaction force (it violates the axiom action = reaction!)." This may be so, but NEWTON also never claimed this to be the case either. Rather he clearly had the imposed forces of his First and Second Law in mind (mark his words "Vis impressa est actio…" in Definitio IV and "… a viribus impressis…" in Lex I as well as "… vi motrici impressae…" from Lex II) and not the temporal change of momentum representing the quantity known as inertia force today. He emphasizes this in his book by means of several examples, e.g., a stone pressing on a finger and vice versa or a horse pulling a stone that prevents the horse's motion since it is pulling back. Nothing can be found at this place about CORIOLIS force & Co. nor does NEWTON claim that they obey the action = reaction principle. This shows even more that D'ALEMBERT's approach has a certain aftertaste.

It is only fair to pose the question what the origin of inertia accelerations really is. To say that they are a consequence of correct differentiation of spatial coordinate transformations w.r.t. time is not a truly satisfying answer. It was the Viennese physicist and natural philosopher Ernst MACH who attempted to give an answer, albeit a sibyllic one. Einstein [6] picks up MACH's ideas and says: "Analogously, I seek in vain for a real something in classical mechanics (or in the special theory of relativity) to which I can attribute the different behaviour of bodies considered with respect to the reference-systems K and K'.[1] Newton saw this objection and attempted to invalidate it, but without success. But E. Mach recognised it most clearly of all, and because of this objection he claimed that mechanics must be placed on a new basis. It can only be got rid of by means of a physics which is conformable to the general principle of relativity, since the equations of such a theory hold for every body of reference, whatever may be its state of motion." In the footnote 1 of the text he becomes even more explicit: "[1] The objection is of importance more especially when the state of motion of the reference-body is of such a nature that it does not require any external agency for its maintenance, e.g. in the case when the reference-body is rotating uniformly." In his German paper of (1914) [5] he says clearly what the physical origin of the forces arising during rotation is, unfortunately without a translation. However, there is an English equivalent to that text in Einstein [7]: "Can gravitation and inertia be identical? This question leads directly to the General Theory of Relativity. It is not possible to regard the earth as free from rotation, if I conceive of the centrifugal force, which acts on all bodies at rest relatively to the earth, as being a "real" field of gravitation, or part of such a field? If this idea can be carried out, then we shall have proved in very truth the identity of gravitation and inertia. ... According to Newton, this interpretation is impossible, because by Newton's law the centrifugal field cannot be regarded as produced by matter, and because in Newton's theory there is no place for a "real" field f the "Koriolis-field" type." In short: The fictitious forces have their origin in the gravitational action of far away masses.

Jean Bernard Léon FOUCAULT was born on September 18, 1819 in Paris and also died there on February 11, 1868. From 1829 on he attended first the Collège Stanislas in Paris. Due to his laziness and rude behavior he was advised to leave school so that he had to get his further education from a private teacher. He quit medical studies due to insurmountable aversion of dissection and dedicated himself to physics from then on as an autodidact. In this field he showed an enormous experimental talent and so he presented publicly in 1851 the famous pendulum named after him. Around 1850 he determined the speed of light by using a cheval mirror construction. In 1855 he invented a typewriter, and so on. His dislike for medicine finally took vengeance when he tragically became almost blind and mute and died of aphasia.

And thus we return to the question regarding the existence of an inertial or GALILEAN system. FOUCAULT's pendulum experiment, which is explained by the presence and the action of the CORIOLIS acceleration, shows very clearly that our planet Earth is everything else but an inertial system. We may be inclined to say that Earth behaves approximately, but not in absolute terms, like an inertial system since the effect is small and does not affect our everyday motions. But if we do not find it on Earth, where else should the inertial frame be? On the sun, in the center of our galaxy, on the way to the Andromeda nebula, or some other place even further away? The sobering answer is: It is nowhere! The notion of an inertial frame is simply a model of our world, a figment of imagination, so to speak, which allows us to describe everyday motions to a good approximation. A better model for an observer accelerated against the Earth is the Euclidean frame, which makes the effects of this state on micro- and macroscopic structures plannable and accessible to engineering by taking CORIOLIS and other inertia forces into account. However, if really large scale structures are of interest, such as planetary systems, galaxies, or even the whole universe, NEWTON's equations provide an insufficient model and have to be replaced by a better one, e.g., by relativistic field equations.

Hans THIRRING was born on March 23, 1888 in Vienna where he also died on March 22, 1976. He studied mathematics, physics and—most useful—gymnastics at the University of Vienna until 1910. In 1911 he became an assistant at the Institute for Theoretical Physics of the university, where he also obtained his Ph.D. in 1911 and presented his habilitation thesis in 1915. In 1921 he became an associate professor and in 1927 finally full professor. Although he did mostly theoretical work he was also open to technology. For example, he invented a method for the production and playback of talkies and started his own business where he exploited that technology. However, in 1938 the Nazis forced him into "early retirement:" His interest in "Jewish physics," such as relativity, his friendship with EINSTEIN and FREUD and his notorious pacifism provided enough reasons. After the war he became the Dean of the Department of Philosophy at the University of Vienna.

This is what THIRRING [12, 13] did in order to explain the existence of inertia forces as the effect of far away rotating masses. In other words he tried to "verify" MACH's idea starting from EINSTEIN's field equations. He considered a thin-walled hollow sphere of homogeneous mass density (total mass $M$, radius $a$). An observer is located in its center turning at a constant angular velocity $\omega$ about the $x_3'$-axis.[3] For this kind of mass distribution he determined approximately the four-dimensional metric tensor (cf., Sect. 13.11) by solution of EINSTEIN's field equations. Once the metric was known he used the geodesic equation and studied the motion of a test mass-point inside of the hollow sphere. Using the nomenclature of Eq. (8.5.9) he derived the following formulae:

[3] Strictly speaking THIRRING's analysis also allows the hollow sphere to rotate with an angular velocity different from $\omega$.

$$\ddot{x}'_1 = 2\omega\left(1 + \frac{GM}{4\pi c^2 a}\right)\dot{x}'_2 + \omega^2\left(1 + \frac{GM}{4\pi c^2 a}\right)x'_1, \tag{8.5.13}$$

$$\ddot{x}'_2 = -2\omega\left(1 + \frac{GM}{4\pi c^2 a}\right)\dot{x}'_1 + \omega^2\left(1 + \frac{GM}{4\pi c^2 a}\right)x'_2, \quad \ddot{z}' = 0.$$

$G = 6.67 \cdot 10^{-11}$ m/kg s$^2$ denotes the gravitational constant and $c = 3 \cdot 10^8$ m/s is the speed of light. The terms on the right hand side correspond w.r.t. their dependence of the angular velocity and the coordinates exactly to the CORIOLIS- and to the centrifugal acceleration—up to relativistically small corrections. We will explore this in detail in the following exercise.

**Exercise 8.5.3: *Rotation about a fixed axis***

Consider the situation shown in Fig. 8.2: A non-inertial system $(\boldsymbol{e}'_1, \boldsymbol{e}'_2)$ rotates counterclockwise about the origin of an inertial system $(\boldsymbol{e}_1, \boldsymbol{e}_2)$ so that the angle $\alpha(t)$ increases steadily. Alternatively we could say that the rotation is counterclockwise about the common axis $\boldsymbol{e}'_3 = \boldsymbol{e}_3$. Show the validity of the following relations:

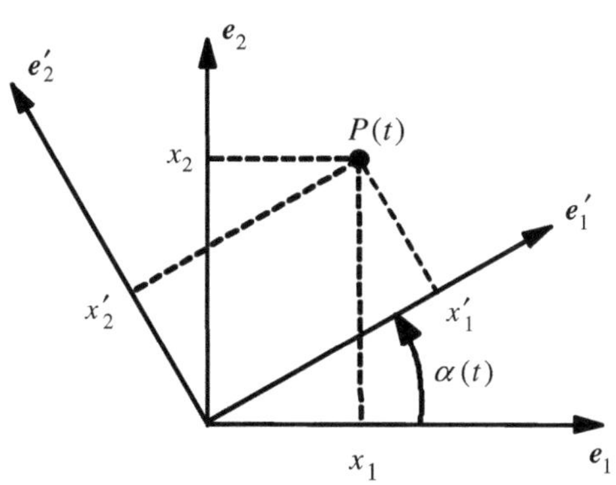


**Fig. 8.2** Counterclockwise spin within the 2D-plane

$$O'_{ij} = \begin{bmatrix} \cos\alpha & \sin\alpha & 0 \\ -\sin\alpha & \cos\alpha & 0 \\ 0 & 0 & 1 \end{bmatrix}, \quad \Omega'_{ij} = \begin{bmatrix} 0 & \omega & 0 \\ -\omega & 0 & 0 \\ 0 & 0 & 0 \end{bmatrix}, \tag{8.5.14}$$

where the absolute value of the angular velocity $\omega = \dot{\alpha}$ has been introduced. Use the result to show that:

$$\boldsymbol{\omega} = (0, 0, \omega), \quad \omega'_1 = 0, \quad \omega'_2 = 0, \quad \omega'_3 = \omega. \tag{8.5.15}$$

Moreover, show that we obtain for the various parts in the equation of motion (8.5.12) (all components are w.r.t. the non-inertial system $\boldsymbol{e}'_i$):

$$\boldsymbol{x}' = \left(x'_1, x'_2, x'_3\right), \quad \dot{\boldsymbol{v}}' = \left(\ddot{x}'_1, \ddot{x}'_2, \ddot{x}'_3\right), \quad \boldsymbol{\omega} \times \boldsymbol{v}' = \omega\left(-\dot{x}'_2, \dot{x}'_1, 0\right), \qquad (8.5.16)$$

$$\boldsymbol{\omega} \times (\boldsymbol{\omega} \times \boldsymbol{x}') = -\omega^2\left(x'_1, x'_2, 0\right), \quad \dot{\boldsymbol{\omega}} \times \boldsymbol{x}' = \dot{\omega}\left(-x'_2, x'_1, 0\right), \quad \boldsymbol{b} = (0, 0, 0).$$

Insert this into the equations of motion (8.5.12) and demonstrate that in Cartesian coordinates of the non-inertial system we have (while neglecting mechanical stresses and imposed volumetric forces):

$$\ddot{x}'_1 = 2\omega\dot{x}'_2 + \omega^2 x'_1 + \dot{\omega}x'_2, \quad \ddot{x}'_2 = -2\omega\dot{x}'_1 + \omega^2 x'_2 - \dot{\omega}x'_2, \quad \ddot{x}'_3 = 0. \qquad (8.5.17)$$

Compare this result to the relativistic equation (8.5.13) and check in particular the corresponding signs of the various terms. Moreover, show that the equations of motion in the inertial frame have the following simple form:

$$\ddot{x}_1 = 0, \quad \ddot{x}_2 = 0, \quad \ddot{x}_3 = 0. \qquad (8.5.18)$$

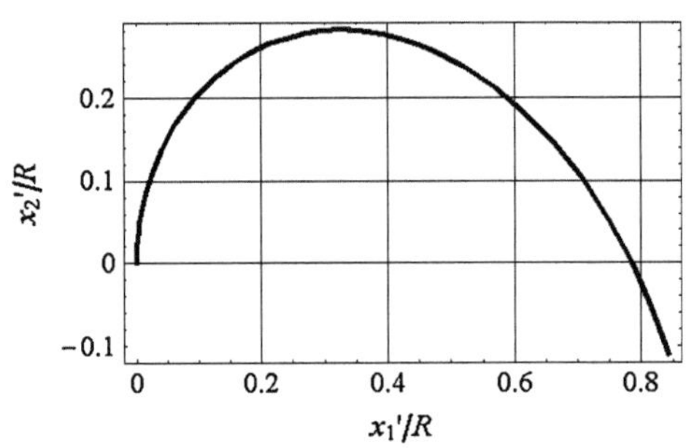


**Fig. 8.3** Trajectory in a moving coordinate frame ($V/(R\omega) = 0.5$)

Solve them analytically by assuming the following initial conditions:

$$x_1(0) = 0, \quad x_2(0) = 0, \quad x_3(0) = 0, \qquad (8.5.19)$$

$$\dot{x}_1(0) = 0, \quad \dot{x}_2(0) = V, \quad \dot{x}_3(0) = 0.$$

Now assume that initially both coordinate systems are located on top of each other in the center of a circular disk of radius $R$. Solve Eq. (8.5.17) numerically and show that (suitably normalized) the trajectory of Fig. 8.3 results. For that purpose assume that the angular velocity is constant. How does this compare to the trajectory in the inertial system?

**Exercise 8.5.4:** ***Planning of a space station***

Use the internet and compile some information regarding the typical dimensions of current and future space stations. Our objective is to create a state of artificial gravity in the periphery by rotation about a center point. In other words gravitation shall be mimicked by centrifugal acceleration, i.e.,

$$\boldsymbol{a}_c = \boldsymbol{\omega} \times (\boldsymbol{\omega} \times \boldsymbol{x}'). \tag{8.5.20}$$

In order to obtain specific numbers, specialize Eq. (8.5.10) to the case of a wheel-like station rotating about its "hub" w.r.t. the fixed stars with a constant angular velocity $\omega_0$:

$$\boldsymbol{\omega} = \omega_0 \boldsymbol{e}_3. \tag{8.5.21}$$

How big is $\omega_0$ and how strong is the corresponding Coriolis force acting on an astronaut who is moving radially forward from the center at pedestrian speed?

## 8.6 Energy Balances in a Rotating System

There are several ways to derive the balance of kinetic energy in a moving system. One possibility is scalar multiplication of the balance of momentum in the moving frame, Eq. (8.5.12), by velocity $\boldsymbol{\upsilon}'$ and to perform a few algebraic manipulations as follows:

$$\rho\left(\frac{1}{2}\upsilon'^2\right)^{\cdot} = \nabla' \cdot (\boldsymbol{\sigma} \cdot \boldsymbol{\upsilon}') - \boldsymbol{\sigma} : \nabla'\boldsymbol{\upsilon}' + \rho\left[\boldsymbol{f} - \boldsymbol{\omega} \times (\boldsymbol{\omega} \times \boldsymbol{x}') - \dot{\boldsymbol{\omega}} \times \boldsymbol{x}' + \ddot{\boldsymbol{b}}\right] \cdot \boldsymbol{\upsilon}'. \tag{8.6.1}$$

Obviously, there are additional supplies of kinetic energy to the well known one, $\boldsymbol{f} \cdot \boldsymbol{\upsilon}'$, namely the power of the centrifugal acceleration, the Euler acceleration, and the relative acceleration.

Alternatively we may start from Eq. (8.5.9), multiply it by $\boldsymbol{\upsilon}'$ and rewrite. The dashes characteristic of the moving frame allow to distinguish very clearly which components we refer to:

$$\begin{aligned}\rho'\left(\frac{1}{2}\upsilon'_i\upsilon'_i\right)^{\cdot} =& \frac{\partial \sigma'_{ji}\upsilon'_i}{\partial x'_j} - \sigma'_{ji}\frac{\partial \upsilon'_i}{\partial x'_j} \\ &+ \rho'\left[f'_i - \Omega'_{il}\Omega'_{lk}(x'_k - b'_k) + \dot{\Omega}'_{ik}(x'_k - b'_k) + \ddot{b}'_i - 2\Omega'_{ik}\dot{b}'_k\right]\upsilon'_i.\end{aligned} \tag{8.6.2}$$

**Exercise 8.6.1:** ***The power of fictitious forces***

Show analogously to the chain of Eqs. (8.3.39–8.3.42) and in Exercise 8.5.1 the equivalence of Eqs. (8.6.1) and (8.6.2). Why is there no power due to the Coriolis force?

In order to derive the balance of internal energy of a moving system we recall Table 3.1, observe the balance of mass, and the definition of the material time derivative so that we find in Cartesian coordinates:

$$\rho \dot{u} = -\frac{\partial q_k}{\partial x_k} + \sigma_{ji}\frac{\partial \upsilon_i}{\partial x_j} + \rho r. \tag{8.6.3}$$

We *postulate* that the specific internal energy and the radiation density are Euclidean scalars:

$$u' = u, \quad r' = r. \tag{8.6.4}$$

Moreover, the heat flux is assumed to behave like a Euclidean vector:

$$q_i' = O_{ik}' q_k. \tag{8.6.5}$$

Note that the last equation follows similar lines of reasoning as Eq. (8.5.6) for the stress tensor: Initially the total heat energy per unit time flowing across the surface of the body is given by:

$$\dot{Q} = -\oint\!\!\!\oint_{\partial V(t)} \hat{q}(\boldsymbol{x}, t; \boldsymbol{n})\mathrm{d}A. \tag{8.6.6}$$

The heat flux density $q = \hat{q}(\boldsymbol{x}, t; \boldsymbol{n})$ is (due to its dependence of the axial normal vector) a rare example of an axial objective tensor of zeroth order, i.e., an axial scalar:

$$q' = \det(\boldsymbol{O}')q. \tag{8.6.7}$$

Cauchy's tetrahedron can be used to map it onto an objective tensor of first order, the heat flux vector, just like the axial traction vector was mapped on the objective stress tensor. We find in complete analogy to Eq. (3.2.7):

$$\hat{q}(\boldsymbol{x}, t; \boldsymbol{n}) = n_j q_j \quad \text{where} \quad q_j = \hat{q}_j\big(\boldsymbol{x}, t; -\boldsymbol{e}_j\big). \tag{8.6.8}$$

The material time derivative of a Euclidean scalar is objective. Moreover, we show easily with Eq. (8.3.18) that:

$$\sigma'_{ji}\frac{\partial \upsilon'_i}{\partial x'_j} = O'_{js}O'_{it}\sigma_{st}\frac{\partial}{\partial x_k}\left(O'_{il}\upsilon_l + \Omega'_{ir}(x'_r - b'_r) + \dot{b}'_i\right)\frac{\partial x_k}{\partial x'_j}$$
$$= O'_{js}O'_{it}\sigma_{st}\left(O'_{il}\frac{\partial \upsilon_l}{\partial x_k} + \Omega'_{ir}O'_{rk}\right)O'_{jk} = \sigma_{kl}\frac{\partial \upsilon_l}{\partial x_k} + O'_{rs}O'_{it}\sigma_{st}\Omega'_{ir} = \sigma_{kl}\frac{\partial \upsilon_l}{\partial x_k}. \tag{8.6.9}$$

Thus Eq. (8.6.3) reads in a Euclidean system:

$$\rho' \dot{u}' = -\frac{\partial q'_k}{\partial x'_k} + \sigma'_{ji}\frac{\partial \upsilon'_i}{\partial x'_j} + \rho' r'. \tag{8.6.10}$$

Consequently the First Law is completely form invariant during change of reference frames. System dependent terms do not occur.

**Exercise 8.6.2:** ***Form invariance of the First Law***

Motivate the objective nature of the fields of the internal energy, the radiation density, and of the heat flux density. Why does the part containing angular velocities vanish in Eq. (8.6.9)?

Hence the form of the total energy balance is determined. By addition of Eqs. (8.6.2) and (8.6.10) we obtain:

$$\rho'\left(u' + \frac{1}{2}\upsilon'_i\upsilon'_i\right)^{\cdot} = -\frac{\partial q'_k}{\partial x'_k} + \frac{\partial \sigma'_{ji}\upsilon'_i}{\partial x'_j}$$
$$+ \rho'\left[\left(f'_i - \Omega'_{il}\Omega'_{lk}(x'_k - b'_k) + \dot{\Omega}'_{ik}(x'_k - b'_k) + \ddot{b}'_i - 2\Omega'_{ik}\dot{b}'_k\right)\upsilon'_i + r'\right]. \tag{8.6.11}$$

## 8.7 Time Derivatives in Moving Systems

In Sect. 8.4 the (material) time derivative of a scalar quantity, namely of the mass density $\rho$, has already been discussed: Eq. (8.4.6). However, in Chap. 10 and 11, dedicated to fluids with memory and to an introduction to plasticity, we will encounter in the constitutive equations time derivatives of Euclidean tensors of higher order, in particular of the stress tensor. Consequently we must pose the question how they transform during change of frames. In particular we expect the constitutive equations to keep their form during the change. Moreover, no frame specific quantities should occur. We now turn to that problem and investigate the time derivative of an objective tensor of first order, i.e., of a Euclidean vector, first:

$$k'_i = O'_{ir} k_r. \tag{8.7.1}$$

The corresponding material time derivative is then given by:

$$\dot{k}'_i = O'_{ir} \dot{k}_r + \dot{O}'_{ir} k_r. \tag{8.7.2}$$

Gustav Andreas Johannes JAUMANN was born on April 18, 1863 in Caransebeş, Hungary and died on July 21, 1924 in the Ötztaler Alps, Austria, during a mountain climbing accident. Initially he studied chemistry at the Technical University of Prague, where he got his Ph.D. in 1890. After that he became a disciple and assistant of Ernst MACH. In the beginning he worked experimentally to turn to theoretical physics later. He finished his habilitation thesis at the German University in Prague where he became an associate professor for experimental physics and physical chemistry in 1893. In 1901 he obtained a chair for physics at the Technical University of Brünn (now Brno). He was a member of the Austrian Academy of Science and also of the German Academy of Scientists Leopoldina. There is an interesting story about the sequence of short-listing for a professorship at the University of Prague. His name was only *secundo loco* after that of Albert EINSTEIN. Rumor has it that JAUMANN was not amused about his "downgrading."

With the definition of the angular velocity:

$$\Omega'_{ij} = \dot{O}'_{ik} O'_{jk}, \tag{8.7.3}$$

this can be rewritten as:

$$\begin{aligned} \dot{k}'_i &= O'_{ir} \dot{k}_r + \dot{O}'_{ir} \delta_{rt} k_t = O'_{ir} \dot{k}_r + \dot{O}'_{ir} O'_{nr} O'_{nt} k_t \\ &= O'_{ir} \dot{k}_r + \Omega'_{in} O'_{nt} k_t = O'_{ir} \dot{k}_r + \Omega'_{in} k'_n. \end{aligned} \tag{8.7.4}$$

Hence the expression:

$$\overset{\circ}{k}'_i = \dot{k}'_i - \Omega'_{in} k'_n = O'_{ir} \dot{k}_r \tag{8.7.5}$$

constitutes an objective tensor of first order. In context with the symbol "∘" for time differentiation it is customary to speak of the so-called JAUMANN time derivative. It is an objective quantity in contrast to the material time derivative of a Euclidean vector. Next we consider an objective tensor $T_{rs}$ of second order:

$$T'_{ij} = O'_{ir} O'_{js} T_{rs}. \tag{8.7.6}$$

James Gardner OLDROYD was born in April, 1921 in Bradford and died on November 22, 1982. He first went to Bradford Grammar School and then attended Trinity College at the University of Cambridge. After his graduation he worked for the Ministry of Supply during World War II. After the war he went to the research laboratories of Courtaulds. In 1953 he became a professor of mathematics at the University of Wales in Swansea. In 1965 he moved to Liverpool University and became Head of Department of Applied Mathematics and Theoretical Physics in 1973 until his death. His main research was dedicated to the visco-elastic behavior of Non-NEWTONIAN fluids.

We will now show that its material time derivative is also *not* objective. For this purpose we first apply the operator $\frac{\mathrm{d}}{\mathrm{d}t} \equiv (\,)^{\cdot}$ to Eq. (8.7.6) and obtain while observing Eq. (8.7.3):

$$
\begin{aligned}
\dot{T}'_{ij} &= O'_{ir}O'_{js}\dot{T}_{rs} + \dot{O}'_{ir}O'_{js}T_{rs} + O'_{ir}\dot{O}'_{js}T_{rs} \\
&= O'_{ir}O'_{js}\dot{T}_{rs} + \dot{O}'_{ir}O'_{pr}O'_{pu}O'_{js}T_{us} + O'_{ir}\dot{O}'_{js}O'_{ps}O'_{pw}T_{rw} \\
&= O'_{ir}O'_{js}\dot{T}_{rs} + \Omega'_{ip}T'_{pj} + \Omega'_{jp}T'_{ip}.
\end{aligned}
\tag{8.7.7}
$$

Clifford Ambrose TRUESDELL III was born on February 18, 1919 in Los Angeles and died on January, 14, 2000 in Baltimore. He first studied physics and mathematics at Caltech. In 1943 he obtained a Ph.D. from Princeton. From 1944 until 1946 he worked at M.I.T.'s Radiation Laboratory and then until 1950 for the Naval Research Laboratory in Washington, D.C. In 1950 he became a professor for mechanics at Indiana University and, finally, in 1961 professor for "Rational Mechanics" at the Johns Hopkins University. His most important achievement are most certainly the various monographs by means of which he established a rational mathematically based way of thinking in mechanics as well as thermodynamics as a fundamental principle.

Marius Sophus LIE was born on December 17, 1842 in Nordfjordeid and died on February, 18, 1899 in Christiania (today Oslo). In Christiania he studied the sciences, obtained a teacher's degree in 1865, and dedicated himself from 1868 onward completely to mathematics. A stipend allowed him to go abroad and so he met the famous German mathematician Felix KLEIN. This acquaintance resulted in several joined papers. In 1872 he became a professor in Christiania and in 1886 successor to KLEIN's chair in Leipzig. LIE's health and psyche were quite delicate. He suffered from pernicious anemia, mental breakdowns and fought regularly with his colleagues about priority issues. In 1894 he was awarded a personal professorship by the Norwegian government in Christiania. However, he returned not before 1898 as a very sick man.

Hence it follows that:

$$\overset{\circ}{T}_{ij} = \dot{T}'_{ij} - \Omega'_{ip}T'_{pj} - \Omega'_{jp}T'_{ip} = O'_{ir}O'_{js}\dot{T}_{rs}. \tag{8.7.8}$$

We conclude that the JAUMANN time derivative of a tensor of second order is also objective.

Finally it should be mentioned that many other objective time derivatives can be defined: OLDROYDT's time derivative, TRUESDELL's time derivative, the LIE derivative, etc.

## 8.8 A Remark on the Form Invariance of Constitutive Equations

Finally in this chapter we want to touch briefly upon the question regarding the form invariance of the constitutive equations mentioned in Chap. 6. In the case of HOOKE's law we start from Eq. (6.2.6). In order to change the frame of reference we assume that LAMÉ's constants, the coefficient of thermal expansion, and the temperature are Euclidean scalars. Moreover, it follows from the orthogonality relations (8.2.5) that the KRONECKER symbol transforms like a Euclidean tensor of second order:

$$\delta'_{ij} = O'_{ir}O'_{js}\delta_{rs}. \tag{8.8.1}$$

By means of the transformations properties of the linear strain tensor and of the stress tensor shown in Eqs. (8.3.14) and (8.5.6) we conclude that the form of HOOKE's law for a Euclidean observer is the same as in an inertial system:

$$\sigma'_{ij} = \lambda'\varepsilon'_{kk}\delta'_{ij} + 2\mu'\varepsilon'_{ij} - 3k'\alpha'\Delta T'\delta'_{ij}. \tag{8.8.2}$$

Similarly we have form invariance for the NAVIER-STOKES law of Eq. (6.3.1) if we, first, assume that the pressure is a Euclidean scalar. In fact it has to be, because it is the trace of the stress tensors and in view of Eqs. (2.6.11), (6.3.2), and the transformations rule for the KRONECKER symbol shown above. Second, we observe the transformation property of the symmetric velocity gradient according to Eq. (8.3.45). Thus:

$$\sigma'_{ij} = -p'\delta'_{ij} + \lambda'd'_{kk}\delta'_{ij} + 2\mu'd'_{ij}. \tag{8.8.3}$$

In view of all that and by observing that the mass density is a Euclidean scalar [cf., Eq. (8.4.4)] it follows that the ideal gas law from Eq. (6.4.1) is also form invariant. We have already mentioned that the internal energy is a Euclidean scalar: Eq. (8.6.4). This in turn is consistent with the representation of the caloric equation of state for the ideal gas in Eq. (6.5.3). As far as FOURIER's law of heat conduction in its form (6.6.1) is concerned it is also form invariant based on Eq. (8.2.22) and if we consider the coefficient of thermal conductivity as a Euclidean scalar:

$$q_i' = -\kappa' \frac{\partial T'}{\partial x_i'}. \tag{8.8.4}$$

## 8.9 Would You Like to Know More?

In all the usual textbooks on continuum mechanics the problem of a change of the frame of reference is a major point of discussion, cf., for example, Greve [8] Sect. 1.4, Bertram [2] in Sect. 4.3 (with additional comments on the Principle of Material Objectivity (PMO), which is not covered in this book), Becker and Bürger [1] in Sect. 1.5, Liu [10], Sects. 1.7 and 3.2 (on the PMO), and in Truesdell [14] on pages 22 pp. and on the P.M.O. on pages 39 pp. The cited text passages also contain remarks on objective time derivatives. In the handbook article by Truesdell and Toupin [16] the problem of a change to a Euclidean frame is treated in Sect. 143. A discussion of Galilean frame invariance can be found in Sect. 171. The problem of the so-called *material frame-indifference* (a synonym for the P.M.O) is extensively outlined in Sect. 19 of Truesdell and Noll [15] and also put into historical context. Even today the notion of material objectivity is subject to fiery controversies and often confused with the concept of objective quantities under Euclidean transformations. More details can be found in the paper by Bertram and Svendsen [3], where also many other references on the subject matter were compiled.

For those interested in the true wording of NEWTONIAN mechanics the various original editions of the Principia are the prime source of information. In fact it is worthwhile to compare the first edition of 1687 with the third one, which shows considerable differences due to numerous comments that NEWTON added in his later years (see the annotated version by Koyré et al. [9]). Here one can also read about NEWTON's ideas on relative motion, absolute space and absolute time, all of which culminates in a famous *gedankenexperiment*, which is known in the literature as NEWTON's bucket. The philosophy behind all this is explained in the reprint of MACH's book [11].

## References

1. Becker E, Bürger W (1975) Kontinuumsmechanik. Teubner Studienbücher Mechanik Band 20. B.G. Teubner, Stuttgart
2. Bertram A (2008) Elasticity and plasticity of large deformations, 2nd edn. Springer, Berlin
3. Bertram A, Svendsen B (2004) Reply to Rivlins's material symmetry revisited or much ado about nothing. GAMM-Mitt 1:88–93

4. d'Alembert JL (1967) Traité de dynamique dans lequel les lois de l'équilibre & du mouvement des corps sont réduites au plus petit nombre possible, & démontrées d'une manière nouvelle, & ou l'on donne un principe général pour trouver le mouvement de plusieurs corps qui agissent les uns sur les autres, d' une manière quelconque. David, Paris
5. Einstein A (1914) Die formale Grundlage der allgemeinen Relativitätstheorie. Sitzungsberichte der Königlich Preußischen Akademie der Wissenschaften (Berlin), pp 1030–1085
6. Einstein A (1920) Relativity. The special and the general theory. A popular exposition, 3rd edn. Methuen & Co. Ltd., London, p. 68
7. Einstein A (1921) A brief outline of the development of the theory of relativity. Nature 106(2677):782–784
8. Greve R (2003) Kontinuumsmechanik—Ein Grundkurs für Ingenieure und Physiker. Springer, Berlin, New York
9. Koyré A, Cohen IB, Whitman A (1972) Isaac Newton's Philosophiae Naturalis Principia Mathematica, 3rd edn (1726), vol I/II with variant readings. Cambridge University Press, Cambridge
10. Liu I-S (2010) Continuum mechanics. Springer, Berlin, New York
11. Mach E (1976) Die Mechanik—historisch-kritisch dargestellt. Wissenschaftliche Buchgesellschaft Darmstadt
12. Thirring H (1918) Über die Wirkungen rotierender ferner Massen in der Einsteinschen Gravitationstheorie. Physikalische Zeitschrift, pp 33–39
13. Thirring H (1921) Berichtigung zu meiner Arbeit: "Über die Wirkungen rotierender ferner Massen in der Einsteinschen Gravitationstheorie". Physikalische Zeitschrift, pp 29–30
14. Truesdell C (1966) The elements of continuum mechanics. Springer, Berlin, New York
15. Truesdell C, Noll W (1965) The non-linear theories of mechanics. In: Flügge S (ed) Encyclopedia of physics, volume III/3. Springer, Berlin, Göttingen
16. Truesdell C, Toupin R (1960) The classical field theories. In: Flügge S (ed.) Encyclopedia of physics, vol III/1, Principles of classical mechanics and field theory. Springer, Berlin, Göttingen

# Chapter 9
# Problems of Linear Elasticity

**Abstract** In this chapter we will investigate some problems of linear elasticity in cylindrical and spherical coordinates. First we consider several circular geometries, namely a disc, cylinders with and without holes, and a sphere, subjected to centrifugal and thermal loads as well as internal and external pressure. We will also discuss the appropriate mathematical tools and procedures that allow us to obtain closed-form solutions, depending on prescribed boundary conditions. Last, but not least, we analyze a more complicated geometry, namely an elliptic hole in a plate subjected to biaxial tensile stress. In the limit of a vanishing half axis the hole degenerates into a pointed slit, the so-called GRIFFITH crack.

*The United States Constitution has proven itself the most marvelously elastic compilation of rules of government ever written.*

Franklin Delano ROOSEVELT

## 9.1 Introduction

The previous sections were dedicated to the mathematical foundations for treating problems of continuum physics namely, in particular, to the balances of mass, momentum, and energy in local and in global form. In combination with suitable constitutive relations, more specifically HOOKE's law and the equations for a NAVIER-STOKES-FOURIER material, we have already solved the resulting field equations for very simple geometries, such as the one-dimensional tensile bar or parallel plate flow. The following chapters are dedicated to more advanced problems. In fact in Chap. 13 we shall even go beyond thermo-mechanics and present the equations of electromagnetism for continuous matter.

W. H. Müller, *An Expedition to Continuum Theory*, Solid Mechanics and Its Applications 210, DOI: 10.1007/978-94-007-7799-6_9,

## 9.2 The Rotating Disc

Consider the circular disc shown in Fig. 9.1 (left) rotating at a constant angular speed $\omega$. Its outer radius is denoted by $R$. Our objective is to determine the internal stresses resulting from the rotation or, in other words, the stresses that arise in the material while counterbalancing the centrifugal forces. We assume that the disc is "very thin" so that stresses will only develop within its plane. In other words, most of the nine components of the stress tensor in cylindrical coordinates will vanish and only $\sigma_{\langle rr\rangle}$, $\sigma_{\langle\vartheta\vartheta\rangle}$, and $\sigma_{\langle r\vartheta\rangle}$ will remain.

Situations where the stress state is considerably reduced due to geometrical constraints occur frequently in solid mechanics. For obvious reasons the present case is known as the state of *plane stress*.

Consequently, the static balance of momentum in cylindrical coordinates, Eq. (5.6.1), consists only of two components:

$$\begin{aligned}\frac{\partial\sigma_{\langle rr\rangle}}{\partial r}+\frac{1}{r}\frac{\partial\sigma_{\langle r\vartheta\rangle}}{\partial\vartheta}+\frac{\sigma_{\langle rr\rangle}-\sigma_{\langle\vartheta\vartheta\rangle}}{r}&=-\rho f_{\langle r\rangle},\\ \frac{\partial\sigma_{\langle r\vartheta\rangle}}{\partial r}+\frac{1}{r}\frac{\partial\sigma_{\langle\vartheta\vartheta\rangle}}{\partial\vartheta}+\frac{2}{r}\sigma_{\langle r\vartheta\rangle}&=-\rho f_{\langle\vartheta\rangle}.\end{aligned}\tag{9.2.1}$$

For the specific body forces on the right hand side we insert the *centrifugal acceleration*. From high school physics it is known that they are proportional to the distance $r$ from the center of rotation:

$$\rho f_{\langle r\rangle}=\rho\omega^2 r,\ \rho f_{\langle\vartheta\rangle}=0.\tag{9.2.2}$$

We now assume that the disc shows linear-elastic material behavior. Consequently, the stress–strain relation is given by HOOKE's law. In particular, HOOKE's law in cylindrical coordinates has already been analyzed in Eq. (6.2.26):

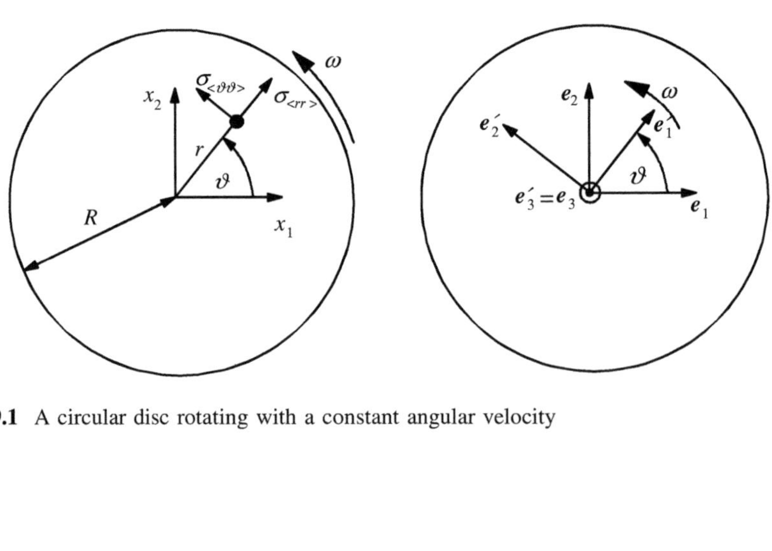


**Fig. 9.1** A circular disc rotating with a constant angular velocity

$$\begin{aligned}
\sigma_{\langle rr\rangle} &= \lambda\big(\varepsilon_{\langle rr\rangle} + \varepsilon_{\langle\vartheta\vartheta\rangle} + \varepsilon_{\langle zz\rangle}\big) + 2\mu\,\varepsilon_{\langle rr\rangle},\\
\sigma_{\langle\vartheta\vartheta\rangle} &= \lambda\big(\varepsilon_{\langle rr\rangle} + \varepsilon_{\langle\vartheta\vartheta\rangle} + \varepsilon_{\langle zz\rangle}\big) + 2\mu\,\varepsilon_{\langle\vartheta\vartheta\rangle},\\
\sigma_{\langle zz\rangle} &= 0 = \lambda\big(\varepsilon_{\langle rr\rangle} + \varepsilon_{\langle\vartheta\vartheta\rangle} + \varepsilon_{\langle zz\rangle}\big) + 2\mu\,\varepsilon_{\langle zz\rangle},\\
\sigma_{\langle r\vartheta\rangle} &= 2\mu\,\varepsilon_{\langle r\vartheta\rangle}, \quad \sigma_{\langle rz\rangle} = 0 = 2\mu\,\varepsilon_{\langle rz\rangle}, \quad \sigma_{\langle\vartheta z\rangle} = 0 = 2\mu\,\varepsilon_{\langle\vartheta z\rangle}.
\end{aligned} \tag{9.2.3}$$

Due to the state of plane stress some of the components of the stress tensor could already be identified as zero. In particular we now eliminate the "annoying" 3D strain component $\varepsilon_{\langle zz\rangle}$ with Eq. $(9.2.3)_3$ from Eq. $(9.2.3)_{1,2}$. Note that $\varepsilon_{\langle zz\rangle}$ is not necessarily zero. Only stress components related to the cylindrical axis $z$ are zero. Moreover, the shear strains $\varepsilon_{\langle rz\rangle}$ and $\varepsilon_{\langle\vartheta z\rangle}$ must vanish due to the state of plane stress. Thus, we obtain:

$$\begin{aligned}
\sigma_{\langle rr\rangle} &= \lambda^*\big(\varepsilon_{\langle rr\rangle} + \varepsilon_{\langle\vartheta\vartheta\rangle}\big) + 2\mu\,\varepsilon_{\langle rr\rangle},\\
\sigma_{\langle\vartheta\vartheta\rangle} &= \lambda^*\big(\varepsilon_{\langle rr\rangle} + \varepsilon_{\langle\vartheta\vartheta\rangle}\big) + 2\mu\varepsilon_{\langle\vartheta\vartheta\rangle},\\
\sigma_{\langle r\vartheta\rangle} &= 2\mu\varepsilon_{\langle r\vartheta\rangle}, \quad \lambda^* = \frac{2\mu}{\lambda + 2\mu}\lambda.
\end{aligned} \tag{9.2.4}$$

We finally use certain relations from Eq. (6.2.27) for the remaining components of the strain tensor in cylindrical coordinates and find:

$$\begin{aligned}
\sigma_{\langle rr\rangle} &= (\lambda^* + 2\mu)\frac{\partial u_{\langle r\rangle}}{\partial r} + \lambda^*\left(\frac{1}{r}\frac{\partial u_{\langle\vartheta\rangle}}{\partial\vartheta} + \frac{u_{\langle r\rangle}}{r}\right),\\
\sigma_{\langle\vartheta\vartheta\rangle} &= \lambda^*\frac{\partial u_{\langle r\rangle}}{\partial r} + (\lambda^* + 2\mu)\frac{1}{r}\left(\frac{\partial u_{\langle\vartheta\rangle}}{\partial\vartheta} + u_{\langle r\rangle}\right),\\
\sigma_{\langle r\vartheta\rangle} &= \mu\left(\frac{1}{r}\left(\frac{\partial u_{\langle r\rangle}}{\partial\vartheta} - u_{\langle\vartheta\rangle}\right) + \frac{\partial u_{\langle\vartheta\rangle}}{\partial r}\right).
\end{aligned} \tag{9.2.5}$$

If the relations (9.2.2) and (9.2.5) are inserted into the balance of momentum (9.2.1) a *coupled* system of partial differential equations for the displacements $u_{\langle r\rangle}$ and $u_{\langle\vartheta\rangle}$ results, which is not easy to solve. However, we have not used all of our knowledge regarding the mathematical form of the displacements yet: Each material point of the disc rotates at a *constant* angular velocity at all times. For reasons of symmetry the radial part of the displacement should therefore not depend on the angle. For the same reason there must be no displacement in angular direction. Thus, it seems natural to propose the following *ansatz* for the displacements:

$$u_{\langle r\rangle} = f(r), \quad u_{\langle\vartheta\rangle} = 0. \tag{9.2.6}$$

$f(r)$ is an unknown function which is yet to be determined from a solution of a differential equation. We have encountered this strategy before in Exercises 7.3.2 and 7.4.1 as well as in Sect. 7.5. It is known as the *semi-inverse method* and widely applied in continuum theory: The structure of the solution to a problem is anticipated by an intelligent guess, which is still general enough so that internal contradictions do not occur. If we now insert Eq. (9.2.6) into Eq. (9.2.5) we obtain:

$$\sigma_{\langle rr\rangle} = (\lambda^* + 2\mu)f'(r) + \lambda^* \frac{f(r)}{r},$$
$$\sigma_{\langle\vartheta\vartheta\rangle} = \lambda^* f'(r) + (\lambda^* + 2\mu)\frac{f(r)}{r}, \quad \sigma_{\langle r\vartheta\rangle} = 0 \tag{9.2.7}$$

and from Eqs. $(9.2.1)_1$ and $(9.2.2)_1$:

$$f''(r) + \frac{f'(r)}{r} - \frac{f(r)}{r^2} = -\frac{\rho\omega^2}{(\lambda^* + 2\mu)} r. \tag{9.2.8}$$

Thus, as we intuitively expect, there are *no shear stresses* in the case of a *stationary* rotating disc. The ordinary differential Eq. (9.2.8) is of the *Euler type*. It is solved as follows (cf., for example [1]):

$$f(r) = f_{\mathrm{h}}(r) + f_{\mathrm{p}}(r). \tag{9.2.9}$$

The indices "h" and "p" refer to the *homogeneous* and the *particular* solutions of which $f(r)$ is additively composed. By inserting the following power functions:

$$f_{\mathrm{h}}(r) = Dr^{\alpha}, \quad f_{\mathrm{p}}(r) = Cr^{s+1} \tag{9.2.10}$$

it follows that:

$$\alpha_1 = 1, \ \alpha_2 = -1, \ s = 2, \quad C = -\frac{\rho\omega^2}{8(\lambda^* + 2\mu)} \tag{9.2.11}$$

and therefore:

$$u_{\langle r\rangle} \equiv f(r) = Ar + \frac{B}{r} - \frac{\rho\omega^2}{8(\lambda^* + 2\mu)} r^3. \tag{9.2.12}$$

The remaining constants $A$ and $B$ are determined from suitable *boundary conditions*. First of all it is intuitively clear that the displacement cannot become singular in any material point $r$ of the disc. Since the disc is not hollow, the point $r = 0$ is part of the system, and therefore:

$$B = 0. \tag{9.2.13}$$

Moreover, the traction vector $t_{\langle i\rangle}$ must be continuous on the free outer rim of the disc. This is a consequence of the local balance of momentum for (non-moving) singular points, cf., Sect. 5.9, Eq. (5.9.5). The traction can be obtained from the stresses:

$$t_{\langle i\rangle} = \sigma_{\langle ji\rangle} n_{\langle j\rangle}. \tag{9.2.14}$$

$n_{\langle j\rangle}$ denotes the unit normal vector of the singular surface. This means in our case:

$$t_{\langle r\rangle} = \sigma_{\langle rr\rangle}, \quad t_{\langle\vartheta\rangle} = 0, \tag{9.2.15}$$

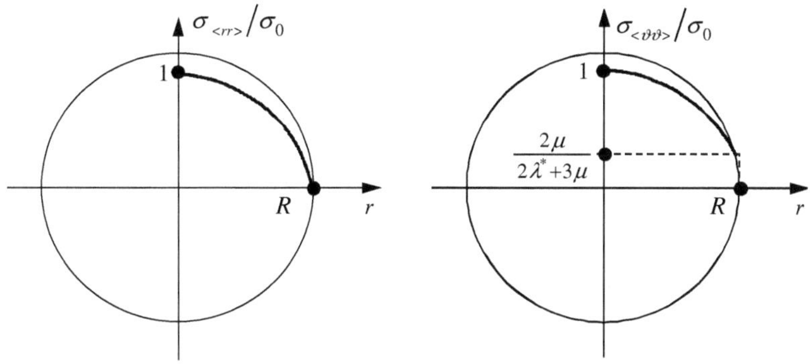


**Fig. 9.2** Radial dependence of the stresses in a disc rotating at a constant angular velocity

because the normal vector in polar coordinates on a circle is given by:

$$n_{\langle r\rangle} = 1, \quad n_{\langle\vartheta\rangle} = 0. \tag{9.2.16}$$

There is no pressure acting on the outer rim. Thus, we obtain by inserting Eq. (9.2.12) into Eq. $(9.2.7)_1$ for the position $r = R$:

$$A = \frac{1}{8}\frac{2\lambda^* + 3\mu}{(\lambda^* + \mu)(\lambda^* + 2\mu)}\rho\omega^2 R^2. \tag{9.2.17}$$

Now all of the constants in (9.2.12) are known. In summary the stresses and the radial displacement of Eqs. (9.2.6/9.2.7) are given by:

$$\begin{aligned}
\sigma_{\langle rr\rangle} &= \sigma_0\left(1 - \left(\frac{r}{R}\right)^2\right), \quad \sigma_{\langle\vartheta\vartheta\rangle} = \sigma_0\left(1 - \frac{2\lambda^* + \mu}{2\lambda^* + 3\mu}\left(\frac{r}{R}\right)^2\right), \\
\sigma_0 &= \frac{1}{4}\frac{2\lambda^* + \mu}{2\lambda^* + 3\mu}\rho\omega^2 R^2, \quad \sigma_{\langle r\vartheta\rangle} = 0, \quad u_{\langle r\rangle} = \frac{\rho\omega^2 R^2}{8(\lambda^* + 2\mu)} r\left(\frac{2\lambda^* + \mu}{2\lambda^* + 3\mu} - \left(\frac{r}{R}\right)^2\right).
\end{aligned} \tag{9.2.18}$$

Figure 9.2 shows how the stresses develop as functions of the radius $r$. Note that in contrast to the radial stress $\sigma_{\langle rr\rangle}$ the hoop stress $\sigma_{\langle\vartheta\vartheta\rangle}$ does *not* reduce to zero at the outer radius $R$ of the disc.

**Exercise 9.2.1:** ***Aftermath on the rotating disc without a central hole***

In the previous section we concentrated on the 2D aspects of the displacements and the strains in a rotating disc. This is reflected in the *ansatz* (9.2.6) and by the fact that it was possible to eliminate the component $\varepsilon_{\langle zz\rangle}$ of the strain tensor by means of the plane stress assumption. Use the results shown in Eq. (9.2.18) and determine $\varepsilon_{\langle zz\rangle}$ explicitly. Integrate this result and calculate the displacement $u_{\langle z\rangle}$ including the proper sign by using Eq. (6.2.27)$_3$. Interpret the sign. Recall that the strain component $\varepsilon_{\langle rz\rangle}$ has to vanish due to the assumption of plane stress. Now calculate $\varepsilon_{\langle rz\rangle}$ directly from Eq. (6.2.27)$_5$ and show that an inconsistency arises. Discuss possible alternative boundary conditions. For example, investigate the case that the average of shear stresses vanishes for a disc of height $-d \leq z \leq +d$.

**Exercise 9.2.2:** ***Rotating disc with a circular hole at its center***

Determine the two constants of integration, $A$ and $B$, of Eq. (9.2.12) for the case that the circular disc of outer radius $R_\mathrm{o}$ contains a circular hole of radius $R_\mathrm{i}$ at its center. Provide analogous formulae to Eq. (9.2.18) for the resulting stresses and for the displacement. In particular, discuss the jump conditions for the stresses (in cylindrical coordinates) starting from the balance in singular points from Sect. 5.8. Try to write the results in a compact and neat form. In other words, normalize suitably (e.g., by $\sigma_0 = \frac{1}{4}\frac{2\lambda^*+3\mu}{\lambda^*+2\mu}\rho\omega^2 R_\mathrm{o}^2$). Discuss the development of the stresses as functions of the radial coordinate $r$: Where are the maxima of the radial and of the hoop stress?

Finally we will show the validity of the equations for the centrifugal force in cylindrical coordinates (9.2.2) by using the general results of Sect. 8.5. We start from Eq. (8.5.9) in Cartesian form and note immediately that in the present situation of a coordinate system circling centrally on a rotating disc w.r.t. the inertial frame we have:

$$b'_i = 0, \tag{9.2.19}$$

since the origin of both systems obviously does not move. Thus, we obtain for the rotation matrix according to Eq. (8.2.3) by computing the scalar products between the unit vectors as shown on the right of Fig. 9.1:

$$O'_{ij}(t) = \begin{bmatrix} \cos\vartheta & \sin\vartheta & 0 \\ -\sin\vartheta & \cos\vartheta & 0 \\ 0 & 0 & 1 \end{bmatrix}. \tag{9.2.20}$$

The change of the angle per unit time or, in other words, the angular velocity is given by:

$$\omega = \dot{\vartheta} = \text{const.} \tag{9.2.21}$$

Hence by Eq. (8.3.17) we find for the matrix of angular velocities and for its time derivative:

$$\Omega'_{ij} = \omega \begin{bmatrix} -\sin\vartheta & \cos\vartheta & 0 \\ -\cos\vartheta & -\sin\vartheta & 0 \\ 0 & 0 & 0 \end{bmatrix} \begin{bmatrix} \cos\vartheta & -\sin\vartheta & 0 \\ \sin\vartheta & \cos\vartheta & 0 \\ 0 & 0 & 1 \end{bmatrix} = \omega \begin{bmatrix} 0 & 1 & 0 \\ -1 & 0 & 0 \\ 0 & 0 & 0 \end{bmatrix} \Rightarrow \dot{\Omega}'_{ij} = 0. \tag{9.2.22}$$

Moreover, we must emphasize that a material point on stationary rotating disc does *not* move w.r.t. the comoving observer:

$$\upsilon'_i = 0. \tag{9.2.23}$$

Thus Eq. (8.5.9) reduces to:

$$\frac{\partial \sigma'_{ji}}{\partial x'_j} = \rho\, \Omega'_{il}\Omega'_{lk}x'_k = \rho\omega^2 \begin{bmatrix} 0 & 1 & 0 \\ -1 & 0 & 0 \\ 0 & 0 & 0 \end{bmatrix} \begin{bmatrix} 0 & 1 & 0 \\ -1 & 0 & 0 \\ 0 & 0 & 0 \end{bmatrix} \begin{bmatrix} x'_1 \\ x'_2 \\ x'_3 \end{bmatrix} = -\rho\omega^2 \begin{bmatrix} x'_1 \\ x'_2 \\ 0 \end{bmatrix}. \tag{9.2.24}$$

Next we transform *mutatis mutandis* the position vector $\boldsymbol{x}'$ (within the plane) to polar coordinates as indicated in Exercise 2.4.5:

$$\underset{(z)}{x}'^1 = r,\ \underset{(z)}{x}'^2 = 0,\ \underset{(z)}{x}'^3 = 0 \Rightarrow x'_{\langle r\rangle} = r,\ x'_{\langle\vartheta\rangle} = 0,\ x'_{\langle z\rangle} = 0. \tag{9.2.25}$$

Thus the result shown in Eqs. (9.2.1/9.2.2) is confirmed incl. the sign on the right hand side.

## 9.3 The Pipeline Problem: A Thick-Walled Hollow Cylinder Under Internal and External Pressure

Consider the situation shown in Fig. 9.3: A long, thick-walled, circular, hollow cylinder of inner radius $R_\mathrm{i}$ and outer radius $R_\mathrm{o}$, the pipeline, is subjected to a (high) internal pressure $p_\mathrm{i}$ and to a (low) external pressure $p_\mathrm{o}$. We want to determine the resulting stresses within the cylinder wall.

Due to its huge length it makes sense to assume that all the relevant fields do *not* depend on the axial coordinate $z$. Moreover, we assume that the displacement in $z$ direction either vanishes or is given by a constant. We say that the cylinder is in a *state of plane strain*.

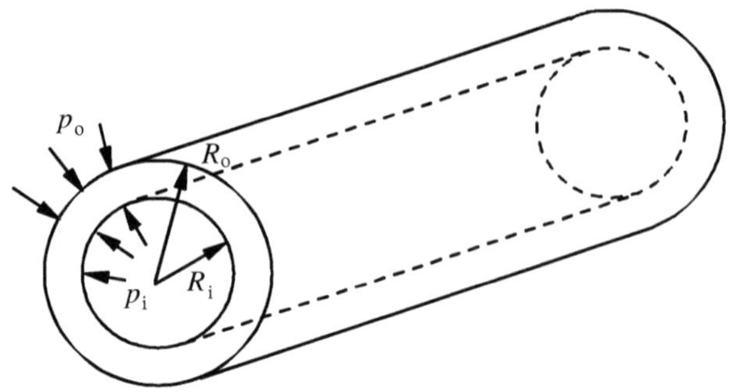


**Fig. 9.3** A thick-walled hollow cylinder subjected to internal and external pressure

Under these circumstances HOOKE's law shown in Eq. (6.2.26) becomes:

$$\begin{aligned} &\sigma_{\langle rr\rangle} = \lambda\left(\varepsilon_{\langle rr\rangle} + \varepsilon_{\langle\vartheta\vartheta\rangle}\right) + 2\mu\,\varepsilon_{\langle rr\rangle}, \quad \sigma_{\langle\vartheta\vartheta\rangle} = \lambda\left(\varepsilon_{\langle rr\rangle} + \varepsilon_{\langle\vartheta\vartheta\rangle}\right) + 2\mu\,\varepsilon_{\langle\vartheta\vartheta\rangle}, \\ &\sigma_{\langle zz\rangle} = \lambda\left(\varepsilon_{\langle rr\rangle} + \varepsilon_{\langle\vartheta\vartheta\rangle}\right), \quad \sigma_{\langle r\vartheta\rangle} = 2\mu\varepsilon_{\langle r\vartheta\rangle}, \quad \sigma_{\langle rz\rangle} = 0, \quad \sigma_{\langle\vartheta z\rangle} = 0. \end{aligned} \tag{9.3.1}$$

We insert the components of the strain tensor in cylindrical coordinates from Eq. (6.2.27):

$$\begin{aligned} \sigma_{\langle rr\rangle} &= (\lambda + 2\mu)\frac{\partial u_{\langle r\rangle}}{\partial r} + \lambda\left(\frac{1}{r}\frac{\partial u_{\langle\vartheta\rangle}}{\partial\vartheta} + \frac{u_{\langle r\rangle}}{r}\right), \\ \sigma_{\langle\vartheta\vartheta\rangle} &= \lambda\frac{\partial u_{\langle r\rangle}}{\partial r} + (\lambda + 2\mu)\left(\frac{1}{r}\frac{\partial u_{\langle\vartheta\rangle}}{\partial\vartheta} + \frac{u_{\langle r\rangle}}{r}\right), \\ \sigma_{\langle r\vartheta\rangle} &= \mu\left(\frac{1}{r}\frac{\partial u_{\langle r\rangle}}{\partial\vartheta} + \frac{\partial u_{\langle\vartheta\rangle}}{\partial r} - \frac{1}{r}u_{\langle\vartheta\rangle}\right), \quad \sigma_{\langle zz\rangle} = \lambda\left(\frac{\partial u_{\langle r\rangle}}{\partial r} + \frac{1}{r}\frac{\partial u_{\langle\vartheta\rangle}}{\partial\vartheta} + \frac{1}{r}u_{\langle r\rangle}\right). \end{aligned} \tag{9.3.2}$$

Note that all but the last of the resulting equations are completely identical to the relations (9.2.5) for plane stress if we only replace $\lambda^*$ by $\lambda$. This simplifies our calculations. For example we may immediately write $(\omega = 0)$ for the displacements in radial and in angular direction by observing Eqs. (9.2.6) and (9.2.12):

$$u_{\langle r\rangle} = Ar + \frac{B}{r}, \quad u_{\langle\vartheta\rangle} = 0. \tag{9.3.3}$$

And the stresses read:

$$\begin{aligned} \sigma_{\langle rr\rangle} &= 2(\lambda + \mu)A - 2\mu\frac{B}{r^2}, \quad \sigma_{\langle\vartheta\vartheta\rangle} = 2(\lambda + \mu)A + 2\mu\frac{B}{r^2}, \\ \sigma_{\langle r\vartheta\rangle} &= 0, \quad \sigma_{\langle zz\rangle} = 2\lambda A. \end{aligned} \tag{9.3.4}$$

The constants of integration, $A$ and $B$, are determined in a slightly different manner than in the previous case of the rotating disc. There are two singular surfaces this time where the traction vector must be continuous, at the inner radius $R_i$ and at the outer radius $R_o$. Now similar arguments as in context with Eqs. (9.2.15) and (9.2.16) do apply. We require that:

$$\sigma_{\langle rr\rangle}\big|_{R_\mathrm{i}} = -p_\mathrm{i}, \quad \sigma_{\langle rr\rangle}\big|_{R_\mathrm{o}} = -p_\mathrm{o}. \tag{9.3.5}$$

By solving the resulting linear system of equations for $A$ and $B$, and by inserting the solution into Eq. (9.3.4) we find:

$$\begin{aligned} \sigma_{\langle rr\rangle} &= \frac{R_\mathrm{i}^2 p_\mathrm{i} - R_\mathrm{o}^2 p_\mathrm{o}}{R_\mathrm{o}^2 - R_\mathrm{i}^2} - (p_\mathrm{i} - p_\mathrm{o})\frac{R_\mathrm{i}^2 R_\mathrm{o}^2}{R_\mathrm{o}^2 - R_\mathrm{i}^2}\frac{1}{r^2}, \\ \sigma_{\langle \vartheta\vartheta\rangle} &= \frac{R_\mathrm{i}^2 p_\mathrm{i} - R_\mathrm{o}^2 p_\mathrm{o}}{R_\mathrm{o}^2 - R_\mathrm{i}^2} + (p_\mathrm{i} - p_\mathrm{o})\frac{R_\mathrm{i}^2 R_\mathrm{o}^2}{R_\mathrm{o}^2 - R_\mathrm{i}^2}\frac{1}{r^2}, \\ \sigma_{\langle r\vartheta\rangle} &= 0, \quad \sigma_{\langle zz\rangle} = \frac{\lambda}{\lambda + \mu}\frac{R_\mathrm{i}^2}{R_\mathrm{o}^2 - R_\mathrm{i}^2}(p_\mathrm{i} - p_\mathrm{o}). \end{aligned} \tag{9.3.6}$$

Next we assume that the cylinder is a thin-walled structure with a wall thickness $t$, so that we may write:

$$R_\mathrm{o} = R_\mathrm{i} + t, \quad R_\mathrm{o}^2 \approx R_\mathrm{i}^2 + 2R_\mathrm{i}t, \quad r \approx \frac{1}{2}(R_\mathrm{o} + R_\mathrm{i}) = R_\mathrm{i} + \frac{1}{2}t. \tag{9.3.7}$$

If these relations are inserted in Eq. (9.3.6) we obtain after expanding w.r.t. the smallness parameter $t/R_\mathrm{i}$:

$$\begin{aligned} \sigma_{\langle rr\rangle} &= -\frac{1}{2}(p_\mathrm{i} + p_\mathrm{o}), \quad \sigma_{\langle \vartheta\vartheta\rangle} = \frac{R_i}{t}(p_\mathrm{i} - p_\mathrm{o}), \\ \sigma_{\langle zz\rangle} &= \frac{\lambda}{\lambda + \mu}\frac{R_\mathrm{i}}{2t}(p_\mathrm{i} - p_\mathrm{o}). \end{aligned} \tag{9.3.8}$$

**Exercise 9.3.1: *Preparations for understanding the "sausage effect"***

Derive Eq. (9.3.8) and discuss the signs. Under which circumstances do we observe tension or compression?

Eq. (9.3.8) is also known as the elementary pressure vessel stress formulae. Obviously for the case of a thin-walled cylinder the radial stress is given by the mean value of the inner and of the outer pressure. This is intuitively clear and it is also indicated by the negative sign that it is a *compressive stress*. In order to interpret the expressions for the hoop stress and for the axial stress, it is helpful to assume that the outer pressure is smaller than the inner one. Under such circumstances both are of tensile nature which, in case of the hoop stress, is also immediately evident. Moreover, it is remarkable that the hoop stress is more than twice as large as the axial one.

The consequences of this observation is known to anyone who has already prepared breakfast sausages in his life or at least watched someone doing it: When the sausage filling starts swelling and the casing finally bursts, the resulting fissure runs (mostly) in axial direction. In other words, the casing is torn apart by the higher hoop stresses.

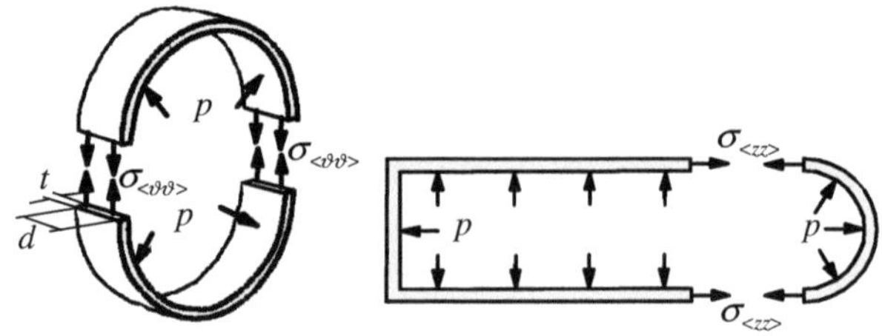


**Fig. 9.4** Deriving the formulae of the pressure vessel code ("sausage effect")

**Exercise 9.3.2:** ***An alternative derivation of the "sausage equations"***

Consider the cartoon drawings shown in Fig. 9.4. Use them for calculating the hoop stress $\sigma_{\langle\vartheta\vartheta\rangle}$: First, determine the force $F$ required to maintain the cylindrical shell in equilibrium and show that:

$$F = 2pdr. \tag{9.3.9}$$

Second, calculate the surface area on which this force is acting to conclude that:

$$\sigma_{\langle\vartheta\vartheta\rangle} = \frac{rp}{t}, \tag{9.3.10}$$

where $p$ denotes the inner pressure in the cylinder. Proceed analogously to find the following expression for the stress $\sigma_{\langle zz\rangle}$:

$$\sigma_{\langle zz\rangle} = \frac{rp}{2t}. \tag{9.3.11}$$

Compare both results with Eq. (9.3.8). Which value of Poisson's ratio guarantees identical results for $\sigma_{\langle zz\rangle}$ and why?

## 9.4 Thermal Stresses in Fiber Reinforced Composites

We now return to the problem that has already been addressed in Chap. 1, i.e., the calculation of the thermal Stresses around a fiber in a composite material. In this context we refer to Fig. 1.7, which shows an idealized model of a fiber-reinforced material. Recall that, in general, the fiber (index 1) as well as the matrix (index 2) possess different elastic and thermal properties, identified by $\lambda_1$, $\mu_1$, $\alpha_1$, and $\lambda_2$, $\mu_2$, $\alpha_2$, respectively. In what follows we will also use the following expressions for the so-called compressibility, $3k_1 = 3\lambda_1 + 2\mu_1$ and $3k_2 = 3\lambda_2 + 2\mu_2$, respectively.

Due to the length of the fibers it makes sense to assume that the presented system is in a state of plane strain. Moreover, in general, each fiber the length of each fiber will be much greater than its distance to the next fiber, i.e., $R_2 - R_1$.

We can immediately read off the solution for the non-vanishing stresses and displacements from Eqs. (9.3.4) and (6.2.24/6.2.26):

$$\left.\begin{aligned}
\sigma^1_{\langle rr\rangle} &= -3k_1\alpha_1(T-T_{\mathrm{R}}) + 2(\lambda_1+\mu_1)A_1 - 2\mu_1\tfrac{B_1}{r^2}\\
\sigma^1_{\langle\vartheta\vartheta\rangle} &= -3k_1\alpha_1(T-T_{\mathrm{R}}) + 2(\lambda_1+\mu_1)A_1 + 2\mu_1\tfrac{B_1}{r^2}\\
\sigma^1_{\langle zz\rangle} &= -3k_1\alpha_1(T-T_{\mathrm{R}}) + 2\lambda_1 A_1,\, u^1_{\langle r\rangle} = A_1 r + \tfrac{B_1}{r}
\end{aligned}\right\}\quad 0 \leq r \leq R_1,$$
$$\left.\begin{aligned}
\sigma^2_{\langle rr\rangle} &= -3k_2\alpha_2(T-T_{\mathrm{R}}) + 2(\lambda_2+\mu_2)A_2 - 2\mu_2\tfrac{B_2}{r^2}\\
\sigma^2_{\langle\vartheta\vartheta\rangle} &= -3k_2\alpha_2(T-T_{\mathrm{R}}) + 2(\lambda_2+\mu_2)A_2 + 2\mu_2\tfrac{B_2}{r^2}\\
\sigma^2_{\langle zz\rangle} &= -3k_2\alpha_2(T-T_{\mathrm{R}}) + 2\lambda_2 A_2,\, u^2_{\langle r\rangle} = A_2 r + \tfrac{B_2}{r}
\end{aligned}\right\},\quad R_1 \leq r \leq R_2, \tag{9.4.1}$$

where it has been observed that, first, the thermal stresses need to be taken into account, which becomes possible by means of the generalized HOOKE's law from Eq. (6.2.24). Second, one needs to distinguish two circular regions with different material parameters. This increases the number of relevant equations and makes each of them longer and more cumbersome. For this reason we refrain from presenting the corresponding relations for the shear stresses (which have to vanish anyway) and for the axial stresses.

Obviously it is necessary to determine *four* constants of integration, namely $A_1$, $A_2$, $B_1$, and $B_2$. For this purpose we need *four* conditions. Following the remarks of the previous sections these are given by:

- a regularity requirement:

$$\sigma^1_{\langle rr\rangle}\Big|_{r=0} < \infty; \tag{9.4.2}$$

- demanding continuity of the traction vector at both interfaces:

$$\sigma^1_{\langle rr\rangle}\Big|_{r=R_1} = \sigma^2_{\langle rr\rangle}\Big|_{r=R_1}, \quad \sigma^2_{\langle rr\rangle}\Big|_{r=R_2} = 0, \tag{9.4.3}$$

- and, finally, continuity of displacement, i.e., a perfect fit of both cylinders along the inner interface:

$$u^1_{\langle r\rangle}\Big|_{r=R_1} = u^2_{\langle r\rangle}\Big|_{r=R_1}. \tag{9.4.4}$$

The last relation can immediately be related to the four constants $A_1, A_2, B_1$, and $B_2$ by using Eq. (9.3.3).

This solves the problem, at least in principle. However, in practice the calculation of the stresses in their most general form leads to rather unwieldy expressions, which most likely will only be of interest to composite specialists. Thus, we shall study two special cases in the following exercises.

**Exercise 9.4.1:** ***Fiber reinforced composites I***

Derive explicit formulae for the resulting thermal stresses in a highly "diluted" fiber reinforced material of equal POISSON numbers for the matrix and for the fiber, i.e., specialize to the case $R_2 \to \infty$:

$$\begin{aligned}
\sigma^1_{\langle rr\rangle} &= \sigma^1_{\langle \vartheta\vartheta\rangle} = \frac{E_1 E_2}{E_1 + E_2(1-2\nu)}(\alpha_2 - \alpha_1)(T - T_R),\\
\sigma^1_{\langle zz\rangle} &= \frac{E_1}{E_1 + E_2(1-2\nu)}(2E_2\nu\alpha_2 - (E_1 + E_2)\alpha_1)(T - T_R),\\
\sigma^2_{\langle rr\rangle} &= -\sigma^2_{\langle \vartheta\vartheta\rangle} = \frac{E_1 E_2}{E_1 + E_2(1-2\nu)}(\alpha_2 - \alpha_1)(T - T_R)\frac{R_1^2}{r^2},\\
\sigma^2_{\langle zz\rangle} &= -E_2\alpha_2(T - T_R).
\end{aligned} \tag{9.4.5}$$

Discuss the sign of the stresses. What results in case of a hole and what for an incompressible fiber? Use the internet and find materials data for typical fiber reinforced composites. Calculate and plot the stresses that arise after the composite is cooled down from fabrication to room temperature.

**Exercise 9.4.2:** ***Fiber reinforced composites II***

Consider the case of a sphere of action of final size radius $R_2$. Assume equal elastic constants for the fiber and for the matrix but different coefficients of thermal expansion and derive compact expressions for all thermal stresses incl. the axial ones, of the form:

$$\sigma^1_{\langle rr\rangle} = \frac{3k\mu}{\lambda + 2\mu}\left(1 - \frac{R_1^2}{R_2^2}\right)(\alpha_2 - \alpha_1)(T - T_R) \tag{9.4.6}$$

(say). Discuss the signs of the stresses.

## 9.5 Transformation Toughened Ceramics

Using ceramic materials in technical applications offers many advantages, at least on first glance: Ceramics are fireproof and can be exposed to extremely high temperatures without suffering from creep, aging, or corrosion. This is important when building engines or turbines, where higher operating temperatures could lead to better efficiencies. The nose of the space shuttles is another example, where the heat resistance of ceramics is used beneficially. It is covered with ceramic tiles that offer a protective shield against the frictional heat during re-entry into the Earth's

**Table 9.1** Fracture toughness of various materials

| Material | $K_{Ic}/(\text{MPa}\sqrt{\text{m}})$ |
|---|---|
| Steel | 50 |
| Cast iron | 13 |
| $Al_2O_3$ | 3–4 |
| $Al_2O_3$ + 10 % $ZrO_2$ | 8–20 |

atmosphere. A few missing tiles can be detrimental as we had to learn the tragic way from the Columbia catastrophe.

As indicated ceramics are also extremely resistant against corrosion, and this is most important for the chemical industry. Moreover, they have excellent tribological properties (also at high temperatures). In other words, they are extremely resistant against wear while showing highest durability. This recommends ceramics for special applications, e.g., as heavy-duty bearings.

However, ceramics also have a well known disadvantage: They are very brittle, i.e., they break easily, and react quite sensitively to impact and tensile stresses because, unlike metals, they do not yield by plastic flow. In order to quantify this statement we compare the fracture toughness of various technically important materials: Table 9.1.

The resistance of a material against (brittle) fracture is characterized by the so-called *fracture toughness* or $K_{Ic}$-value for short. Without going into the details we may say that the higher the toughness the more resistant against fracture the material will be. Indeed, high strength steel has a much higher $K_{Ic}$-value than brittle cast iron (say). Note that the $K_{Ic}$ of a typical ceramic, for example Alumina (or $Al_2O_3$ in chemical terms), is still much lower than that cast iron, namely almost by an order of magnitude. This makes sense since ceramics are typically very brittle and quite susceptible to fracture.

However, if a certain amount of Zirconia particles ($ZrO_2$) of micrometer size are added to Alumina powder we observe that the fracture toughness of the sintered compound has grown considerably, even beyond the $K_{Ic}$ value of cast iron. With a grain of salt we may claim that we have invented the "ceramic steel."

Naturally the question arises why an addition of Zirconia leads to such a dramatic increase of the fracture toughness. The reason for the effect is hidden in the pronounced *polymorphic* nature of $ZrO_2$. The crystal structure of polymorphic solids depends on temperature and on the applied pressure. At atmospheric pressure and above 1,480 °K (ca. 1,200 °C) Zirconia exists in *tetragonal* form. Below that it has a *monoclinic* crystal lattice. Moreover, the volumes of the unit cell of these two crystal structures are very different. The monoclinic phase is approximately 3 % more voluminous than the tetragonal[1] one. Thus, if a piece of Zirconia

[1] Strictly speaking, during the tetragonal to monoclinic phase transition the volume of Zirconia increases *and* we observe a shear as well: The right-angled, tetragonal unit cell transforms into a slightly inclined monoclinic configuration. In other words the angle $\beta$ of Fig. 9.5 is not quite 90°. Consequently shear strains will arise and a more complex state of stress will result. However, in our calculations we will ignore this effect.

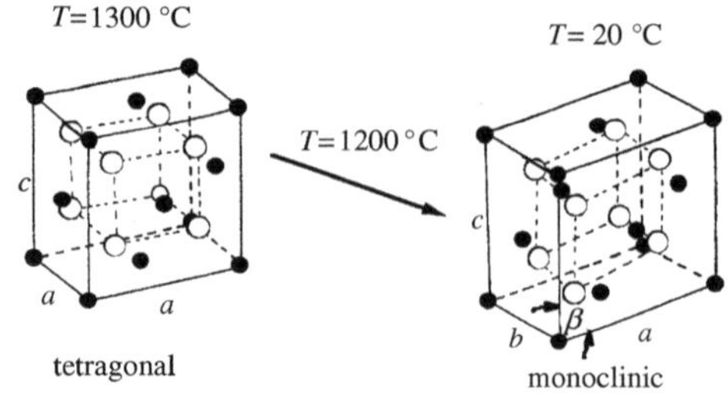


**Fig. 9.5** Zirconia and its polymorphic crystal structures

is cooled down from high to room temperature, we observe at roughly 1,480 K a spontaneous volume expansion. In other words we are confronted with a phase transition similarly to the one when fluid water turns into ice.

But indeed, the temperature at which Zirconia transforms from its tetragonal into the monoclinic phase depends strongly upon the applied pressure. If Zirconia is inserted in a high pressure cell, a so-called Bridgman anvil, and exposed to pressures of several GPa, the phase transition temperature decreases from 1,480 K down to room temperature, cf., Fig. 9.6.

Percy Williams Bridgman was born on April 21, 1882 in Cambridge (Mass.) and died on August 20, 1961 in Randolph (N. H.). He entered Harward University in 1900 to study physics. After graduation he worked and taught at Harward where he became a full professor in 1919. His research focused on experimental high pressure physics. However, he also provided the theoretical background of thermodynamics in order to understand the phase transitions that take place at such extreme conditions. His greatest achievement is probably the development of the "anvil" named after him, by means of which he applied pressures above 10 GPa, i.e., 100,000 atmospheres to samples for the first time. All of this got him the Nobel prize in 1946.

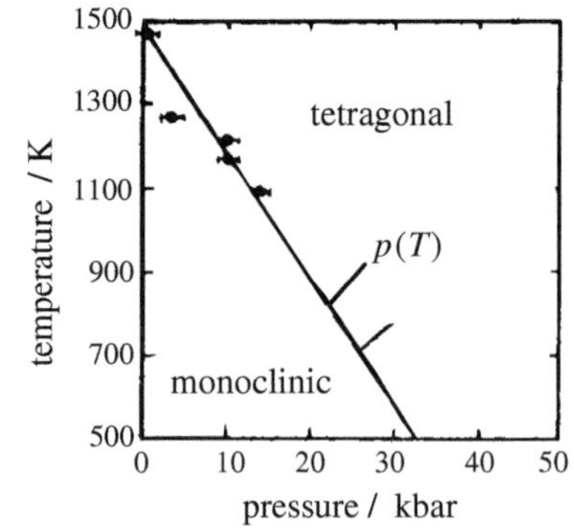


**Fig. 9.6** Phase diagram of $ZrO_2$ (after Whitney [14])

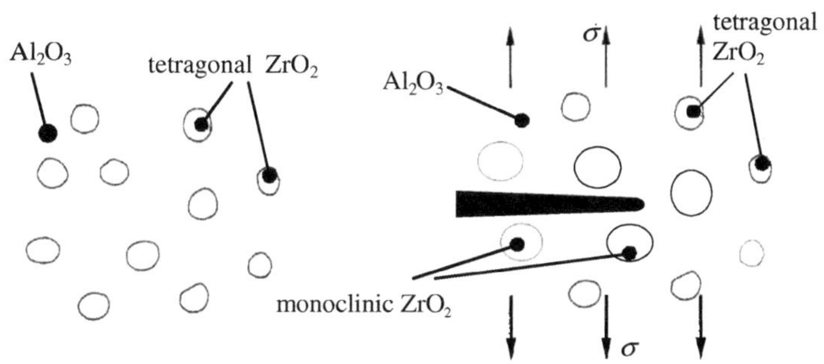


**Fig. 9.7** Transformation toughening in a Zirconia containing ceramic

This is exactly what should happen if $ZrO_2$ particles are sintered into a $Al_2O_3$ matrix. After cooling below 1,480 K the Zirconia particles are basically ready to switch from their tetragonal phase into the monoclinic one. However, this is not automatically possible due to the increase in volume, which makes it necessary to push the matrix aside. This, however, creates a counter pressure which is high enough to stabilize the $ZrO_2$ particles in their tetragonal phase way down below room temperature: Fig. 9.7 (left).

Now if a crack enters such a metastable system under the influence of an external load, cf., Fig. 9.7 (right), it reduces in his vicinity the stabilizing effect of the matrix on the Zirconia particles. In the neighborhood of the crack the particles will then spontaneously transform from their tetragonal into the monoclinic phase. During the process they will increase their volume, push the matrix aside, and generate a compressive zone around the crack. In order to move through that zone an extra amount of energy is required, which in macroscopical terms results in an increased fracture toughness of Zirconia containing ceramics.

In order to guarantee an optimal increase of fracture toughness, it is obviously necessary that the Zirconia particles are in a metastable, tetragonal state and do not transform before the crack is running through the material. It is therefore of interest to predict at which temperature a Zirconia particle of given size will transform in an undamaged matrix of given stiffness. In what follows we will try to provide an estimate for that temperature.

To this end we idealize the situation and consider a spherical Zirconia particle, denoted by "1," of radius $R_1$ in an infinitely large hollow sphere of matrix, denoted by "2," cf., Fig. 1.7. In order to calculate the stresses in that system we make use of the balance of the static balance of momentum in its form (5.5.2) to (5.6.4) in combination with Hooke's law from Eq. (6.2.28), and the strain tensor from Eq. (6.2.29). In other words, all equations are written in spherical coordinates. Moreover, for reasons of symmetry it is reasonable to assume that:

$$u_{\langle r\rangle} = f(r), \quad u_{\langle\varphi\rangle} = 0, \quad u_{\langle\vartheta\rangle} = 0. \tag{9.5.1}$$

If this semi-inverse *ansatz* is inserted we obtain for the strain tensor:

$$\begin{aligned} &\varepsilon_{\langle rr\rangle} = f'(r), \quad \varepsilon_{\langle\vartheta\vartheta\rangle} = \frac{f(r)}{r}, \quad \varepsilon_{\langle\varphi\varphi\rangle} = \frac{f(r)}{r}, \\ &\varepsilon_{\langle r\varphi\rangle} = 0, \quad \varepsilon_{\langle r\vartheta\rangle} = 0, \quad \varepsilon_{\langle\varphi\vartheta\rangle} = 0; \end{aligned} \tag{9.5.2}$$

and for the stress tensor:

$$\begin{aligned} &\sigma_{\langle rr\rangle} = (\lambda + 2\mu)f'(r) + 2\lambda\frac{f(r)}{r}, \\ &\sigma_{\langle\vartheta\vartheta\rangle} = \sigma_{\langle\varphi\varphi\rangle} = 2(\lambda + \mu)\frac{f(r)}{r} + \lambda f'(r), \\ &\sigma_{\langle r\varphi\rangle} = 0, \quad \sigma_{\langle r\vartheta\rangle} = 0, \quad \sigma_{\langle\varphi\vartheta\rangle} = 0; \end{aligned} \tag{9.5.3}$$

and from the radial component of the balance of momentum (the other two components are identically satisfied):

$$f''(r) + 2\frac{f'(r)}{r} - 2\frac{f(r)}{r^2} = 0. \tag{9.5.4}$$

This ordinary differential equation is solved by using the power function $(9.2.10)_1$. Thus we obtain for the displacement in radial direction:

$$u_{\langle r\rangle} = Ar + \frac{B}{r^2} \tag{9.5.5}$$

and, consequently, for the stresses different from zero:

$$\sigma_{\langle rr\rangle} = (3\lambda + 2\mu)A - 4\mu\frac{B}{r^3}, \quad \sigma_{\langle\vartheta\vartheta\rangle} = \sigma_{\langle\varphi\varphi\rangle} = (3\lambda + 2\mu)A + 2\mu\frac{B}{r^3}. \tag{9.5.6}$$

As before $A$ and $B$ denote two constants of integration.

**Exercise 9.5.1:** ***Displacement and stresses in case of spherical symmetry***

Explain in detail the various steps required in order to obtain Eqs. (9.5.5/9.5.6) for the displacements and stresses for the case of perfect spherical symmetry. Moreover, recall the definition of the displacement:

$$\underset{(z)}{u^i} = z^i - Z^i \tag{9.5.7}$$

with the current and with the reference positions $z^i$ and $Z^i$, respectively. Evaluate this relation for the case of spherical coordinates and derive an expression for the *current* radial distance $r$ as a function of the radial distance $R$ in the reference configuration by using Eq. (9.5.5). In the same context discuss the question if it is necessary to distinguish between $r$ and $R$ in Eqs. (9.5.5/9.5.6) for small displacements and strains.

These equations need to be written down for a solid sphere made of (monoclinic) Zirconia, denoted by an index 1 of radius $R_{\mathrm{m}}$ in the reference configuration, and for a hollow sphere made of matrix material, denoted by the index 2, with an inner radius suitable for a tetragonal solid sphere of Zirconia of radius $R_{\mathrm{t}}$ in the reference configuration, respectively. In order to obtain concise formulae that still allow for a modeling of the effect, we will assume that the outer radius of the matrix sphere is infinitely large. This leads to the following equations containing four unknown constants of integration $A_1$, $A_2$, $B_1$, and $B_2$ (also see the analogous cylindrical problem depicted in Fig. 1.7):

$$r_1 = \left(1 + A_1 + \frac{B_1}{R^3}\right)R, \quad \sigma^1_{\langle rr\rangle} = (3\lambda_1 + 2\mu_1)A_1 - 4\mu_1 \frac{B_1}{R^3},$$
$$\sigma^1_{\langle\vartheta\vartheta\rangle} = \sigma^1_{\langle\varphi\varphi\rangle} = (3\lambda_1 + 2\mu_1)A_1 + 2\mu_1 \frac{B_1}{R^3}, \quad 0 \le R \le R_{\mathrm{m}} \tag{9.5.8}$$

and:

$$r_2 = \left(1 + A_2 + \frac{B_2}{R^3}\right)R, \quad \sigma^2_{\langle rr\rangle} = (3\lambda_2 + 2\mu_2)A_2 - 4\mu_2 \frac{B_2}{R^3},$$
$$\sigma^2_{\langle\vartheta\vartheta\rangle} = \sigma^2_{\langle\varphi\varphi\rangle} = (3\lambda_2 + 2\mu_2)A_2 + 2\mu_2 \frac{B_2}{R^3}, \quad R_{\mathrm{t}} \le R \le R_2 \to \infty. \tag{9.5.9}$$

In order to determine the four constants four boundary conditions are required. These include requirements for

- regularity in the origin at $r = 0$, for example:

$$\sigma^1_{\langle rr\rangle}\Big|_{r=0} < \infty; \tag{9.5.10}$$

- continuity of the traction vector at the transition between Zirconia sphere and the matrix:

$$\sigma^1_{\langle rr\rangle}\Big|_{r=R_{\mathrm{m}}} = \sigma^2_{\langle rr\rangle}\Big|_{r=R_{\mathrm{t}}}; \tag{9.5.11}$$

- a perfect fit; in other words, the cavity must be extended and the solid sphere of monoclinic Zirconia must be compressed until the volume expansion due to the phase transition of roughly $\zeta = 3\,\%$ has been compensated:

$$r_1\big|_{r=R_{\mathrm{m}}} = r_2\big|_{r=R_{\mathrm{t}}}; \tag{9.5.12}$$

- the decay of the stresses to zero at infinity:

$$\sigma^2_{\langle rr\rangle}\Big|_{R_2\to\infty} = 0. \tag{9.5.13}$$

If these four equations are evaluated with Eqs. (9.5.8/9.5.9) and if, in addition, the balance of mass is observed:

$$\frac{4\pi}{3}\rho_t R_t^3 = \frac{4\pi}{3}\rho_m R_m^3, \tag{9.5.14}$$

with the mass densities $\rho_t$ and $\rho_m$ of the tetragonal and of the monoclinic Zirconia in the reference configuration, respectively, we obtain with the smallness parameter:

$$\zeta = \frac{\rho_t - \rho_m}{3\rho_m} > 0 \tag{9.5.15}$$

immediately:

$$\begin{aligned} A_2 = B_1 = 0, \quad \frac{B_2}{R_t^3} &= \frac{3\lambda_1 + 2\mu_1}{3\lambda_1 + 2\mu_1 + 4\mu_2}\zeta, \\ A_1 &= -\frac{4\mu_2}{3\lambda_1 + 2\mu_1 + 4\mu_2}\zeta. \end{aligned} \tag{9.5.16}$$

Thus the stresses of interest are given by:

$$\begin{aligned} \sigma^1_{\langle rr\rangle} = \sigma^1_{\langle\vartheta\vartheta\rangle} = \sigma^1_{\langle\varphi\varphi\rangle} &= -\frac{4\mu_2(3\lambda_1 + 2\mu_1)}{3\lambda_1 + 2\mu_1 + 4\mu_2}\zeta, \\ \sigma^2_{\langle rr\rangle} &= -\frac{4\mu_2(3\lambda_1 + 2\mu_1)}{3\lambda_1 + 2\mu_1 + 4\mu_2}\frac{R_t^3}{R^3}\zeta, \\ \sigma^2_{\langle\vartheta\vartheta\rangle} = \sigma^2_{\langle\varphi\varphi\rangle} &= \frac{2\mu_2(3\lambda_1 + 2\mu_1)}{3\lambda_1 + 2\mu_1 + 4\mu_2}\frac{R_t^3}{R^3}\zeta. \end{aligned} \tag{9.5.17}$$

Note that the sphere of Zirconia is subjected to an *isotropic homogeneous* state of stress, i.e., under a "pressure" that can be linked to the phase diagram shown in Fig. 9.7. In order to compute the transition temperature $T$ of the Zirconia embedded in the matrix we put:

$$-\frac{1}{3}\left(\sigma^1_{\langle rr\rangle} + \sigma^1_{\langle\vartheta\vartheta\rangle} + \sigma^1_{\langle\varphi\varphi\rangle}\right) = p(T), \tag{9.5.18}$$

where $p(T)$ is the straight line shown in Fig. 9.7 that divides the region of tetragonal phase from the one of monoclinic stability. From measurements it is known that:

$$\begin{aligned} p(T) &= -\frac{10^{-2}\,\text{GPa}}{3.02\ \ \text{K}} \cdot T + 4.89\,\text{GPa}, \\ \lambda_1 = 31\ \text{GPa}, \quad \mu_1 = 66\ \text{GPa}, &\quad \lambda_2 = 102\ \text{GPa}, \quad \mu_2 = 110\ \text{GPa}. \end{aligned} \tag{9.5.19}$$

Hence we find:

$$T = 128\ \text{K}. \tag{9.5.20}$$

**Exercise 9.5.2: *Transition temperature of a Zirconia inclusion***

Confirm in detail the solution for the stresses shown in Eq. (9.5.17) and calculate the constants of integration by means of the boundary and jump conditions (9.5.10–9.5.13). In particular study the arguments that led to the continuity condition (9.5.12).

Also verify the transition temperature of Eq. (9.5.20) and explain why we may use the experimental result of Fig. 9.7 for the transformed *sphere* of Zirconia in our model. Would this also work with a Zirconia inclusion of arbitrary shape?

And exactly this has been observed: It has been shown experimentally that depending on their size Zirconia particles embedded in Alumina need to be cooled down up to temperatures of liquid nitrogen. It turns out that the smaller the particle, the lower the transformation temperature, i.e., the more difficult it becomes to accomplish the phase transition. Nevertheless, we may argue that our estimate concerns notably small particles since the sphere of Zirconia was embedded in an infinitely large matrix. However, the presented method can be extended to particles embedded in a matrix of finite size and the afore-mentioned size effect can be predicted from the resulting equations.

**Exercise 9.5.3: *Extended Hooke's law***

The concept of a reference configuration is typically not used in context with problems of linear elasticity. In fact, following the book by Mura [2] we can avoid it in Eqs. (9.5.7–9.5.9) by arguing as follows. In general Hooke's law relates stresses with *elastic* strains, i.e., in Cartesian coordinates (for simplicity) we may write:

$$\sigma_{ij} = C_{ijkl}\, \varepsilon^{\text{el}}_{kl}. \tag{9.5.21}$$

The elastic strains are an *additive* part of the total strains:

$$\varepsilon_{ij} = \varepsilon^{\text{el}}_{ij} + \varepsilon'_{ij}, \tag{9.5.22}$$

where $\varepsilon'_{ij}$ stands for all non-elastic strain contributions. The latter are, for example, thermal strains, for which we may write in the isotropic case:

$$\varepsilon'_{ij} = \alpha(T - T_{\text{R}})\delta_{ij}. \tag{9.5.23}$$

Strains related to phase transformations that are accompanied by a change of volume or shape as indicated in Eq. (9.5.15), or plastic strains are further examples of $\varepsilon'_{ij}$. Use this relation and state the corresponding HOOKE's law. Show that it leads to the same results as given by Eq. (9.5.17) by evaluating the condition

$$u^1_{\langle r\rangle}\Big|_{r_\mathrm{i}} = u^2_{\langle r\rangle}\Big|_{r_\mathrm{i}}, \tag{9.5.24}$$

and by using the inner radius $r_\mathrm{i}$ of the hollow matrix as a measure of distance. Discuss the pros and cons of this method.

## 9.6 Compression of a Sphere

In this section we will address the problem of the broken sphere in a bushing shown in Figs. 1.1 and 1.2. In contrast to the transformed Zirconia sphere this problem does not show perfect spherical symmetry any more. However, there is some symmetry left, namely w.r.t. the polar axis of the sphere. Hence it makes sense to look for a solution for the stresses, strains, and for the displacements by starting from the static balance of momentum, Hooke's law, and kinematic conditions in spherical coordinates as presented in Exercises 5.4.2 and 6.2.5. As we shall see the semi-inverse method can still be used, although the *ansatz* will be more complex and basically consist of products of unknown functions of the radius $r$ and of the polar angle $\vartheta$. The azimuthal angle $\varphi$ will *not* occur because of the symmetry with respect to the polar axis. During the solution we will discover a special class of mathematical functions, which are very useful when tackling problems of spherical symmetry, the so-called LEGENDRE *polynomials*. In fact, this whole section is inspired by the paper of Hiramatsu and Oka [3] who, unlike us, in the end specialized the solution the case of a sphere symmetrically compressed at the poles.

In this spirit we assume static conditions, neglect body forces, and allow no dependence on $\varphi$ so that the balance of momentum of Eq. (5.4.7) reduces to:

$$\begin{aligned}
&\frac{\partial\sigma_{\langle rr\rangle}}{\partial r} + \frac{1}{r}\frac{\partial\sigma_{\langle r\vartheta r\rangle}}{\partial\vartheta} + \frac{1}{r}\left[2\sigma_{\langle rr\rangle} - \sigma_{\langle\vartheta\vartheta\rangle} - \sigma_{\langle\varphi\varphi\rangle} + \sigma_{\langle r\vartheta\rangle}\cot\vartheta\right] = 0,\\
&\frac{\partial\sigma_{\langle r\vartheta\rangle}}{\partial r} + \frac{1}{r}\frac{\partial\sigma_{\langle\vartheta\vartheta\rangle}}{\partial\vartheta} + \frac{1}{r}\left[3\sigma_{\langle r\vartheta\rangle} + \left(\sigma_{\langle\vartheta\vartheta\rangle} - \sigma_{\langle\varphi\varphi\rangle}\right)\cot\vartheta\right] = 0,\\
&\frac{\partial\sigma_{\langle r\varphi\rangle}}{\partial r} + \frac{1}{r}\frac{\partial\sigma_{\langle\vartheta\varphi\rangle}}{\partial\vartheta} + \frac{1}{r}\left(3\sigma_{\langle r\varphi\rangle} + 2\sigma_{\langle\vartheta\varphi\rangle}\cot\vartheta\right) = 0.
\end{aligned} \tag{9.6.1}$$

We now turn to HOOKE's law including the strains and the displacements. After simplifying the kinematic relations of Eq. (6.2.29) by omitting derivatives w.r.t. $\varphi$ and inserting the strains into Eq. (6.2.28) we find that:

$$\sigma_{\langle rr\rangle} = \lambda\Delta + 2\mu\frac{\partial u_{\langle r\rangle}}{\partial r}, \quad \sigma_{\langle\vartheta\vartheta\rangle} = \lambda\Delta + 2\mu\left(\frac{\partial u_{\langle\vartheta\rangle}}{\partial\vartheta} + \frac{1}{r}u_{\langle r\rangle}\right),$$
$$\sigma_{\langle\varphi\varphi\rangle} = \lambda\Delta + \frac{2\mu}{r}\left(u_{\langle r\rangle} + u_{\langle\vartheta\rangle}\cot\vartheta\right), \quad \sigma_{\langle r\vartheta\rangle} = \mu\left[\frac{\partial u_{\langle\vartheta\rangle}}{\partial r} - \frac{1}{r}\left(u_{\langle\vartheta\rangle} - \frac{\partial u_{\langle r\rangle}}{\partial\vartheta}\right)\right],$$
$$\sigma_{\langle\vartheta\varphi\rangle} = \frac{\mu}{r}\left(\frac{\partial u_{\langle\varphi\rangle}}{\partial\vartheta} - u_{\langle\varphi\rangle}\cot\vartheta\right), \quad \sigma_{\langle r\varphi\rangle} = \mu\left(\frac{\partial u_{\langle\varphi\rangle}}{\partial r} - \frac{1}{r}u_{\langle\varphi\rangle}\right),$$
(9.6.2)

where the following abbreviation has been used:

$$\Delta = \frac{1}{r^2\sin\vartheta}\left[\frac{\partial}{\partial r}\left(r^2 u_{\langle r\rangle}\sin\vartheta\right) + \frac{\partial}{\partial\vartheta}\left(r u_{\langle\vartheta\rangle}\sin\vartheta\right)\right]. \tag{9.6.3}$$

We now insert Eq. (9.6.2) into Eq. $(9.6.1)_{1,2}$ and a system of coupled partial differential equations for $u_{\langle r\rangle}$ and $u_{\langle\vartheta\rangle}$ results:

$$\begin{aligned}(\lambda + 2\mu)\frac{\partial\Delta}{\partial r} - \frac{2\mu}{r}\frac{\partial\Omega}{\partial\vartheta} - \frac{2\mu}{r}\Omega\cot\vartheta = 0,\\ (\lambda + 2\mu)\frac{1}{r}\frac{\partial\Delta}{\partial\theta} + 2\mu\frac{\partial\Omega}{\partial r} + 2\mu\frac{\Omega}{r} = 0,\end{aligned} \tag{9.6.4}$$

where a second abbreviation has been used:

$$2\Omega = \frac{\partial u_{\langle\vartheta\rangle}}{\partial r} + \frac{u_{\langle\vartheta\rangle}}{r} - \frac{1}{r}\frac{\partial u_{\langle r\rangle}}{\partial\vartheta}. \tag{9.6.5}$$

Equation $(9.6.1)_3$ will later serve us to determine $u_{\langle\varphi\rangle}$ and is ignored for the time being. By cross-differentiation w.r.t. $r$ and $\vartheta$ and mutual addition and subtraction, Eq. (9.6.4) can be decoupled and the following second order partial differential equations for $\Delta$ and $\Omega$ result:

$$\begin{aligned}&\frac{\partial^2\Delta}{\partial r^2} + \frac{2}{r}\frac{\partial\Delta}{\partial r} + \frac{1}{r^2}\frac{\partial^2\Delta}{\partial\vartheta^2} + \frac{\cot\vartheta}{r^2}\frac{\partial\Delta}{\partial\vartheta} = 0,\\ &\frac{\partial^2\Omega}{\partial r^2} + \frac{2}{r}\frac{\partial\Omega}{\partial r} + \frac{1}{r^2}\frac{\partial^2\Omega}{\partial\vartheta^2} + \frac{\cot\vartheta}{r^2}\frac{\partial\Omega}{\partial\vartheta} - \frac{1}{(r\sin\vartheta)^2}\Omega = 0.\end{aligned} \tag{9.6.6}$$

Now we apply the semi-inverse method: We look for solutions in terms of products of functions that depend exclusively on $r$ and $\vartheta$. In fact this idea goes back to Johann Bernoulli and is a.k.a. *method of separation* in the mathematical literature:

$$\Delta(r,\vartheta) = f_n(r)\,g_n(\vartheta), \quad \Omega(r,\vartheta) = F_n(r)\,G_n(\vartheta). \tag{9.6.7}$$

Insertion into Eq. (9.6.6) yields:

$$
\begin{aligned}
r^2 \frac{f_n''}{f_n} + 2r \frac{f_n'}{f_n} &= -\left[\frac{g_n''}{g_n} + \cot\vartheta \frac{g_n'}{g_n}\right], \\
r^2 \frac{F_n''}{F_n} + 2r \frac{F_n'}{F_n} &= -\left[\frac{G_n''}{G_n} + \cot\vartheta \frac{G_n'}{G_n} - \frac{1}{\sin^2\vartheta}\right].
\end{aligned}
\tag{9.6.8}
$$

The dashes refer to differentiation w.r.t. the corresponding argument. Since $r$ and $\vartheta$ are independent of each other the left and right hand side of Eq. (9.6.8) must be constant. For reasons that will become evident shortly we choose this constant as $n(n+1)$ where $n = 0, 1, 2, \ldots$. We attempt a solution by power functions of the form $f_n(r) \sim r^\alpha$ and $F_n(r) \sim r^\beta$ for the differential equations on the left hand side and find that $\alpha_{1,2} = \beta_{1,2} = -\frac{1}{2} \pm \left(n + \frac{1}{2}\right)$. Hence:

$$
f_n(r) = A_n r^n - \frac{B_n}{r^{n+1}}, \quad F_n(r) = a_n r^n - \frac{b_n}{r^{n+1}}. \tag{9.6.9}
$$

Recall that infinitely many choices of $n$ are possible. Thus, we have found infinitely many solutions of power functions in $r$, positive as well as negative ones. If $n$ had not been chosen as integer the $r$-dependence of the displacements would not result in power functions and thus contradict the idea of a power series. In fact an infinite sum of functions with positive *and* negative powers is known as a LAURENT series. With a grain of salt it can be considered as an extension of the well known TAYLOR series including also contributions that become singular for $r = 0$. As we shall see such power functions in $r$ are perfectly sufficient in order to solve the axially symmetric boundary value problem. It will not lead to contradictions. This is in agreement with our afore-mentioned principle that the semi-inverse *ansatz* should always be kept as simple as possible. Finally note that the minus in front of the second terms in (9.6.9) is arbitrary. We are just following the convention established in Hiramatsu and Oka [3].

**233.**

ANALYSE MATHÉMATIQUE. — *Rapport sur un Mémoire de* M. LAURENT, *qui a pour titre :* Extension du théorème de M. Cauchy relatif à la convergence du développement d'une fonction suivant les puissances ascendantes de la variable $x$.

C. R., T. XVII, p. 938 (30 octobre 1843).

L'Académie nous a chargés, M. Liouville et moi, de lui rendre compte d'un Mémoire de M. Laurent relatif à l'extension d'un théorème que l'un de nous a donné dans le Mémoire présenté à l'Académie de Turin le 11 octobre 1831, et dont il a fourni une démonstration nouvelle dans ses *Exercices d'Analyse et de Physique mathématique*. Le théorème en question peut s'énoncer comme il suit :

*$x$ désignant une variable réelle ou imaginaire, une fonction réelle ou imaginaire de $x$ sera développable en une série convergente ordonnée suivant les puissances ascendantes de cette variable, tant que le module de la variable conservera une valeur inférieure à la plus petite de celles pour lesquelles la fonction ou sa dérivée cesse d'être finie ou continue.*

En examinant attentivement la première démonstration de ce théorème, M. Laurent a reconnu, comme il le dit lui-même, que l'analyse employée par l'auteur pouvait conduire à un théorème plus général, relatif au développement d'une fonction en une série ordonnée suivant les puissances entières positives, nulle et négatives de la variable. Déjà, dans une des séances de l'Académie, le rapporteur avait montré qu'un semblable développement, lorsqu'il peut s'effectuer entre deux

Pierre Alphonse LAURENT was born on July 18, 1813 in Paris where he also died on September 2, 1854. Following the French tradition he was a mathematician *and* an engineer. The latter passion explains his engagement in setting up the army port of Le Havre. He is known as the discoverer of LAURENT series, an expansion of a function into an infinite power series, with positive and negative exponents thus generalizing the TAYLOR series expansion. His result was contained in a memoir submitted for the Grand Prize of the Académie des Sciences in 1843. Rumor has it that his submission arrived after due date. Consequently the paper was not published and never considered for the prize. Laurent died at the very young age of 41. Tragically his work was not really published until after his death. However, Cauchy reported on its contents as early as 1843 (see inset).

We now turn to the right hand sides of Eq. (9.6.8) and obtain the following ordinary differential equations of second order:

$$\begin{aligned} &g''(\vartheta) + \cot\vartheta\, g'(\vartheta) + n(n+1)g(\vartheta) = 0, \\ &G''(\vartheta) + \cot\vartheta\, G'(\vartheta) + \left[n(n+1) - \frac{1}{\sin^2\vartheta}\right]G(\vartheta) = 0. \end{aligned} \tag{9.6.10}$$

The first one is LEGENDRE's differential equation in standard form. It is solved by certain polynomials in $\vartheta$, the so-called LEGENDRE polynomials, $P_n(\cos\vartheta)$, defined by:

$$\frac{d^2 P_n}{d\vartheta^2} + \cot\vartheta\, \frac{dP_n}{d\vartheta} + n(n+1)P_n = 0. \tag{9.6.11}$$

We will investigate the specific form of these polynomials in Exercise 9.6.1. If we differentiate $(9.6.10)_1$ by $\vartheta$ it is easy to show that the second differential equation is solved by $dP_n/d\vartheta$. Thus, the general solutions for $\varDelta$ and $\Omega$ read:

$$\varDelta = \sum_{n=0}^{\infty}\left(A_n r^n - \frac{B_n}{r^{n+1}}\right)P_n, \quad \Omega = \sum_{n=0}^{\infty}\left(a_n r^n - \frac{b_n}{r^{n+1}}\right)\frac{dP_n}{d\vartheta}. \tag{9.6.12}$$

However, $\varDelta$ and $\Omega$ are not independent but linked to each other via Eq. (9.6.4). Thus, if we insert (9.6.12) into $(9.6.4)_1$ (say) and observe Eq. (9.6.11), we finally find that:

$$2\Omega = -\frac{\lambda + 2\mu}{\mu}\sum_{n=0}^{\infty}\left(\frac{A_n}{n+1}r^n + \frac{B_n}{n}\frac{1}{r^{n+1}}\right)\frac{dP_n}{d\vartheta}. \tag{9.6.13}$$

The term $n = 0$ in this equation does not really present a singularity problem since we shall see in Exercise 9.6.1 that $dP_0/d\vartheta = 0$. The solutions shown in Eqs. $(9.6.12)_1$ and (9.6.13) will now help us to find the general solution for the displacements $u_{\langle r\rangle}$ and $u_{\langle\vartheta\rangle}$. We use the definitions (9.6.3) and (9.6.5) to find uncoupled differential equations for both displacements. First for $u_{\langle\vartheta\rangle}$:

$$\begin{aligned} &\frac{\partial^2 u_{\langle\vartheta\rangle}}{\partial r^2} + \frac{1}{r^2}\frac{\partial^2 u_{\langle\vartheta\rangle}}{\partial\vartheta^2} + \frac{4}{r}\frac{\partial u_{\langle\vartheta\rangle}}{\partial r} + \frac{\cot\vartheta}{r^2}\frac{\partial u_{\langle\vartheta\rangle}}{\partial\vartheta} + \frac{1}{r^2}\left(2 - \frac{1}{\sin^2\vartheta}\right)u_{\langle\vartheta\rangle} \\ &= \frac{\partial(2\Omega)}{\partial r} + \frac{1}{r}\frac{\partial\varDelta}{\partial\vartheta} + \frac{3}{r}(2\Omega). \end{aligned} \tag{9.6.14}$$

As before we look for solutions of the product type:

$$u_{\langle\vartheta\rangle}(r,\vartheta) = f_n(r)\, g_n(\vartheta). \tag{9.6.15}$$

By inserting this in the homogeneous equivalent to (9.6.14) we find after separation for the homogeneous part of the $r$-solution:

$$f_n^{\text{hom}}(r) = C_n r^{n-1} - \frac{D_n}{r^{n+2}}. \tag{9.6.16}$$

The $\vartheta$-dependence resulting from Eq. (9.6.14) leads again to a differential equation of the type $(9.6.10)_2$. However, for the complete solution we also need one particular solution to cover the inhomogeneous part on the right hand side of Eq. (9.6.14). It can be determined from the $r$-dependencies of Eqs. $(9.6.12)_1$ and (9.6.13). Thus after some algebraic manipulations we find that in total:

$$u_{\langle\vartheta\rangle} = \sum_{n=0}^{\infty} \left[ -\frac{(n+3)\lambda + (n+5)\mu}{(n+1)(2n+3)2\mu} A_n r^{n+1} - \frac{(n-2)\lambda + (n-4)\mu}{n(2n-1)2\mu} \frac{B_n}{r^n} + C_n r^{n-1} - \frac{D_n}{r^{n+2}} \right] \frac{\mathrm{d}P_n}{\mathrm{d}\vartheta}. \tag{9.6.17}$$

Now that we have found $u_{\langle\vartheta\rangle}$ the radial displacement $u_{\langle r\rangle}$ can be obtained by integration from Eqs. (9.6.5) or (9.6.3):

$$\begin{aligned} \frac{\partial u_{\langle r\rangle}}{\partial\vartheta} &= -2r\Omega + \frac{\partial}{\partial r}\left(r u_{\langle\vartheta\rangle}\right), \\ \frac{\partial}{\partial r}\left(r^2 u_{\langle r\rangle}\right) &= r^2 \Delta - r\left(\frac{\partial u_{\langle\vartheta\rangle}}{\partial\vartheta} + u_{\langle\vartheta\rangle}\cot\vartheta\right). \end{aligned} \tag{9.6.18}$$

If we insert Eqs. $(9.6.12)_1$, (9.6.13), and (9.6.17) we obtain almost the same results from both equations. There is a subtle difference though: The first integration can be performed up to an unknown function in $\vartheta$ and the second one up to an unknown function of $r$. However, since both results must be equal, these functions degenerate into a constant, which is equal to zero, since we do not allow rigid body translations to occur. Therefore, the final result reads:

$$u_{\langle r\rangle} = \sum_{n=0}^{\infty} \left[ -\frac{n\lambda + (n-2)\mu}{(2n+3)2\mu} A_n r^{n+1} + \frac{(n+1)\lambda + (n+3)\mu}{(2n-1)2\mu} \frac{B_n}{r^n} \\ n C_n r^{n-1} + (n+1)\frac{D_n}{r^{n+2}} \right] P_n. \tag{9.6.19}$$

We now turn to the third displacement $u_{\langle\varphi\rangle}$ which can be obtained by combining Eqs. $(9.6.1)_3$ and $(9.6.2)_{5,6}$:

$$\frac{\partial^2 u_{\langle\varphi\rangle}}{\partial r^2} + \frac{2}{r}\frac{\partial u_{\langle\varphi\rangle}}{\partial r} + \frac{1}{r^2}\frac{\partial^2 u_{\langle\varphi\rangle}}{\partial\vartheta^2} + \frac{\cot\vartheta}{r^2}\frac{\partial u_{\langle\varphi\rangle}}{\partial\vartheta} - \frac{1}{(r\sin\vartheta)^2} u_{\langle\varphi\rangle} = 0. \tag{9.6.20}$$

As before a product *ansatz* is used:

$$u_{\langle\vartheta\rangle}(r,\vartheta) = f_n(r)\, g_n(\vartheta) \tag{9.6.21}$$

leading to:

$$\left[f_n'' + \frac{2}{r}f_n' - n(n+1)f_n\right] g_n + \frac{1}{r^2} f_n \left[g_n'' + \cot\vartheta\, g_n' + \left(n(n+1) - \frac{1}{\sin^2\vartheta}\right) g_n\right] = 0. \tag{9.6.22}$$

The radial part is solved by $f_n(r) \sim r^\alpha \Rightarrow \alpha_{1,2} = -\frac{1}{2} \pm \left(n + \frac{1}{2}\right)$ and the angular one is satisfied by putting $g_n = \mathrm{d}P_n/\mathrm{d}\vartheta$. Thus, we conclude that:

$$u_{\langle\varphi\rangle} = \sum_{n=1}^{\infty} \left(E_n r^n + \frac{F_n}{r^{n+1}}\right) \frac{\mathrm{d}P_n}{\mathrm{d}\vartheta}. \tag{9.6.23}$$

It is interesting to note that the solutions for perfect radial symmetry from Eqs. (9.5.1) and (9.5.5) can be obtained from the terms for $n = 0$ in Eqs. (9.6.17/9.6.19/9.6.23) after neglecting the rigid body part. It is now only a matter of algebra to obtain the stresses by inserting the formulae for the displacements into Eq. (9.6.2):

$$\begin{aligned}
\sigma_{\langle rr\rangle} =& \sum_{n=0}^{\infty} \left[ -\frac{(n^2 - n - 3)\lambda + (n+1)(n-2)\mu}{2n+3} A_n r^n - \frac{(n^2 + 3n - 1)\lambda + n(n+3)\mu}{2n-1} \frac{B_n}{r^{n+1}} \right. \\
& \left. + n(n-1)2\mu C_n r^{n-2} - (n+1)(n+2)2\mu \frac{D_n}{r^{n+3}} \right] P_n, \\
\sigma_{\langle\vartheta\vartheta\rangle} =& \sum_{n=0}^{\infty} \left[ \frac{(n+3)\lambda - (n-2)\mu}{2n+3} A_n r^n - \frac{(n-2)\lambda - (n+3)\mu}{2n-1} \frac{B_n}{r^{n+1}} \right. \\
& \left. + 2\mu n C_n r^{n-2} + 2\mu(n+1) \frac{D_n}{r^{n+3}} \right] P_n \\
& + \sum_{n=1}^{\infty} \left[ -\frac{(n+3)\lambda + (n+5)\mu}{(n+1)(2n+3)} A_n r^n - \frac{(n-2)\lambda + (n-4)\mu}{n(2n-1)} \frac{B_n}{r^{n+1}} \right. \\
& \left. + 2\mu C_n r^{n-2} - 2\mu \frac{D_n}{r^{n+3}} \right] \frac{\mathrm{d}^2 P_n}{\mathrm{d}\vartheta^2}, \\
\sigma_{\langle\varphi\varphi\rangle} =& \sum_{n=0}^{\infty} \left[ \frac{(n+3)\lambda - (n-2)\mu}{2n+3} A_n r^n - \frac{(n-2)\lambda - (n+3)\mu}{2n-1} \frac{B_n}{r^{n+1}} \right. \\
& \left. + 2\mu n C_n r^{n-2} + 2\mu(n+1) \frac{D_n}{r^{n+3}} \right] P_n \\
& + \sum_{n=1}^{\infty} \left[ -\frac{(n+3)\lambda + (n+5)\mu}{(n+1)(2n+3)} A_n r^n - \frac{(n-2)\lambda + (n-4)\mu}{n(2n-1)} \frac{B_n}{r^{n+1}} \right. \\
& \left. + 2\mu C_n r^{n-2} - 2\mu \frac{D_n}{r^{n+3}} \right] \frac{\mathrm{d}P_n}{\mathrm{d}\vartheta} \cot\vartheta, \\
\sigma_{\langle r\vartheta\rangle} =& \sum_{n=1}^{\infty} \left[ -\frac{n(n+2)\lambda + (n^2 + 2n - 1)\mu}{(n+1)(2n+3)} A_n r^n + \frac{(n^2 - 1)\lambda + (n^2 - 2)\mu}{n(2n-1)} \frac{B_n}{r^{n+1}} \right. \\
& \left. + 2\mu(n-1) C_n r^{n-2} + 2\mu(n+2) \frac{D_n}{r^{n+3}} \right] \frac{\mathrm{d}P_n}{\mathrm{d}\vartheta}, \\
\sigma_{\langle\vartheta\varphi\rangle} =& \mu \sum_{n=1}^{\infty} \left(E_n r^{n-1} + \frac{F_n}{r^{n+2}}\right) \left(\frac{\mathrm{d}^2 P_n}{\mathrm{d}\vartheta^2} - \frac{\mathrm{d}P_n}{\mathrm{d}\vartheta} \cot\vartheta\right), \\
\sigma_{\langle r\varphi\rangle} =& \mu \sum_{n=1}^{\infty} \left((n-1) E_n r^{n-1} - (n+2) \frac{F_n}{r^{n+2}}\right) \frac{\mathrm{d}P_n}{\mathrm{d}\vartheta}.
\end{aligned} \tag{9.6.24}$$

**Exercise 9.6.1: *A briefing on LEGENDRE polynomials***

Use a suitable mathematical textbook, for example, Butkov [4], and show that the LEGENDRE differential equation (9.6.11) can be solved by polynomials following RODRIGUES' generating formula:

$$P_n(x) = \frac{1}{2^n n!}\frac{\mathrm{d}^n}{\mathrm{d}x^n}\left(x^2-1\right)^n, \quad x = \cos\vartheta. \tag{9.6.25}$$

Use this formula to determine and plot the first five LEGENDRE polynomials:

$$\begin{aligned} &P_0 = 1, \quad P_1 = x, \quad P_1 = \frac{3}{2}\left(x^2 - \frac{1}{3}\right), \quad P_3 = \frac{5}{2}x\left(x^2 - \frac{3}{5}\right), \\ &P_4 = \frac{35}{8}\left(x^4 - \frac{6}{7}x^2 + \frac{3}{35}\right), \quad P_5 = \frac{63}{8}x\left(x^4 - \frac{10}{9}x^2 + \frac{5}{21}\right). \end{aligned} \tag{9.6.26}$$

Moreover, show that the LEGENDRE polynomials are orthogonal to each other:

$$\int_{-1}^{+1} P_m(x)P_n(x)\,\mathrm{d}x = \frac{2}{2m+1}\delta_{mn} \tag{9.6.27}$$

and that the following integration rule holds:

$$\int_0^t P_m(x)\,\mathrm{d}x = -\frac{1}{2m+1}\left[P_{m-1}(t) - P_{m+1}(t)\right]. \tag{9.6.28}$$

Also prove the following symmetry properties:

$$P_{2m}(x) = P_{2m}(-x), \quad P_{2m+1}(x) = -P_{2m+1}(-x), \quad m = 0, 1, 2, \ldots. \tag{9.6.29}$$

Benjamin Olinde RODRIGUES was born on October 6, 1795 in Bordeaux, and died on December 17, 1851 in Paris. He was the oldest of eight children of a Jewish family, originally from Portugal or Spain. After their move to Paris he graduated in mathematics in 1816 at the recently established University of Paris. In his thesis he presented his generating formula for the polynomials that were named after him. However, he was smart enough to leave mathematics and to start a very profitable career as a banker. Interestingly, and probably counter-productive to his banking passion, he was also an ardent activist of the socialist movement that had just started in France.

**Exercise 9.6.2:** ***A test of patience***

Study the derivation of Eqs. (9.6.1–9.6.24) carefully and in all detail. What purpose do the various terms with negative powers of $r$ in the final equations for the displacements and for the stresses serve? For the answer note why and how in the following sections all of the coefficients proceeding these terms are eliminated.

We now need to take care of the six times infinitely many constants of integration $A_n$, $B_n$, $C_n$, $D_n$, $E_n$, and $F_n$ in the general solutions for the stresses and for the displacements. As usual this can be done by adjusting them to the requirements of our problem or, in other words, to the boundary conditions. First, all of the stresses and displacements should not be singular at any point. This concerns notably the origin $r = 0$ which is part of a solid sphere. Therefore, we require that:

$$B_n = 0, \; D_n = 0, \; F_n = 0, \quad n = 0, 1, 2, \ldots. \tag{9.6.30}$$

Second, recall that the traction vector must be continuous at the outer radius, $R$, of the sphere. Since no loads are applied in angular direction we have:

$$\sigma_{\langle r\vartheta\rangle}\Big|_{r=R} = 0, \quad \sigma_{\langle r\varphi\rangle}\Big|_{r=R} = 0. \tag{9.6.31}$$

This leads to:

$$C_n = \frac{n(n+2)\lambda + (n^2 + 2n - 1)\mu}{(n-1)(n+1)(2n+3)2\mu} R^2 A_n, \quad E_n = 0, \quad n = 2, 3, \ldots, \quad A_1 = 0. \tag{9.6.32}$$

However, in radial direction a constant pressure $p_0$ is applied symmetrically along a certain equatorial region of size $\vartheta_0$ due to the misfit with the cylindrical bushing:

$$\sigma_{\langle rr\rangle}\Big|_{r=R} = \begin{cases} -p_0, & \text{if } -\vartheta_0 + \frac{\pi}{2} \leq \vartheta \leq \vartheta_0 + \frac{\pi}{2} \\ 0, & \text{else} \end{cases}. \tag{9.6.33}$$

We assume that the pressure is known. The trick is now to expand the radial stress on the outer surface in a series of LEGENDRE polynomials and to compare corresponding terms with Eq. $(9.6.24)_1$ evaluated at the outer radius $R$. Thus, we write:

$$\sigma_{\langle rr\rangle}\Big|_{r=R} = \sum_{n=0}^{\infty} a_n P_n(\cos\vartheta). \tag{9.6.34}$$

Now we multiply by $P_m \mathrm{d}(\cos\vartheta)$, integrate from $\vartheta = 0$ to $\vartheta = \pi$, and observe Eqs. (9.6.27–9.6.29). The final result is:

$$\begin{aligned} &a_0 = -p_0 \sin\vartheta_0, \quad a_{2m} = p_0\left[P_{2m-1}(\sin\vartheta_0) - P_{2m+1}(\sin\vartheta_0)\right], \\ &a_{2m-1} = 0, \quad m = 1, 2, \ldots. \end{aligned} \tag{9.6.35}$$

By comparison between Eq. (9.6.24)$_1$, evaluated at $r = R$, and Eqs. (9.6.34) and (9.6.35) we find that:

$$
\begin{aligned}
A_0 &= -\frac{3\sin\vartheta_0}{3\lambda + 2\mu} p_0, \quad A_{2n-1} = 0, \\
A_{2n} &= \frac{(2n+1)(4n+3)[P_{2n-1}(\sin\vartheta_0) - P_{2n+1}(\sin\vartheta_0)]}{(8n^2+8n+3)\lambda + (4n^2+2n+1)2\mu} \frac{1}{R^{2n}} p_0, \quad n = 1, 2, \ldots.
\end{aligned}
\tag{9.6.36}
$$

All coefficients can now be inserted into the equations for the non-vanishing stresses:

$$
\begin{aligned}
\frac{\sigma_{\langle rr\rangle}}{p} =& -1 + \sum_{n=1}^{\infty} \big[-(2n+1)\big((4n^2-2n-3)\alpha + (2n^2-n-1)\beta\big)\rho^2 \\
&+ n\big(8n(n+1)\alpha + (4n^2+4n-1)\beta\big)\big] \frac{\rho^{2(n-1)}\alpha_{2n}}{(8n^2+8n+3)\alpha + (4n^2+2n+1)\beta} P_{2n}, \\
\frac{\sigma_{\langle\vartheta\vartheta\rangle}}{p} =& -1 + \sum_{n=1}^{\infty} \Big\{\Big[(2n+1)((2n+3)\alpha - (n-1)\beta)\rho^2 \\
&+ \frac{n}{2n-1}\big(8n(n+1)\alpha + (4n^2+4n-1)\beta\big)\Big] P_{2n} \\
&+ \Big[-(2(2n+3)\alpha + (2n+5)\beta)\frac{\rho^2}{2} \\
&+ \frac{1}{2n-1}\big(8n(n+1)\alpha + (4n^2+4n-1)\beta\big)\Big] \frac{1}{2}\frac{\mathrm{d}^2 P_{2n}}{\mathrm{d}\vartheta^2}\Big\} \\
&\times \frac{\rho^{2(n-1)}}{(8n^2+8n+3)\alpha + (4n^2+2n+1)\beta}\alpha_{2n}, \\
\frac{\sigma_{\langle\varphi\varphi\rangle}}{p} =& -1 + \sum_{n=1}^{\infty} \Big\{\Big[(2n+1)((2n+3)\alpha - (n-1)\beta)\rho^2 \\
&+ \frac{n}{2n-1}\big(8n(n+1)\alpha + (4n^2+4n-1)\beta\big)\Big] P_{2n} \\
&+ \Big[-(2(2n+3)\alpha + (2n+5)\beta)\frac{\rho^2}{2} \\
&+ \frac{1}{2n-1}\big(8n(n+1)\alpha + (4n^2+4n-1)\beta\big)\Big] \frac{1}{2}\frac{\mathrm{d}P_{2n}}{\mathrm{d}\vartheta}\cot\vartheta\Big\} \\
&\times \frac{\rho^{2(n-1)}}{(8n^2+8n+3)\alpha + (4n^2+2n+1)\beta}\alpha_{2n}, \\
\frac{\sigma_{\langle r\vartheta\rangle}}{p} =& \sum_{n=1}^{\infty} \frac{8n(n+1)\alpha + (4n^2+4n-1)\beta}{(8n^2+8n+3)\alpha + (4n^2+2n+1)\beta}\big(1-\rho^2\big)\rho^{2(n-1)}\alpha_{2n}\frac{1}{2}\frac{\mathrm{d}P_{2n}}{\mathrm{d}\vartheta},
\end{aligned}
\tag{9.6.37}
$$

and the following abbreviations were used:

$$p = p_0 \sin\vartheta_0, \quad \rho = \frac{r}{R}, \quad \alpha = \frac{\nu}{(1+\nu)(1-2\nu)}, \quad \beta = \frac{\nu}{1+\nu},$$
$$\alpha_{2n} = \frac{P_{2n-1}(\sin\vartheta_0) - P_{2n+1}(\sin\vartheta_0)}{\sin\vartheta_0}. \tag{9.6.38}$$

We now consider the special case of a sphere subjected to a (constant) line force $q_0$ along its equator, i.e., the case $\vartheta_0 \to 0$ so that $pR\sin\vartheta_0 \approx pR\mathrm{d}\vartheta \to q_0$. L'HOPITAL's rule is used to calculate the following relevant limit for a numerical evaluation of the stress formulae shown in Eq. (9.6.37):

$$\begin{aligned}\lim_{x\to 0}\alpha_n &= \lim_{x\to 0}\frac{P_{2n-1}(x) - P_{2n+1}(x)}{x} = (4n+1)P_{2n}(0)\\ &= \frac{(-1)^{n+1}(2n)!(4n+1)}{2^{2n}(n!)^2}.\end{aligned} \tag{9.6.39}$$

In the same context it is helpful to know the following relations for LEGENDRE polynomials:

$$\begin{aligned}P_n''(x) =& \frac{n}{(1-x^2)^2}\left[(n-2)x^2P_n(x) - (2n-3)xP_{n-1}(x) + (n+1)P_{n-2}(x)\right]\\ &- \frac{n}{1-x^2}\left[P_{n-1}(x) - xP_n(x)\right],\\ \frac{1}{2}\frac{\mathrm{d}^2P_{2n}}{\mathrm{d}\vartheta^2} =& \frac{1}{2}\left(P_{2n}''(x)\left(1-x^2\right) - xP_{2n}'(x)\right),\\ \frac{1}{2}\cot\vartheta\frac{\mathrm{d}P_{2n}}{\mathrm{d}\vartheta} =& -\frac{1}{2}xP_{2n}'(x), \quad P_n'(x) = \frac{n}{1-x^2}\left[P_{n-1}(x) - xP_n(x)\right], \quad x = \cos\vartheta.\end{aligned} \tag{9.6.40}$$

Guillaume François Antoine Marquis DE L'HÔPITAL was born in Paris in 1661, where he also died on February 2, 1704. As a typical aristocrat of his days he was forced to overcome his boredom on a daily basis. So he indulged himself in mathematical problems. His name is linked eternally to his rule for calculating limits involving indeterminate forms, such as "0/0" or "$\infty/\infty$." Since calculus was a "hot" subject at that time it is unlikely that he was the first one to have used "his" rule. However, it appeared in print for the first time in his treatise on infinitesimal calculus, entitled *Analyse des Infiniment Petits pour l'Intelligence des Lignes Courbes.*

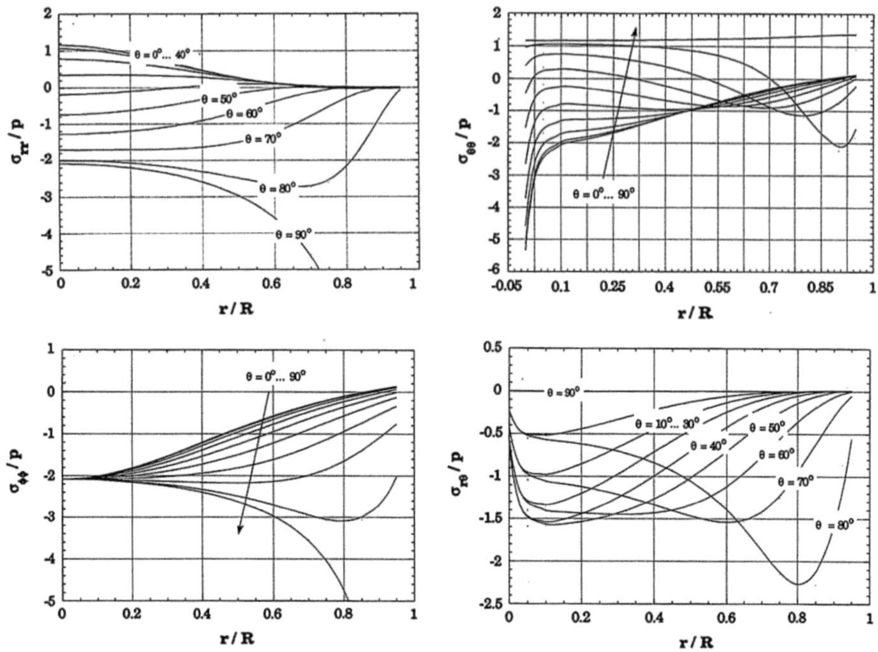


**Fig. 9.8** Stresses in a sphere compressed along its equator

Figure 9.8 shows all the stresses as a function of (normalized) radius. POISSON's number for glass was used, i.e., $\nu = 0.256$, and $\vartheta_0 = 0$. As expected, compressive stresses are dominant. However, there are also zones of tensile stress. Consider for example the angular stress $\sigma_{\vartheta\vartheta}$ in the equatorial plane, i.e., at $\vartheta = 90°$. The corresponding values are all positive and almost constant w.r.t. $r$. However, a thorough examination shows that for the limit $r \to R$ the tensile stress turns rapidly into a compressive one. Nevertheless this explains the tensile fracture shown in Fig. 1.2.

**Exercise 9.6.3:** ***Auxiliary formulae for LEGENDRE polynomials***

Prove Eqs. (9.6.39/9.6.40). Use the results to program the formulae for the stresses shown in Eq. (9.6.37). Discuss the transition from a pressure load in the equatorial zone to a line force along the equator.

## 9.7 The GRIFFITH Crack Model

The first two-dimensional model of a sharp crack in a brittle solid was proposed and mathematically analyzed during the 20 s of the last century by A. A. GRIFFITH.

As indicated in Fig. 9.8 GRIFFITH considers a sequence of ellipses with decreasing minor axis. It was mentioned in Sect. 2.3 that these ellipses can mathematically be described by elliptic coordinates, $(z^1, z^2)$. Recall that they are related to Cartesian coordinates $(x_1, x_2)$ as follows:

$$\begin{aligned} & x_1 = c \cosh(z^1) \cos(z^2), \quad x_2 = c \sinh(z^1) \sin(z^2), \\ & z^1 \in [0, \infty), \quad z^2 \in [0, 2\pi). \end{aligned} \tag{9.7.1}$$

If the angle $z^2$ is eliminated the well known equation of an ellipse in Cartesian coordinates results:

$$\frac{x_1^2}{c^2 \cosh^2(z^1)} + \frac{x_2^2}{c^2 \sinh^2(z^1)} = 1, \tag{9.7.2}$$

and we realize that the major and the minor axes, $a$ and $b$, are given by:

$$a = c \cosh(z^1), \quad b = c \sinh(z^1). \tag{9.7.3}$$

Obviously the ellipse degenerates to a slit of infinite sharpness in the limit $z^1 \to 0$, which is known as the GRIFFITH crack. The crack length is given by the parameter $2c$. We are interested in the stresses developing in the immediate neighborhood in front of the crack tip at $x_1 = c + r$ if a biaxial tensile stress $\sigma$ is applied at infinity in $x_1$ as well as in $x_2$ direction (cf., Fig. 9.9). Thus we have:

$$c + r = c \cosh(\Delta z^1), \quad \frac{r}{c} \ll 1. \tag{9.7.4}$$

If the hyperbolic function is expanded in a TAYLOR series it follows that:

$$\Delta z^1 = \sqrt{2\frac{r}{c}}. \tag{9.7.5}$$

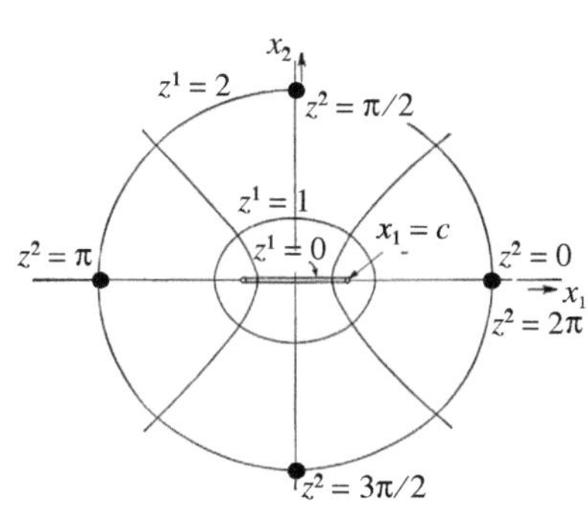


**Fig. 9.9** Nomenclature used for elliptic coordinates

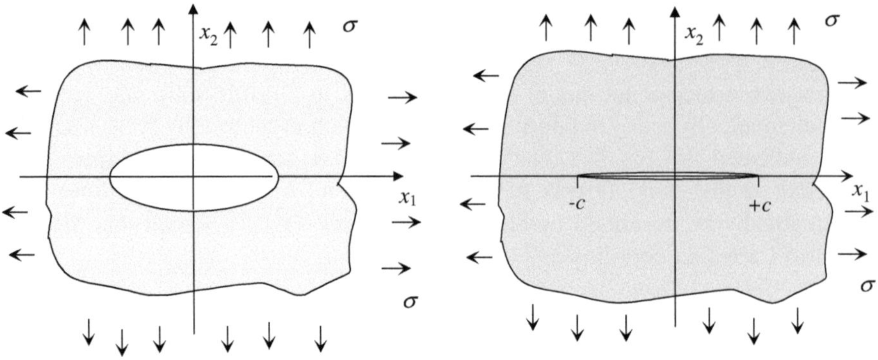


**Fig. 9.10** Transition from an *elliptic hole* to a GRIFFITH crack

In order to calculate the stresses in the vicinity of the GRIFFITH crack we will first determine the stresses around an elliptical hole in a most general manner. The periphery of the hole is characterized by $z^1 = \alpha =$ constant. Then, in a second step, we will study the limit case $z^1 \rightarrow 0$, i.e., the crack problem, cf., Fig. 9.10.

The stresses around an elliptic hole can be characterized best by working with elliptic coordinates from the very beginning on. As a first step in that direction we specialize the general form of the static balance of momentum from Eq. (5.5.1) to such coordinates. By taking the definition of the covariant derivative for tensors in account, Eqs. (4.4.3/4.4.5), it follows in two dimensions:

$$\begin{aligned} &\frac{\partial \sigma^{11}}{\partial z^1} + \frac{\partial \sigma^{12}}{\partial z^2} + \sigma^{11}\left(2\Gamma^1_{11} + \Gamma^2_{12}\right) + \sigma^{12}\left(3\Gamma^1_{12} + \Gamma^2_{22}\right) + \sigma^{22}\Gamma^1_{22} = 0, \\ &\frac{\partial \sigma^{12}}{\partial z^1} + \frac{\partial \sigma^{22}}{\partial z^2} + \sigma^{11}\Gamma^2_{11} + \sigma^{12}\left(3\Gamma^2_{12} + \Gamma^1_{11}\right) + \sigma^{22}\left(2\Gamma^2_{22} + \Gamma^1_{12}\right) = 0. \end{aligned} \tag{9.7.6}$$

From Eq. (9.7.1) we find the coefficients of the metric tensor by observing Eqs. (2.2.8) and (2.4.7) (also see Exercise 4.2.4):

$$\begin{aligned} g_{ij} &= \frac{c^2}{2}\begin{pmatrix} \cosh(2z^1) - \cos(2z^2) & 0 \\ 0 & \cosh(2z^1) - \cos(2z^2) \end{pmatrix}, \\ g^{kl} &= \frac{2}{c^2}\begin{pmatrix} \frac{1}{\cosh(2z^1)-\cos(2z^2)} & 0 \\ 0 & \frac{1}{\cosh(2z^1)-\cos(2z^2)} \end{pmatrix}. \end{aligned} \tag{9.7.7}$$

Recall that the elliptic coordinates $(z^1, z^2)$ are orthogonal so that only the components along the diagonal of the metric tensor are different from zero. Moreover, it is curious to note that the diagonal metric coefficients are identical. Eq. (4.2.4) allows now to calculate the corresponding CHRISTOFFEL symbols (also see Exercise 4.2.4):

$$\begin{aligned} \Gamma^1_{11} &= -\Gamma^1_{22} = \Gamma^2_{12} = \frac{\sinh(2z^1)}{\cosh(2z^1) - \cos(2z^2)}, \\ \Gamma^1_{12} &= -\Gamma^2_{11} = \Gamma^2_{22} = \frac{\sin(2z^2)}{\cosh(2z^1) - \cos(2z^2)}. \end{aligned} \tag{9.7.8}$$

By means of these relations and the following auxiliary formulae easily confirmed by differentiation:

$$2\Gamma^1_{11} = \frac{1}{g_{11}} \frac{\partial g_{11}}{\partial z^1}, \quad 2\Gamma^1_{12} = \frac{1}{g_{11}} \frac{\partial g_{11}}{\partial z^2}. \tag{9.7.9}$$

and by observing the definition of physical tensor components in Eq. (2.6.5), the two-dimensional balance of momentum in contravariant form of Eq. (9.7.6) can alternatively be written as:

$$\begin{aligned} &\frac{\partial \sigma_{\langle 11 \rangle}}{\partial z^1} + \frac{\partial \sigma_{\langle 12 \rangle}}{\partial z^2} + \Gamma^1_{11} \sigma_{\langle 11 \rangle} + 2\Gamma^1_{12} \sigma_{\langle 12 \rangle} + \Gamma^1_{22} \sigma_{\langle 22 \rangle} = 0, \\ &\frac{\partial \sigma_{\langle 12 \rangle}}{\partial z^1} + \frac{\partial \sigma_{\langle 22 \rangle}}{\partial z^2} + \Gamma^2_{11} \sigma_{\langle 11 \rangle} + 2\Gamma^1_{11} \sigma_{\langle 12 \rangle} + \Gamma^2_{22} \sigma_{\langle 22 \rangle} = 0. \end{aligned} \tag{9.7.10}$$

**Exercise 9.7.1:** ***Balance of momentum in physical elliptic coordinates***

Verify Eq. (9.7.10), i.e., the two-dimensional balance of momentum in physical elliptic coordinates. Moreover, rewrite HOOKE's law in these coordinates.

It is easily verified by inserting that the following expressions for the stresses into satisfy the momentum balance identically:

$$\begin{aligned} \frac{\sigma_{\langle 11 \rangle}}{\sigma} &= \frac{\sinh(2z^1)[\cosh(2z^1) - \cosh(2\alpha)]}{[\cosh(2z^1) - \cos(2z^2)]^2}, \\ \frac{\sigma_{\langle 12 \rangle}}{\sigma} &= \frac{\sin(2z^2)[\cosh(2z^1) - \cosh(2\alpha)]}{[\cosh(2z^1) - \cos(2z^2)]^2}, \\ \frac{\sigma_{\langle 22 \rangle}}{\sigma} &= \frac{\sinh(2z^1)[\cosh(2z^1) + \cosh(2\alpha) - 2\cos(2z^2)]}{[\cosh(2z^1) - \cos(2z^2)]^2}. \end{aligned} \tag{9.7.11}$$

It is straightforward to proof that these relations fulfill the requirements for a hole that is free of tractions. In other words the traction vector vanishes at positions $z^1 = \alpha = \text{constant}$:

$$\frac{\sigma_{\langle 11 \rangle}}{\sigma} = 0, \quad \frac{\sigma_{\langle 12 \rangle}}{\sigma} = 0. \tag{9.7.12}$$

Moreover, we obtain for $z^1 \to \infty$:

$$\sigma_{\langle 11\rangle} \to \sigma, \quad \sigma_{\langle 12\rangle} \to 0, \quad \sigma_{\langle 22\rangle} \to \sigma. \tag{9.7.13}$$

In summary, we may say that Eq. (9.7.11) is the solution for the stresses around an elliptic hole in a plate under biaxial tension.

**Exercise 9.7.2:** ***Stress fields around a* G*RIFFITH crack I***

Confirm the last statement by starting from Eq. (9.7.11). Moreover, show that the components $\sigma_{\langle 11\rangle}$ and $\sigma_{\langle 12\rangle}$ are the relevant components of the traction vector at the periphery of the hole. Write down and evaluate the jump conditions for the flux of momentum.

We will now investigate the behavior of the stress component $\sigma_{\langle 22\rangle}$ in front of a GRIFFITH crack by using the equations shown in (9.7.11). The stress $\sigma$ is tensile, i.e., positive. Thus we conclude from Eq. $(9.7.11)_3$ that the stress $\sigma_{\langle 22\rangle}$ leads to an opening of the plate material along the $x_1$-axis. In other words, the flanks of the GRIFFITH crack are driven apart.

By means of a TAYLOR expansion and by observing Eqs. $(9.7.11)_3$ and (9.7.5) we conclude that:

$$\lim_{\alpha\to 0} \frac{\sigma_{\langle 22\rangle}}{\sigma}\bigg|_{\substack{z^1=\Delta z^1\\ z^2=0}} \approx \frac{1}{\Delta z^1}, \tag{9.7.14}$$

and thus:

$$\sigma_{\langle 22\rangle} \approx \frac{K_{\mathrm{I}}}{\sqrt{2\pi r}}, \quad K_{\mathrm{I}} = \sigma\sqrt{\pi c}. \tag{9.7.15}$$

The quantity $K_{\mathrm{I}}$ is also known as *stress intensity factor* of a GRIFFITH crack in an infinite plane subjected to biaxial tension. As its name indicates, $K_{\mathrm{I}}$ characterizes the intensity of the $1/\sqrt{r}$ singularity characteristic of the stresses in front of the crack tip. Obviously the opening mode stress becomes infinitely large at the crack tip. Thus, we should suspect that a GRIFFITH crack cannot be stable. However, this is an artifact of the linear elastic analysis. In fact, it is not the stresses that decide about stability or instability of a crack. Rather it is the release rate of elastic energy in competition with the energy required to form new surfaces that controls crack instability and propagation. It can be shown that the elastic energy release rate is proportional to the square of the stress intensity factor. Catastrophic fracture occurs if a critical value of the stress intensity factor is reached, the so-called fracture toughness $K_{\mathrm{Ic}}$, which we have already introduced phenomenologically in Sect. 9.5.

**Exercise 9.7.3: *Stress fields around a GRIFFITH crack II***

Confirm Eqs. (9.7.14/9.7.15) in detail. Also compute the (regular) higher order terms following the $1/\sqrt{r}$ singularity Depict the stresses given by Eq. (9.7.11), in particular, focus on the stress singularity and the far field stresses, some of which should approach the value $\sigma$. Which ones and why?

**Exercise 9.7.4: *Fracture toughness***

Use the internet or a suitable materials science textbook to obtain numerical values for critical stress intensity factors/fracture toughness for brittle as well as for ductile materials. Motivate the validity of GRIFFITH' fracture criterion:

$$K_{\mathrm{I}}\big|_{\mathrm{c}} = K_{\mathrm{Ic}} = \sqrt{2E\gamma}, \tag{9.7.16}$$

where $E$ denotes YOUNG's modulus and $\gamma$ the specific surface energy of the material. Also try to obtain some data for surface energies of brittle materials.

## 9.8 Would You Like to Know More?

The theory of elasticity is an extensive mathematical playground. This holds already for 2D elastic problems, i.e., the world of plane strain and plane stress. The books by Sokolnikoff [5] or by Hahn [6] give a nice overview on the whole topic. Moreover, the two classic monographs on the mathematical treatment of plane elastic are the books by Muskhelishvili [7] and Milne-Thomson [8], which bring the use of complex functions to perfection. Another classic on linear elasticity is the book by Love [9] where, in particular, the question regarding the so-called *generalized plane stress and strain* from Exercise 9.2.1 is answered, at least implicitly. General solutions for cylindrical and circular problems can be found in the book by Lurie [10].

A more profound introduction to (linear) fracture mechanics can be obtained from the textbooks by Anderson [11], Hahn [12], and by Gross and Seelig [13]. Finally the afore-mentioned book by Mura [2] is a real treasure chest, if one is interested in the calculation of stresses and strains inside and outside of inclusions of different shapes.

## References

1. Kneschke A (1965) Differentialgleichungen und Randwertprobleme, Band I, Gewöhnliche Differentialgleichungen. B.G. Teubner, Leipzig
2. Mura T (1987) Micromechanics of defects in solids, 2nd edn. Martinus Nijhoff Publishers, Dordrecht
3. Hiramatsu Y, Oka Y (1966) Determination of the tensile strength of rock by a compression test of an irregular test piece. Int. J. Rock. Min. Sci. 3:89–99
4. Butkov E (1968) Mathematical physics. Addison-Wesley Publishing Company, Reading Massachusetts, Menlo Park, California, London, Sydney, Manila
5. Sokolnikoff IS (1956) Mathematical theory of elasticity. McGraw-Hill Book Company Inc., New York, Toronto, London
6. Hahn HG (1986) Elastizitätstheorie. B.G. Teubner, Stuttgart
7. Muskhelishvili NI (1963) Some basic problems of the mathematical theory of elasticity. P. Noordhoff Ltd, Groningen
8. Milne-Thomson LM (1968) Plane elastic systems. Ergebnisse der angewandten Mathematik, 2nd edn. Springer, Berlin, Heidelberg, New York
9. Love AEH (1944) A treatise on the mathematical theory of elasticity, 4th edn. Dover Publications, New York
10. Lurie AI (2005) Theory of elasticity. Foundations of engineering mechanics. Springer, Berlin, Heidelberg
11. Anderson TL (2005) Fracture mechanics: fundamentals and applications, 3rd edn. CRC Press, Boca Raton
12. Hahn HG (1976) Bruchmechanik. B.G. Teubner, Stuttgart
13. Gross D, Seelig T (2011) Fracture mechanics: with an introduction to micromechanics, 2nd edn. Springer, Heidelberg, Dordrecht, London, New York
14. Whitney ED (1965) Electrical resistivity and diffusionless phase transformations of Zirconia at high temperatures and ultrahigh pressures. J Electrochem Soc 112(1):91–94

# Chapter 10
# Selected Problems for Newtonian and Maxwellian Fluids

**Abstract** This chapter is dedicated to the art of modeling, in particular modeling problems involving viscous gases and fluids. We start with the previously introduced NAVIER–STOKES constitutive law in context with transient flow through a channel. We shall see that this constitutive relations results in a parabolic partial differential equation for the velocity of the fluid, which immediately leads to the artefact of infinite speeds of propagation. This can be avoided by application of the MAXWELL fluid model, which accounts for memory effects. Further examples of gas dynamics in spherical coordinates include expanding and contracting stars as well as the whole universe, which are modeled by means of classical continuum theory.

*Very simple was my explanation, and plausible enough —as most wrong theories are!*
H. G. WELLS, The Time Machine

## 10.1 Some Comments on Modeling, in Particular Modeling of Transient Channel Flow

Consider in analogy to Fig. 7.4 the plane-parallel shear flow between two infinitely large plates as shown in Fig. 10.1. However, this time the upper plate at position $y = H$ is fixed and the lower one at $y = 0$ moves at a *time-dependent* speed $V(t)$, which is a big difference when compared to the stationary case of Exercise 7.4.1. The switch of plates has computational reasons only. Note that the time-dependent speed adds a new quality to the problem: We are no longer exclusively interested

For the solutions to the problems presented in this chapter I would like to give credit to Prof. Dr. W. Dreyer and Dr. W. Weiss of the Weierstrass Institut in Berlin and also Prof. emer. Dr. rer. nat. Dr. h.c. I. Müller, formerly with Technische Universität Berlin.

W. H. Müller, *An Expedition to Continuum Theory*, Solid Mechanics and Its Applications 210, DOI: 10.1007/978-94-007-7799-6_10,

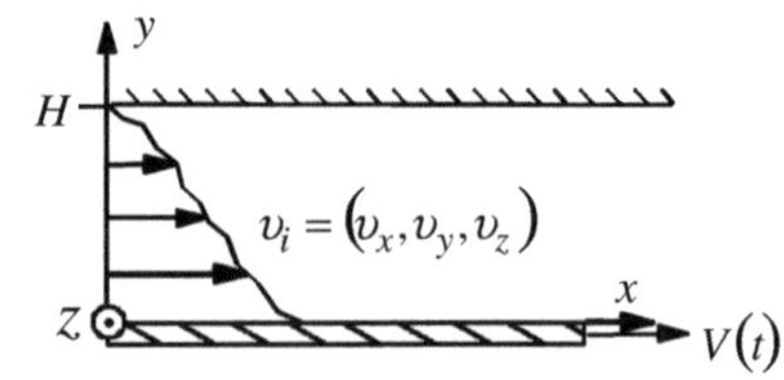


**Fig. 10.1** Transient parallel plate flow

in the final stationary state. Rather we focus on *transient flow*. In particular, we will show how the "information" regarding the movement of the lower plate is transferred to the layers of fluid above. For physical reasons we expect this to take some time. However, we shall see that in order to mimic finite speeds of propagation, proper modeling is required. In particular the choice of constitutive equation is crucial. To keep the resulting formulae simple the origin of the coordinate system is located at the lower plate. In the spirit of the semi-inverse method we assume for the resulting velocity field:

$$\upsilon_x = \upsilon(y, t), \upsilon_y = 0, \upsilon_z = 0. \tag{10.1.1}$$

We neglect the influence of gravity, do not apply any pressure gradients to propagate the flow, and assume that the fluid is incompressible. Thus according to the mass balance of Eq. (3.8.13) the divergence of the velocity field must vanish, $\partial\upsilon_k/\partial x_k = 0$. This is in full accord with the *ansatz* of (10.1.1). We obtain for the non-vanishing components of the stress tensor according to the NAVIER–STOKES constitutive law (6.3.1):

$$\sigma_{xx} = \sigma_{yy} = \sigma_{zz} = -p, \quad \sigma_{xy} = \sigma_{yx} = \mu\frac{\partial\upsilon}{\partial y}. \tag{10.1.2}$$

Recall that a fluid behaving according to this equation is also known as a *Newtonian fluid* in the literature. However, we are about to see that this kind of constitutive relation leads to a *parabolic* partial differential equation for the velocity component $\upsilon(y, t)$, similarly as FOURIER's law for the heat flux leads to the traditional parabolic equation of heat conduction for the temperature field. In other words, the information, that the lower plate starts moving at time $t = 0$ and any other change in its movement, $V(t)$!, will *instantaneously* be transferred through the fluid to the upper plate, no matter how far away it is located. This unphysical behavior can be avoided by using another constitutive relation for the shear component of the stress tensor. For example, we write for a so-called *MAXWELL fluid*, a.k.a. *non-Newtonian fluid with memory*:

$$\lambda\frac{\partial\sigma_{xy}}{\partial t} + \sigma_{xy} = \mu\frac{\partial\upsilon}{\partial y}. \tag{10.1.3}$$

As in Eq. (10.1.2) $\mu$ denotes the shear viscosity but $\lambda$ is a relaxation time. Typical values for the material data are

(a) Newtonian fluid, e.g., water:

$$\rho = 10^3 \frac{\text{kg}}{\text{m}^3}, \mu = 10^{-2} \frac{\text{Ns}}{\text{m}^2}. \tag{10.1.4}$$

(b) MAXWELL fluid, e.g., 6.5 wt% of polyisobutylene in decahydronaphthalene:

$$\rho = 10^3 \frac{\text{kg}}{\text{m}^3}, \quad \mu = 5.6 \frac{\text{Ns}}{\text{m}^2}, \quad \lambda = 0.0427 \text{ s}. \tag{10.1.5}$$

Obviously, the relaxation time is very short. In other words, the propagation of a disturbance is rather fast. In the case of water the relaxation time is even smaller by several decades. Thus in that case the influence of a finite speed of propagation of disturbances should be (negligibly) small.

In combination with the *ansatz* (10.1.1) and Eqs. (10.1.2/10.1.3) for the $y$ and $z$ components of the balance of momentum shown in Eq. $(3.8.14)_1$, we conclude that the pressure can only be a function of position $x$ and of time $t$—independent of the constitutive relations. On the other hand the $x$ component of the balance of momentum reads:

$$\rho \frac{\partial \upsilon}{\partial t} = -\frac{\partial p}{\partial x} + \frac{\partial \sigma_{xy}}{\partial y} \quad \Rightarrow \quad \rho \frac{\partial \upsilon}{\partial t} - \frac{\partial \sigma_{xy}}{\partial y} = -\frac{\partial p}{\partial x}. \tag{10.1.6}$$

According to the *ansatz* (10.1.1) and the constitutive Eqs. (10.1.2/10.1.3) the function on the left can only depend on $y$ (at all times $t$). However, as it was just pointed out, the function on the right can only depend on $x$ (at all times $t$). Hence both must be equal and we conclude that they can only be a constant. Thus, physically speaking, the pressure gradient in flow direction $x$ must be constant. But since no pressure gradient was applied this constant must be zero and we have:

$$\rho \frac{\partial \upsilon}{\partial t} = \frac{\partial \sigma_{xy}}{\partial y}. \tag{10.1.7}$$

If we insert Eq. (10.1.2) for Newtonian fluids the afore-mentioned *parabolic* Partial Differential Equation (PDE) results:

$$\frac{\partial \upsilon}{\partial t} = D \frac{\partial^2 \upsilon}{\partial y^2}, D = \frac{\mu}{\rho}. \tag{10.1.8}$$

On the other hand we obtain by combination of Eqs. (10.1.3) and (10.1.6) for MAXWELL fluids:

$$\rho \frac{\partial \upsilon}{\partial t} = \mu \frac{\partial^2 \upsilon}{\partial y^2} - \lambda \frac{\partial^2 \sigma_{xy}}{\partial y \partial t}. \tag{10.1.9}$$

If we now differentiate the field equation (10.1.6) w.r.t. time and recall that the fluid is *incompressible* the following *hyperbolic* equation is finally obtained:

$$\frac{\partial^2 \upsilon}{\partial t^2} + \frac{1}{\lambda}\frac{\partial \upsilon}{\partial t} = c^2 \frac{\partial^2 \upsilon}{\partial y^2}, \quad c^2 = \frac{D}{\lambda}. \tag{10.1.10}$$

The quantity $c$ has the dimension of a velocity. By using the data from above we find that:

$$c = \left(\frac{D}{\lambda}\right)^{1/2} = \left(\frac{\mu}{\rho\lambda}\right)^{1/2} = 0.36\frac{\mathrm{m}}{\mathrm{s}}. \tag{10.1.11}$$

**Exercise 10.1.1:** ***Newtonian versus Maxwell fluid: Nature of the resulting field equations***

Consider the following general form of a linear PDE of second order of the two variables $x$ and $y$:

$$A\frac{\partial^2 u}{\partial x^2} + 2B\frac{\partial^2 u}{\partial x \partial y} + C\frac{\partial^2 u}{\partial y^2} + \alpha\frac{\partial u}{\partial x} + \beta\frac{\partial u}{\partial y} + \gamma u = f. \tag{10.1.12}$$

In older textbooks it is customary to use the following quantity $\delta$ for a classification of the PDE:

$$\delta = AC - B^2. \tag{10.1.13}$$

We say that the PDE is of the *elliptic* type if $\delta > 0$, of *parabolic* type if $\delta = 0$, and of *hyperbolic* type if $\delta > 0$. Use that scheme and show that Eqs. (10.1.8) and (10.1.10) represent PDEs of the parabolic and of the hyperbolic type, respectively. Interpret $c$ as a wave propagation speed (of which physical quantity?). Moreover, consult modern mathematics textbooks or the internet and find alternative classification schemes for PDEs.

In order to solve Eqs. (10.1.8) and (10.1.10), respectively, we need initial and boundary conditions. The boundary conditions read in both cases:

$$\upsilon(0,t) = V(t), \quad \upsilon(H,t) = 0, \quad t \geq 0 \tag{10.1.14}$$

and the initial conditions:

$$\upsilon(y,0) = 0, \quad \frac{\partial \upsilon}{\partial t}(y,0) = 0, \quad t = 0, \tag{10.1.15}$$

where the second one is only required for the Maxwell case, which led to a PDE of second order in time. As plate velocity $V(t)$ we choose the step function shown in Fig. 10.2:

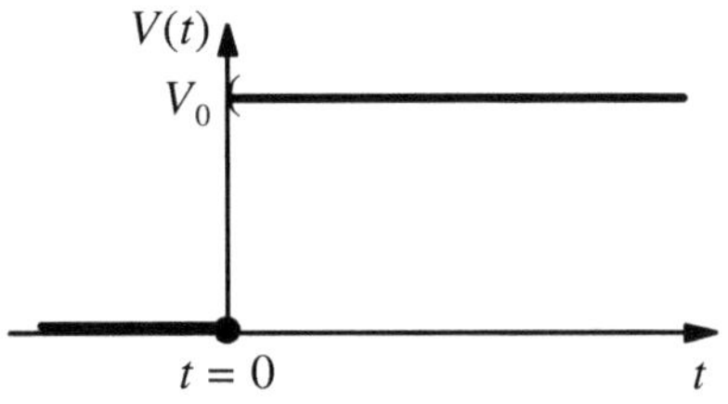


**Fig. 10.2** The step function prescribed on the lower plate

$$V(t) = \begin{cases} 0, & t \leq 0 \\ V_0, & t > 0 \end{cases} \tag{10.1.16}$$

For $t \to \infty$ *stationary* conditions must result. These have already been analyzed in Exercise 7.4.1 and we may write:

$$\upsilon_s(y) = V_0\left(1 - \frac{y}{H}\right) \text{ where } \upsilon_s = \lim_{t \to \infty} \upsilon(y, t). \tag{10.1.17}$$

If $\upsilon_s(y)$ is split off from the full solution the calculations simplify considerably. Thus we define a reduced velocity:

$$w(y, t) = \upsilon(y, t) - \upsilon_s(y) \tag{10.1.18}$$

and obtain the following PDEs:

$$\frac{\partial w}{\partial t} = D\frac{\partial^2 w}{\partial y^2} \text{ and } \frac{\partial^2 w}{\partial t^2} + \frac{1}{\lambda}\frac{\partial w}{\partial t} = c^2\frac{\partial^2 w}{\partial y^2}. \tag{10.1.19}$$

Note that the initial and boundary conditions change as well:

$$w(0, t) = 0, \quad w(H, t) = 0 \tag{10.1.20}$$

and:

$$w(y, 0) = -\upsilon_s(y), \quad \frac{\partial w}{\partial t}(y, 0) = 0. \tag{10.1.21}$$

In the next two chapters we will explain in detail that the parabolic and the hyperbolic PDE lead to a totally different time behavior of the evolving parallel flow. In fact the parabolic equation carries in itself the artifact of infinite propagation of disturbances which, from a technological point-of-view, may or may not be important. We will also see that a numerical solution in form of a FOURIER series is required, which unless converted into a D'ALEMBERT type of expression must be cut off after a finite number of terms for practical reasons. This may bring in other artifacts that we must judge very critically. The problem of a numerical solution will become even more evident in Sect. 10.4, which deals with the transient as well as with the stationary behavior of gas clouds, i.e., models for expanding stars if not for the whole universe. We should always keep in mind that a good model should capture as many aspects of reality as possible. However, models will never be able to mimic all aspects of reality simultaneously. In the end, the final answers can only be obtained by asking questions to mother nature, as some physicists so aptly put it, i.e., by performing experiments. Models are work of art, like paintings, and therefore only feeble substitutes for our material world.

## 10.2 Transient Channel Flow of a Navier-Stokes-Fluid

Following Bernoulli we attempt to solve the first PDE in (10.1.19) by *separation of variables*:

$$w(y,t) = \psi(y)\chi(t). \tag{10.2.1}$$

We obtain:

$$\frac{1}{D}\frac{\dot{\chi}}{\chi} = \frac{\psi''}{\psi} = \text{const.} = -k^2, \tag{10.2.2}$$

since the left side depends only on time $t$ and the right side only on position $y$. The minus sign can be motivated physically: We expect that the motion will be damped down when time increases. We do not expect excitation. The general solution of the two Ordinary Differential Equations (ODEs) reads:

$$\dot{\chi} + Dk^2\chi = 0 \quad \Rightarrow \quad \chi = \exp\left(-Dk^2 t\right) \tag{10.2.3}$$

and:

$$\psi'' + k^2\psi = 0 \quad \Rightarrow \quad \psi = A\sin(ky) + B\cos(ky). \tag{10.2.4}$$

Johann Bernoulli was born in Basel on August 6, 1667 where he also died on January 1, 1748. He was one of the “senior members” of the Bernoulli’s—a family clan of mathematicians. Originally he was supposed to become a merchant. However, he decided to turn into an intellectual instead. So he began to study at the University of Basel in 1683, where he graduated as a magister in 1685. After that he turned to medicine. His older brother Jacob finally introduced him to mathematics and, in particular, to the new technique of calculus. Initially they worked together very closely. However, scientific envy eventually turned into big dispute and animosity, with his brother and with his son Daniel. In 1693 he started an extensive correspondence with Leibniz, where the method of separation of variables is also mentioned. As a fierce supporter of Leibniz’ work and ideas he became one of the spearheads continuing the feud of continental mathematicians against the Newtonians after Leibniz’ death.

By evaluating the boundary conditions (10.1.20) we find that:

$$B = 0, k = \frac{n\pi}{H}, \quad n = 1,2,3,\ldots. \tag{10.2.5}$$

Consequently, the expression

$$w_n(y,t) = A_n \exp\left(-D\frac{n^2\pi^2}{H^2}t\right)\sin\left(\frac{n\pi}{H}y\right) \tag{10.2.6}$$

is a possible solution. According to the principle of linear superposition we conclude that:

$$w(y,t) = \sum_{n=1}^{\infty} A_n \exp\left(-D\frac{n^2\pi^2}{H^2}t\right)\sin\left(\frac{n\pi}{H}y\right). \tag{10.2.7}$$

The constants $A_n$ must be adjusted to meet the (first) initial condition from Eq. (10.1.21):

$$w(y,0) = -V_0\left(1-\frac{y}{H}\right) = \sum_{n=1}^{\infty} A_n \sin\left(\frac{n\pi}{H}y\right). \tag{10.2.8}$$

We multiply this equation by $\sin(\frac{m\pi}{H}y)$ and integrate between 0 and $H$, i.e., the region on which the problem has been defined initially. After evaluating the corresponding integrals the coefficients $A_n$ result:

$$A_n = -\frac{2V_0}{n\pi}. \tag{10.2.9}$$

Thus the solution for the transient parallel plate flow of a Newtonian fluid reads:

$$w(y,t) = -\frac{2V_0}{\pi}\sum_{n=1}^{\infty}\frac{1}{n}\sin\left(\frac{n\pi}{H}y\right)\exp\left(-D\frac{n^2\pi^2}{H^2}t\right), \tag{10.2.10}$$

or with Eqs. (10.1.17) and (10.1.18):

$$\upsilon(y,t) = V_0\left(1-\frac{y}{H}-\frac{2}{\pi}\sum_{n=1}^{\infty}\frac{1}{n}\sin\left(\frac{n\pi}{H}y\right)\exp\left(-D\frac{n^2\pi^2}{H^2}t\right)\right). \tag{10.2.11}$$

We realize that the disturbance spreads infinitely fast from the lower plate at $y=0$ to the position $y=H$ since:

$$\upsilon(y,t) \neq 0, \forall y \in [0,H), \quad \forall t > 0. \tag{10.2.12}$$

A graphic representation of the solution at different times is shown in Fig. 10.3. For the distance $H$ we chose the value 1 mm.

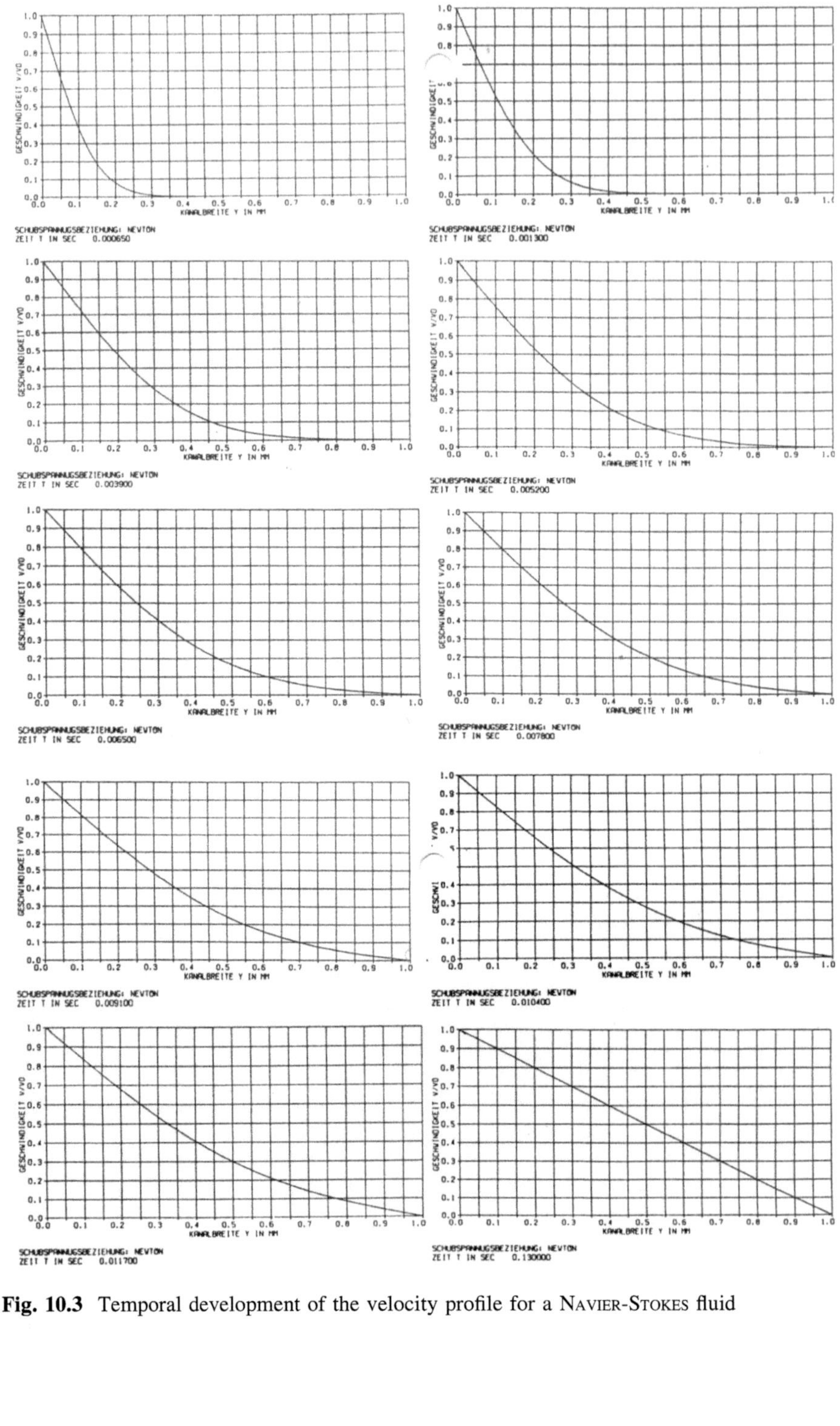


**Fig. 10.3** Temporal development of the velocity profile for a Navier-Stokes fluid

**Exercise 10.2.1:** ***Continuation of a function and its application to the transient flow of a NAVIER-STOKES fluid***

Prove Eq. (10.2.11) and put it in a physical context. Moreover, show that the right hand side in the following relation:

$$1 - \frac{y}{H} = \frac{2}{\pi}\sum_{n=1}^{\infty}\frac{1}{n}\sin\left(\frac{n\pi}{H}y\right) \tag{10.2.13}$$

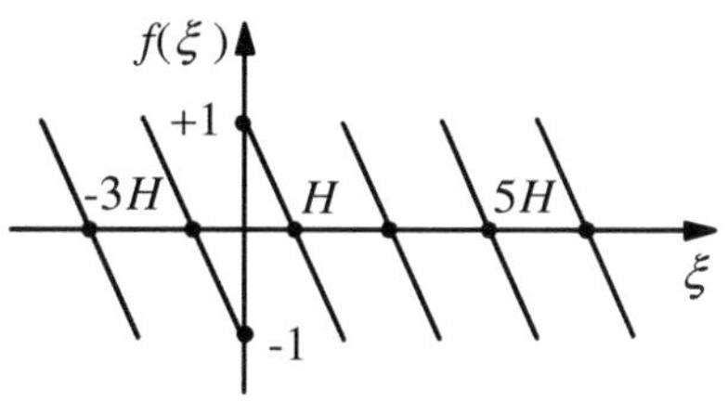


**Fig. 10.4** Saw tooth function: odd periodic continuation of a straight line

represents the FOURIER series of the function $f(\xi)$ shown in Fig. 10.4.

$$f(\xi) = \frac{1}{H}(H - [\xi - 2nH]) \text{ for } 2nH < \xi < (2n+2)H \tag{10.2.14}$$

and

$$f(\xi) = -\frac{1}{H}(H + [\xi + 2nH]) \text{ for } -2(n+1)H < \xi < -2nH, \tag{10.2.15}$$

where $n = 0, 1, 2, \ldots$. Interpret this function as the periodic continuation of the straight line $1 - \frac{y}{H}$.

## 10.3 Transient Channel Flow of a MAXWELL Fluid

We now turn to the second PDE shown in (10.1.19). For our discussion it is most helpful to consider a special case first, a.k.a. the *non-dissipative wave propagation*: We neglect the term $\partial w/\partial t$ and obtain the classical wave equation:

$$\frac{\partial^2 w}{\partial t^2} = c^2\frac{\partial^2 w}{\partial y^2}, \tag{10.3.1}$$

which will now be solved in combination with the aforementioned initial and boundary conditions. By using the Eq. (10.2.1) it follows that:

$$\frac{1}{c^2}\frac{\ddot{\chi}}{\chi} = \frac{\psi''}{\psi} = -k^2, \tag{10.3.2}$$

and consequently:

$$\ddot{\chi} + c^2k^2\chi = 0 \quad \Rightarrow \quad \chi = C\sin(ckt) + D\cos(ckt), \tag{10.3.3}$$

and:

$$\psi'' + k^2\psi = 0 \quad \Rightarrow \quad \psi = A\sin(ky) + B\cos(ky). \tag{10.3.4}$$

By evaluating the boundary conditions (10.1.20) we conclude that:

$$B = 0, k = \frac{n\pi}{H}, n = 1, 2, 3, \ldots \tag{10.3.5}$$

and obtain:

$$w(y,t) = \sum_{n=1}^{\infty} \left[A_n \sin\left(\frac{n\pi}{H}ct\right) + B_n \cos\left(\frac{n\pi}{H}ct\right)\right] \sin\left(\frac{n\pi}{H}y\right) \tag{10.3.6}$$

for the solution of the PDE (10.3.1), with $A_n = AC$ and $B_n = AD$. The remaining coefficients will be adjusted to the initial conditions from Eq. (10.1.21). We obtain in complete analogy to Eq. (10.2.9) of the Navier–Stokes case:

$$\sum_{n=1}^{\infty} B_n \sin\left(\frac{n\pi}{H}y\right) = -\upsilon_s(y) \quad \Rightarrow \quad B_n = -\frac{2V_0}{n\pi} \tag{10.3.7}$$

and

$$\sum_{n=1}^{\infty} \frac{n\pi}{H} cA_n \sin\left(\frac{n\pi}{H}y\right) = 0 \quad \Rightarrow \quad A_n = 0. \tag{10.3.8}$$

Thus the solution for non-dissipative wave propagation reads:

$$w(y,t) = -\frac{2V_0}{\pi}\sum_{n=1}^{\infty}\frac{1}{n}\cos\left(\frac{n\pi}{H}ct\right)\sin\left(\frac{n\pi}{H}y\right), \tag{10.3.9}$$

or alternatively with Eq. (10.1.18):

$$\upsilon(y,t) = V_0\left[1 - \frac{y}{H} - \frac{2}{\pi}\sum_{n=1}^{\infty}\frac{1}{n}\cos\left(\frac{n\pi}{H}ct\right)\sin\left(\frac{n\pi}{H}y\right)\right]. \tag{10.3.10}$$

By means of the trigonometric theorem

$$\sin(\alpha)\cos(\beta) = \frac{1}{2}\left[\sin(\alpha - \beta) + \sin(\alpha + \beta)\right] \tag{10.3.11}$$

this can also be written as:

$$\upsilon(y,t) = V_0\left(1 - \frac{y}{H} - \frac{1}{\pi}\left[\sum_{n=1}^{\infty}\frac{1}{n}\sin\left(\frac{n\pi}{H}(y - ct)\right) + \sum_{n=1}^{\infty}\frac{1}{n}\sin\left(\frac{n\pi}{H}(y + ct)\right)\right]\right) \tag{10.3.12}$$

If we define:

$$\xi_{\pm} = y \pm ct, \tag{10.3.13}$$

it becomes obvious that the two sums in Eq. (10.3.12) correspond to FOURIER series of the saw tooth function shown in Fig. 10.4. Thus by replacing the FOURIER series we obtain the following compact form of the solution:

$$\upsilon(y,t) = V_0\left(1 - \frac{y}{H} - \frac{1}{2}[f(\xi_-) + f(\xi_+)]\right) \tag{10.3.14}$$

with:

$$f(\xi_{\pm}) = \frac{1}{H}(H - [\xi_{\pm} - 2nH]) \text{ for } 2nH < \xi_{\pm} < 2(n+1)H, \tag{10.3.15}$$

and:

$$f(\xi_{\pm}) = -\frac{1}{H}(H + [\xi_{\pm} + 2nH]) \text{ for } -2(n+1)H < \xi < -2nH, \tag{10.3.16}$$

where $n = 0, 1, 2, \ldots$. This way of writing is known as D'ALEMBERT's solution for the wave equation in the literature.

In order to convince ourselves that the solution really describes a propagating wave that starts as a disturbance at $y = 0$ and moves at a finite speed $c = \sqrt{D/\lambda} = \sqrt{\mu/\rho\lambda}$ into a fluid at rest, we study the following scheme:

| $y = \frac{H}{2}$ | $ct = 0$ | $\frac{H}{4}$ | $\frac{2H}{4}$ | $\frac{3H}{4}$ | $\frac{4H}{4}$ | $\frac{5H}{4}$ | $\frac{6H}{4}$ | $\frac{7H}{4}$ | $\frac{8H}{4}$ |
|---|---|---|---|---|---|---|---|---|---|
| $\xi_- = \frac{2H}{4} - ct$ | $\frac{2H}{4}$ | $\frac{H}{4}$ | $0$ | $-\frac{H}{4}$ | $-\frac{2H}{4}$ | $-\frac{3H}{4}$ | $-\frac{4H}{4}$ | $-\frac{5H}{4}$ | $-\frac{6H}{4}$ |
| $\xi_+ = \frac{2H}{4} + ct$ | $\frac{2H}{4}$ | $\frac{3H}{4}$ | $\frac{4H}{4}$ | $\frac{5H}{4}$ | $\frac{6H}{4}$ | $\frac{7H}{4}$ | $\frac{8H}{4}$ | $\frac{9H}{4}$ | $\frac{10H}{4}$ |
| $f(\xi_-)$ | $\frac{2}{4}$ | $\frac{3}{4}$ | $\frac{4}{4}$ | $-\frac{3}{4}$ | $-\frac{2}{4}$ | $-\frac{1}{4}$ | $0$ | $\frac{1}{4}$ | $\frac{2}{4}$ |
| $f(\xi_+)$ | $\frac{2}{4}$ | $\frac{1}{4}$ | $0$ | $-\frac{1}{4}$ | $-\frac{2}{4}$ | $-\frac{3}{4}$ | $-\frac{4}{4}$ | $\frac{3}{4}$ | $\frac{2}{4}$ |
| $\frac{v(H/2,t)}{V_0}$ | 0 | 0 | 0 | 1 | 1 | 1 | 1 | 0 | 0 |

Obviously, the velocity $\upsilon(H/2,t)$ of the fluid at the center of the fluid, $y = H/2$, is equal to zero until $ct = H/2$. At that time $\upsilon$ jumps to $V_0 \neq 0$, where it remains until the time $ct = 6H/4$ has been exceeded. This corresponds to the amount of time it takes until the reflection (i.e., the negative) of the first wave at the wall $y = H$ has reached the position $y = H/2$ and eliminates the original disturbance from $y = 0$ completely. The temporal development at all positions $y$ is shown in the sequence of plots in Fig. 10.5.

With the same methods we will now tackle the general MAXWELL case with damping. In other words, we will solve the second PDE from (10.1.19) with the initial and boundary conditions shown in Eqs. (10.1.20) and (10.1.21). Separation of variables according to Eq. (10.2.1) yields:

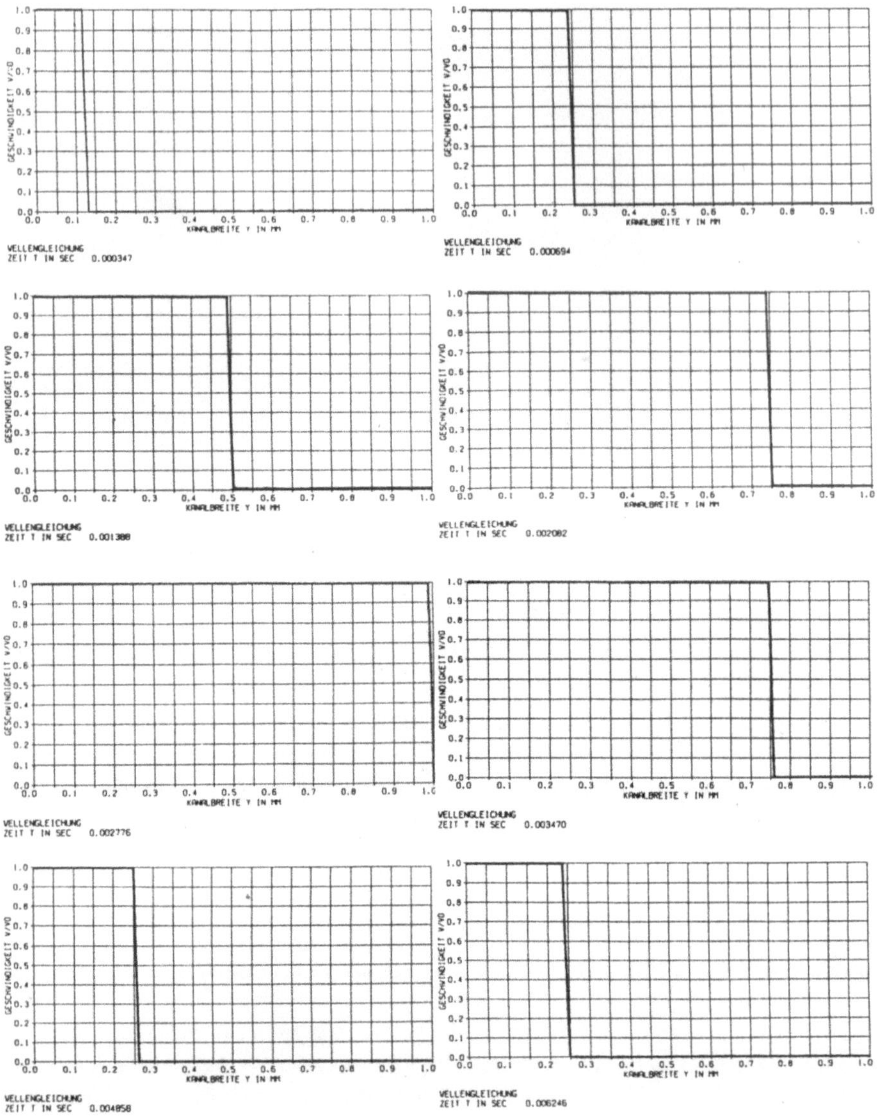


**Fig. 10.5** Temporal development of the solution for a propagating wave w/o damping

$$\frac{1}{D}\frac{\lambda\ddot{\chi}+\dot{\chi}}{\chi}=\frac{\psi''}{\psi}=-k^2 \tag{10.3.17}$$

and thus:

$$\frac{1}{c^2}\ddot{\chi}+\frac{1}{D}\dot{\chi}+k^2\chi=0, \tag{10.3.18}$$

which we will solve with the following *ansatz*:

$$\chi = \exp(mt). \tag{10.3.19}$$

We obtain the following characteristic equation:

$$m^2 + \frac{c^2}{D} m + c^2 k^2 = 0, \tag{10.3.20}$$

which has two solutions:

$$m_{\pm} = -\frac{c^2}{2D} \pm \sqrt{\left(\frac{c^2}{2D}\right)^2 - c^2 k^2} = -\frac{1}{2\lambda}\left(1 \mp \sqrt{1 - 4\lambda^2 \frac{n^2\pi^2}{H^2} c^2}\right), \tag{10.3.21}$$

i.e.,

$$\chi = A_n^+ \exp(m_+ t) + A_n^- \exp(m_- t). \tag{10.3.22}$$

Here we have already made use of the solution of the second remaining ODE from (10.3.17). As in the previous case we must write:

$$\psi'' + k^2 \psi = 0 \quad \Rightarrow \quad \psi = A \sin(ky) + B \cos(ky), \tag{10.3.23}$$

and, once more, by evaluating the boundary conditions (10.1.20) we have:

$$B = 0, k = \frac{n\pi}{H}, n = 1, 2, 3, \ldots \tag{10.3.24}$$

and thus:

$$w(y,t) = \sum_{n=1}^{\infty} \left[A_n^+ \exp(m_+ t) + A_n^- \exp(m_- t)\right] \sin\left(\frac{n\pi}{H} y\right). \tag{10.3.25}$$

Note that the constant $A$ has been chosen as 1. The relation (10.3.25) will now be adjusted to meet the initial conditions shown in Eq. (10.1.21). We obtain two equations for the unknown coefficients $A_n^{\pm}$:

$$\sum_{n=1}^{\infty} \left(A_n^+ + A_n^-\right) \sin\left(\frac{n\pi}{H} y\right) = -\upsilon_s(y) = -V_0\left(1 - \frac{y}{H}\right) \tag{10.3.26}$$

and:

$$\sum_{n=1}^{\infty} \left(m_+ A_n^+ + m_- A_n^-\right) \sin\left(\frac{n\pi}{H} y\right) = 0. \tag{10.3.27}$$

The FOURIER series from Eq. (10.2.8) allows us to write:

$$A_n^+ + A_n^- = -V_0 \frac{2}{\pi n}, m_+ A_n^+ + m_- A_n^- = 0 \tag{10.3.28}$$

or:

$$A_n^+ = \frac{m_-}{m_+ - m_-}\frac{2V_0}{\pi n}, A_n^- = -\frac{m_+}{m_+ - m_-}\frac{2V_0}{\pi n}. \tag{10.3.29}$$

These formulae are inserted in Eq. (10.3.25) and the representation (10.1.18) is observed. Hence the following equation for the velocity field results:

$$\begin{aligned} \upsilon(y,t) = V_0\left(1-\frac{y}{H}\right) - \frac{2V_0}{\pi}\exp\left(-\frac{t}{2\lambda}\right)\sum_{n=1}^{\infty}\frac{1}{n}\left(\cosh\left(\frac{t}{2\lambda}\sqrt{1-4\lambda^2\frac{n^2\pi^2}{H^2}c^2}\right)+\right. \\ \left.\left(\sqrt{1-4\lambda^2\frac{n^2\pi^2}{H^2}c^2}\right)^{-1}\sinh\left(\frac{t}{2\lambda}\sqrt{1-4\lambda^2\frac{n^2\pi^2}{H^2}c^2}\right)\right)\sin\left(\frac{n\pi}{H}y\right). \end{aligned} \tag{10.3.30}$$

We will now investigate the solution somewhat further and simplify appropriately. In particular, we are interested to know as to whether a change in velocity at the lower plate will propagate infinitely fast, as in the NAVIER-STOKES case, or moves at a finite speed, as in the previously discussed case of a traveling wave.

First note that Eq. (10.3.30) will lead to solutions of the wave type only if:

$$\sqrt{1-4\lambda^2\frac{n^2\pi^2}{H^2}c^2} = \mathrm{i}\sqrt{4\lambda^2\frac{n^2\pi^2}{H^2}c^2-1}\ \text{with}\ 4\lambda^2\frac{n^2\pi^2}{H^2}c^2 \geq 1. \tag{10.3.31}$$

For $n = 1$ and with the data for polyisobutylene in decahydronaphthalene it follows that:

$$4\lambda^2\frac{\pi^2}{H^2}c^2 = 4\pi^2\frac{D\lambda}{H^2} = \frac{9.4\cdot 10^{-3}\ \mathrm{m}^2}{H^2} \geq 1. \tag{10.3.32}$$

This relation is satisfied only if $H \leq 9.7$ cm. Otherwise we suspect that the velocity $\upsilon(y,t)$ at an arbitrary point $y$ is different from zero for $t > 0$. In what follows we restrict ourselves to the case $H < 9.7$ cm. With the identities

$$\cosh(\mathrm{i}\alpha) = \cos(\alpha), \sinh(\mathrm{i}\alpha) = \mathrm{i}\sin(\alpha) \tag{10.3.33}$$

the solution for the velocity field (10.3.30) becomes:

$$\begin{aligned} \upsilon(y,t) = V_0\left(1-\frac{y}{H}\right) - \frac{2V_0}{\pi}\exp\left(-\frac{t}{2\lambda}\right)\sum_{n=1}^{\infty}\frac{1}{n}\left(\cos\left(\frac{t}{2\lambda}\sqrt{4\lambda^2\frac{n^2\pi^2}{H^2}c^2-1}\right)\right. \\ \left.+\frac{1}{\sqrt{4\lambda^2\frac{n^2\pi^2}{H^2}c^2-1}}\sin\left(\frac{t}{2\lambda}\sqrt{4\lambda^2\frac{n^2\pi^2}{H^2}c^2-1}\right)\right)\sin\left(\frac{n\pi}{H}y\right). \end{aligned} \tag{10.3.34}$$

**Exercise 10.3.1:** ***Velocity profile for a MAXWELL fluid***

Prove Eqs. (10.3.30/10.3.34) and explain concisely the underlying assumptions.

This relation can be simplified further for the following special case. First, we have:

$$4\pi^2 \frac{\lambda D}{H^2} = \frac{9.4 \cdot 10^{-3}\,\mathrm{m}^2}{H^2} = \begin{cases} 9400 & \text{for } H = 1\text{ mm} \\ 94 & \text{for } H = 1\text{ cm} \\ 0.94 & \text{for } H = 10\text{ cm} \end{cases} \tag{10.3.35}$$

Consequently the term "1" in the square roots of Eq. (10.3.34) can be neglected for $H \leq 1$ mm. Thus by the addition theorem shown in (10.3.11) and the following one:

$$\sin(\alpha)\sin(\beta) = \frac{1}{2}[\cos(\alpha - \beta) - \cos(\alpha + \beta)] \tag{10.3.36}$$

it follows that:

$$\begin{aligned} \upsilon(y,t) = V_0\left(1 - \frac{y}{H}\right) - \frac{V_0}{\pi}\exp\left(-\frac{t}{2\lambda}\right)\Bigg\{&\sum_{n=1}^{\infty}\frac{1}{n}\left(\sin\left[\frac{n\pi}{H}(y - ct)\right] + \sin\left[\frac{n\pi}{H}(y + ct)\right]\right) \\ &+ \frac{1}{2\pi}\frac{H}{\sqrt{\lambda D}}\sum_{n=1}^{\infty}\frac{1}{n^2}\left(\cos\left[\frac{n\pi}{H}(y - ct)\right] - \cos\left[\frac{n\pi}{H}(y + ct)\right]\right)\Bigg\}. \end{aligned} \tag{10.3.37}$$

In this special case the FOURIER series can be rewritten by means of the function $f(\xi)$ introduced in Exercise 10.2.1, and a quadratic function $g(\xi)$ that is defined piecewise as follows:

$$g(\xi) = \frac{1}{12}\left(2 - \frac{6}{H}[\xi - 2nH] + \frac{3}{H^2}[\xi - 2nH]^2\right) \tag{10.3.38}$$

for $2nH < \xi < 2(n+1)H$ and

$$g(\xi) = \frac{1}{12}\left(2 + \frac{6}{H}[\xi + 2nH] + \frac{3}{H^2}[\xi + 2nH]^2\right) \tag{10.3.39}$$

for $-2(n+1)H < \xi < -2nH$. It can be expanded into the following FOURIER series:

$$g(\xi) = \frac{1}{\pi^2}\sum_{n=1}^{\infty}\frac{1}{n^2}\cos\left(\frac{n\pi}{H}\xi\right). \tag{10.3.40}$$

Thus the solution rewritten in D'ALEMBERT form reads:

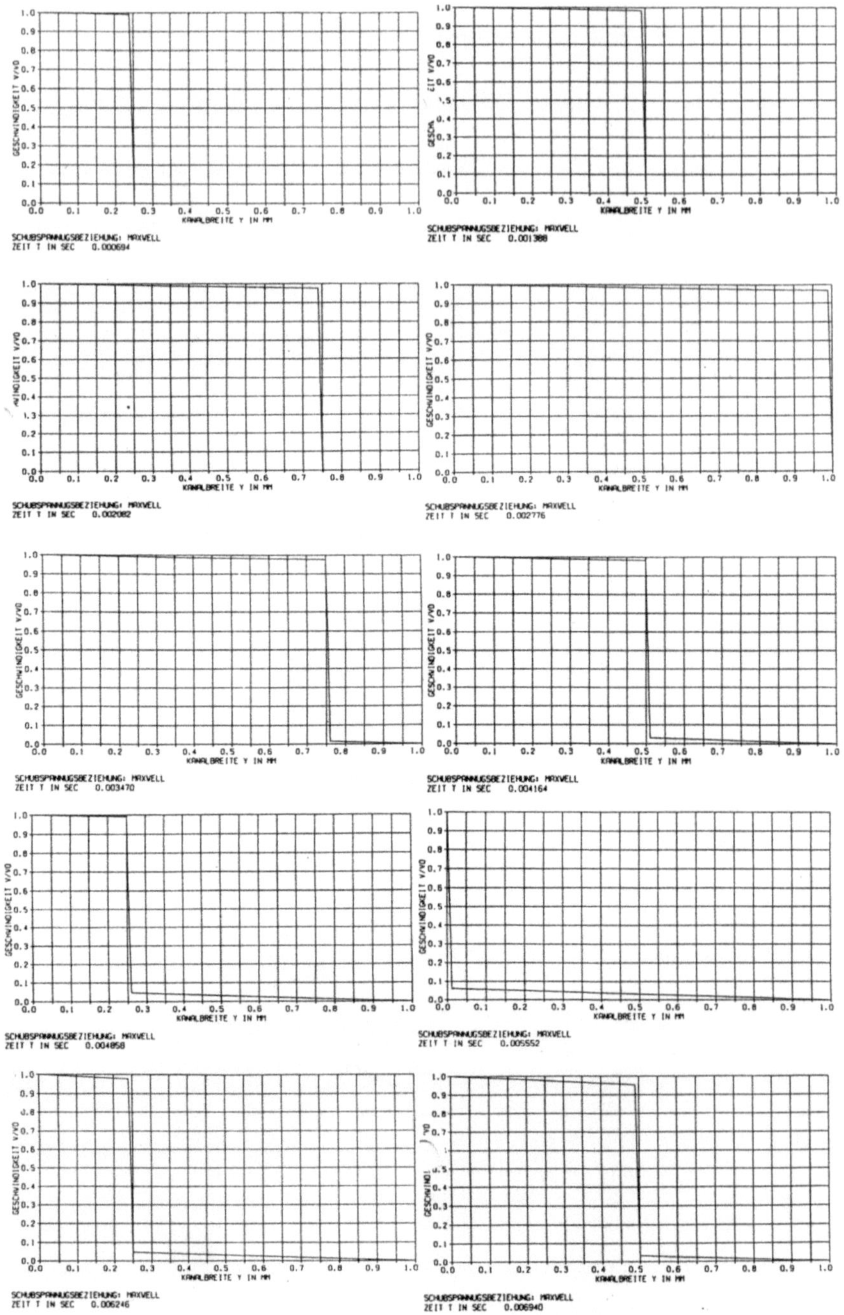


**Fig. 10.6** Temporal development of the velocity profile for a MAXWELL fluid

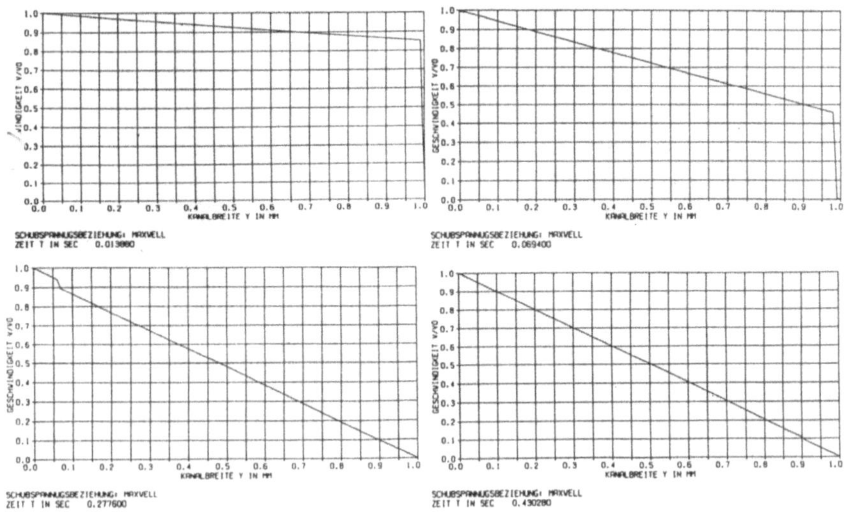


**Fig. 10.6** continued

$$v(y,t) = V_0\left(1 - \frac{y}{H}\right) - \frac{V_0}{2}\exp\left(-\frac{t}{2\lambda}\right)\left\{f(\xi_-) + f(\xi_+) + \frac{H}{\sqrt{\lambda D}}(g(\xi_-) - g(\xi_+))\right\}. \tag{10.3.41}$$

Figure 10.6 shows the result of a numerical evaluation of the various terms in Eq. (10.3.37) for the velocity profile of transient plan-parallel flow of a MAXWELL fluids at various times $t$. It becomes evident that from a certain point in time the profile cannot be distinguished from the stationary one any more.

**Exercise 10.3.2:** ***FOURIER series for the D'ALEMBERT solution of the transient flow of a MAXWELL fluid***

Prove Eq. (10.3.40) and explain concisely its relevance for obtaining the solution shown in Eq. (10.3.41).

## 10.4 Expanding and Contracting Stars and Universes: An Approach Based on Classical Continuum Physics

Physicists love to pooh-pooh theories by saying that they are "non-relativistic" or claim that without quantum mechanics "these problems cannot be understood." Whilst it may be true that both relativity as well as quantum mechanics provide us with better quantitative models of reality we have to keep in mind that they are also just models after all. In fact, classical theories allow us frequently to get first useful insights of what the problem is all about. An example of this kind is the application of classical continuum concepts to the problem of an expanding or contracting "sphere of gas" which may serve as a (coarse) model for the stars or even for the whole universe.

We will assume perfect spherical symmetry, i.e., no angular dependence. This means that the velocity field in spherical coordinates is given by:

$$\upsilon_{\langle r\rangle} = \upsilon_{\langle r\rangle}(r), \quad \upsilon_{\langle\vartheta\rangle} = 0, \quad \upsilon_{\langle\varphi\rangle} = 0. \tag{10.4.1}$$

Thus the balance of mass (5.2.10) reduces to:

$$\frac{\partial\rho}{\partial t} + \upsilon_{\langle r\rangle}\frac{\partial\rho}{\partial r} + \rho\frac{\partial\upsilon_{\langle r\rangle}}{\partial r} + \frac{2\rho}{r}\upsilon_{\langle r\rangle} = 0. \tag{10.4.2}$$

Before we turn to the balance of momentum we reduce the form of the NAVIER-STOKES stress tensor. If we insert the *ansatz* (10.4.1) into Eqs. (6.3.6/6.3.9) we find that all shear components vanish and obtain for the normal stresses:

$$\sigma_{\langle rr\rangle} = -p + (\lambda + 2\mu)\frac{\partial\upsilon_{\langle r\rangle}}{\partial r} + 2\lambda\frac{\upsilon_{\langle r\rangle}}{r}, \tag{10.4.3}$$

$$\sigma_{\langle\vartheta\vartheta\rangle} = \sigma_{\langle\varphi\varphi\rangle} = -p + \lambda\frac{\partial\upsilon_{\langle r\rangle}}{\partial r} + 2(\lambda + \mu)\frac{\upsilon_{\langle r\rangle}}{r}.$$

Note that for a self-gravitating sphere the specific volume force acts only in $r$-direction:

$$f_{\langle r\rangle} = f_{\langle r\rangle}(r), \quad f_{\langle\vartheta\rangle} = 0, \quad f_{\langle\varphi\rangle} = 0. \tag{10.4.4}$$

Thus, if we insert these relations in the balance of momentum shown in Eq. (5.4.7), and if we observe once more Eq. (10.4.1), we see that the $\vartheta$ and $\varphi$ components are identically satisfied. However, the $r$ component reads:

$$\rho\left(\frac{\partial\upsilon_{\langle r\rangle}}{\partial t} + \upsilon_{\langle r\rangle}\frac{\partial\upsilon_{\langle r\rangle}}{\partial r}\right) - (\lambda + 2\mu)\left(\frac{\partial^2\upsilon_{\langle r\rangle}}{\partial r^2} + \frac{2}{r}\frac{\partial\upsilon_{\langle r\rangle}}{\partial r} - \frac{2\upsilon_{\langle r\rangle}}{r^2}\right) = -\frac{\partial p}{\partial r} + \rho f_{\langle r\rangle}. \tag{10.4.5}$$

Equations (10.4.2) and (10.4.5) are two equations for the two unknowns, $\rho$ and $\upsilon_{\langle r\rangle}$. However, they are not field equations yet, unless we specify the pressure and the body force in terms of the unknowns. We start with the body force. According to NEWTON's law of gravity we find for an infinitesimal contribution to the gravitational acceleration between a mass element $\rho(\bar{r})\mathrm{d}V$ and a test mass $M$ *outside* of the sphere (cf., Fig. 10.7 left):

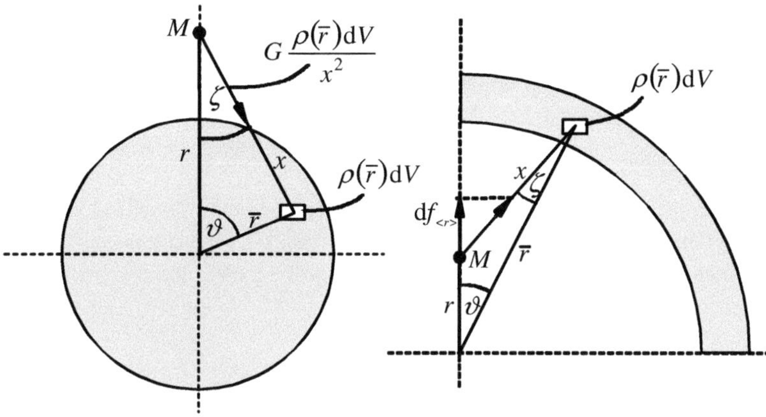


**Fig. 10.7** Test mass $M$ outside and within radially heterogeneous spherical regions

$$\mathrm{d}f_{\langle r\rangle} = -G\frac{\rho(\bar{r})\mathrm{d}V\cos\zeta}{x^2}, \tag{10.4.6}$$

where $G = 6.673 \cdot 10^{-11}\frac{\mathrm{m}^3}{\mathrm{kgs}^2}$ denotes the gravitational constant. Simple trigonometry in combination with Eq. $(4.5.13)_2$ leads to:

$$f_{\langle r\rangle} = -G\int\limits_{\bar{r}=0}^{R}\int\limits_{\vartheta=0}^{\pi}\int\limits_{\varphi=0}^{2\pi}\frac{\rho(\bar{r})\,\bar{r}^2\sin\vartheta\,(r-\bar{r}\cos\vartheta)\mathrm{d}\varphi\mathrm{d}\vartheta\mathrm{d}\bar{r}}{\left(r^2+\bar{r}^2-2r\bar{r}\cos\vartheta\right)^{3/2}}. \tag{10.4.7}$$

With suitable substitutions (see the following exercise) the integrations can relatively easily be performed and the following astonishing result is obtained:

$$f_{\langle r\rangle} = -G\frac{m(R)}{r^2}, m(R) = 4\pi\int\limits_{\bar{r}=0}^{R}\rho(\bar{r})\,\bar{r}^2\mathrm{d}\bar{r}, r \geq R. \tag{10.4.8}$$

In words: The gravitational acceleration outside of a radially heterogeneous sphere of radius $R$ and mass $m(R)$ is equal to that of a point mass of equal size located in its center. Moreover, as we shall see in the next exercise, a test mass within a radially heterogeneous spherical shell experiences no gravitational force at all.

**Exercise 10.4.1:** ***Gravitational pull within and outside spherical regions***

Perform the integrations shown in Eq. (10.4.7). To this end use the following substitutions and integral:

$$y = \cos\vartheta, a = \frac{\bar{r}}{r}, \int\frac{(1-ay)\mathrm{d}y}{\left(1+a^2-2ay\right)^{3/2}} = \frac{y-a}{\sqrt{1+a^2-2ay}}. \tag{10.4.9}$$

Also use Fig. 10.7 (left) in combination with the sine and cosine rule to explain how Eq. (10.4.7) was obtained. Why is it possible to orient the test mass $M$ the way it is shown in the figure? What happens to the gravitational forces $\mathrm{d}f'_{\langle r \rangle}$ i.e., the pull within planes of constant angle $\vartheta$ perpendicular to the vertical axis?

Consider now the situation shown on the right hand side of Fig. 10.7: The test mass $M$ is situated *within the hole* of a radially heterogeneous *spherical* shell. Show that the total gravitational pull can be expressed by the following integral:

$$f_{\langle r \rangle} = G \int_{\bar{r}=0}^{R} \int_{\vartheta=0}^{\pi} \int_{\varphi=0}^{2\pi} \frac{\rho(\bar{r})\, \bar{r}^2 \sin\vartheta \, \cos(\zeta + \vartheta) \mathrm{d}\varphi \mathrm{d}\vartheta \mathrm{d}\bar{r}}{r^2 + \bar{r}^2 - 2r\bar{r}\cos\vartheta}. \tag{10.4.10}$$

Perform the integrations by using the substitutions of Eq. (10.4.9) and the following formula:

$$\int \frac{-(1 - by)y + (1 - y^2)b}{(1 + b^2 - 2by)^{3/2}} \mathrm{d}y = \frac{-1 + by}{b^2\sqrt{1 + b^2 - 2by}}, \, b = \frac{1}{a}, \tag{10.4.11}$$

and show that the test mass is free of forces, even for the case of a radially heterogeneous, hollow sphere.

Now consider a homogeneous sphere of a constant mass density $\rho_0$. Argue that the gravitational acceleration of a test mass at radial position $r$ is given by:

$$\boldsymbol{f} = f_{\langle r \rangle} \boldsymbol{e}_r, \quad f_{\langle r \rangle} = \begin{cases} -\frac{4\pi}{3} G\rho_0 r & \text{if } 0 \leq r \leq R \\ -G\frac{m(R)}{r^2}, m(R) = \frac{4\pi}{3}\rho_0 R^3 & \text{if } R \leq r < \infty. \end{cases} \tag{10.4.12}$$

Recall the statement made in context with Eq. (7.2.10) according to which conservative forces can be obtained by differentiation of a potential, $\varphi$. Show that the following formulae hold in the present case:

$$\boldsymbol{f} = -\nabla\varphi, \quad \varphi = \varphi(r) = \begin{cases} \frac{4\pi}{6} G\rho_0 r^2 + C_1 & \text{if } 0 \leq r \leq R \\ -G\frac{m(R)}{r} + C_2 & \text{if } R \leq r < \infty, \end{cases} \tag{10.4.13}$$

where the del operator is given by Eq. (5.6.5). Determine both constants, $C_1$ and $C_2$, from the requirement that the potential is continuous at the sphere's surface, i.e., at $R$, and vanishes at infinity so that:

$$\varphi = \varphi(r) = \begin{cases} -\frac{1}{2} G\frac{m(R)}{R} \left[3 - \left(\frac{r}{R}\right)^2\right] & \text{if } 0 \leq r \leq R \\ -G\frac{m(R)}{r} & \text{if } R \leq r < \infty \end{cases}. \tag{10.4.14}$$

Note that Eqs. (10.4.12/10.4.14) are mathematically identical to the restoring force and the potential of a linear Hookean spring. In other words: A test mass $M$ falling into a hole through the center of the Earth will behave like a harmonic oscillator. Show that the spring constant $c$ and the duration $T$ of the oscillation of the test mass are given by:

$$c = \frac{4\pi}{3} G \rho_{\mathrm{E}}, \quad T = \sqrt{\frac{3\pi}{G \rho_{\mathrm{E}}}}. \tag{10.4.15}$$

Recall that the average value of the Earth's radius is ca. 6370 km. Show that the average mass density of Earth is therefore given by

$$\rho_{\mathrm{E}} = \frac{3g}{4\pi G R} \approx 5500 \frac{\mathrm{kg}}{\mathrm{m}^3} \tag{10.4.16}$$

and that it would take roughly 2500 s, i.e., 42 min to send a person to the other side of the globe by free frictionless fall through a radial hole. Why is the acceleration of this person during the fall always less or equal to 1 g or, in other words, why are the accelerations harmless? What is the maximum velocity that could be reached? Moreover, find arguments why a considerable portion of the Earth must consist of iron.

As it was pointed out in Sect. 6.4 the pressure is given by the so-called thermal equation of state as a function of density and temperature, $p = p(\rho, T)$. Thus we conclude that Eq. (10.4.5) can be rewritten as follows:

$$\begin{aligned} &\rho \left( \frac{\partial \upsilon_{\langle r \rangle}}{\partial t} + \upsilon_{\langle r \rangle} \frac{\partial \upsilon_{\langle r \rangle}}{\partial r} \right) - (\lambda + 2\mu) \left( \frac{\partial^2 \upsilon_{\langle r \rangle}}{\partial r^2} + \frac{2}{r} \frac{\partial \upsilon_{\langle r \rangle}}{\partial r} - \frac{2\upsilon_{\langle r \rangle}}{r^2} \right) \\ &+ \frac{\partial p}{\partial \rho} \frac{\partial \rho}{\partial r} + \frac{\partial p}{\partial T} \frac{\partial T}{\partial r} = -G\rho \frac{m(r)}{r^2}. \end{aligned} \tag{10.4.17}$$

If we now use the thermal equation of state for an ideal gas, Eq. (6.4.1), this turns into:

$$\begin{aligned} &\rho \left( \frac{\partial \upsilon_{\langle r \rangle}}{\partial t} + \upsilon_{\langle r \rangle} \frac{\partial \upsilon_{\langle r \rangle}}{\partial r} \right) - (\lambda + 2\mu) \left( \frac{\partial^2 \upsilon_{\langle r \rangle}}{\partial r^2} + \frac{2}{r} \frac{\partial \upsilon_{\langle r \rangle}}{\partial r} - \frac{2\upsilon_{\langle r \rangle}}{r^2} \right) \\ &+ \frac{R}{M} \left( T \frac{\partial \rho}{\partial r} + \rho \frac{\partial T}{\partial r} \right) = -G\rho \frac{m(r)}{r^2}. \end{aligned} \tag{10.4.18}$$

Similar to the arguments in Sect. 7.5 we are now going to turn the system of coupled PDEs, Eqs. (10.4.2) and (10.4.18), into a system of ODEs by choosing a suitable semi-inverse *ansatz*. Basically we claim that the density and temperature fields are homogeneous but time-dependent. We assume perfect spherical symmetry for the velocity field or, in other words, the angular components vanish. Moreover, the radial velocity follows HUBBLE's law with a time-dependent HUBBLE's constant:

$$\rho = \rho(t), \quad T = T(t), \quad \frac{\upsilon_{\langle r\rangle}}{r} \equiv \frac{\dot{r}}{r} = H(t), \quad \upsilon_{\langle\vartheta\rangle} = \upsilon_{\langle\varphi\rangle} = 0. \tag{10.4.19}$$

In the equation for the radial velocity we have used the explicit form for its physical component shown in Eq. $(5.2.7)_1$. It can be integrated easily:

$$r(t) = r(0)\exp\left(\int_{\tilde{t}=0}^{\tilde{t}=t} H(\tilde{t})\mathrm{d}\tilde{t}\right), \quad 0 \le r(t) \le R(t), \ 0 \le r(0) \le R(0). \tag{10.4.20}$$

In order to perform the integration, HUBBLE's constant must be a known function of time and, indeed, it is, *after* we have solved the aforementioned PDEs. We shall turn to that problem shortly. Note that Eq. (10.4.20) holds for *all* radii of a material sphere, in particular for its outer radius $R$:

$$R(t) = R(0)\exp\left(\int_{\tilde{t}=0}^{\tilde{t}=t} H(\tilde{t})\mathrm{d}\tilde{t}\right). \tag{10.4.21}$$

However, there is another simple way of determining the time dependence of the outer radius. Due to the homogeneity of the mass density the mass balance for the sphere can be integrated easily:

$$\frac{4\pi}{3}\rho(t)R^3(t) = \frac{4\pi}{3}\rho(0)R^3(0) \quad \Rightarrow \quad \frac{R(t)}{R(0)} = \left[\frac{\rho(t)}{\rho(0)}\right]^{-1/3}. \tag{10.4.22}$$

Thus we may say that the time dependence of mass density is the key to the evolution of the outer radius. Of course, it is necessary to compute $\rho(t)$, which is also subject of the PDEs, as we shall see now. The solution of the two PDEs benefits considerably from the semi-inverse *ansatz*. They turn into coupled ODEs, which can be examined by using standard numerical techniques:

$$\dot{\rho} + 3\rho H = 0, \quad \dot{H} + H^2 + \frac{4\pi}{3}G\rho = 0. \tag{10.4.23}$$

The following dimensionless variables are now used:

$$\bar{t} = H(0)t, \quad \bar{\rho}(\bar{t}) = \frac{\rho(t)}{\rho(0)}, \quad \bar{H}(\bar{t}) = \frac{H(t)}{H(0)}, \tag{10.4.24}$$

and we obtain:

$$\frac{\mathrm{d}\bar{\rho}}{\mathrm{d}\bar{t}} + 3\bar{\rho}\bar{H} = 0, \quad \frac{\mathrm{d}\bar{H}}{\mathrm{d}\bar{t}} + \bar{H}^2 + \gamma\bar{\rho} = 0, \quad \gamma = \frac{4\pi}{3}\frac{G\rho(0)}{H^2(0)}. \tag{10.4.25}$$

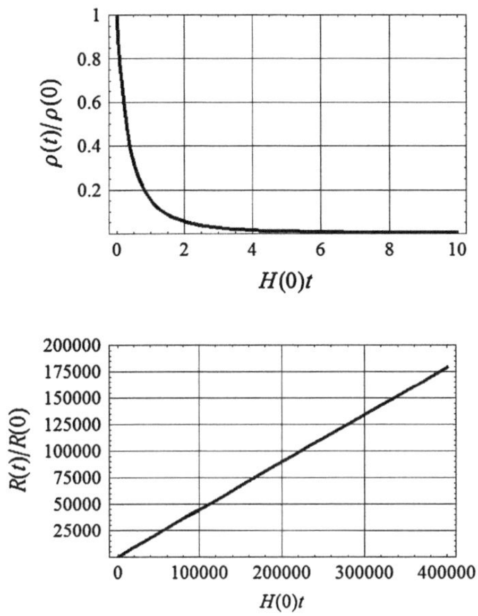


**Fig. 10.8** Mass density, HUBBLE's constant, and outer radius over time for $\gamma = 0.4$ (see text)

Obviously the initial conditions read:

$$\bar{\rho}(0) = 1, \quad \bar{H}(0) = 1. \tag{10.4.26}$$

The temporal behavior of the solution depends strongly on the choice for the value of the constant $\gamma$. For example, if we choose a value below a certain threshold, say 0.4, the density and HUBBLE's constant decrease monotonically over time, as shown in Fig. 10.8.

This is also reflected in a monotonic, almost linear increase of the outer radius according to Eqs. (10.4.21/10.4.22): Fig. 10.8 (bottom left). However, if we take a value above 0.5 instead, say $\gamma = 0.51$, the plots change considerably: Fig. 10.9 (bottom left). The radius increases initially. It reaches a maximum and then starts to decrease. Note that it even goes below the value $\bar{R}(\bar{t}) = R(t)/R(0) = 1$, cf., Fig. 10.9 (bottom right), and one is tempted to interpret this behavior in terms of a big crunch theory for the universe. Note that the mass density and HUBBLE's constant behave accordingly: Fig. 10.9 (top left and right). Interestingly, HUBBLE's constant has to become negative for a collapsing star or universe.

We proceed to investigate this effect in more detail. In fact, it can be explained by the principle of *mechanical* energy conservation and is based on a competition between kinetic and potential energy. But why is the total mechanical energy conserved in the present case? We know from Eq. (3.9.3) that, in general, kinetic

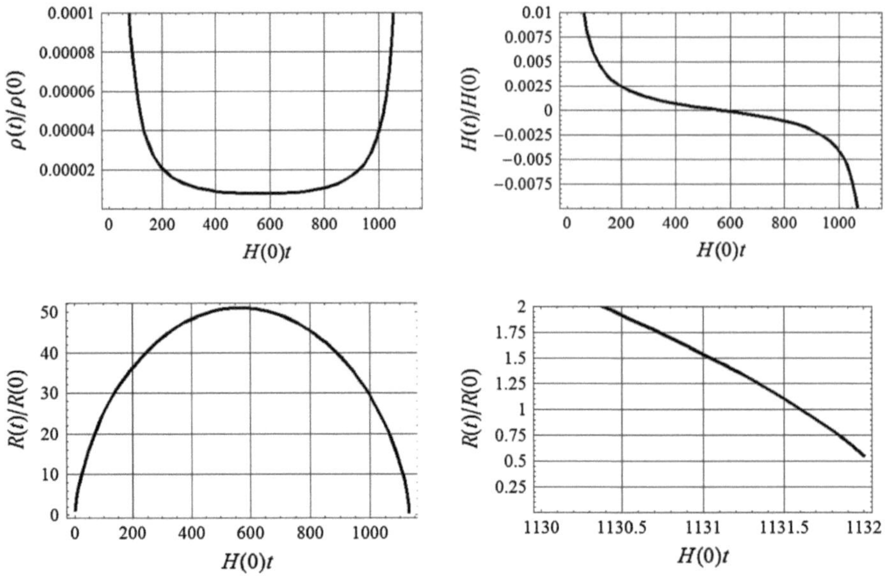


**Fig. 10.9** Mass density, HUBBLE's constant, and outer radius over time for $\gamma = 0.51$ (see text)

energy is *not* conserved due to the term $-\iiint_{V(t)} \boldsymbol{\sigma} : \nabla \boldsymbol{\upsilon}\, \mathrm{d}V$. This term can be integrated for the case of perfect spherical symmetry by using the *ansatz* from Eq. (10.4.19) in combination with Eqs. (10.4.3), $(4.5.13)_2$, and $(5.7.21)_2$. The final result reads:

$$-\iiint_{V(t)} \boldsymbol{\sigma} : \nabla \boldsymbol{\upsilon}\, \mathrm{d}V = [-p + (\lambda + 2\mu)H(t)]H(t)\, 4\pi R^3(t). \tag{10.4.27}$$

Interestingly the same result, but with the opposite sign, is obtained for the term $\oiint_{\partial V} \boldsymbol{n} \cdot \boldsymbol{\sigma} \cdot \boldsymbol{\upsilon} \mathrm{d}A$, if we only observe Eq. $(4.5.14)_2$. Thus the two terms involving the stress tensor cancel each other. The general formula for the kinetic energy $\iiint_{V(t)} \frac{\rho}{2}\upsilon^2 \mathrm{d}V$ can also easily be integrated:

$$\begin{aligned} E_{\mathrm{kin}}(t) &= \iiint_{V(t)} \frac{\rho}{2}\upsilon^2 \mathrm{d}V = \frac{3}{10} m(R) R^2(t) H^2(t), \\ m(R) &= \frac{4\pi}{3}\rho(t)R^3(t) \equiv \frac{4\pi}{3}\rho(0)R^3(0). \end{aligned} \tag{10.4.28}$$

Moreover, as we shall prove later, the result for the gravitational field shown in Eq. (10.4.12) can be used to prove the following formulae:

$$\iiint\limits_{V(t)} \boldsymbol{f} \cdot \boldsymbol{\upsilon}\, \mathrm{d}V = \frac{\mathrm{d}}{\mathrm{d}t}\left[\frac{3}{5}G\frac{m^2(R)}{R(t)}\right], \quad E_{\text{pot}}(t) = -\frac{3}{5}G\frac{m^2(R)}{R(t)}. \tag{10.4.29}$$

We may interpret the potential energy as the total gravitational energy stored in a sphere of gas of perfect radial symmetry. Moreover, we conclude that the balance of kinetic energy can be rewritten in the following form:

$$\frac{\mathrm{d}}{\mathrm{d}t}\left[E_{\text{pot}}(t) + E_{\text{kin}}(t)\right] = 0. \tag{10.4.30}$$

In other words, the mechanical energy is surprisingly conserved under the present assumptions even though the gas was treated as a viscous NAVIER-STOKES fluid:

$$E_{\text{pot}}(t) + E_{\text{kin}}(t) = E_{\text{pot}}(0) + E_{\text{kin}}(0). \tag{10.4.31}$$

The definitions for kinetic and potential energy can now be used to interpret the factor $2\gamma$ originally defined in Eq. $(10.4.25)_3$. A simple calculation shows that it is nothing else but the ratio between the total gravitational and the kinetic energy initially assigned to the sphere of gas:

$$\frac{\left|E_{\text{pot}}(0)\right|}{E_{\text{kin}}(0)} = 2\frac{Gm(0)}{R^3(0)H^2(0)} \equiv 2\gamma. \tag{10.4.32}$$

If the initial amount of kinetic energy is very large, i.e., for small values of $\gamma$, gravity is not strong enough and the sphere will continue to expand, and vice versa. It is exactly that kind of behavior we see in Figs. 10.8 and 10.9. A threshold value is reached when both energies are equally strong, i.e., $\gamma = \frac{1}{2}$. It is also quite instructive to ask under which circumstances the expanding star will come to a standstill. In that case the kinetic energy must vanish and Eq. (10.4.31) yields:

$$E_{\text{pot}}(0) + E_{\text{kin}}(0) = E_{\text{pot}}(t) \quad \Rightarrow \quad R(t) = R(0)\frac{2\gamma}{2\gamma - 1}. \tag{10.4.33}$$

Since the radius must always be positive (*plus* infinity is also acceptable) we conclude that in order to reach a state of rest it is required that $\gamma \geq \frac{1}{2}$. Recall that in the original definition $\gamma$ was only required to be positive. Therefore choices $0 \leq \gamma < \frac{1}{2}$ will lead to continuously expanding stars or, in other words, a never-ending blast.

**Exercise 10.4.2:** ***Energy competition during the explosion of a sphere of gas***

Prove the previous statements and in particular Eqs. (10.4.27–10.4.33).

We finish our discussions with a proof of Eq. (10.4.29). The starting points of our line of arguments form Eqs. $(10.4.12)_2$ and (10.4.14), which we apply to the case of a gravitating expanding star of current radius $R(t)$ and total mass $m(t) =$

$\frac{4\pi}{3}\rho(t)R^3(t)$, thus acknowledging Eq. $(10.4.19)_1$ for a time-dependent density and a time-dependent external radius. Only the fields within the star, i.e., $0 \leq r \leq R(t)$ are relevant for the proof and, consequently, we write:

$$f_{\langle r\rangle} = -\frac{4\pi}{3} G\rho(t) r, \quad \varphi = -\frac{2\pi}{3} G\rho(t)\left[3R^2(t) - r^2\right]. \tag{10.4.34}$$

Recall that due to the perfect spherical symmetry this is the only non-vanishing component of the body force. Eq. $(10.4.13)_1$ allows us to write:

$$\boldsymbol{f} = -\nabla\varphi \quad \Leftrightarrow \quad f_{\langle r\rangle}(r,t) = -\frac{\partial}{\partial r}\varphi(r,t). \tag{10.4.35}$$

These results will now be used to obtain the *total* potential energy of the gravitating mass. To begin with, the total potential energy is given by:

$$E_{\text{pot}}(t) = \frac{1}{2}\iiint\limits_{V(t)} \rho\varphi \,\mathrm{d}V. \tag{10.4.36}$$

The surprising factor $\frac{1}{2}$ is required because the temporal change of potential energy must equal the power related to gravity:

$$\iiint\limits_{V(t)} \boldsymbol{f}\cdot\boldsymbol{\upsilon}\,\mathrm{d}V = \frac{\mathrm{d}}{\mathrm{d}t}E_{\text{pot}}(t). \tag{10.4.37}$$

Indeed, the following sequence of transformations proves the correctness of Eq. (10.4.36):

$$\begin{aligned}\frac{\mathrm{d}}{\mathrm{d}t}\iiint\limits_{V(t)} \rho\varphi\,\mathrm{d}V &= \iiint\limits_{V(t)} \rho\frac{\mathrm{d}\varphi}{\mathrm{d}t}\mathrm{d}V = \iiint\limits_{V(t)} \rho\left(\frac{\partial\varphi}{\partial t} + \boldsymbol{\upsilon}\cdot\nabla\varphi\right)\mathrm{d}V \\ &= 2\iiint\limits_{V(t)} \rho\boldsymbol{\upsilon}\cdot\nabla\varphi\,\mathrm{d}V = 2\iiint\limits_{V(t)} \rho\boldsymbol{\upsilon}\cdot\boldsymbol{f}\mathrm{d}V,\end{aligned} \tag{10.4.38}$$

if we make use of the following auxiliary formula:

$$\iiint\limits_{V(t)} \rho\frac{\partial\varphi}{\partial t}\mathrm{d}V = \iiint\limits_{V(t)} \rho\boldsymbol{\upsilon}\cdot\nabla\varphi\,\mathrm{d}V, \tag{10.4.39}$$

which holds only in the resent case of the perfectly symmetric star. The auxiliary formula is easy to proof, if we start from Eq. $(10.4.34)_2$ and differentiate:

$$\begin{aligned}\frac{\partial\varphi}{\partial t} &= -\frac{2\pi}{3}G\dot{\rho}\left(3R^2 - r^2\right) - 4\pi G\rho R\dot{R} = \frac{3}{2}\frac{4\pi}{3}G\rho H\left(R^2 - r^2\right), \\ \frac{\partial\varphi}{\partial r} &= \frac{4\pi}{3}G\rho\, r,\end{aligned} \tag{10.4.40}$$

where Eqs. (10.4.21) and $(10.4.23)_1$ have been used. This can now be integrated over the total volume of the star and we find:

$$\iiint\limits_{V(t)} \rho \frac{\partial \varphi}{\partial t} \mathrm{d}V = 4\pi \int\limits_{r=0}^{R(t)} \rho \frac{\partial \varphi}{\partial t} r^2 \mathrm{d}r = \frac{(4\pi)^2}{15} G \rho^2 H R^5, \tag{10.4.41}$$

$$\iiint\limits_{V(t)} \rho \upsilon \cdot \nabla \varphi \,\mathrm{d}V = 4\pi \int\limits_{r=0}^{R(t)} \rho \dot{r} \frac{\partial \varphi}{\partial r} r^2 \mathrm{d}r = 4\pi \int\limits_{r=0}^{R(t)} \rho H \frac{\partial \varphi}{\partial r} r^3 \mathrm{d}r = \frac{(4\pi)^2}{15} G \rho^2 H R^5,$$

where Eq. $(10.4.19)_3$ has been used. Now we are ready to perform the remaining integration in Eq. (10.4.36):

$$\begin{aligned} E_{\mathrm{pot}}(t) &= \frac{1}{2} \iiint\limits_{V(t)} \rho \varphi \,\mathrm{d}V = -\frac{4\pi^2}{3} G \rho^2 \int\limits_{r=0}^{r=R(t)} \left(3R^2 - r^2\right) r^2 \mathrm{d}r \\ &= -\frac{(4\pi)^2}{3 \cdot 5} G \rho^2 R^2 = -\frac{3}{5} G \frac{m^2(R)}{R}. \end{aligned} \tag{10.4.42}$$

This concludes the cumbersome proof of Eq. (10.4.29) without any artificial tricks as they are frequently used, e.g., by calculating the work required for transporting spherical shells of mass successively from infinity to form the total volume of the expanding star.

**Exercise 10.4.3:** ***Potential energy of Newtonian gravity***

Study the alternative proofs of Eq. (10.4.29) shown on the World Wide Web (for example). Also show that the result still holds for radially *and* time dependent mass densities $\rho = \rho(r,t)$ for $0 \leq r \leq R(t)$.

**Exercise 10.4.4:** ***Temperature development during the explosion of a sphere of gas***

From the proof above we have learned that the balance of kinetic energy is unaffected by the dissipative stress terms (cf., Eq. (10.4.27) and the corresponding text). Thus the sum of kinetic and potential energy does not change over time and the corresponding balance decouples completely from the balance of internal energy. However, all of this does not mean that the temperature stays constant, too: The total internal energy of the sphere of gas is conserved (in the case of vanishing bulk viscosity), but the sphere expands, and so the temperature should decrease over time since the internal energy is distributed over a greater volume. We will explore this in more detail: Show the local balance of internal energy, Eq. (3.9.7), results in the following ODE for the dimensionless temperature, $\bar{T} = T(t)/T(0)$:

$$\bar{\rho}\frac{\mathrm{d}\bar{T}}{\mathrm{d}\bar{t}} = \frac{3}{\zeta}\left(-\bar{\rho}\,\bar{T} + \frac{(3\lambda + 2\mu)H(0)}{\rho(0)R/MT(0)}\bar{H}\right)\bar{H}. \tag{10.4.43}$$

Assume that the star can be treated as an ideal gas, and use Eq. (10.4.3) to evaluate the stresses. Note the different signs of the two terms in the parentheses. What is their effect regarding the change in temperature? Show that the dissipative term in Eq. (10.4.43) vanishes for the case of vanishing bulk viscosity, $\mu' = 0$. Interpret the *ansatz* (10.4.1) as pure volumetric deformation and relate it to the "compressibility factor" $3\lambda + 2\mu$ in Eq. (10.4.43) (in the same context also discuss the analogue for the Hookean solid shown in Eq. $(6.2.6)_2$).

**Exercise 10.4.5:** ***The isothermal barometric equation revisited***

Neglect the mass of the atmosphere and show that we may write for the gravitational field of the Earth:

$$f_{\langle r\rangle} = -\frac{g}{(1+\frac{\Delta r}{r_\mathrm{E}})^2} = -g\left(1 - 2\frac{\Delta r}{r_\mathrm{E}} + 3\frac{\Delta r^2}{r_\mathrm{E}^2} \mp \cdots\right), \quad \Delta r = r - r_\mathrm{E}. \tag{10.4.44}$$

$g = G\frac{m_\mathrm{E}}{r_\mathrm{E}^2} = 9.81\,\frac{\mathrm{m}}{\mathrm{s}^2}$ stands for the gravitational acceleration at ground level, $m_\mathrm{E}$ and $r_\mathrm{E}$ denote the mass and the radius of the Earth, respectively. Use the stationary momentum balance and the ideal gas law to show that the mass density of an *isothermal* atmosphere decreases exponentially:

$$\rho(r) = \rho(r_\mathrm{E})\exp\left[-C_\mathrm{E}\left(1 - \frac{1}{1+\frac{\Delta r}{r_\mathrm{E}}}\right)\right], \quad C_E = \frac{Gm_\mathrm{E}}{\frac{R}{M}Tr_\mathrm{E}}, \quad \Delta r = r - r_\mathrm{E}. \tag{10.4.45}$$

Expand into a TAYLOR series and rediscover the classical expression for the *isothermal barometric equation*:

$$\rho(r) \approx \rho(r_\mathrm{E})\exp\left[-\frac{g}{\frac{R}{M}T}\Delta r\right] \quad \Leftrightarrow \quad p(r) \approx p(r_\mathrm{E})\exp\left[-\frac{g}{\frac{R}{M}T}\Delta r\right]. \tag{10.4.46}$$

Explain why the pressure at ground level must equal the weight of the atmosphere per unit surface, which is given by the following integral:

$$p(r_{\mathrm{E}}) = \int\limits_{r=r_{\mathrm{E}}}^{\infty} \rho(r_{\mathrm{E}}) \left| f_{\langle r \rangle} \right| \mathrm{d}r. \tag{10.4.47}$$

Now replace the integrand by the stationary balance of momentum, perform the integration and show that an identity results if we choose:

$$p(\infty) = 0. \tag{10.4.48}$$

Note that this requirement is consistent with the boundary condition at infinity: We expect no gas at infinity, i.e., vacuum conditions. No tractions are applied at infinity and thus Eq. (10.4.48) must hold. Recall the ideal gas equation and conclude that this requires $\rho(\infty) = 0$. Show that this disagrees with the exact solution for the mass density from Eq. (10.4.45):

$$\bar{\rho}(\bar{r}) = \frac{\rho(r)}{\rho(r_{\mathrm{E}})} = A \exp\left[\frac{C_{\mathrm{E}}}{\bar{r}}\right] \quad \Rightarrow \quad \bar{\rho}(\infty) = A, \quad A = \exp(-C_{\mathrm{E}}). \tag{10.4.49}$$

but *not* with the approximation of constant gravity $g$ shown in Eq. (10.4.46). Calculate the value for $C_{\mathrm{E}}$ and $A$ using the data of Exercise 10.4.1 at 0 °C and show that $C_{\mathrm{E}} \approx 797$, $A \approx 7.1 \cdot 10^{-347}$ for $M = 29$. Conclude that in this model of decreasing ($r^{-2}$) gravitational acceleration the density of the atmosphere decreases. However, note that at infinity it does not become zero. It only approaches an extremely small, but finite value.

Now calculate the total mass of the atmosphere by integrating the mass density between $r_{\mathrm{E}} \leq r < \infty$. Show that in the case of constant gravity $g$ the total gas mass is finite and can be expressed in closed-form:

$$m_{\mathrm{g}} = \frac{4\pi r_{\mathrm{E}}^4 p(r_{\mathrm{E}})}{G m_{\mathrm{E}}} \left(1 + 2\frac{1 + C_{\mathrm{E}}}{C_{\mathrm{E}}^2}\right). \tag{10.4.50}$$

Use the data from Exercise 10.4.1 ($p(r_{\mathrm{E}}) \approx 1$ bar) and show that the mass of Earth's atmosphere is roughly $m_{\mathrm{g}} = 5.4 \cdot 10^{18}$ kg. Note that the mass of the Earth is a million times greater. This means that it is justified to neglect the gravitational effect of the gas, at least in the case of planet Earth. Next use the exact formula for the density shown in Eq. (10.4.49) and prove that:

$$\bar{m}_{\mathrm{g}}(\bar{r}) = \frac{m_{\mathrm{g}}(r)}{B m_{\mathrm{E}}} = A \left[\exp\left(\frac{C_{\mathrm{E}}}{\bar{s}}\right) \bar{s}\left(C_{\mathrm{E}}^2 + C_{\mathrm{E}}\bar{s} + 2\bar{s}^2\right) - C_{\mathrm{E}}^3 \mathrm{Ei}\left(\frac{C_{\mathrm{E}}}{\bar{s}}\right)\right]_{\bar{s}=1}^{\bar{s}=\bar{r}} \tag{10.4.51}$$

with a normalized radius $\bar{r} = r/r_{\mathrm{E}}$, the exponential integral function known as "Ei," and $B = \frac{1}{2}\rho(r_{\mathrm{E}})/\rho_{\mathrm{E}}$. Show by expansion of Eq. (10.4.51) and by means of a simple argument based on the fact of a constant mass density at

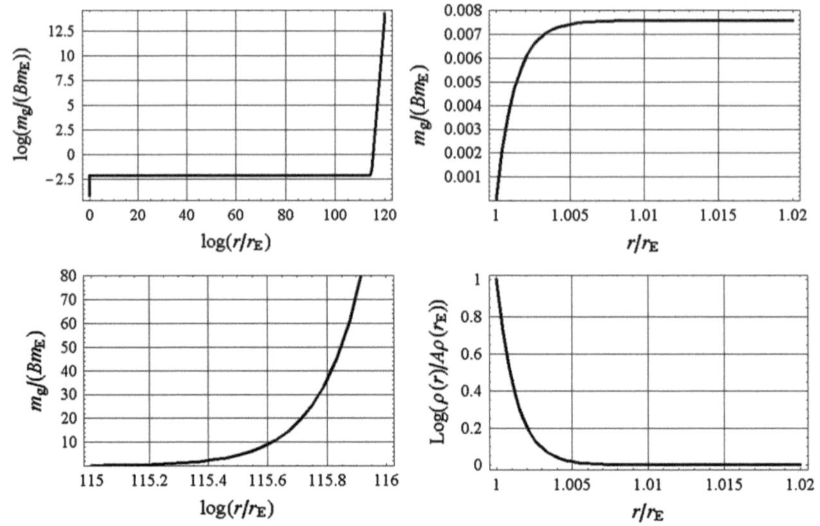


**Fig. 10.10** Gas mass and gas mass density for $r^{-2}$ varying gravity

infinity that $\bar{m}_g(\bar{r} \to \infty) \sim \bar{r}^3$, i.e., the mass of the gas diverges for $r \to \infty$. Conclude that the assumption of a negligible gas mass, which formed the basis of the exact form of the gravitational acceleration shown in Eq. (10.4.44), does not hold. Discuss the behavior of mass and mass density shown in the sequence of plots of Fig. 10.10.

So far we have developed a very simple model of an exploding star, i.e., we have studied the dynamics of a spherical Navier-Stokes gas cloud. We will now turn to stationary states of gas clouds and investigate the balance between internal pressure and gravity. This may serve as a crude model of the current state of huge gas planets, such as Jupiter or Saturn. To this end we require $\boldsymbol{\upsilon} = \mathbf{0}$ in the momentum balance (10.4.18):

$$\frac{R}{M}\left(T\frac{\partial \rho}{\partial r} + \rho\frac{\partial T}{\partial r}\right) = -G\rho\frac{m(r)}{r^2}. \tag{10.4.52}$$

For the mass relevant for the gravitational pull at the position $r$ we may write:

$$m(r) = 4\pi \int_{r'=0}^{r'=r} \rho(r')\, r'^2 \mathrm{d}r' \quad \Rightarrow \quad \frac{\mathrm{d}m}{\mathrm{d}r} = 4\pi\rho(r)\, r^2. \tag{10.4.53}$$

We also assume that the planet is isothermal and obtain the following system of coupled ODEs:

$$\frac{\mathrm{d}m}{\mathrm{d}r} = 4\pi\rho(r)\,r^2, \quad \frac{\mathrm{d}\rho}{\mathrm{d}r} + \frac{G}{\frac{R}{M}T}\rho\frac{m(r)}{r^2} = 0. \tag{10.4.54}$$

This system can be uncoupled by eliminating the mass. We obtain the following non-linear second order ODE for the mass density:

$$r\rho\frac{\mathrm{d}^2\rho}{\mathrm{d}r^2} - r\left(\frac{\mathrm{d}\rho}{\mathrm{d}r}\right)^2 + 2\rho\frac{\mathrm{d}\rho}{\mathrm{d}r} + \frac{4\pi G}{\frac{R}{M}T}r\rho^3 = 0. \tag{10.4.55}$$

Note that the differential equations hold for all material points of the stationary gas cloud. Due to the assumed perfect spherical symmetry this means that they hold within $0 \le r \le R$ where $R$ denotes the outer radius of the cloud. However, the point $r = 0$ will cause trouble since the ODEs degenerate at that point. Moreover, Exercise 10.4.5 taught us that we should be prepared for a semi-infinite domain, i.e., the case $R \to \infty$. Consequently, a "natural" length scale parameter for normalization is not available. Therefore we choose an arbitrary radius $r_0 \in (0, \infty)$ and use it to define a dimensionless radius $\bar{r} = r/r_0$. Moreover, we expect a finite value $\rho(0)$ for the mass density at $r = 0$ and define a dimensionless density by $\bar{\rho}(\bar{r}) = \rho/\rho(0)$. Thus Eq. (10.4.55) reads:

$$\bar{r}\bar{\rho}\frac{\mathrm{d}^2\bar{\rho}}{\mathrm{d}\bar{r}^2} - \bar{r}\left(\frac{\mathrm{d}\bar{\rho}}{\mathrm{d}\bar{r}}\right)^2 + 2\bar{\rho}\frac{\mathrm{d}\bar{\rho}}{\mathrm{d}\bar{r}} + 3C\bar{r}\bar{\rho}^3 = 0, \quad 0 \le \bar{r} < \infty \tag{10.4.56}$$

where:

$$C = \frac{Gm_0}{\frac{R}{M}Tr_0}, \quad m_0 = \frac{4\pi}{3}\rho(0)r_0^3. \tag{10.4.57}$$

The mass-like quantity $m_0$ was introduced only formally and has no immediate physical meaning, unlike its analogue in Eq. $(10.4.45)_2$. The boundary conditions to be observed in context with Eq. (10.4.56) read:

$$\bar{\rho}(0) = 1, \quad \left.\frac{\mathrm{d}\bar{\rho}}{\mathrm{d}\bar{r}}\right|_{\bar{r}=0} = 0. \tag{10.4.58}$$

The first one follows from the definition of the normalized mass density, and the second one from Eq. $(10.4.54)_2$ since there is no mass at $r = 0$. In fact the vicinity of $\bar{r} = 0$ presents a problem during the numerical solution of the non-linear boundary value problem (10.4.56/10.4.58). Therefore we solve it for $\bar{r} \in [\varepsilon, \infty)$, with $\varepsilon = 10^{-4}$ and explore the situation for $\bar{r} \in [0, \varepsilon]$ analytically by using the following quadratic *ansatz*:

$$\bar{\rho}(\bar{r}) = \alpha + \beta\,\bar{r} + \gamma\,\bar{r}^2. \tag{10.4.59}$$

If this is inserted in (10.4.55/57) we find that:

$$\alpha = 1, \beta = 0, \gamma = -\frac{C}{2} \quad \Rightarrow \quad \bar{\rho}(\bar{r}) = 1 - \frac{C}{2}\,\bar{r}^2. \tag{10.4.60}$$

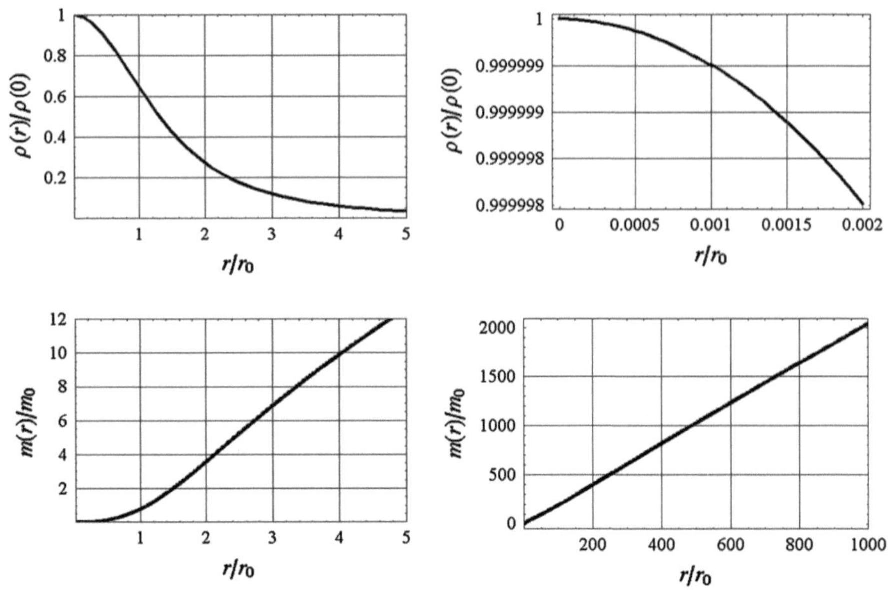


**Fig. 10.11** Mass density and mass distribution in a sphere of gas (see text)

We also rewrite Eq. $(10.4.53)_{1,2}$ in normalized form:

$$\bar{m}(\bar{r}) = 3 \int_{\bar{s}=0}^{\bar{s}=\bar{r}} \bar{\rho}(\bar{s})\, \bar{s}^2 \mathrm{d}\bar{s} \equiv -\frac{1}{C}\frac{\bar{r}^2}{\bar{\rho}}\frac{\mathrm{d}\bar{\rho}}{\mathrm{d}\bar{r}}. \qquad (10.4.61)$$

The radial development of mass density and mass is shown in Fig. 10.11. Obviously mass is accumulated linearly at large distances. In view of the last equation we conclude that the mass density varies like $\bar{r}^{-2}$. If such an *ansatz* is inserted in the ODE (10.4.55) we find that:

$$\bar{\rho}(\bar{r}) = \frac{2C}{3}\frac{1}{\bar{r}^2}, \quad \bar{r} \gg 1. \qquad (10.4.62)$$

This leads us to conclude that stability between internal pressure and gravity is possible for a gas cloud of *infinite* size containing *infinite* mass. To hear the word infinite twice is a bit disturbing and we should try to improve the model, especially since it is hard to believe that the mass of a gas planet like Jupiter is infinite. Maybe a rigid core of sufficiently high mass $m_c = \frac{4\pi}{3}\rho_c r_c^3$ with a (constant) mass density $\rho_c$ and a radius $r_c$ would increase the gravitational pull so much that the mass density would decrease faster and, after integration, lead to a *finite* amount of gas distributed within an infinitely large spherical shell? As a matter of fact we have already studied such situations in Exercise 10.4.5. There we have noticed that a *constant* gravitational field $g$ reaching out to infinity would indeed lead to an atmosphere of finite mass: Eq. (10.4.50). However, the gravitational field of a solid core is *not* constant. Rather it follows NEWTON's law of gravity and decreases like $\bar{r}^{-2}$. This results in an atmosphere of infinite mass: Eq. (10.4.51). However, in view of Figs. $10.10_{1,2}$ we may argue that the accumulated mass seems to saturate initially. The density and the pressure at the onset of saturation are governed by the factor $A = \exp(-C_{\mathrm{E}})$ and thus extremely small. Consequently we may consider the corresponding position as the "extension" of the planet incl. its "atmosphere" and disregard the later increase of mass as an artifact of the model. In fact, we neglected the contribution of the atmosphere's mass to gravity in order to obtain the closed-form result shown in Eq. (10.4.51). If we now *include* the mass of the atmosphere the stabilizing effect should even be more pronounced. However, this case can only be studied numerically. We proceed to discuss the details.

We define a dimensional radius in a "natural" manner by $\bar{r} = r/r_{\mathrm{c}}$. However, in the case of mass density we have a choice, $\bar{\rho}(\bar{r}) = \rho/\rho_{\mathrm{c}}$ or $\bar{\rho}(\bar{r}) = \rho/\rho(r_{\mathrm{c}})$. We will choose the latter way of normalization. However, then the differential equation for the normalized density differs slightly from Eq. (10.4.56):

$$\bar{r}\bar{\rho}\frac{\mathrm{d}^2\bar{\rho}}{\mathrm{d}\bar{r}^2} - \bar{r}\left(\frac{\mathrm{d}\bar{\rho}}{\mathrm{d}\bar{r}}\right)^2 + 2\bar{\rho}\frac{\mathrm{d}\bar{\rho}}{\mathrm{d}\bar{r}} + 3\alpha C_{\mathrm{c}}\bar{r}\bar{\rho}^3 = 0, \quad 1 \le \bar{r} < \infty \tag{10.4.63}$$

where:

$$C_{\mathrm{c}} = \frac{G m_{\mathrm{c}}}{\frac{R}{M} T r_{\mathrm{c}}}, \quad m_{\mathrm{c}} = \frac{4\pi}{3}\rho_{\mathrm{c}}\, r_{\mathrm{c}}^3, \quad \alpha = \frac{\rho(r_{\mathrm{c}})}{\rho_{\mathrm{c}}}, \tag{10.4.64}$$

and the boundary conditions read:

$$\bar{\rho}(1) = 1, \quad \left.\frac{\mathrm{d}\bar{\rho}}{\mathrm{d}\bar{r}}\right|_{\bar{r}=1} = -C_{\mathrm{c}}. \tag{10.4.65}$$

The latter one follows from the relation for the (normalized) mass of the core-gas system, which can be obtained by integration or by differentiation:

$$\begin{aligned}\bar{m}(\bar{r}) &= 1 + 3\alpha \int\limits_{\bar{s}=1}^{\bar{s}=\bar{r}} \bar{\rho}(\bar{s})\, \bar{s}^2 \mathrm{d}\bar{s} \equiv -\frac{1}{C_{\mathrm{c}}}\frac{\bar{r}^2}{\bar{\rho}}\frac{\mathrm{d}\bar{\rho}}{\mathrm{d}\bar{r}}, \quad 1 \le \bar{r} < \infty, \\ \bar{m}(\bar{r}) &= \bar{r}^3, \quad 0 \le \bar{r} < 1.\end{aligned} \tag{10.4.66}$$

Solving the ODE (10.4.63/10.4.65) for the density and then performing the integration for the mass shown in (10.4.66), both numerically, is, in principle, a feasible way, at least for small normalized radii $\bar{r}$. However, we would like to study the behavior at very large values of $\bar{r}$ and find out as to whether the mass saturates. Thus it is much more advisable to eliminate the mass density from Eqs. (10.4.54) and to derive the following ODE for the mass instead:

$$\frac{\mathrm{d}^2 m}{\mathrm{d}r^2} - \left(\frac{2}{r} - \frac{G}{\frac{R}{M}T}\frac{m}{r^2}\right)\frac{\mathrm{d}m}{\mathrm{d}r} = 0, \quad r_{\mathrm{c}} \le r < \infty. \tag{10.4.67}$$

Two boundary conditions are required. The first stems from the fact that at the radius $r_c$ of the core, the mass of the rigid core is present:

$$m(r_c) = \frac{4\pi}{3}\rho_c r_c^3 \equiv m_c. \tag{10.4.68}$$

The second boundary condition follows from Eq. $(10.4.54)_1$:

$$\left.\frac{dm}{dr}\right|_{r_c} = \frac{3m_c}{r_c}\frac{\rho(r_c)}{\rho_c}. \tag{10.4.69}$$

If we normalize mass by the mass of the core,

$$\bar{m} = \frac{m}{m_c}, \tag{10.4.70}$$

we may rewrite Eqs. (10.4.67, 10.4.69) in dimensionless form:

$$\begin{gathered}\frac{d^2\bar{m}}{d\bar{r}^2} - \left(\frac{2}{\bar{r}} - C_c\frac{\bar{m}}{\bar{r}^2}\right)\frac{d\bar{m}}{d\bar{r}} = 0, \quad 1 \le \bar{r} < \infty, \\ \bar{m}(1) = 1, \quad \left.\frac{d\bar{m}}{d\bar{r}}\right|_{\bar{r}=1} = 3\frac{\rho(r_c)}{\rho_c}.\end{gathered} \tag{10.4.71}$$

The plot on the top left in Fig. 10.12 shows what happens if we increase the value of $C_c$ from 10 to 12 and finally to 15 (for $\rho(r_c)/\rho_c = 0.5$): The curves start to saturate if the mass of the core is increasing. In other words the core has a stabilizing effect. However, does the mass still saturate at extremely high values of $\bar{r}$ or is the situation similar as in Fig. 10.10 of Exercise 10.4.5? This was examined in the plot on the top right for $C_c = 40$. It is fair to say that high values of $C_c$ do stabilize the mass at saturation level even at very high values of $\bar{r}$. However, the numerics starts to get jumpy, and thus it is not quite clear as to whether the mass will remain constant if $\bar{r} \to \infty$. It is also fair to say that for that particular choice of parameters the mass of the atmosphere is just a measly few percent when compared to that of the core. This does not appear to be realistic in view of real gas planets. However, this can be adjusted if we choose $\rho(r_c)/\rho_c = 2$ (say), i.e., make the gas much denser than the matter of the solid core. The plot on the middle left in Fig. 10.12 shows that the atmosphere can then easily be 50 % of the mass of the core and more (curves were plotted for $C_c$ equal to 10, 12, and 15). Increasing the ratio of densities has also a stabilizing effect as shown in the plot on the middle right of Fig. 10.12 ($C_c = 40$).

Still we are not sure what happens if $\bar{r} \to \infty$. In this context it is most instructive to investigate the long range behavior of the coupled system of PDEs (10.4.54) in dimensionless form

$$\frac{d\bar{m}}{d\bar{r}} = 3\frac{\rho(r_c)}{\rho_c}\bar{r}^2\bar{\rho}(\bar{r}), \quad \frac{d\bar{\rho}}{d\bar{r}} = -C_c\frac{\bar{m}}{\bar{r}^2}\bar{\rho}(\bar{r}) \tag{10.4.72}$$

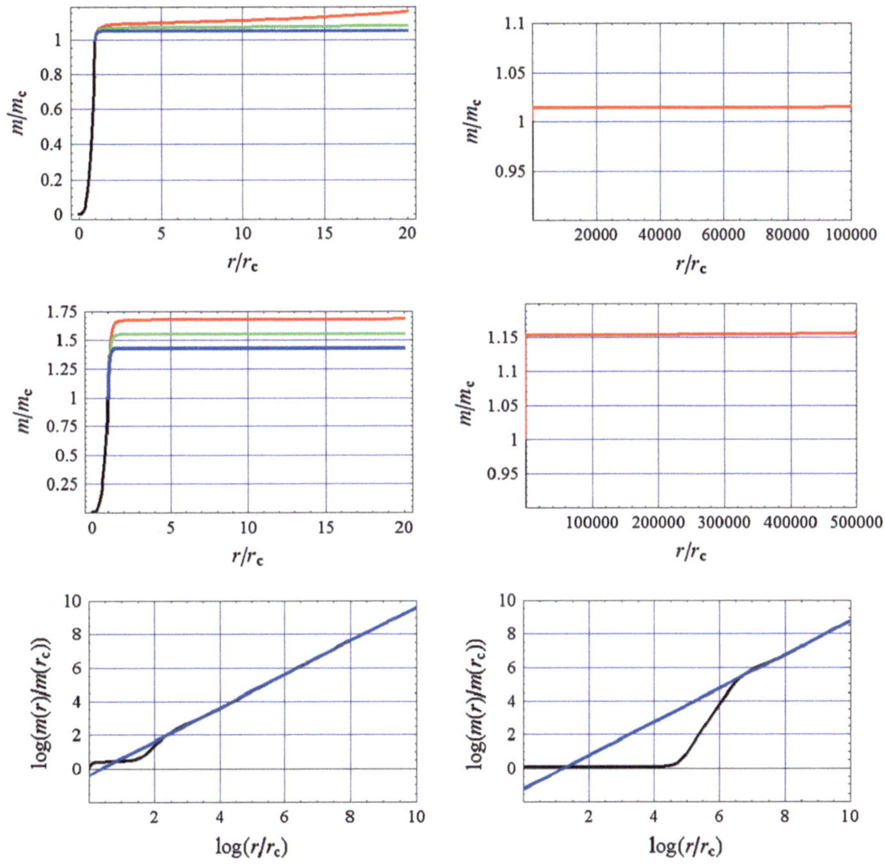

Fig. 10.12 Mass distribution in a gas planet with a solid core (see text)

by means of a power *ansatz*:

$$\bar{m} = A\bar{r}^{\alpha}, \quad \bar{\rho} = B\bar{r}^{\beta}. \tag{10.4.73}$$

After insertion we find the following relations:

$$\frac{\alpha A}{3\frac{\rho(r_c)}{\rho_c}B} = \bar{r}^{\beta-\alpha+3}, \quad -\frac{\beta}{C_c A} = \bar{r}^{\alpha-1}. \tag{10.4.74}$$

The constancy and independence of radius of the left hand sides must still be guaranteed if $\bar{r} \to \infty$. Thus we conclude that:

$$\alpha = 1, \quad \beta = -2. \tag{10.4.75}$$

This means that the mass grows linearly[1] and the density decreases quadratically with increasing distance:

$$\bar{m} = \frac{2}{C_c}\bar{r}, \quad \bar{\rho} = \frac{2}{3C_c \frac{\rho(r_c)}{\rho_c}} \frac{1}{\bar{r}^2}. \tag{10.4.76}$$

Thus the situation is very similar to that discussed in context with Eq. (10.4.51). The last row of double-logarithmic plots in Fig. 10.12 (for $\rho(r_c)/\rho_c = 0.5$) shows the development of mass together with the asymptotic evolution for $C_c = 5$ (left) and $C_c = 35$ (right). The stabilizing influence of and increasing value of $C_c$, i.e., increasing core mass is clearly visible. However, in the end the mass diverges unless $C_c \to \infty$.

**Exercise 10.4.6:** ***Gas planets with rigid cores***

Show that the mass within the rigid core grows like

$$\bar{m}_c(\bar{r}) = \bar{r}^3, \quad 0 \le \bar{r} \le 1. \tag{10.4.77}$$

Also determine the constants $A$ and $B$ in Eq. (10.4.73). Gather information from the World Wide Web and find out more about the core-aggregation hypothesis used to explain the genesis of gas giants like Jupiter or Saturn.

## 10.5 Would You Like to Know More?

Newtonian as well as non-Newtonian fluid mechanics are extensive fields of teaching and research in their own right. In order to get a thorough overview of the subject matter, the textbook by the famous fluid mechanics professor Ludwig PRANDTL in the new edition by Oertel [5] can be recommended. The theoretical foundations for (a non-Newtonian) fluid are also described in great detail in Chap. F of Truesdell and Noll [6] or can be found in the monograph by Coleman et al. [2].

The correlation between the choice of constitutive relations and finite speed of disturbances is most extensively discussed in the book by Müller and Ruggieri [4]. However, this book is not recommended for easy reading. It is reserved for the advanced expert. The book makes use not only of the arguments of continuum theory but also of concepts of kinetic theory.

Methods of solution for PDEs, in particular those by BERNOULLI and D'ALEMBERT, can be found in almost any standard textbook on higher mathematics. However, the book by Butkov [1] illuminates the topic from a more applied point of view.

[1] Note that the third order increase of mass (*cf.*, Exercise 10.4.3) was reduced drastically due to the gravitational action of the atmosphere.

Finally, those who wish to learn more about the stationary or dynamic behavior of stars and gas planets are recommended to consult the monograph by Kippenhahn et al. [3] and the literature cited therein.

## References

1. Butkov E (1968) Mathematical physics. Addison-Wesley Publishing Company, Reading, MA
2. Coleman BD, Markovitz H, Noll W (1966) Viscometric flows of non-newtonian fluids—theory and experiment. Springer, Berlin, Heidelberg (Springer Tracts in Natural Philosophy)
3. Kippenhahn R, Weigert A, Weiss A (2013) Stellar structure and evolution, 2nd edn. Springer, Berlin, Heidelberg
4. Müller I, Ruggieri T (1998) Rational extended thermodynamics. Springer, New York (Springer Tracts in Natural Philosophy)
5. Oertel H (2002) Prandtl—Führer durch die Strömungslehre. 11. Auflage. Vieweg, Braunschweig
6. Truesdell C, Noll W (1965) The non-linear theories of mechanics. In: Flügge S (ed) Encyclopedia of physics, vol III/3. Springer, Berlin, Göttingen

# Chapter 11
# Introduction to Time-Independent Plasticity Theory

**Abstract** In this chapter we study irreversible deformation in solids by means of time-independent plasticity theory. In order to establish fundamental concepts of plasticity we examine the autofrettage process of a hollow metallic sphere, the pressure vessel, which is subjected to an enormous internal pressure. For computing the corresponding plastic strain and (eigen-) stresses we use the yield criterion of VON MISES in combination with the stress-strain relations for perfect radial symmetry, which have been established in the previous chapters on spherical coordinates. The second part of this section is dedicated to a more formal treatment of time-independent plasticity where we derive and use the PRANDTL-REUSS equations.

*I remember back when I was a kid there was a comic strip called Plastic Man. His body was elastic and he could make his extremities as long as he wanted. As a youngster I didn't fully appreciate. But I'm now thinking Plastic Man was probably pretty popular with the ladies.*

BEN AFFLECK, American actor

## 11.1 An Important Problem in Plasticity: Autofrettage of a Hollow Spherical Vessel

In what follows we consider a hollow sphere made of metal (Fig. 11.1), the pressure vessel, which is subjected to a very high internal pressure. In fact, the pressure is supposed to be high enough to induce plastic deformation in the metal. Recall Exercise 2.6.5 where it was mentioned that the VON MISES equivalent stress $\sigma_{\text{Mises}}$ must reach the following threshold for this purpose:

W. H. Müller, *An Expedition to Continuum Theory*, Solid Mechanics and Its Applications 210, DOI: 10.1007/978-94-007-7799-6_11,

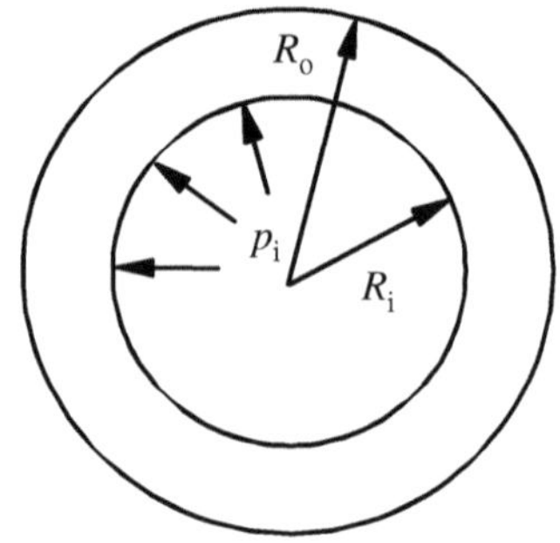


**Fig. 11.1** A hollow sphere subjected to internal pressure

$$\sigma^2_{\text{Mises}} \equiv \frac{3}{2} \underset{(x)}{\Sigma}_{ij} \underset{(x)}{\Sigma}_{ij} = \sigma^2_{\text{y}} \tag{11.1.1}$$

with a (material dependent) yield stress $\sigma_{\text{y}}$ and the stress deviator $\underset{(x)}{\Sigma}_{ij}$:

$$\underset{(x)}{\Sigma}_{ij} = \underset{(x)}{\sigma}_{ij} - \frac{1}{3} \underset{(x)}{\sigma}_{kk}\, \delta_{ij} \tag{11.1.2}$$

If the inner pressure $p$ increases monotonously we intuitively expect that from a certain minimum pressure $p_{\min}$ on a plastified spherical zone will form, starting from the inner surface at radius $R_{\text{i}}$. If we increase the pressure further up to the value $p_\rho$, the plastified zone will grow in a radially symmetric manner and, finally, occupy a spherical shell of outer radius $\rho$. Our objective consists of finding equations for the pressures $p_{\min}$ and $p_\rho$ and to derive a equation for the growth of $\rho$.

## 11.2 The Radially Symmetric Solution

Recall the results from Sect. 9.5. The linear-elastic solution for the non-vanishing stresses and strains of a completely spherically symmetric problem in physical spherical coordinates reads:

$$\begin{aligned} \sigma_{\langle rr\rangle} &= 3kA - 4\mu\frac{B}{r^3}, \quad \sigma_{\langle\vartheta\vartheta\rangle} = \sigma_{\langle\varphi\varphi\rangle} = 3kA + 2\mu\frac{B}{r^3}, \\ u_{\langle r\rangle} &= Ar + \frac{B}{r^2}, \quad 3k = 3\lambda + 2\mu. \end{aligned} \tag{11.2.1}$$

We now observe the boundary conditions of the problem, i.e., the radial stress must equal the internal pressure $p$ at radius $R_{\text{i}}$, and the outer pressure vanishes at radius $R_{\text{o}}$, and find:

$$\sigma_{\langle rr\rangle} = -p\frac{\left(\frac{R_\mathrm{o}}{r}\right)^3 - 1}{\left(\frac{R_\mathrm{o}}{R_\mathrm{i}}\right)^3 - 1}, \quad \sigma_{\langle\vartheta\vartheta\rangle} = \sigma_{\langle\varphi\varphi\rangle} = p\frac{\frac{1}{2}\left(\frac{R_\mathrm{o}}{r}\right)^3 + 1}{\left(\frac{R_\mathrm{o}}{R_\mathrm{i}}\right)^3 - 1},$$
$$u_{\langle r\rangle} = \frac{p}{E}\left[(1-2\nu)r + \frac{(1+\nu)R_\mathrm{o}^3}{2r^2}\right]\frac{1}{\left(\frac{R_\mathrm{o}}{R_\mathrm{i}}\right)^3 - 1}, \quad R_\mathrm{i} \le r \le R_\mathrm{o}. \tag{11.2.2}$$

**Exercise 11.2.1:** ***Linear-elastic stress–strains relations for a hollow sphere under pressure***

Verify Eqs. (11.2.1/11.2.2) and state all assumptions required for their derivation. Use results of previous exercises for that purpose. In particular show that:

$$3kA = p\frac{1}{\left(\frac{R_\mathrm{o}}{R_\mathrm{i}}\right)^3 - 1}, \quad 3kA = 4\mu\frac{B}{R_\mathrm{o}^3}. \tag{11.2.3}$$

We will now evaluate the VON MISES flow condition with these results. We start by calculating the stress deviator. Equation 2.6.11 allows us to write in spherical coordinates:

$$\Sigma_{\langle ij\rangle} = \frac{p}{2}\frac{\left(\frac{R_\mathrm{o}}{r}\right)^3}{\left(\frac{R_\mathrm{o}}{R_\mathrm{i}}\right)^3 - 1}\begin{bmatrix} -2 & 0 & 0 \\ 0 & 1 & 0 \\ 0 & 0 & 1 \end{bmatrix}, \tag{11.2.4}$$

since:

$$\frac{1}{3}\left(\sigma_{\langle rr\rangle} + \sigma_{\langle\vartheta\vartheta\rangle} + \sigma_{\langle\varphi\varphi\rangle}\right) = 3kA = p\frac{1}{\left(\frac{R_\mathrm{o}}{R_\mathrm{i}}\right)^3 - 1}. \tag{11.2.5}$$

This result will now be inserted into the flow condition from Eq. (2.6.20). We obtain:

$$\sigma_\mathrm{Mises}^2 = \frac{3}{2}\Sigma_{\langle ij\rangle}\Sigma_{\langle ij\rangle} = \frac{3}{2}p^2\left[\frac{\left(\frac{R_\mathrm{o}}{r}\right)^3}{\left(\frac{R_\mathrm{o}}{R_\mathrm{i}}\right)^3 - 1}\right]^2\left(1 + \frac{1}{2}\right)$$
$$\Rightarrow \sigma_\mathrm{Mises} = \frac{3}{2}p\frac{\left(\frac{R_\mathrm{o}}{r}\right)^3}{\left(\frac{R_\mathrm{o}}{R_\mathrm{i}}\right)^3 - 1} = \sigma_\mathrm{Mises}(p, r).$$

We first answer the question at which radial position $r$ the VON MISES equivalent stress will assume a maximum for a given pressure $p$. This happens at $r = R_\mathrm{i}$ where the nominator in Eq. (11.2.6) assumes its maximum. This is where the critical value for the material-dependent flow stress $\sigma_\mathrm{y}$ will be reached first. Thus from Eq. (11.2.6) it follows for the smallest pressure $p_\mathrm{min}$ required for first plastic flow:

$$
\begin{aligned}
&\sigma_{\mathrm{Mises}}(p = p_{\min}, r = R_{\mathrm{o}}) = \sigma_{\mathrm{y}} \\
&\Rightarrow p_{\min} = \frac{2}{3}\sigma_{\mathrm{y}}\left(\frac{R_{\mathrm{i}}}{R_{\mathrm{o}}}\right)^3\left[\left(\frac{R_{\mathrm{o}}}{R_{\mathrm{i}}}\right)^3 - 1\right] = \frac{2}{3}\sigma_{\mathrm{y}}\left[1 - \left(\frac{R_{\mathrm{i}}}{R_{\mathrm{o}}}\right)^3\right].
\end{aligned}
\tag{11.2.7}
$$

From Eq. $(11.2.2)_3$ the corresponding displacements at the inner and at the outer radius can be determined:

$$
\begin{aligned}
u_{\langle r\rangle}^{\min}\Big|_{R_{\mathrm{i}}} &= u_{\langle r\rangle}(p = p_{\min}, r = R_{\mathrm{i}}) = \frac{p_{\min}}{E}\left[1 - 2\nu + \frac{1+\nu}{2}\left(\frac{R_{\mathrm{o}}}{R_{\mathrm{i}}}\right)^3\right]\frac{R_{\mathrm{i}}}{\left(\frac{R_{\mathrm{o}}}{R_{\mathrm{i}}}\right)^3 - 1} \\
&= \frac{2\sigma_{\mathrm{y}}}{3E}\left[(1 - 2\nu)\left(\frac{R_{\mathrm{o}}}{R_{\mathrm{i}}}\right)^3 + \frac{1+\nu}{2}\right]R_{\mathrm{i}}, \\
u_{\langle r\rangle}^{\min}\Big|_{R_{\mathrm{o}}} &= u_{\langle r\rangle}(p = p_{\min}, r = R_{\mathrm{o}}) = \frac{p_{\min}}{E}\left[1 - 2\nu + \frac{1+\nu}{2}\right]\frac{R_{\mathrm{o}}}{\left(\frac{R_{\mathrm{o}}}{R_{\mathrm{i}}}\right)^3 - 1} \\
&= \frac{\sigma_{\mathrm{y}}}{E}(1 - \nu)\left(\frac{R_{\mathrm{i}}}{R_{\mathrm{o}}}\right)^3 R_{\mathrm{o}}.
\end{aligned}
\tag{11.2.8}
$$

Note that for the case of total spherical symmetry the flow condition can alternatively be written as follows:

$$
\sigma_{\mathrm{y}} = \sigma_{\langle\vartheta\vartheta\rangle} - \sigma_{\langle rr\rangle}. \tag{11.2.9}
$$

For a proof we start from Eq. (11.2.1) and write:

$$
\frac{1}{3}\sigma_{\langle kk\rangle} = \frac{1}{3}\left(\sigma_{\langle rr\rangle} + \sigma_{\langle\vartheta\vartheta\rangle} + \sigma_{\langle\varphi\varphi\rangle}\right) = \frac{1}{3}\left(\sigma_{\langle rr\rangle} + 2\sigma_{\langle\vartheta\vartheta\rangle}\right) = 3kA. \tag{11.2.10}
$$

Thus the stress deviator in spherical coordinates reads according to Eq. (11.1.2):

$$
\Sigma_{\langle ij\rangle} = \frac{1}{3}\left(\sigma_{\langle\vartheta\vartheta\rangle} - \sigma_{\langle rr\rangle}\right)\begin{bmatrix} 2 & 0 & 0 \\ 0 & -1 & 0 \\ 0 & 0 & -1 \end{bmatrix}, \tag{11.2.11}
$$

and because of Eq. (11.1.1) we finally find that:

$$
\begin{aligned}
\sigma_{\mathrm{Mises}}^2 &= \frac{3}{2}\Sigma_{\langle ij\rangle}\Sigma_{\langle ij\rangle} = \frac{3}{2}\left(\sigma_{\langle\vartheta\vartheta\rangle} - \sigma_{\langle rr\rangle}\right)^2\left(\frac{4}{9} + \frac{1}{9} + \frac{1}{9}\right) \\
&= \left(\sigma_{\langle\vartheta\vartheta\rangle} - \sigma_{\langle rr\rangle}\right)^2 = \sigma_{\mathrm{y}}^2.
\end{aligned}
\tag{11.2.12}
$$

With Eq. (11.2.1) it follows that:

$$
\sigma_{\mathrm{y}} = 6\mu\frac{B}{r^3} = 4\mu\frac{B}{R_{\mathrm{i}}^3}\frac{3}{2}\left(\frac{R_{\mathrm{i}}}{r}\right)^3. \tag{11.2.13}
$$

Thus because of Eq. (11.2.3) we may write:

$$\sigma_{\mathrm{y}} = \frac{3}{2}\, p \frac{\left(\frac{R_{\mathrm{i}}}{r}\right)^3}{1 - \left(\frac{R_{\mathrm{i}}}{R_{\mathrm{o}}}\right)^3}, \tag{11.2.14}$$

which is identical to Eq. (11.2.6). We now turn to the next problem, namely the mathematical description of the development and growth of the plastified zone due to the steady increase of the pressure beyond the critical initial value $p_{\mathrm{min}}$. For *reasons of symmetry* we expect that a plastified spherical shell grows radially starting from the inside at $r = R_{\mathrm{i}}$ to the position $r = \rho$. We would like to calculate the required pressure $p_\rho$ and the stress distribution within the hollow sphere. To this end we first note that the sphere remains in a linear elastic state in the region beyond $r = \rho$. The solution shown in Eq. (11.2.1) can be used to compute the corresponding stresses:

$$\begin{aligned} \sigma_{\langle rr\rangle} &= 3k\tilde{A}^* - 4\mu\frac{B^*}{r^3}, \quad \sigma_{\langle\vartheta\vartheta\rangle} = \sigma_{\langle\varphi\varphi\rangle} = 3k\tilde{A}^* + 2\mu\frac{B^*}{r^3}, \\ u_{\langle r\rangle} &= \tilde{A}^* r + \frac{B^*}{r^2}, \quad 3k = 3\lambda + 2\mu, \quad \rho \le r \le R_{\mathrm{o}}. \end{aligned} \tag{11.2.15}$$

The pressure along the outer radius is still zero, thus:

$$4\mu\frac{B^*}{R_{\mathrm{o}}^3} = 3k\tilde{A}^* = A^*. \tag{11.2.16}$$

By inserting this result into Eq. (11.2.15) for the stresses we obtain:

$$\begin{aligned} \sigma_{\langle rr\rangle} &= A^*\left[1 - \left(\frac{R_{\mathrm{o}}}{r}\right)^3\right] = -A^*\left[\left(\frac{R_{\mathrm{o}}}{r}\right)^3 - 1\right], \\ \sigma_{\langle\vartheta\vartheta\rangle} &= \sigma_{\langle\varphi\varphi\rangle} = A^*\left[\frac{1}{2}\left(\frac{R_{\mathrm{o}}}{r}\right)^3 + 1\right]. \end{aligned} \tag{11.2.17}$$

The unknown constant $A^*$ can be determined from the requirement that at the position $r = \rho$ plastic flow conditions have just been reached. The situation shows still perfect spherical symmetry so that the von Mises flow criterion in the form shown in Eq. (11.2.9) can be used:

$$\sigma_{\mathrm{y}} = \sigma_{\langle\vartheta\vartheta\rangle} - \sigma_{\langle rr\rangle} = A^*\frac{3}{2}\left(\frac{R_{\mathrm{o}}}{\rho}\right)^3 \Rightarrow A^* = \frac{2}{3}\sigma_{\mathrm{y}}\left(\frac{\rho}{R_{\mathrm{o}}}\right)^3. \tag{11.2.18}$$

In combination with Eq. (11.2.17) we obtain:

$$\begin{aligned} \sigma_{\langle rr\rangle} &= -\frac{2}{3}\sigma_{\mathrm{y}}\left(\frac{\rho}{R_{\mathrm{o}}}\right)^3\left[\left(\frac{R_{\mathrm{o}}}{r}\right)^3 - 1\right], \\ \sigma_{\langle\vartheta\vartheta\rangle} &= \sigma_{\langle\varphi\varphi\rangle} = \frac{2}{3}\sigma_{\mathrm{y}}\left(\frac{\rho}{R_{\mathrm{o}}}\right)^3\left[\frac{1}{2}\left(\frac{R_{\mathrm{o}}}{r}\right)^3 + 1\right], \quad \rho \le r \le R_{\mathrm{o}}. \end{aligned} \tag{11.2.19}$$

Note that $\sigma_{\langle rr\rangle}$ is a negative, i.e., compressive stress, whereas $\sigma_{\langle\vartheta\vartheta\rangle}$ and $\sigma_{\langle\varphi\varphi\rangle}$ are of tensile nature. The displacements within the elastic region can now also be obtained from Eq. (11.2.15)$_3$:

$$\begin{aligned} u_{\langle r\rangle} &= \left(\tilde{A}^* + \frac{B^*}{r^3}\right) r = \left(\frac{A^*}{3k} + 4\mu \frac{B^*}{R_o^3}\frac{1}{4\mu}\left(\frac{R_o}{r}\right)^3\right) r \\ &= A^*\left(\frac{1}{3k} + \frac{1}{4\mu}\left(\frac{R_o}{r}\right)^3\right) r, \end{aligned} \tag{11.2.20}$$

and with Eq. (11.2.18):

$$u_{\langle r\rangle} = \frac{2\sigma_y}{3E}\left(\frac{\rho}{R_o}\right)^3\left(1 - 2\nu + \frac{1+\nu}{2}\left(\frac{R_o}{r}\right)^3\right) r, \quad \rho \le r \le R_o, \tag{11.2.21}$$

since:

$$3k = 3\lambda + 2\mu = \frac{E}{1 - 2\nu}, \quad \mu = \frac{E}{2(1+\nu)}. \tag{11.2.22}$$

However, that is not all there is. In order to determine the stresses in the plastified region $R_i \le r \le \rho$ we argue as follows. In this region we will also have *perfect spherical symmetry*. Thus:

$$\sigma_{\langle rr\rangle} \neq 0, \quad \sigma_{\langle\vartheta\vartheta\rangle} = \sigma_{\langle\varphi\varphi\rangle} \neq 0, \quad \sigma_{\langle r\vartheta\rangle} = \sigma_{\langle r\varphi\rangle} = \sigma_{\langle\vartheta\varphi\rangle} = 0 \tag{11.2.23}$$

and:

$$\sigma_{\langle rr\rangle} = f(r), \quad \sigma_{\langle\vartheta\vartheta\rangle} = \sigma_{\langle\varphi\varphi\rangle} = g(r). \tag{11.2.24}$$

Recall the balance of momentum in spherical coordinates from Eq. (5.6.4). Obviously the $\varphi$ and $\vartheta$ components are identically satisfied. The $r$ component reduces to:

$$\frac{\partial\sigma_{\langle rr\rangle}}{\partial r} + \frac{1}{r}\left(2\sigma_{\langle rr\rangle} - \sigma_{\langle\vartheta\vartheta\rangle} - \sigma_{\langle\varphi\varphi\rangle}\right) = 0, \tag{11.2.25}$$

if bulk forces are neglected, which can be written as:

$$\frac{\mathrm{d}\sigma_{\langle rr\rangle}}{\mathrm{d}r} = \frac{2}{r}\left(\sigma_{\langle\vartheta\vartheta\rangle} - \sigma_{\langle rr\rangle}\right). \tag{11.2.26}$$

The expression shown in the parentheses of the right hand side of this differential equation is equal to the yield stress $\sigma_y$ according to the von Mises flow rule of Eq. (11.2.9). Thus we can immediately integrate:

$$\sigma_{\langle rr\rangle} = 2\sigma_y \ln(r) + C. \tag{11.2.27}$$

The constant of integration, $C$, follows from the continuity relation for the traction at the interface to the elastic zone at $r = \rho$ [cf., Eq. (11.2.19)]:

$$\sigma_{\langle rr\rangle}\Big|_{r=\rho} = 2\sigma_y \ln(\rho) + C = -\frac{2}{3}\sigma_y\left(\frac{\rho}{R_o}\right)^3\left[\left(\frac{R_o}{\rho}\right)^3 - 1\right]$$
$$\Rightarrow \sigma_{\langle rr\rangle} = -\sigma_y\left\{2\ln\left(\frac{\rho}{r}\right) + \frac{2}{3}\left[1 - \left(\frac{\rho}{R_o}\right)^3\right]\right\}, \quad R_i \le r \le \rho. \tag{11.2.28}$$

Note that the radial stress is always compressive and becomes even more negative with decreasing distance $r$ (cf., Fig. 11.2 top, left). By using the VON MISES flow criterion it follows within the region $R_i \le r \le \rho$:

$$\sigma_{\langle\vartheta\vartheta\rangle} = \sigma_{\langle\varphi\varphi\rangle} = \sigma_y + \sigma_{\langle rr\rangle} = \sigma_y\left\{1 - 2\ln\left(\frac{\rho}{r}\right) - \frac{2}{3}\left[1 - \left(\frac{\rho}{R_o}\right)^3\right]\right\}. \tag{11.2.29}$$

Note that the angular stresses are *tensile* at radial distances $r$ close to $\rho$ (cf., Fig. 11.2 top, right). They increase with increasing distance $r$. In order to find a relation for the pressure $p_\rho$ required to initiate plastic flow up to the radius $\rho$, we simply evaluate Eq. (11.2.28) at $r = R_i$:

$$\sigma_{\langle rr\rangle}\Big|_{R_i} = -\sigma_y\left\{2\ln\left(\frac{\rho}{R_i}\right) + \frac{2}{3}\left[1 - \left(\frac{\rho}{R_o}\right)^3\right]\right\} = -p_\rho \tag{11.2.30}$$

since due to the continuity of the traction at that position the radial stress must be equal to the negative pressure. Consequently:

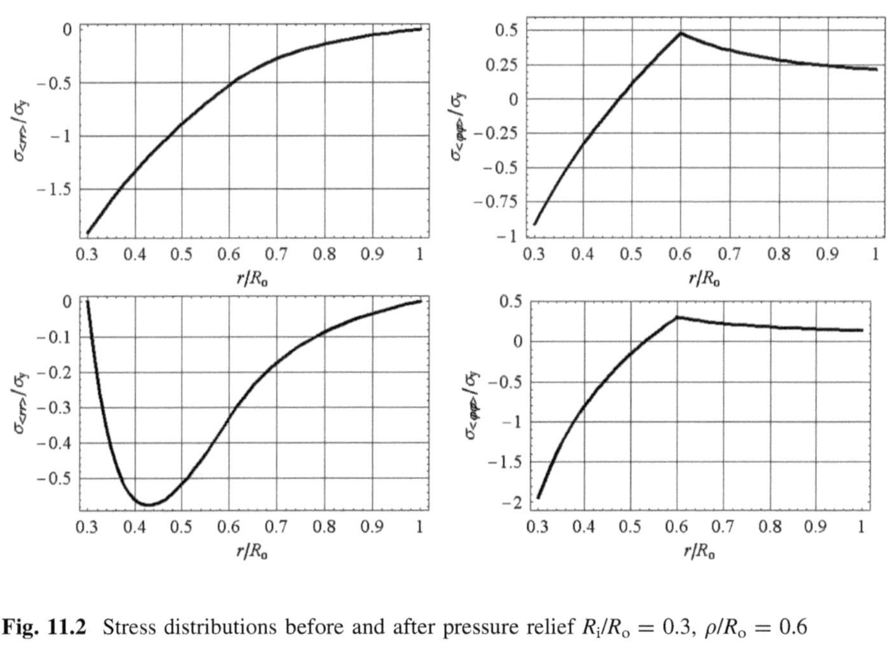


**Fig. 11.2** Stress distributions before and after pressure relief $R_i/R_o = 0.3$, $\rho/R_o = 0.6$

$$p_\rho = \sigma_\mathrm{y}\left\{2\ln\left(\frac{\rho}{R_\mathrm{i}}\right) + \frac{2}{3}\left[1 - \left(\frac{\rho}{R_\mathrm{o}}\right)^3\right]\right\}. \tag{11.2.31}$$

In summary: If we increase the pressure monotonously from $p_\mathrm{min}$ to $p_\rho$ the following stresses will arise within the hollow sphere:

(a) within the plastified region:

$$\begin{aligned}\sigma_{\langle rr\rangle} &= -\sigma_\mathrm{y}\left\{2\ln\left(\frac{\rho}{r}\right) + \frac{2}{3}\left[1 - \left(\frac{\rho}{R_\mathrm{o}}\right)^3\right]\right\},\\ \sigma_{\langle\vartheta\vartheta\rangle} = \sigma_{\langle\varphi\varphi\rangle} &= \sigma_\mathrm{y}\left\{1 - 2\ln\left(\frac{\rho}{r}\right) - \frac{2}{3}\left[1 - \left(\frac{\rho}{R_\mathrm{o}}\right)^3\right]\right\}, \quad R_\mathrm{i} \le r \le \rho;\end{aligned} \tag{11.2.32}$$

(b) within the linear-elastic region:

$$\begin{aligned}\sigma_{\langle rr\rangle} &= -\frac{2}{3}\sigma_\mathrm{y}\left(\frac{\rho}{R_\mathrm{o}}\right)^3\left[\left(\frac{R_\mathrm{o}}{r}\right)^3 - 1\right],\\ \sigma_{\langle\vartheta\vartheta\rangle} = \sigma_{\langle\varphi\varphi\rangle} &= \frac{2}{3}\sigma_\mathrm{y}\left(\frac{\rho}{R_\mathrm{o}}\right)^3\left[\frac{1}{2}\left(\frac{R_\mathrm{o}}{r}\right)^3 + 1\right], \quad \rho \le r \le R_\mathrm{o}.\end{aligned} \tag{11.2.33}$$

We will now reduce the pressure $p_\rho$ down to zero again. This will lead to a redistribution of the stresses. During this process equilibrium of forces and continuity of the traction at the interfaces must be satisfied. We claim that in the end the stresses are as follows

(a) within the inner plastified region:

$$\begin{aligned}\sigma_{\langle rr\rangle} &= \frac{2}{3}\sigma_\mathrm{y}\left(-\frac{p_\rho}{p_\mathrm{min}}\left[1 - \left(\frac{R_\mathrm{i}}{r}\right)^3\right] + 3\ln\left(\frac{r}{R_\mathrm{i}}\right)\right) < 0,\\ \sigma_{\langle\vartheta\vartheta\rangle} = \sigma_{\langle\varphi\varphi\rangle} &= \frac{2}{3}\sigma_\mathrm{y}\left(\frac{3}{2} + 3\ln\left(\frac{r}{R_\mathrm{i}}\right) - \frac{p_\rho}{p_\mathrm{min}}\left[1 + \frac{1}{2}\left(\frac{R_\mathrm{i}}{r}\right)^3\right]\right), \quad R_\mathrm{i} \le r \le \rho;\end{aligned} \tag{11.2.34}$$

(b) in the outer region:

$$\begin{aligned}\sigma_{\langle rr\rangle} &= -\frac{2}{3}\sigma_\mathrm{y}\left(\left(\frac{\rho}{R_\mathrm{i}}\right)^3 - \frac{p_\rho}{p_\mathrm{min}}\right)\left(\left(\frac{R_\mathrm{i}}{r}\right)^3 - \left(\frac{R_\mathrm{i}}{R_\mathrm{o}}\right)^3\right),\\ \sigma_{\langle\vartheta\vartheta\rangle} = \sigma_{\langle\varphi\varphi\rangle} &= \frac{2}{3}\sigma_\mathrm{y}\left(\left(\frac{\rho}{R_\mathrm{i}}\right)^3 - \frac{p_\rho}{p_\mathrm{min}}\right)\left(\frac{1}{2}\left(\frac{R_\mathrm{i}}{r}\right)^3 + \left(\frac{R_\mathrm{i}}{R_\mathrm{o}}\right)^3\right), \quad \rho \le r \le R_\mathrm{o}.\end{aligned} \tag{11.2.35}$$

Obviously the contraction of the outer region compresses the inner one: Fig. 11.2, second row. Compressive residual stresses dominate, which is advantageous since they strengthen the whole structure. This technique is actually used for the benefit of pressure vessels and known as *autofrettage*.

**Exercise 11.2.2:** ***Distribution of the residual stresses***

In order to verify the results for the residual stresses due to the autofrettage process proceed as follows. First specify all continuity conditions at the interfaces and boundaries. Use the solution shown in Eqs. (11.2.34/ 11.2.35) and show that they are all satisfied. Second, analyze the differential equations for static equilibrium of forces in both regions and prove that they are identically satisfied by the solution for the stresses. In this context make use of the perfect spherical symmetry.

Hill [1] claims in his famous textbook on plasticity that the solution can alternatively be obtained by subtraction of the elastic stress field resulting from Eq. (11.2.2) for the choice $p = p_\rho$ from the stresses shown in Eqs. (11.2.32) and (11.2.33). Confirm this statement and explain why it is legitimate to calculate the stresses in this manner after pressure relief?

## 11.3 The Prandtl-Reuss Equations

The fundamental equations of time-independent plasticity according to Prandtl and Reuss will be formulated as rate equations (in terms of stress or strain increments, identifiable by a dot above the corresponding symbol). We start by assuming that the linear strain rates can be decomposed *additively* into elastic and plastic parts (in what follows we shall exclusively use Cartesian coordinates; for simplicity we will not emphasize this choice by the identifier $(x)$ beneath the formula symbols):

$$\dot{\varepsilon}_{ij} = \dot{\varepsilon}_{ij}^{\mathrm{el}} + \dot{\varepsilon}_{ij}^{\mathrm{pl}}. \tag{11.3.1}$$

Ludwig PRANDTL was born on February 4, 1875 in Freising near Munich and died on August 15, 1953 in Göttingen. He studies engineering in Munich. After his graduation he first becomes the assistant and later also the son-in-law of the famous mechanics professor August FÖPPL. In 1900 he finishes his Ph.D. thesis on stability theory. PRANDTL starts working for the company MAN, but only for a very short while. In the same year he becomes a professor of engineering mechanics at the Technische Hochschule in Hannover. A few years later in 1904 he is appointed as a professor for applied mechanics at the University of Göttingen. In 1925 he becomes the director of the local Kaiser–Wilhelm Institute for fluid mechanics, which was turned into a Max-Planck-Institute after WW II. Special credit is due to PRANDTL for developing fundamental concepts of fluid mechanics, e.g., the boundary layer theory. His name lives on in the so-called PRANDTL number, which characterizes the transition from laminar to turbulent flow.

HOOKE's law is also written in incremental form:

$$\dot{\sigma}_{ij} = C_{ijkl}\dot{\varepsilon}^{\mathrm{el}}_{kl} = C_{ijkl}\left(\dot{\varepsilon}_{kl} - \dot{\varepsilon}^{\mathrm{pl}}_{kl}\right). \tag{11.3.2}$$

Moreover, we postulate a so-called *associated flow rule* according to which the plastic strain rates can be obtained from a gradient w.r.t. the stress components from a scalar function $\phi$:

$$\dot{\varepsilon}^{\mathrm{pl}}_{kl} = \dot{\lambda}\frac{\partial\phi}{\partial\sigma_{kl}}. \tag{11.3.3}$$

Endre A. REUSS was born on July 1, 1900 in Budapest where he also died on May 10, 1968. He worked at the Department of Applied Mechanics of the Technical University of Budapest. His name is not only known in context with plasticity but also from homogenization theory, which is used to determine effective material properties of heterogeneous matter.

$\dot{\lambda}$ is a currently unknown factor which implicitly carries the time dependence (indicated by the dot which is generally not related to a total differential in time). In a one-dimensional tensile test we may write after plastic flow has set in, i.e., after the initial yield stress, $\sigma_{\mathrm{y},0}$, has been surpassed:

$$\dot{\sigma} = \dot{\lambda}h \text{ with } \quad \sigma \geq \sigma_{\mathrm{y},0}, \tag{11.3.4}$$

$h$ denotes another scalar, currently unknown function. The VON MISES yield criterion from Eq. (11.1.1) will now be used as a special but explicit example of a flow function. We put:

$$\phi = \phi\left(\vec{\sigma}, \sigma_{\mathrm{y}}\right) = \frac{1}{2}\Sigma_{rs}\Sigma_{rs} - \frac{1}{3}\sigma_{\mathrm{y}}^2 = 0. \tag{11.3.5}$$

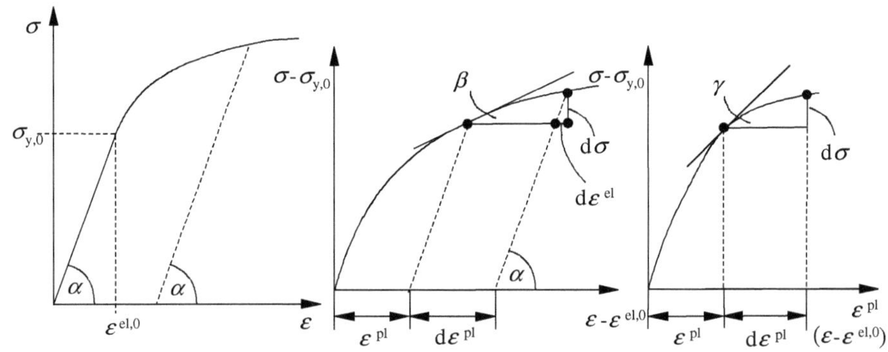


**Fig. 11.3** Non-linear stress-strain curve

$\sigma_y$ is the current yield stress an must not be confused with the initial one, $\sigma_{y,0}$. The difference is shown in Fig. 11.3. Only for the case of ideal plasticity, where there is no hardening, both are equal. In addition, we must guarantee that the so-called *consistency condition* is satisfied. This is a capricious expression for the fact that the flow condition is identically satisfied at each incremental loading step or at each point in quasi-time:

$$\dot{\phi} = 0 \Rightarrow \dot{\phi}(\sigma, \sigma_y) = \frac{\partial \phi}{\partial \sigma_{ij}} \dot{\sigma}_{ij} + \frac{\partial \phi}{\partial \sigma_y} \dot{\sigma}_y \equiv 0. \tag{11.3.6}$$

In particular, we obtain for the case of VON MISES plasticity, Eq. (11.1.1), and by observing the definition for the stress deviator shown in Eq. (11.1.2):

$$\frac{\partial \phi}{\partial \sigma_{ij}} = \frac{\partial \phi}{\partial \Sigma_{rs}} \frac{\partial \Sigma_{rs}}{\partial \sigma_{ij}} = \Sigma_{rs} \left( \delta_{ri} \delta_{sj} - \frac{1}{3} \delta_{ni} \delta_{nj} \delta_{rs} \right) = \Sigma_{ij} \tag{11.3.7}$$

and from Eq. (11.3.5) that:

$$\frac{\partial \phi}{\partial \sigma_y} = -\frac{2}{3} \sigma_y. \tag{11.3.8}$$

Stress-strain curves of metals, which are experimentally determined in one-dimensional tensile tests are not only used to determine *YOUNG's modulus E*, i.e., an *elastic* property, but also the so-called *tangent modulus* $E_T$, which is a measure of the hardening capacity of a metal. Suggestively speaking, *Young*'s modulus is nothing else but the slope of the initial "line" of the stress-strain curve, i.e., characteristic of the *Hookean* reversible region (Fig. 11.3, left). The tangent modulus corresponds also to a slope, but of the non-linear part of the stress-strain curve. In general there is not just one value for the slope since that part is *curved*. However, it is frequently idealized by a straight line. Then we speak of *linear hardening* or of a *bilinear elastic-plastic* stress-strain model. Strictly speaking, the tangent modulus is also not exactly equal to the slope directly visible in a recorded stress-strain curve, but in practice it is very close to this value provided plastic deformation is dominant. For its proper characterization one first eliminates the elastic strain from the curve. Thus only the non-linear stress-strain part remains starting from the initial yield stress $\sigma_{\mathrm{y},0}$ to an arbitrary (current yield stress) value $\sigma = \sigma_{\mathrm{y}}$. Consequently we obtain a plot of $\sigma - \sigma_{\mathrm{y},0}$ versus $\varepsilon - \varepsilon^{\mathrm{el},0}$ (Fig. 11.3, center) and we define:

$$E = \frac{\sigma_{\mathrm{y},0}}{\varepsilon^{\mathrm{el},0}}. \tag{11.3.9}$$

Figure 11.3 (center) also shows what happens if we start unloading at a stress value above the point of initial yield. We then return to zero stress level, basically along a straight line parallel to the HOOKEAN branch. The abscissa is intersected at a strain $\varepsilon^{\mathrm{pl}} \neq 0$. This is the remaining plastic strain after monotonous loading and the total strain $\varepsilon$ is simply the sum of the elastic and of the plastic strain parts $\varepsilon = \varepsilon^{\mathrm{el}} + \varepsilon^{\mathrm{pl}}$, in other words the integrated, one-dimensional form of Eq. (11.3.1). As indicated in Fig. 11.3 (center) we obtain for a neighboring point an increment of plastic strain $\mathrm{d}\varepsilon^{\mathrm{pl}}$. The corresponding elastic (reversible) strain increment is given by $\mathrm{d}\varepsilon^{\mathrm{el}} = E\,\mathrm{d}\sigma$. If we now subtract in each point of the curve $\left(\varepsilon - \varepsilon^{\mathrm{el},0}, \sigma - \sigma_{\mathrm{y},0}\right)$ the elastic bit $E\left(\sigma - \sigma_{\mathrm{y},0}\right)$ from the ordinate value we can "renormalize" and generate the curve $\left(\varepsilon - \varepsilon^{\mathrm{el},0} - E\left(\sigma - \sigma_{\mathrm{y},0}\right), \sigma - \sigma_{\mathrm{y},0}\right) = \left(\varepsilon^{\mathrm{pl}}, \sigma - \sigma_{\mathrm{y},0}\right)$. For this purpose we make use of Eq. (11.3.9). Thus we obtain the third viewgraph in Fig. 11.3 (right), which shows stress exclusively as a function of plastic strain. This allows us to define the (plastic) tangent modulus correctly as the slope of the function shown in Fig. 11.3 (right). Note that there is a connection to the slope $\beta$ of the function shown in Fig. 11.3 (center):

$$\begin{gathered} E_{\mathrm{T}} \sim \tan(\gamma) \Rightarrow E_{\mathrm{T}} = \frac{\mathrm{d}\sigma}{\mathrm{d}\varepsilon^{\mathrm{pl}}} = \frac{\dot{\sigma}}{\dot{\varepsilon}^{\mathrm{pl}}} \text{ and} \\ \tan(\beta) \sim \frac{\mathrm{d}\sigma}{\mathrm{d}\varepsilon^{\mathrm{el}} + \mathrm{d}\varepsilon^{\mathrm{pl}}} = \frac{\mathrm{d}\sigma/\mathrm{d}\varepsilon^{\mathrm{pl}}}{1 + \dfrac{\mathrm{d}\sigma/\mathrm{d}\varepsilon^{\mathrm{pl}}}{\mathrm{d}\sigma/\mathrm{d}\varepsilon^{\mathrm{el}}}} = \frac{E_{\mathrm{T}}}{1 + E_{\mathrm{T}}/E} = \frac{E_{\mathrm{T}}E}{E + E_{\mathrm{T}}}. \end{gathered} \tag{11.3.10}$$

Thus the difference between $\tan(\beta)$ and $\tan(\gamma)$ becomes obsolete if $E_{\mathrm{T}}/E \ll 1$. If we now observe that:

$$\Sigma_{rs}\sigma_{rs} = \Sigma_{rs}\left(\Sigma_{rs} + \frac{1}{3}\sigma_{nn}\delta_{rs}\right) = \Sigma_{rs}\Sigma_{rs} + \frac{1}{3}\sigma_{nn}\Sigma_{rr} = \Sigma_{rs}\Sigma_{rs} \tag{11.3.11}$$

and assume that the power dissipated due to plastic deformation in the one-dimensional case is equal to that in 3D (recall that $\sigma \equiv \sigma_{\mathrm{y}}$ is the currently applied one-dimensional tension during plastic deformation, i.e., the current yield stress):

$$\sigma\,\dot{\varepsilon}^{\mathrm{pl}} \equiv \sigma_{\mathrm{y}}\,\dot{\varepsilon}^{\mathrm{pl}} = \sigma_{ij}\dot{\varepsilon}_{ij}^{\mathrm{pl}}, \tag{11.3.12}$$

the quantity $h$ from Eq. (11.3.4) can be identified:

$$\begin{aligned}\dot{\sigma} &= E_{\mathrm{T}}\dot{\varepsilon}^{\mathrm{pl}} = E_{\mathrm{T}}\frac{\sigma_{ij}\dot{\varepsilon}_{ij}^{\mathrm{pl}}}{\sigma_{\mathrm{y}}} = \dot{\lambda}E_{\mathrm{T}}\frac{\sigma_{ij}\frac{\partial\phi}{\partial\sigma_{ij}}}{\sigma_{\mathrm{y}}} = \dot{\lambda}E_{\mathrm{T}}\frac{\sigma_{ij}\Sigma_{ij}}{\sigma_{\mathrm{y}}} \\ &= \dot{\lambda}E_{\mathrm{T}}\frac{\Sigma_{ij}\Sigma_{ij}}{\sigma_{\mathrm{y}}} = \dot{\lambda}\frac{2}{3}\frac{E_{\mathrm{T}}\sigma_{\mathrm{y}}^2}{\sigma_{\mathrm{y}}} = \dot{\lambda}\frac{2}{3}E_{\mathrm{T}}\sigma_{\mathrm{y}} \Rightarrow h = \frac{2}{3}E_{\mathrm{T}}\sigma_{\mathrm{y}}\end{aligned} \tag{11.3.13}$$

Consequently we now may rewrite Eq. (11.3.4) as follows:

$$\dot{\sigma} = \dot{\lambda}\frac{2}{3}E_{\mathrm{T}}\sigma_{\mathrm{y}}. \tag{11.3.14}$$

This relation shows us indirectly that we are dealing with a quasistatic, i.e., time-independent theory. A last potential time- (or better rate-) dependence is contained in the quantity $\dot{\lambda}$, which we shall identify now. We start from the consistency condition (11.3.6) and insert successively the newly derived Eqs. (11.3.3/11.3.7/11.3.8/11.3.14):

$$\begin{aligned}0 &= \frac{\partial\phi}{\partial\sigma_{ij}}\dot{\sigma}_{ij} + \frac{\partial\phi}{\partial\sigma_{\mathrm{y}}}\dot{\sigma} = \Sigma_{ij}\dot{\sigma}_{ij} - \tfrac{2\sigma_{\mathrm{y}}}{3}\dot{\sigma} = \Sigma_{ij}C_{ijkl}\left(\dot{\varepsilon}_{ij} - \dot{\varepsilon}_{ij}^{\mathrm{pl}}\right) \\ &- \tfrac{2\sigma_{\mathrm{y}}}{3}\dot{\sigma} = \Sigma_{ij}C_{ijkl}\left(\dot{\varepsilon}_{kl} - \dot{\lambda}\frac{\partial\phi}{\partial\sigma_{kl}}\right) - \dot{\lambda}\frac{4E_{\mathrm{T}}}{9}\sigma_{\mathrm{y}}^2 = \Sigma_{ij}C_{ijkl}\left(\dot{\varepsilon}_{kl} - \dot{\lambda}S_{kl}\right) - \dot{\lambda}\frac{4E_{\mathrm{T}}}{9}\sigma_{\mathrm{y}}^2.\end{aligned} \tag{11.3.15}$$

Consequently it follows for the unknown function $\dot{\lambda}$:

$$\dot{\lambda} = \frac{\Sigma_{rs}C_{rstu}\dot{\varepsilon}_{tu}}{\Sigma_{op}C_{opmn}\Sigma_{mn} + \frac{4}{9}E_{\mathrm{T}}\sigma_{\mathrm{y}}^2} \Rightarrow \dot{\varepsilon}_{kl}^{\mathrm{pl}} = \frac{\Sigma_{rs}C_{rstu}\Sigma_{kl}\dot{\varepsilon}_{tu}}{\Sigma_{op}C_{opmn}\Sigma_{mn} + \frac{4}{9}E_{\mathrm{T}}\sigma_{\mathrm{y}}^2}. \tag{11.3.16}$$

If we insert this in Eq. (11.3.2) we obtain the following equation, which connects stress and strain rates:

$$\dot{\sigma}_{ij} = \left(C_{ijkl} - \frac{C_{ijrs}\Sigma_{rs}\Sigma_{uv}C_{uvkl}}{\Sigma_{op}C_{opmn}\Sigma_{mn} + \frac{4}{9}E_{\mathrm{T}}\sigma_{\mathrm{y}}^2}\right)\dot{\varepsilon}_{kl}. \tag{11.3.17}$$

These are the Prandtl-Reuss equations for a material showing *anisotropic* linear-elastic behavior. We write for short:

$$\dot{\sigma}_{ij} = C_{ijkl}^{\mathrm{ep}}\,\dot{\varepsilon}_{kl} \tag{11.3.18}$$

introducing the so-called elastic-plastic stiffness matrix:

$$C_{ijkl}^{\mathrm{ep}} = C_{ijkl} - \frac{C_{ijrs}\Sigma_{rs}\Sigma_{uv}C_{uvkl}}{\Sigma_{op}C_{opmn}\Sigma_{mn} + \frac{4}{9}E_{\mathrm{T}}\sigma_{\mathrm{y}}^2}. \tag{11.3.19}$$

Equation (11.3.19) must be solved incrementally during each loading step while equilibrium of forces, i.e., the incremental static balance of momentum is observed:

$$\frac{\partial \dot{\sigma}_{ji}}{\partial x_j} = 0. \tag{11.3.20}$$

Thus we are confronted with a mathematical problem that—in a certain way—bears a certain similarity to the LAMÉ-NAVIER equations from Sect. 6.2. The equations can be simplified if we assume *isotropic* linear-elastic behavior. To this end we write:

$$C_{ijkl} = \lambda \delta_{ij}\delta_{kl} + \mu\left(\delta_{ik}\delta_{jl} + \delta_{il}\delta_{jk}\right) \tag{11.3.21}$$

and find (recall that the trace of the deviator vanishes, $\Sigma_{rr} = 0$, as well as its symmetry, $\Sigma_{ij} = \Sigma_{ji}$):

$$\begin{aligned} C_{ijrs}\Sigma_{rs}\Sigma_{uv}C_{uvkl} &= \lambda\delta_{ij}\Sigma_{rr}\Sigma_{uv}C_{uvkl} + \mu\left(\Sigma_{ij} + \Sigma_{ji}\right)\mu(\Sigma_{kl} + \Sigma_{lk}) \\ &= 4\mu^2\Sigma_{ij}\Sigma_{kl} = \Sigma_{op}C_{opmn}\Sigma_{mn} = \lambda\Sigma_{oo}\Sigma_{mn} + \mu(\Sigma_{nm} + \Sigma_{mn})\,\Sigma_{mn} \\ &= 2\mu\Sigma_{mn}\Sigma_{mn} = \frac{4}{3}\mu\sigma_{\mathrm{y}}^2. \end{aligned} \tag{11.3.22}$$

Thus:

$$\dot{\sigma}_{ij} = \left(C_{ijkl} - \frac{9\mu^2\Sigma_{ij}\Sigma_{kl}}{(3\mu + E_{\mathrm{T}})\sigma_{\mathrm{y}}^2}\right)\dot{\varepsilon}_{kl} = \left(C_{ijkl} - \frac{3\mu\Sigma_{ij}\Sigma_{kl}}{\left(1 + \frac{E_{\mathrm{T}}}{3\mu}\right)\sigma_{\mathrm{y}}^2}\right)\dot{\varepsilon}_{kl}. \tag{11.3.23}$$

Note that according to VON MISES we may also write for the yield stress [see Eq. (11.1.1)], where the right hand side is a.k.a. *equivalent stress*:

$$\sigma_{\mathrm{y}}\left(\varepsilon^{\mathrm{pl}}\right) = \sqrt{\frac{3}{2}\Sigma_{ij}\Sigma_{ij}} \equiv \sigma_{\mathrm{eq}}. \tag{11.3.24}$$

A similar relation can be found for the *equivalent plastic strain rate* by means of Eqs. (11.3.7/11.3.10/11.3.14/11.3.13):

$$\begin{aligned} &\dot{\varepsilon}_{kl}^{\mathrm{pl}}\dot{\varepsilon}_{kl}^{\mathrm{pl}} = \dot{\lambda}^2\frac{\partial\phi}{\partial\sigma_{kl}}\frac{\partial\phi}{\partial\sigma_{kl}} = \dot{\lambda}^2\Sigma_{kl}\Sigma_{kl} = \dot{\lambda}^2\frac{2}{3}\sigma_{\mathrm{y}}^2 = \frac{2}{3}\frac{\dot{\sigma}^2}{E_{\mathrm{T}}^2\sigma_{\mathrm{y}}^2}\sigma_{\mathrm{y}}^2 = \frac{3}{2}\left(\dot{\varepsilon}^{\mathrm{pl}}\right)^2 \\ &\Rightarrow \dot{\varepsilon}^{\mathrm{pl}} = \sqrt{\frac{2}{3}\dot{\varepsilon}_{kl}^{\mathrm{pl}}\dot{\varepsilon}_{kl}^{\mathrm{pl}}}. \end{aligned} \tag{11.3.25}$$

Note that the factor $\frac{2}{3}$ is exactly the inverse of the one shown in Eq. (11.1.1) for the *equivalent plastic strain rate*. This expression can easily be integrated w.r.t. "time:"

$$\varepsilon_{\mathrm{eq}}^{\mathrm{pl}} = \int \dot{\varepsilon}^{\mathrm{pl}}\,\mathrm{d}t = \int \sqrt{\frac{2}{3}\dot{\varepsilon}_{kl}^{\mathrm{pl}}\dot{\varepsilon}_{kl}^{\mathrm{pl}}}\,\mathrm{d}t. \tag{11.3.26}$$

It can be used to calculate the accumulated equivalent plastic strain during a loading process.

**Exercise 11.3.1:** ***The one-dimensional plastically deformed tensile beam***

A slender metallic beam subjected to a one-dimensional tensile state of stress (in $x_1$ direction) is loaded beyond initial yield by a monotonously increasing, time-dependent stress rate $\dot{\sigma}$. Assume that the material is elastically isotropic and show that:

$$\begin{aligned}\dot{\varepsilon}_{11} &= \frac{1}{3k}\frac{\lambda + \mu - \frac{\mu/3}{1+E_{\mathrm{T}}/(3\mu)}}{\mu\left(1 - \frac{1}{1+E_{\mathrm{T}}/(3\mu)}\right)}\dot{\sigma}, \\ \dot{\varepsilon}_{22} = \dot{\varepsilon}_{33} &= -\frac{1}{3k}\frac{\lambda + \frac{2\mu/3}{1+E_{\mathrm{T}}/(3\mu)}}{2\mu\left(1 - \frac{1}{1+E_{\mathrm{T}}/(3\mu)}\right)}\dot{\sigma}.\end{aligned} \tag{11.3.27}$$

Explain by integrating explicitly with respect to time for the case $E_{\mathrm{T}} =$ const. why it is customary to speak of time-independent incremental plasticity. Determine the strain rates also for an elastic ideal-plastic material for which, by definition, $E_{\mathrm{T}} = 0$. Moreover, show that the plastic strain rates and the equivalent plastic strain are given by:

$$\dot{\varepsilon}_{ij}^{\mathrm{pl}} = \frac{\dot{\sigma}}{2E_{\mathrm{T}}}\begin{bmatrix} 2 & 0 & 0 \\ 0 & -1 & 0 \\ 0 & 0 & -1 \end{bmatrix}, \quad \dot{\varepsilon}^{\mathrm{pl}} = \frac{\dot{\sigma}}{E_{\mathrm{T}}}. \tag{11.3.28}$$

Integrate these equations explicitly w.r.t. time for the case $E_{\mathrm{T}} =$ const.

**Exercise 11.3.2:** ***Plastic strain rates for a pressurized spherical shell***

Use the PRANDTL-REUSS relation (11.3.16) and show that the trace of the plastic strain rates vanishes:

$$\mathrm{tr}\ \dot{\varepsilon}^{\mathrm{pl}} = 0. \tag{11.3.29}$$

What does this have to do with incompressibility? Specialize the PRANDTL–REUSS relation to the case of the plastifying hollow sphere subjected to an internal pressure from Sect. 11.2 and show that within the region $R_{\mathrm{i}} \leq r \leq \rho$:

$$\dot{\varepsilon}_{\langle rr\rangle}^{\mathrm{pl}} = \frac{2}{3}\left[\frac{\partial \dot{u}_{\langle r\rangle}}{\partial r} - \frac{\dot{u}_{\langle r\rangle}}{r}\right], \quad \dot{\varepsilon}_{\langle\vartheta\vartheta\rangle}^{\mathrm{pl}} = \dot{\varepsilon}_{\langle\varphi\varphi\rangle}^{\mathrm{pl}} = -\frac{1}{3}\left[\frac{\partial \dot{u}_{\langle r\rangle}}{\partial r} - \frac{\dot{u}_{\langle r\rangle}}{r}\right]. \tag{11.3.30}$$

To this end recall that it was assumed that the material of the sphere was ideally plastic:

$$E_{\mathrm{T}} = 0 \tag{11.3.31}$$

and kinematic relations of the form (6.2.29) must be used but for the total *strain rates*. Confirm the general expression shown in Eq. (11.3.28) by evaluating it with Eq. (11.3.29). Moreover, show that the equivalent plastic strain for the pressurized hollow sphere is given by:

$$\begin{aligned}\varepsilon_{\mathrm{eq}}^{\mathrm{pl}} &= \int\limits_{t(p_{\min})}^{t(p_\rho)} \sqrt{\frac{2}{3}\dot{\varepsilon}_{\langle kl\rangle}^{\mathrm{pl}}\dot{\varepsilon}_{\langle kl\rangle}^{\mathrm{pl}}}\,\mathrm{d}t = \sqrt{\frac{2}{3}} \int\limits_{t(p_{\min})}^{t(p_\rho)} \sqrt{\left(\dot{\varepsilon}_{\langle rr\rangle}^{\mathrm{pl}}\right)^2 + \left(\dot{\varepsilon}_{\langle \vartheta\vartheta\rangle}^{\mathrm{pl}}\right)^2 + \left(\dot{\varepsilon}_{\langle \varphi\varphi\rangle}^{\mathrm{pl}}\right)^2}\,\mathrm{d}t \\ &= \varepsilon_{\mathrm{eq}}^{\mathrm{pl}} = \frac{2}{3} \int\limits_{t(p_{\min})}^{t(p_\rho)} \left[\frac{\partial \dot{u}_{\langle r\rangle}}{\partial r} - \frac{\dot{u}_{\langle r\rangle}}{r}\right] \mathrm{d}t,\; R_{\mathrm{i}} \le r \le \rho.\end{aligned} \tag{11.3.32}$$

How can this result be evaluated further? Recall HOOKE's law for the isotropic case and show by assuming additive decomposition of the strain rates according to Eq. (11.3.1) that:

$$\begin{aligned}\sigma_{\langle ij\rangle} &= \lambda \varepsilon_{\langle kk\rangle}^{\mathrm{el}}\,\delta_{\langle ij\rangle} + 2\mu\varepsilon_{\langle ij\rangle}^{\mathrm{el}} \\ &= \lambda\left[\varepsilon_{\langle kk\rangle} - \varepsilon_{\langle kk\rangle}^{\mathrm{pl}}\right]\delta_{\langle ij\rangle} + 2\mu\left[\varepsilon_{\langle ij\rangle} - \varepsilon_{\langle ij\rangle}^{\mathrm{pl}}\right].\end{aligned} \tag{11.3.33}$$

Calculate the trace of this equation and derive the following differential equation for the radial component of the displacement:

$$\frac{\mathrm{d}u_{\langle r\rangle}}{\mathrm{d}r} + 2\frac{u_{\langle r\rangle}}{r} = \frac{1-2\nu}{E}\left[\sigma_{\langle rr\rangle} + 2\sigma_{\langle \vartheta\vartheta\rangle}\right], \quad R_{\mathrm{i}} \le r \le \rho. \tag{11.3.34}$$

Recall that complete spherical symmetry was assumed and observe the kinematic relations (6.2.29). Integrate this equation by using the expressions for the stresses within the plastic region, Eq. (11.2.33), and show by adjusting the resulting constant of integration to the solution for the radial displacement in the elastic region from Eq. (11.2.21) that within $R_{\mathrm{i}} \le r \le \rho$:

$$u_{\langle r\rangle} = \frac{2(1-2\nu)\sigma_{\mathrm{y}}}{3E}\left[\frac{3}{2}\frac{1-\nu}{1-2\nu}\left(\frac{\rho}{r}\right)^3 - \left[1 - \left(\frac{\rho}{R_{\mathrm{o}}}\right)^3\right] + \ln\left(\frac{r}{\rho}\right)^3\right] r. \tag{11.3.35}$$

## 11.4 Would You Like to Know More?

Plasticity theory, in particular time-dependent, so-called visco-plastic deformation, a.k.a. "creep," which is not discussed in this book, is a most active field of teaching and research. A very good introduction to time-independent plasticity can be found in the classic textbook by Hill [1]. The problem of autofrettage in pressure vessels is more extensively discussed in Chap. 6 of Betten [2]. This book can also be used to get a first insight to time-dependent plasticity: Chap. 10. Khan and Huang [3] present plasticity from a continuum mechanics point-of-view. The monograph by Houlsby and Puzrin [4] goes well beyond the description of plastic deformation of metals. It presents many applications from soil mechanics and emphasizes thermodynamic arguments. In the same context the book by Maugin [5] may be consulted.

## References

1. Hill R (1998) The mathematical theory of plasticity, Oxford classic texts in the physical sciences. Clarendon Press, Oxford
2. Betten J (2001) Kontinuumsmechanik—Elastisches und inelastisches Verhalten isotroper und anisotroper Stoffe, 2nd edn. Springer, Berlin, Heidelberg
3. Khan AS, Huang S (1995) Continuum theory of plasticity. Wiley, New York, Chichester, Brisbane, Toronto, Singapore
4. Houlsby GT, Puzrin AM (2006) Principles of hyperplasticity—an approach to plasticity theories based on thermodynamic principles. Springer, London
5. Maugin GA (1992) The thermomechanics of plasticity and fracture, Cambridge texts in applied mathematics. Cambridge University Press, Cambridge

# Chapter 12
# Entropy

**Abstract** This chapter is dedicated to a very delicate subject, entropy. More advanced textbooks on materials theory use the so-called entropy principle for further reduction of the possible form of constitutive relations. Moreover, entropy is very useful when it comes to the quantification of the degree of irreversibility of a physical process. In order to obtain some familiarity with this rather abstract quantity entropy is introduced as just another field obeying a balance law. Thus entropy is directly in line with other additive quantities that were used previously in this book, namely mass, momentum, and energy. Moreover, potential "insecurity" in context with the notion of entropy is reduced by computing it for a few simple but illuminating cases: We shall understand why it is a measure of the state of disorder of a system. Similarly we shall calculate the entropy production in simple cases and understand that it is a measure of irreversibility or of how difficult it is to reverse a "natural" process. At the end of the chapter we present an introduction to the theory of irreversible processes according to ECKART. This way we will rediscover the constitutive equations of NAVIER-STOKES-FOURIER. This may serve as a starting point for further studies of entropy principles.

*In the end it is unavoidable to meet the Grim Reaper.*
*However, by means of entropy we can assess his strength,*
*reveal at least one of his secrets, and boldly*
*hurl defiance: "We'll be watching you!"*

Anonymus

## 12.1 Entropy as a Balanceable Quantity

Entropy and the Second Law of thermodynamics belong to the concepts of science with mystical quality. On the one hand side this may be due to the fact that entropy and the corresponding entropy production are relatively abstract quantities, for which we have no gut feeling, as in the case of mass, velocity, or even internal

W. H. Müller, *An Expedition to Continuum Theory*, Solid Mechanics and Its Applications 210, DOI: 10.1007/978-94-007-7799-6_12,

energy and heat flux. On the other hand side the concept of entropy is related to a very fundamental, if not frightening property of our physical world, namely to its irreversibility and mortality. Clearly, this attracts very quickly the attention of philosophers, prophets, esoterics, and other witch doctors.

Of course, we shall not follow their line of reasoning. On the contrary: In this chapter we will show that entropy is a very useful tool for the rational engineer, a tool that can be understood and mastered easily, if it is presented within a proper mathematical setting. Entropy will help us, first, to reduce the number of calorimetric measurements considerably, which are required to determine the dependence of the internal energy on density and temperature. Second, entropy poses constraints regarding the possible form and dependence of constitutive equations, like the heat flux and the stress tensor on the state variables. Third, entropy puts us in a position to quantify the degree of irreversibility of a physical process.

Constantin CARATHÉODORY was born on September 13, 1873 in Berlin and died (despite his fundamental contributions to the interpretation of entropy) on February 2, 1950 in Munich. He was born into a family of diplomats. Nevertheless he was much more interested in mathematics and engineering. He got his first engineering degree from the École Militaire de Belgique in Brussels in 1985. Immediately after that he went as an military civil engineer to Mytilene on Lesbos, which was part of the Osmanic Empire in those days. There he contributed to the local infrastructure by building better streets and thoroughfares. Moreover, he conducted measurements of the entrance of the Cheops pyramid, which he also published. To the surprise of his family he suddenly declared that From now on he would exclusively dedicate his time to mathematics. For this purpose he went to the Universities of Berlin and Göttingen in 1901. He advanced quickly in academia and was appointed to several chairs in Germany and abroad. Judging from his biography it is not too surprising that he could communicate fluently in Greek, French, German, English, Italian, and Turkish.

Nevertheless we must find a suitable way of how to introduce entropy, in particular in a book on continuum theory, which follows certain formalisms. For example, we do not want to start with a discussion on machine cycles, in particular not with the CARNOT process, as it is customarily done in a course on technical thermodynamics, when entropy is presented for the first time. However, we do expect that the reader has at least some knowledge of such things and, if not, acquires that knowledge by self study. For this purpose some references to suitable literature will be provided later. In this book we will follow a way suggested by Carl ECKART [1], which is based on the scientific writings of Constantin CARATHÉODORY. We will start from the First Law of Thermodynamics in local regular form according to Eq. (6.5.5). Analogously to Eq. (11.1.2) we decompose the stress tensor into an isotropic part, consisting of the mechanical pressure, and into a stress deviator:

$$\sigma_{ij} = -(p+\pi)\delta_{ij} + \Sigma_{ij}, \quad p_{\mathrm{me}} = -\frac{1}{3}\sigma_{kk} \equiv p + \pi. \tag{12.1.1}$$

Note that the mechanical pressure is given by the sum of the thermodynamic pressure, $p$, the state variable, so to speak, and of the dynamic pressure, which covers all irreversible isotropic stress related parts (cf., also Exercise 7.4.2). We find that:

$$\rho \frac{\mathrm{d}u}{\mathrm{d}t} = -\frac{\partial q_j}{\partial x_j} - (p+\pi)\frac{\partial \upsilon_i}{\partial x_i} + \Sigma_{ji}\frac{\partial \upsilon_i}{\partial x_j}. \tag{12.1.2}$$

Nicolas Léonard Sadi CARNOT was born on June 1, 1796 in Paris where he also died on August 24, 1832. In 1812 he enrolls at a very young age as a student at the École Polytechnique. However, in 1814 he quits to become a military engineer. But now his republican convictions lead to problems with his army colleagues. Thus in 1819 he tenders his (preliminary) resignation and dedicates his life to science, in particular to the theoretical foundations of (steam) engines and their maximum efficiency. This culminates in his famous paper *Réflexions sur la puissance motrice du feu et sur les machines propres à développer cette puissance.* By the end of the year 1826 CARNOT reenters the military. However, in 1828 his health forces him to retire for good. In June 1832 he is taken ill with scarlet fever and he finally dies at age 36 during a cholera epidemic.

If we now observe the balance of mass in Eq. (6.5.7) this result can be rewritten as follows:

$$\rho\left(\frac{\mathrm{d}u}{\mathrm{d}t} + p\frac{\mathrm{d}\upsilon}{\mathrm{d}t}\right) = -\frac{\partial q_j}{\partial x_j} - \pi\frac{\partial \upsilon_i}{\partial x_i} + \Sigma_{ji}\frac{\partial \upsilon_i}{\partial x_j}. \tag{12.1.3}$$

We now assume that the specific internal energy depends on *only two* variables, as we did previously in Sect. 6.5. However, this time and for reasons which will become immediately obvious, we choose two "mechanical" quantities as variables, namely the thermodynamic pressure and the specific volume:

$$u = u(p,\upsilon) \Rightarrow \frac{\mathrm{d}u}{\mathrm{d}t} = \left.\frac{\partial u}{\partial p}\right|_{\upsilon}\frac{\mathrm{d}p}{\mathrm{d}t} + \left.\frac{\partial u}{\partial \upsilon}\right|_{p}\frac{\mathrm{d}\upsilon}{\mathrm{d}t}. \tag{12.1.4}$$

Thus the expression in parentheses on the left hand side of Eq. (12.1.3) becomes:

$$\frac{\mathrm{d}u}{\mathrm{d}t} + p\frac{\mathrm{d}\upsilon}{\mathrm{d}t} = \left.\frac{\partial u}{\partial p}\right|_{\upsilon}\frac{\mathrm{d}p}{\mathrm{d}t} + \left(\left.\frac{\partial u}{\partial \upsilon}\right|_{p} + p\right)\frac{\mathrm{d}\upsilon}{\mathrm{d}t}. \tag{12.1.5}$$

The objective is now to write this expression as a single total differential. This is always possible thanks to a mathematical theorem due to EULER, which is known as the *method of the integrating factor*. Recall that EULER considered the following differential equation:

$$f(x,y) + g(x,y)\frac{\mathrm{d}y}{\mathrm{d}x} = 0, \tag{12.1.6}$$

where $f(x, y)$ and $g(x, y)$ are arbitrary, continuously differentiable functions of the variables $x$ and $y$. He now multiplied the equation with a currently unknown function $\mu(x, y)$—the so-called integrating factor—, such that:

$$\mu(x, y)f(x, y) + \mu(x, y)g(x, y)\frac{\mathrm{d}y}{\mathrm{d}x} = 0, \tag{12.1.7}$$

and required that:

$$\frac{\partial \varphi}{\partial x} = \mu(x, y)f(x, y), \quad \frac{\partial \varphi}{\partial y} = \mu(x, y)g(x, y). \tag{12.1.8}$$

Euler called the function $\varphi(x, y)$ the *potential*. This was done for good reasons since then Eq. (12.1.7) can be transformed into the following form:

$$\frac{\partial \varphi}{\partial x}\mathrm{d}x + \frac{\partial \varphi}{\partial y}\mathrm{d}y \equiv \mathrm{d}\varphi, \tag{12.1.9}$$

which can immediately be integrated. From a mathematical point-of-view finding the integrating factor may present a certain problem. In principle we could find it by solving the following PDE:

$$\frac{\partial \mu f}{\partial y} = \frac{\partial \mu g}{\partial x}, \tag{12.1.10}$$

which follows from Eq. (12.1.8) if Schwarz' theorem is observed. However, it is impossible to find a general solution to this PDE. On the other hand for a specific choice of $f$ and $g$ we only need one particular solution $\mu(x, y)$. Thus special functional forms of $\mu(x, y)$ are tried, for example:

$$\mu = \mu(x), \ \mu = \mu(y), \ \mu = \mu(xy), \ \mu = \mu\left(\frac{x}{y}\right), \text{ etc.} \tag{12.1.11}$$

These are useful if they convert the PDE of Eq. (12.1.10) into an ordinary differential equation. Obviously this method is successful only if the functions $f(x, y)$ and $g(x, y)$ are explicitly known. However, this is unfortunately not the case for our physics-based problem (12.1.5)—with the exception of the ideal gas, of course. We shall now stop our mathematical excursion and apply Euler's technique as far as possible to the solution of (12.1.5). To this end we identify:

$$x \to p, \ y \to \upsilon, \ f \to \left.\frac{\partial u}{\partial p}\right|_{\upsilon}, \ g \to \left.\frac{\partial u}{\partial \upsilon}\right|_{p} + p. \tag{12.1.12}$$

We multiply Eq. (12.1.5) by the integrating factor $1/T = 1/T(p, \upsilon)$ and denote the corresponding potential by the symbol $s = s(p, \upsilon)$:

$$\frac{\mathrm{d}s}{\mathrm{d}t} = \frac{1}{T}\left[\left.\frac{\partial u}{\partial p}\right|_{\upsilon}\frac{\mathrm{d}p}{\mathrm{d}t} + \left(\left.\frac{\partial u}{\partial \upsilon}\right|_{p} + p\right)\frac{\mathrm{d}\upsilon}{\mathrm{d}t}\right]. \tag{12.1.13}$$

Hermann Amandus SCHWARZ was born on January 25, 1843 in Hermsdorf, Silesia and died on November 30, 1921 in Berlin. He started to study chemistry at what is today's Technical University of Berlin. However, he switched to mathematics quite soon. He finished his Ph.D. in 1864 and became a research assistant at the University of Halle. In 1869 he went to the ETH in Zürich. During 1875–1892 he worked at the University in Göttingen and, finally, became the successor to WEIERSTRASS' chair at the University of Berlin. He stayed there until 1917. Rumor has it that SCHWARZ started from very specialized problems but successively extended the methods of solution so that they finally went far beyond the special case.

The choice of symbols is not accidental. Indeed, $T$ corresponds to absolute temperature and $s$ is the specific entropy, as we shall see in Exercise 12.1.1. We insert the result in Eq. (12.1.3) and find:

$$\rho T \frac{\mathrm{d}s}{\mathrm{d}t} = -\frac{\partial q_j}{\partial x_j} - \pi \frac{\partial \upsilon_i}{\partial x_i} + \Sigma_{ji} \frac{\partial \upsilon_i}{\partial x_j}, \tag{12.1.14}$$

or:

$$\rho \frac{\mathrm{d}s}{\mathrm{d}t} = -\frac{\partial q_j / T}{\partial x_j} - \frac{q_j}{T^2} \frac{\partial T}{\partial x_j} - \frac{\pi}{T} \frac{\partial \upsilon_i}{\partial x_i} + \frac{\Sigma_{ji}}{T} \frac{\partial \upsilon_i}{\partial x_j}. \tag{12.1.15}$$

**Exercise 12.1.1:** ***The integrating factor for the case of an ideal gas***

Statistical mechanics of multi-particle systems allows to calculate the so-called *grand canonical sum* for the case of ideal gas particles analytically. It can then be used to determine the internal energy, $U$, as a function of the pressure, $p$, and of the gas volume, $V$, up to a constant (see, e.g., [2], §1.9):

$$U = \zeta p V + U_0, \quad \zeta = \left\{ \begin{matrix} \frac{3}{2} & & \text{1-} \\ \frac{5}{2} & \text{for} & \text{2-} \\ 3 & & \text{multi-} \end{matrix} \right\} \text{atomic particles.} \tag{12.1.16}$$

$\zeta$ is a number characteristic of the type and of the corresponding degrees of freedom of the ideal gas particles (see also Sect. 6.5). Use this result in context with Eq. (12.1.13) and determine the integrating factor $T^{-1}(p, \upsilon)$ for the case of ideal gases. More specifically, inspired by Eq. (12.1.11), use power functions of the following kind:

$$\frac{1}{T} = a p^b, \ \frac{1}{T} = a \upsilon^b, \ \frac{1}{T} = a (p\upsilon)^b, \ \frac{1}{T} = a \left(\frac{p}{\upsilon}\right)^b. \tag{12.1.17}$$

Show that all of these choices will lead to contradictions with the exception of the third one for $b = -1$. Thus we conclude for *mathematical reasons* that:

$$pv = aT. \qquad (12.1.18)$$

Recall now the thermal equation of state for ideal gases (6.4.1), which was established empirically by the experiments of BOYLE, MARIOTTE, GAY-LUSSAC, and AMONTONS. Conclude that the integrating factor is indeed (absolute) temperature and the remaining factor $a$ is given by:

$$a = \frac{R}{M}. \qquad (12.1.19)$$

Robert BOYLE was born on February 4, 1627 in Lismore (Scotland) and died on January 10, 1692 in London. He was the son of an Irish nobleman who took a strong interest in the then new natural philosophy. He worked experimentally in chemistry as well as in physics. He was the first chemist to produce a gas other than air, namely hydrogen, by treating some metals with acids, but he did not name the gas. His contribution to thermodynamics is BOYLE's law: In a gas at constant temperature the volume is inversely proportional to pressure. BOYLE was a zealous religious advocate—he had met God in a particularly intense thunderstorm while traveling in Switzerland—and in his lectures he defended Christianity against "notorious infidels," namely atheists, theists, pagans, jews, and muslims.

Edme MARIOTTE was born in Til-Châtel ca. 1620 and died in Paris on May 12, 1684. He was at least as devout as BOYLE. More specifically, he was a priest and Prior of St. Martin sous Beaune. He rediscovered BOYLE's law and he extended it by noting that, at a constant pressure, the volume of a gas grows with temperature. BOYLE had not made that observation or, at least, he did not mention it. In that early age of natural science most scientists did not limit their attention to a single field of research and so MARIOTTE was also a keen meteorologist and physiologist. He discovered the circulation of the Earth's water supply and the "blind spot" of the eye.

Guillaume Amontons was born on Aug. 31, 1663, in Paris where he also died on October 11, 1705. He was an experimental physicist, known for his work on temperature and friction and inventor of scientific instruments for the corresponding measurements. In particular, he developed an air-pressure thermometer in 1702 and published two relevant papers on thermometry. In context with the ideal gas law he was the one measuring a change in temperature in terms of a proportional change in pressure at a constant mass and volume (of air). By doing this he anticipated the concept of absolute zero temperature.

In view of Eq. (3.7.3) and the balance of mass $(3.8.3)_1$, Eq. (12.1.15) must be interpreted as the local balance of entropy in regular points:

$$\frac{\partial \rho s}{\partial t} + \frac{\partial}{\partial x_j}\left(\rho\, s\, \upsilon_i + \frac{q_j}{T}\right) = -\frac{q_j}{T^2}\frac{\partial T}{\partial x_j} - \frac{\pi}{T}\frac{\partial \upsilon_i}{\partial x_i} + \Sigma_{ji}\frac{\partial \upsilon_i}{\partial x_j}. \tag{12.1.20}$$

Alternatively we may integrate Eq. (12.1.15) w.r.t. a material system, observe Eq. (3.8.16) and GAUSS' theorem (3.4.2). This will lead to the following global balance:

$$\begin{aligned} \frac{\mathrm{d}}{\mathrm{d}t}\iiint\limits_{V(t)} \rho\, s\, \mathrm{d}V = &- \oiint\limits_{\partial V(t)} \frac{q_j}{T} n_j\, \mathrm{d}A + \\ &\iiint\limits_{V(t)} \left(-\frac{q_j}{T^2}\frac{\partial T}{\partial x_j} - \frac{\pi}{T}\frac{\partial \upsilon_i}{\partial x_i} + \Sigma_{ji}\frac{\partial \upsilon_i}{\partial x_j}\right) \mathrm{d}V. \end{aligned} \tag{12.1.21}$$

Obviously the first term after the equal sign represents the *non-convective flux of entropy*. The second term is the *entropy production*. There is no volume supply of entropy because we cannot control the development of the heat flux, of temperature, of the stress deviator, or of the velocity gradient within the body (see Sect. 3.3 where the difference between a production and a supply was explained). All of these will simply develop inside the system after initial and boundary conditions have been specified.

The "derived" entropy balances (12.1.15, 12.1.20 and 12.1.21) resulted basically from the assumption that the thermodynamic process is governed by two mechanical *state variables*, the thermodynamic pressure $p$ and the (specific) volume $\upsilon$. Two comments are now in order. First, we may use the integrating factor $T(p, \upsilon)$ as a state variable and replace one of the two mechanical variables by a thermodynamic quantity, namely the temperature $T$. Thus possible sets of state variables are either $(p, T)$ or $(\upsilon, T)$. These choices have certain advantages, which we shall learn to appreciate during the rational setup of measurements required to determine the internal energy of real materials in the vicinity of thermodynamic equilibrium. Of course, in order to give a physical meaning to such a substitution, we must establish rules of measurement for the temperature. In the case of pressure and volume we did not even mention these since they are easy to define. However, how to measure temperature is neither a trivial task nor immediately evident.

Second, it is by no means obvious that the specific internal energy depends only on two state variables. This choice may be sufficient if processes are concerned that are close to thermodynamic equilibrium. Indeed, the local state of a one-component material in thermodynamic equilibrium can be described completely by only two variables, for example volume and pressure. But there are more complicated materials and processes associated with them than that.

In fact, this simple choice of variables is pertinent to the so-called $p\,\mathrm{d}\upsilon$-thermodynamics, which is, in many cases, perfectly sufficient for quantification of the performance of fast operating engineering machines (in this context also see the remarks in Sects. 7.2 and 7.5). It is surprising that it works, because during such processes there will certainly be gradients of pressure, velocity, or temperature, which might be important for the description of a current state and of the whole process. After all, note that these processes cannot simply be reversed: They have a certain direction in time and reversing them is an extreme, sometimes impossible effort. In other words, they are irreversible. Indeed, there are technically important processes that cannot be described within the framework of $p\,\mathrm{d}\upsilon$-thermodynamics. Heat conduction problems are just one example.

It seems obvious that, in non-equilibrium, the internal energy must depend on additional variables beyond pressure and volume (or temperature). However, the question is which variables this should be, what kind of dependencies are relevant and, finally, how can these dependencies be assessed experimentally, in particular in context with non-mechanical, diva-like quantities, such as temperature. Indeed, from a mathematical point-of-view it would be possible without major problems to generalize and extend the aforementioned procedure of the integrating factor to more than two variables. However, this could easily degenerate into a physically unmotivated, rigid formalism, totally useless for engineering applications.

Two phenomenological methods have been established in the literature in order to avoid such problems. The first one is known as *Thermodynamics of Irreversible Processes*, or TIP for short. In fact we just presented some of its arguments following the early paper of Eckart [1]. One of TIP's essential features is the so-called *hypothesis of local thermodynamic equilibrium*: Even if in the considered processes velocity and temperature gradients, i.e., non-equilibrium phenomena, are present, it is sufficient to characterize the state of a material particle and the state functions required for its description (e.g., its specific internal energy) just by pressure and by specific volume (or alternatively by temperature), at least if we consider a heat conducting, viscous gas or liquid.[1] In a moment we shall see how the entropy relations (12.1.15 / 12.1.20 / 12.1.21) of TIP can be used to find constraints regarding the possible mathematical form of constitutive quantities like the specific internal energy, the heat flux, and the stress (deviator).

Thus in TIP entropy is introduced not for sheer academic pleasure. Rather entropy provides us a *service* and helps to reduce the amount of measurements required to determine constitutive relations. Moreover, as we shall see, it also allows for a quantitative interpretation of the state of order or rather disorder of a system and quantifies how much it takes to reverse a given thermodynamic processes. This is particularly useful when characterizing technical systems, where waffling and hand-waving would not do us any good.

[1] In order to describe the behavior of solids we have to introduce strain or its trace. Moreover, it may be advisable to work with other than linear strain measures. However, we shall not go into details here.

Bernard David COLEMAN was born in 1930. He obtained his B.S. at Indiana University in 1951. In 1952 he completed his M.S. and two years later his Ph.D., both from Yale. He was appointed to various chairs in mathematics, biology, and chemistry. Since July 1988 he is the J. Willard GIBBS Professor of Thermomechanics at Rutgers University. One of his essential contributions to physics concerns a rational exploitation of the entropy principle, known as COLEMAN-NOLL-procedure in the pertinent literature.

The second method—also aiming at a unified description of thermodynamic processes—concerns the so-called rational materials theories of thermodynamic non-equilibrium. They were presented by pioneers like COLEMAN, TRUESDELL, or NOLL for the first time, and they can be found in compiled form in the books by Haupt [3] or Müller [4, 5]. However, in this book intended for beginners we shall not elaborate on these theories and say only this: One of the main foundations of Rational Thermodynamics is the so-called *entropy principle*, which can be considered as a generalization of Eq. (12.1.21). First, the entropy principle postulates that entropy is an additive quantity which can be balanced. This is expressed mathematically by the following global equation:

$$\frac{\mathrm{d}}{\mathrm{d}t}\iiint\limits_{V(t)} \rho\, s\, \mathrm{d}V = -\oint\!\!\!\oint\limits_{\partial V(t)} \phi_j n_j\, \mathrm{d}A + \iiint\limits_{V(t)} z\, \mathrm{d}V + \iiint\limits_{V(t)} \sigma\, \mathrm{d}V. \tag{12.1.22}$$

Walter NOLL was born on January 7, 1925 in Berlin. In 1946 he began to study engineering at the Technical University of Berlin. He finished his studies with a diploma degree in 1951. After that he became a scientific assistant for four years at the Institute of Technical Mechanics under Istvan SZABO. However, his enthusiasm for mathematics prevailed and so he decided in 1953 to move to the U.S. and join Clifford TRUESDELL's group at Indiana University in Bloomington. There he obtained a Ph.D. in Applied Mathematics in 1954. In 1955 he decided to emigrate completely, spent that year at the University of Southern California and finally accepted a professorship at the Mathematics Department of Carnegie Mellon University in 1956, He became an emeritus in 1993. His scientific achievements are at the borderline between mathematics and mechanics where he contributed in particular to the rational foundations of the latter.

All the quantities in that equation, the specific entropy, $s$, the entropy flux vector, $\phi_j$, the volume supply of entropy, $z$, and also the entropy production, $\sigma$, are (unknown) constitutive functions of a suitable set of state variables. In other words, the entropy flux is not a priori given by $q_j/T$ and the volume supply is not automatically equal to $\rho r/T$ ($r$: specific radiation density). Rather these results might follow under certain assumptions from exploitation of the entropy principle. The second important requirement during the exploitation consists of the entropy production being positive-semidefinite for all thermodynamically admissible processes:

$$\sigma \geq 0. \tag{12.1.23}$$

We may call this the Second Law of thermodynamics. Finally, the third requirement concerns the continuity of the entropy flux at heat conducting walls at rest:

$$[\![\phi_j]\!] e_j = 0. \tag{12.1.24}$$

This relation is very useful when defining (absolute) temperature as a measureable quantity.

**Exercise 12.1.2:** ***The entropy of the ideal gas***

Start from Eqs. (12.1.5 / 12.1.13) and combine them to the so-called GIBBS's equation:

$$T\,\mathrm{d}s = \mathrm{d}u + p\,\mathrm{d}\upsilon, \tag{12.1.25}$$

where the specific entropy, $s$, the specific internal energy, $u$, and the pressure, $p$, are functions of two state variables, namely $(p, \upsilon)$, or $(p, T)$, or $(\upsilon, T)$, respectively. In particular, choose the thermal and the caloric equations of state for the ideal gas according to Eqs. (6.4.1) and (6.5.2), and integrate Eq. (12.1.25) between a reference state "0" and the current state. Show that the specific entropy of the ideal gas is given by:

$$s - s_0 = \zeta\frac{R}{M}\ln\left(\frac{T}{T_0}\right) + \frac{R}{M}\ln\left(\frac{\upsilon}{\upsilon_0}\right),\quad s - s_0 = (\zeta + 1)\frac{R}{M}\ln\left(\frac{T}{T_0}\right) - \frac{R}{M}\ln\left(\frac{p}{p_0}\right)$$
$$s - s_0 = \zeta\frac{R}{M}\ln\left(\frac{p}{p_0}\right) + (\zeta + 1)\frac{R}{M}\ln\left(\frac{\upsilon}{\upsilon_0}\right), \tag{12.1.26}$$

depending on the choice of state variables each version is particularly suited for describing a given problem.

## 12.2 Entropy as a Measure of (Dis-)Order and (Ir-)Reversibility

We have just learned that the concept of entropy will put us in a position to restrict the possible form of constitutive relations. Before we explore this any further we shall see that entropy also offers a quantitative measure to characterize the state of (dis-)order in a systems as well as the degree of (ir-)reversibility of a process. We shall demonstrate this by means of simple examples.

First, we shall consider the vessel shown in Fig. 12.1, which is divided by a sliding wall into two volumes of equal size, $V$. One half is filled with an ideal gas,

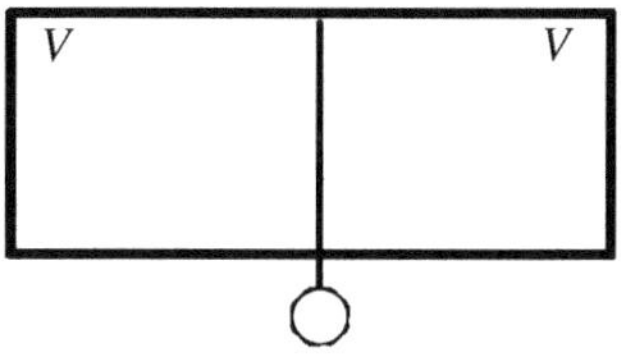


**Fig. 12.1** Two chambers one filled with an ideal gas, separated by a sliding wall

the other one is empty. At the start of the process the gas is characterized by an initial pressure $p_s$ and an initial temperature $T_s$. According to the thermal equation of state for ideal gases (6.4.1) we may then write for its specific volume:

$$\upsilon_s = \frac{R}{M}\frac{T_s}{p_s}. \tag{12.2.1}$$

We now remove the slide. Turbulent flow will set in and after some time the ideal gas will come to rest again due to internal friction. This process is obviously irreversible: The gas will not "by itself" reorganize and occupy its original space. We want to calculate by how much entropy has grown after this process. The mass of the gas is constant and therefore we may as well calculate the change of the specific entropy and make use of Eq. (12.1.26). In principle we can use any of the three representations. We simply replace symbols referring to the current state by quantities characteristic of the end of the process and the 0-state by the corresponding quantities at the start. The latter ones are prescribed, the former ones still need to be determined. To this end we apply the global energy balance (also see the remarks in Sect. 7.2 and in Exercise 7.2.1) to the total volume and find that in the end:

$$T_e = T_s, \tag{12.2.2}$$

i.e., the temperature does not change. We may interpret this result to that effect that on the one hand side the ideal gas will cool down if its volume is increased. On the other hand side the internal friction will heat it up. Both effects are equally strong. This is also known as the JOULE-THOMSON effect in the literature. The ideal gas law (6.4.1) allows us now to calculate the final pressure:

$$p_e = \frac{1}{2}p_s. \tag{12.2.3}$$

Thus the specific entropy will increase. Equation $(12.1.26)_2$ tells us by how much:

$$s_e - s_s = \zeta\frac{R}{M}\ln(2). \tag{12.2.4}$$

The *increase* of entropy makes immediate sense due to the apparent irreversibility of the process. Intuitively we also expect a decrease of the amount of order. Indeed, the amount of order must decrease because after removal of the slide the gas has twice as much space than before. This also explains the factor of 2 in Eq. (12.2.4). We can also interpret the effect probabilistically: BOLTZMANN developed the following famous formula according to which entropy can be calculated from

$$S = k\,\ln(W), \tag{12.2.5}$$

where the probability $W$ is related to the number of possible realizations of system states.

William THOMSON, 1. Baron KELVIN, a.k.a. LORD KELVIN or KELVIN OF LARGS was born on June 26, 1824 in Belfast, Ireland and died on December 17, 1907 in Netherhall near Largs, Scotland. THOMSON attended Glasgow University already at the young age of ten. In 1841 he went to Cambridge to obtain B.A. in 1845. He became a Professor of Natural Philosophy at the University of Glasgow in 1846, where he remained for the rest of his scientific career. Thomson was an extremely versatile physicist working theoretically as well as experimentally. Particularly worth mentioning are his attempts to estimate the age of the Earth and his contributions to setting up the first transatlantic telegraph cable. He published more than 600 papers, was elected to the Royal Society in 1851 and remained its president from 1890 until 1895.

We will now extend the argument slightly and imagine that the second half of the vessel is now also filled with an ideal gas of the same type. For simplicity we assume that the sliding wall is permeable to heat. In other words the temperatures in both sections will initially be equal and given by $T_s$. The pressure in the second half differs by a positive factor $\alpha$ from the other one and is initially equal to $\alpha p_s$. We pull the slide, both gases will mix turbulently due to the difference in pressure. After a while they will come to rest due to internal friction. As before we ask by how much the state of order, i.e., entropy, has changed. As before it is intuitively clear that the state of order must have decreased, i.e., entropy must have increased, because we do not observe that the gases will rearrange spontaneously into two regions with different pressures.

For the computation it is, first, important to note that the masses in both halves of the vessel are different: If we apply the ideal gas law to the initial state in both sections, we conclude that the mass in the section with the pressure $\alpha\, p_s$ must differ by a factor $\alpha$ from the mass $m$ in the other half. If we apply the global energy balance to the total chamber we obtain the result (12.2.2) again, i.e., the temperature does not change as a consequence of the mixing. The final pressure is obtained from the ideal gas law:

$$p_e = \frac{1+\alpha}{2} p_s. \tag{12.2.6}$$

Note that for $\alpha = 0$ the previous case is recovered. Equation $(12.1.26)_3$ results in the following change of entropy if the two different masses are taken into account:

$$\Delta S = S_e - S_s = m\frac{R}{M}\left[\ln\left(\frac{2}{1+\alpha}\right) + \alpha\ln\left(\frac{2\alpha}{1+\alpha}\right)\right]. \tag{12.2.7}$$

It is not immediately obvious that the expression is always positive for all values $0 \le \alpha < \infty$. The practical engineer shuns and despises the general mathematical proof and solves the problem graphically: As can be seen in Fig. 12.2 the

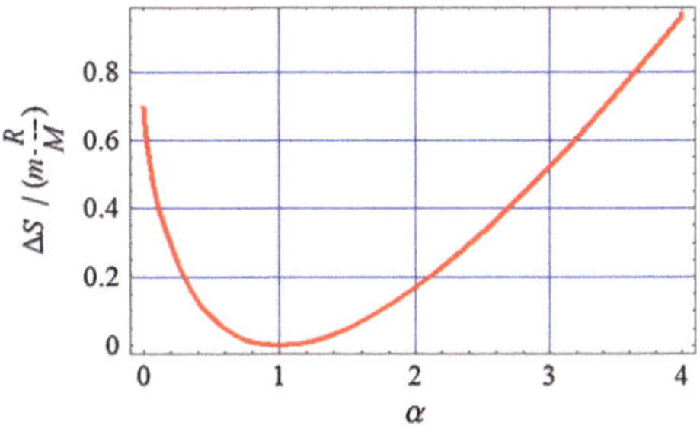


**Fig. 12.2** Growth of entropy after mixing of two ideal gases of different initial pressures (see text)

growth of entropy is *always* positive. For the value $\alpha = 1$ it is exactly equal to zero, as expected for the same pressure in both sections.

Now we imagine that both sections of the vessel are filled with two ideal gases of different type 1 and 2, respectively. For simplicity the starting pressure and the starting temperature are assumed as equal. Application of the ideal gas law shows that the masses in both parts must be related to the molecular weights as follows:

$$\frac{m_1}{m_2} = \frac{M_1}{M_2}. \tag{12.2.8}$$

After pulling the slide both gases will diffuse into each other and mix homogeneously. This time the reason for mixing is neither a difference in pressure nor in temperature. An in-depth analysis within the framework of a theory of mixtures would show that the driving force is based on different, so-called *chemical potentials* in both halves. This enforces homogenization. We will not get into this any further and argue phenomenologically instead: Nature prefers states of higher disorder and the corresponding increase in entropy can be calculated from Eq. $(12.1.26)_3$:

$$\Delta S = S_e - S_s = (N_1 + N_2)k \ln(2). \tag{12.2.9}$$

During the derivation the thermal equation of state in the form $(6.4.10)_1$, Eq. (12.2.8), and

$$m_{1,2} = N_{1,2} M_{1,2} \mu_H \tag{12.2.10}$$

were taken into account. $N_{1,2}$ denotes the number of particles and $\mu_H$ refers to the mass of the hydrogen atom. Obviously entropy grows linearly depending on the number of particles available for mixing. This makes immediate sense because the possibility to increase disorder will grow with increasing number of particles. On the other hand Eq. (12.2.9) carries an intrinsic problem: It holds in exactly the same form if two gases of the same type, the same temperature, and at the same pressure are initially present. If we now pull the slide we cannot distinguish on a macroscopic level between the initial and the final state. Consequently, there should be no change in entropy. However, Eq. (12.2.9) claims there is one. This dilemma is known as the GIBBS paradox in literature. Rumor has it that quantum mechanics arguments are required to explain it. But even if this were possible, it somehow discredits Eq. (12.1.26) for computing the change in entropy. Nevertheless, it should not shatter our belief in the usefulness and interpretability of entropy. In order to gain further reassurance we will now tend to two more complex examples.

The first one concerns the calculation of the entropy difference for the heavy oscillating piston under the influence of gravity from Sect. 7.2. For simplicity we shall neglect the mass of the gas, $m_g$, the external pressure, $p_0$, and the specific heat, $c_\varepsilon$, of the piston in Eq. (7.2.22). Under these circumstances we obtain:

$$z_e = \frac{m_p g + p_s A\zeta}{(\zeta+1)m_p g} z_s, \quad T_e = \frac{m_p g + p_p A\zeta}{(\zeta+1)p_p A} T_s. \tag{12.2.11}$$

This can be inserted into Eq. $(12.1.26)_1$ to compute the difference between the specific entropies, which due to the homogeneity of the situation is proportional to the difference in entropy. We define the following positive factor:

$$0 \leq x = \frac{m_p g}{p_s A} < \infty, \tag{12.2.12}$$

and find:

$$\Delta S = S_e - S_s = Nk\left[\zeta \ln\left(\frac{x+\zeta}{\zeta+1}\right) + \ln\left(\frac{x+\zeta}{\zeta+1}\frac{1}{x}\right)\right]. \tag{12.2.13}$$

Again it is hard to see that the difference is always positive for all possible values of $x$, independently of the choice of $\zeta$. As before, we provide a "proof" by means of a graphical solution, which is shown in Fig. 12.3.

We realize that the curves are always positive and almost coincide for the three possible choices of $\zeta$. Moreover, the entropy difference is equal to zero for $x = 1$. This makes sense because in this case the piston will not move at all. Consequently a slight difference between the weight of the piston and the inner pressure will lead to almost no increase of entropy. This confirms the conclusions from our previous analyses by means of the adiabatic equation in from Sects. 7.2 and 7.5. Recall that the adiabatic equation (6.5.29) is a consequence of $p\,\mathrm{d}v$-thermodynamics, which as explained above is reversible *a priori*.

The second example analyzes the growth of entropy after the turbulent mixing process of two ideal gases separated in two chambers that are initially at different pressure but the same temperature. The details of the situation are described in Exercise 7.2.1 and the results are presented in what follows.

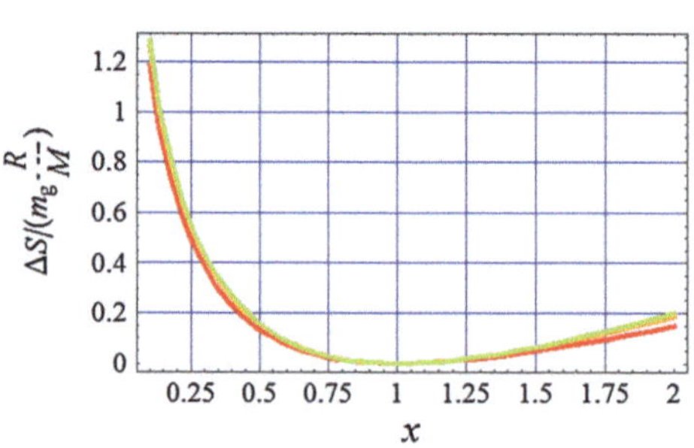


**Fig. 12.3** Growth of entropy of an ideal gas in an adiabatically sealed cylinder compressed by a falling piston for the three possible choices of $\zeta$ (see text)

**Exercise 12.2.1:** ***Growth of entropy after equilibrating the pressure between two separate gas chambers***

Use the results compiled in Eq. (7.2.29) from Exercise 7.2.1, define the following positive, dimensionless parameters

$$0 \leq x = \frac{V_2^s}{V_1^s} < \infty, \quad 0 \leq y = \frac{N_2}{N_1} < \infty \tag{12.2.14}$$

in combination with Eq. $(12.1.26)_1$ and show that the difference in entropy is given by:

$$\Delta S = S_e - S_s = N_1 k \left[ \ln\frac{1+x}{1+y} + y \ln\frac{y(1+x)}{x(1+y)} \right], \tag{12.2.15}$$

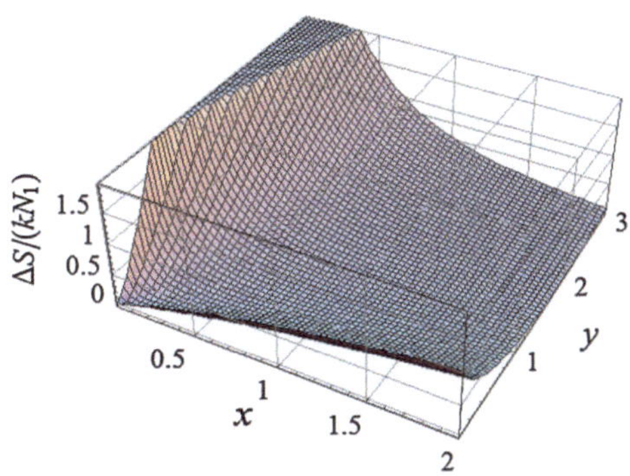


**Fig. 12.4** Growth of entropy after equilibrating the pressure acting between the two chambers

depending on the choice of state variables.

Prepare a plot (see Fig. 12.4) and show that this expression is positive for all possible choices of $x$ and $y$. Discuss the circumstances for which the entropy production vanishes. What can be said about the case of *adiabatic* dividing piston?

## 12.3 Properties of the Global Entropy Inequality: The Concept of Availability

We now start from the global balance of entropy shown in Eq. (12.1.22), assume that the entropy flux is given by $q_j/T$, and that there is no entropy supply due to radiation. Then it follows that:

$$\frac{\mathrm{d}}{\mathrm{d}t}\iiint\limits_{V(t)} \rho\, s\,\mathrm{d}V + \oiint\limits_{\partial V(t)} \frac{q_j}{T} n_j\,\mathrm{d}A = \iiint\limits_{V(t)} \sigma\,\mathrm{d}V \geq 0. \tag{12.3.1}$$

First, we consider a system with an adiabatic hull for which we may write by definition:

$$q_j \equiv 0 \quad \text{on} \quad \partial V(t). \tag{12.3.2}$$

Then it follows from Eq. (12.3.1) that the entropy can only grow after an internal (irreversible) process in the system has ended. Note that we do not need to know the details of the internal process. Rather we may simply write:

$$\Delta S = S_\mathrm{e} - S_\mathrm{s} = \iiint\limits_{V(t_\mathrm{e})} \rho\, s\,\mathrm{d}V - \iiint\limits_{V(t_\mathrm{s})} \rho\, s\,\mathrm{d}V = \int\limits_{t_\mathrm{s}}^{t_\mathrm{e}} \iiint\limits_{V(t)} \sigma\,\mathrm{d}V\mathrm{d}t \geq 0. \tag{12.3.3}$$

In the previous section we have already provided a few specific examples for this kind of situation. Moreover, we have calculated the entropy production,

$$\Sigma = \int\limits_{t_\mathrm{s}}^{t_\mathrm{e}} \iiint\limits_{V(t)} \sigma\,\mathrm{d}V\mathrm{d}t, \tag{12.3.4}$$

explicitly. We now turn to systems that are not adiabatically sealed. However, the temperature on their surface is assumed as constant and the state of stress on the surface is given by a pressure, which is constant as well, for the whole duration of the process:

$$T = T_0 = \text{const.} \quad \text{and} \quad \sigma_{ij} = -p_0\delta_{ij} = \text{const.} \quad \text{on} \quad \partial V(t). \tag{12.3.5}$$

Equation (12.3.1) leads us to conclude that:

$$-\frac{\mathrm{d}}{\mathrm{d}t}\iiint\limits_{V(t)} T_0 \rho\, s\,\mathrm{d}V - \oiint\limits_{\partial V(t)} q_j n_j\,\mathrm{d}A = -T_0 \iiint\limits_{V(t)} \sigma\,\mathrm{d}V \leq 0. \tag{12.3.6}$$

We eliminate the heat flux with the energy balance (3.9.5), which we may rewrite for the present case as [in particular observe the relations (7.2.8) and (7.2.11)]:

$$\frac{\mathrm{d}}{\mathrm{d}t}\left(U + E_\mathrm{kin} + E_\mathrm{pot} + p_0 V\right) = -\oiint\limits_{\partial V(t)} q_j n_j\,\mathrm{d}A \tag{12.3.7}$$

with:

$$U = \iiint\limits_{V(t)} \rho\, u\,\mathrm{d}V, \quad E_\mathrm{kin} = \iiint\limits_{V(t)} \frac{\rho}{2}\upsilon^2\,\mathrm{d}V, \quad E_\mathrm{pot} = \iiint\limits_{V(t)} \rho\varphi\,\mathrm{d}V. \tag{12.3.8}$$

The following result is obtained:

$$\frac{\mathrm{d}}{\mathrm{d}t}\left(U + E_{\mathrm{kin}} + E_{\mathrm{pot}} + p_0 V - \iiint\limits_{V(t)} T_0 \rho\, s\, \mathrm{d}V\right) = -T_0 \iiint\limits_{V(t)} \sigma\, \mathrm{d}V \leq 0. \tag{12.3.9}$$

It is customary to introduce the so-called *availability* of the system by:

$$\mathfrak{A} = U + E_{\mathrm{kin}} + E_{\mathrm{pot}} + p_0 V - \iiint\limits_{V(t)} T_0 \rho\, s\, \mathrm{d}V. \tag{12.3.10}$$

Obviously the availability of such a system tends to a minimum:

$$\Delta\mathfrak{A} = \mathfrak{A}_{\mathrm{e}} - \mathfrak{A}_{\mathrm{s}} = -T_0 \Sigma \leq 0. \tag{12.3.11}$$

For chemical experiments with open vessels subjected to a constant environmental pressure, $p_0$, where neither kinetic nor potential energies are an issue, the availability reduces to the following expression:

$$\mathfrak{A} = \iiint\limits_{V(t)} \rho\, (u - T_0 s + p_0 \upsilon)\, \mathrm{d}V. \tag{12.3.12}$$

This is easily confused with the so-called GIBBS free energy (sometimes also termed free enthalpy), which is defined as follows:

$$G = \iiint\limits_{V(t)} \rho\, (u - T s + p \upsilon)\, \mathrm{d}V. \tag{12.3.13}$$

Note the subtle difference: In here the *local* temperature and pressure *field inside* of the system must be used and *not* the corresponding values on the surface of the system, which are assumed not to change during the process. Nevertheless, chemical engineers often claim in a sloppy manner that the GIBBS free energy in an open system assumes a minimum.

Frequently chemists also consider closed systems, i.e., sealed reaction vessels of a fixed volume. If we again neglect kinetic and potential energies during a process performed with such a system, Eq. (12.3.10) simplifies to:

$$\mathfrak{A} = \iiint\limits_{V} \rho\, (u - T_0 s)\, \mathrm{d}V. \tag{12.3.14}$$

Erroneously this expression is often confused with the so-called HELMHOLTZ free energy, which is defined as follows:

$$F = \iiint\limits_{V} \rho\, (u - T s)\, \mathrm{d}V, \tag{12.3.15}$$

and despite its obvious difference in comparison with Eq. (12.3.14) we often hear chemical engineers say that the HELMHOLTZ free energy of a closed systems must turn into a minimum.

## 12.4 Reduction of the Constitutive Equations for a Viscous Heat-Conducting Fluid

As we shall realize now the concept of entropy, in combination with the GIBBS equation, can be used to reduce the amount of calorimetric measurements required for determination of the specific internal energy as a function of two state variables considerably. For calorimetric measurements it is most appropriate to include temperature in the pair of variables. We therefore choose specifically the pair $(T, \upsilon)$. Indeed, one does not measure $u = u(T, \upsilon)$ directly. Rather it is determined from its derivatives, as already indicated by the remarks in Sect. 6.5. We write:

$$u = u(T, \upsilon) \Rightarrow \mathrm{d}u(T, \upsilon) = \left.\frac{\partial u}{\partial T}\right|_{\upsilon} \mathrm{d}T + \left.\frac{\partial u}{\partial \upsilon}\right|_{T} \mathrm{d}\upsilon. \tag{12.4.1}$$

Thus we will obtain the specific internal energy for viscous, heat-conducting gases and fluids by (numerical) integration up to an additive constant, if we only determine the derivatives on the right hand side on the second equation experimentally:

$$u(T, \upsilon) = \int \left.\frac{\partial u}{\partial T}\right|_{\upsilon} \mathrm{d}T + \int \left.\frac{\partial u}{\partial \upsilon}\right|_{T} \mathrm{d}\upsilon + \text{const.} \tag{12.4.2}$$

In fact, the derivatives can be related to the specific heats from Sect. 6.5. These are known by measuring the change in temperature after energy in form of heat has been added to the system in a controlled manner. Indeed, the first derivative in Eq. (12.4.2) is nothing else but the specific heat at a constant volume, which has already been introduced in Eq. (6.5.12). We will relate the second derivative with the specific heats at a constant volume and at a constant pressure in combination with the pressure, i.e., the thermal equation of state. In order to learn how, we start from the First Law for "slow processes" according to Eq. (6.5.10) and replace $\mathrm{d}\upsilon$ by the thermal equation of state $[p = p(T, \upsilon) \Rightarrow \upsilon = \upsilon(p, T)]$, so that:

$$\mathrm{d}\upsilon(p, T) = \left.\frac{\partial \upsilon}{\partial T}\right|_{p} \mathrm{d}T + \left.\frac{\partial \upsilon}{\partial p}\right|_{T} \mathrm{d}p. \tag{12.4.3}$$

This leads to:

$$-\frac{1}{\rho}\frac{\partial q_j}{\partial x_j}\mathrm{d}t = \left(c_\upsilon + \left(\left.\frac{\partial u}{\partial \upsilon}\right|_{T} + p\right)\left.\frac{\partial \upsilon}{\partial T}\right|_{p}\right)\mathrm{d}T + \left(\left.\frac{\partial u}{\partial \upsilon}\right|_{T} + p\right)\left.\frac{\partial \upsilon}{\partial p}\right|_{T}\mathrm{d}p. \tag{12.4.4}$$

For $\mathrm{d}p = 0$, the left hand side obviously represents the specific heat provided at a constant pressure, i.e., during isobaric processes. Simple algebraic manipulations allow us now to write:

$$c_p = c_\upsilon + \left(\left.\frac{\partial u}{\partial \upsilon}\right|_T + p\right)\left.\frac{\partial \upsilon}{\partial T}\right|_p \Rightarrow \left.\frac{\partial u}{\partial \upsilon}\right|_T = \frac{c_p - c_\upsilon}{\left.\frac{\partial \upsilon}{\partial T}\right|_p} - p. \tag{12.4.5}$$

Equation (12.4.2) may thus be rewritten as follows:

$$u(T,\upsilon) = \int c_\upsilon \mathrm{d}T + \int \left[\frac{c_p - c_\upsilon}{\left.\frac{\partial \upsilon}{\partial T}\right|_p} - p\right]\mathrm{d}\upsilon + \text{const.} \tag{12.4.6}$$

This equation shows explicitly that the internal energy can indeed be determined if only the thermal equation of state and the specific heats at a constant volume and at a constant pressure are known. Note that the specific heats must be known *for every set of data* $\upsilon$, $T$. In other words, until now there is no other way but measuring the specific heats for every pair by means of calorimetry applied to every gas or fluid. This is a substantial effort and from the experimental point-of-view by no means trivial, because we have to pay attention that the heat is really supplied to the substance of interest and does not simply vanish in the container, the environment, etc.

However, if we now make use of the concept of entropy in combination with the Gibbs equation, the amount of measurements required can be reduced dramatically. We start from Eq. (12.1.25) and write:

$$\begin{gathered} \mathrm{d}s(T,\upsilon) = \frac{1}{T}[\mathrm{d}u(T,\upsilon) + p\,\mathrm{d}\upsilon] \Rightarrow \\ \left.\frac{\partial s}{\partial T}\right|_\upsilon \mathrm{d}T + \left.\frac{\partial s}{\partial \upsilon}\right|_T \mathrm{d}\upsilon = \frac{1}{T}\left[\left.\frac{\partial u}{\partial T}\right|_\upsilon \mathrm{d}T + \left(\left.\frac{\partial u}{\partial \upsilon}\right|_T + p\right)\mathrm{d}\upsilon\right]. \end{gathered} \tag{12.4.7}$$

By comparing the corresponding terms on both sides we find that:

$$\left.\frac{\partial s}{\partial T}\right|_\upsilon = \frac{1}{T}\left.\frac{\partial u}{\partial T}\right|_\upsilon, \quad \left.\frac{\partial s}{\partial \upsilon}\right|_T = \frac{1}{T}\left(\left.\frac{\partial u}{\partial \upsilon}\right|_T + p\right). \tag{12.4.8}$$

We differentiate the first expression w.r.t. $\upsilon$ and the second one w.r.t. $T$. According to Schwarz' theorem for continuously differentiable functions the sequence of both derivatives does not matter. We assume that entropy and internal energy do have that property and conclude that:

$$\left.\frac{\partial u}{\partial \upsilon}\right|_T = -p + T\left.\frac{\partial p}{\partial T}\right|_\upsilon. \tag{12.4.9}$$

In view of Eq. (12.4.5) we conclude that it is no longer required to know and measure both specific heats for all specific volumes at a constant temperature, because $\left.\frac{\partial u}{\partial \upsilon}\right|_T$ can already be determined completely from the thermal equation of state $p = p(\upsilon, T)$, which we assume to be known.

However, there is more to be concluded from the concept of entropy and from the GIBBS equation. If we differentiate Eq. (12.4.9) by $T$ we find that:

$$\frac{\partial^2 u}{\partial T \partial \upsilon} = T \left.\frac{\partial^2 p}{\partial T^2}\right|_{\upsilon} \equiv \left.\frac{\partial c_{\upsilon}}{\partial \upsilon}\right|_{T} \Rightarrow c_{\upsilon}(\upsilon, T) = \int T \left.\frac{\partial^2 p}{\partial T^2}\right|_{\upsilon} \mathrm{d}\upsilon + f(T). \tag{12.4.10}$$

This shows that the dependence of the specific heat at a constant volume, $c_{\upsilon}$, of the specific volume results by integration from the known thermal equation of state. Thus we conclude that it is sufficient to measure $c_{\upsilon}$ at *one* specific volume $\upsilon$ and *all* temperatures $T$ in order to fix the last unknown function $f(T)$, which depends only of the temperature, $T$.

Finally in this section we turn to the right hand side of Eq. (12.1.15). In particular we will now examine the entropy production, $\sigma$, for which we may write according to Eq. (12.1.23):

$$T\sigma = -\frac{q_j}{T}\frac{\partial T}{\partial x_j} - \pi \frac{\partial \upsilon_i}{\partial x_i} + \Sigma_{ji}\frac{\partial \upsilon_i}{\partial x_j} \geq 0. \tag{12.4.11}$$

The requirement of the positive-semi-definite of the inequality is guaranteed within the framework of TIP by relating so-called *fluxes* (in our case the heat flux, the dynamic pressure, and the stress deviator) in a *linear* manner to the so-called driving *forces* (in our case the temperature gradient, the divergence of velocity, and the deviatoric parts of the velocity gradients):

$$q_j = -\kappa \frac{\partial T}{\partial x_j}, \quad \pi = -\lambda \frac{\partial \upsilon_i}{\partial x_i}, \quad \Sigma_{ji} = \mu \left( \frac{\partial \upsilon_i}{\partial x_j} + \frac{\partial \upsilon_j}{\partial x_i} - \frac{2}{3}\frac{\partial \upsilon_k}{\partial x_k}\delta_{ij} \right). \tag{12.4.12}$$

In the last relation it has been guaranteed that the stress deviator is symmetric and trace free. We have already encountered the heat conduction coefficient, $\kappa$, the (quasi) bulk viscosity, $\lambda$, and the shear viscosity, $\mu$, in Eqs. (6.6.1) and (6.3.1). According to the entropy principle they must all be *positive* parameters that potentially depend on temperature. In summary, the entropy principle puts us in a position to *reduce* the possible form of constitutive relations considerably.

On the other hand Eq. (12.4.11) can also be used to find out how large the local entropy production is, i.e., to determine the intensity of local irreversibility. Of course, for this purpose we must know the constitutive equations for the heat flux and for the stress deviator as well as the temperature and the velocity field for the corresponding system as functions of time and space. The latter may be hard to achieve, in particular, if turbulent flow is concerned, as we discussed in the Examples of Sects. 12.2 and 12.5. Interestingly it was possible to calculate the total entropy production, i.e., the dissipation integrated w.r.t. space and time *in closed form* [see Eq. (12.3.4)] just from the initial and final state of the system. Both were homogeneous *states of equilibrium*. Specific results are compiled in Eqs. (12.2.4, 12.2.7, 12.2.9 / 12.2.13).

However, for *stationary* processes it is sometimes also possible to compute $\sigma$ in closed form. An example of this is provided by the stationary parallel flow between plates from Exercise 7.4.1. Without anticipating the proof required in the exercise we note that velocity field for laminar flow in Cartesian coordinates is given by:

$$\upsilon_i = \left(V\frac{x_2}{h}, 0, 0\right). \tag{12.4.13}$$

Then the stress deviator and the velocity gradient can be calculated from the NAVIER–STOKES constitutive relation:

$$\Sigma_{ji} = \begin{pmatrix} 0 & \mu\frac{V}{h} & 0 \\ \mu\frac{V}{h} & 0 & 0 \\ 0 & 0 & 0 \end{pmatrix}, \quad \frac{\partial \upsilon_i}{\partial x_j} = \begin{pmatrix} 0 & \frac{V}{h} & 0 \\ 0 & 0 & 0 \\ 0 & 0 & 0 \end{pmatrix}. \tag{12.4.14}$$

This allows us already to determine the mechanical part of the entropy production:

$$\Sigma_{ji}\frac{\partial \upsilon_i}{\partial x_j} = \mu\left(\frac{V}{h}\right)^2. \tag{12.4.15}$$

In order to obtain the thermal one we need to turn to the heat conduction equation which follows from the First Law (7.6.1) and solve it for the present case. There is no radiation $r$, the heat flux is given by FOURIER's law, and the time derivative of the specific internal energy vanishes completely because of stationarity and the *ansatz* (7.4.7) for the velocity:

$$\frac{\mathrm{d}u}{\mathrm{d}t} = \frac{\partial u}{\partial t} + \upsilon_i\frac{\partial u}{\partial x_i} = 0. \tag{12.4.16}$$

Since the internal energy depends on the density and on the temperature it can only be a function of height, $x_2$. However, $\upsilon_2$ is equal to zero and thus:

$$-\kappa\frac{\mathrm{d}^2 T}{\mathrm{d}x_2^2} = \mu\left(\frac{V}{h}\right)^2. \tag{12.4.17}$$

Obviously isothermal conditions contradict this differential equation. We assume that the upper as well as the lower plate are kept at a constant temperature level, $T_0$. It is easily verified that the corresponding solution for the temperature must be:

$$T = T_0 + \frac{\mu}{2\kappa}V^2\frac{x_2}{h}\left(1 - \frac{x_2}{h}\right). \tag{12.4.18}$$

Now we are in a position to determine the local entropy production $\sigma$ from Eq. (12.4.11):

$$T\sigma = \left[\frac{\mu}{4\kappa}\frac{V^2}{T}\left(1 - \frac{2x_2}{h}\right)^2 + 1\right]\mu\left(\frac{V}{h}\right)^2 \geq 0. \tag{12.4.19}$$

Note that in contrast to the mechanical bit the thermal part of the entropy production depends on position. In order to find out at which height the local entropy production assumes a maximum, it is advisable to introduce the following parameters:

$$0 \leq \bar{x} = \frac{x_2}{h} \leq 1, \quad 0 \leq \alpha = \frac{\mu}{2\kappa} \frac{V^2}{T_0} < \infty, \quad \sigma_0 = \frac{\mu}{T_0} \left( \frac{V}{h} \right)^2. \tag{12.4.20}$$

Thus we obtain a dimensionless temperature:

$$\bar{T}(\bar{x}) \equiv \frac{T(x)}{T_0} = 1 + \alpha \bar{x}(1 - \bar{x}), \tag{12.4.21}$$

and a dimensionless entropy production:

$$\bar{\sigma}(\bar{x}) \equiv \frac{\sigma(x)}{\sigma_0} = \frac{1 + \frac{\alpha}{2} + \alpha \bar{x}(\bar{x} - 1)}{[1 + \alpha \bar{x}(1 - \bar{x})]^2}. \tag{12.4.22}$$

Figure 12.5 shows the latter function vs. $\bar{x}$ for $\alpha = 0.5$, 1.5, and 10. It is obviously symmetric, as it should, due to the choice of the same temperatures at the lower and at the upper plate. It assumes a minimum in the center, and it tends to maximum values when we get closer to the plates. This seems reasonable because this is where the temperature gradient is particularly strong, whereas in the center it is equal to zero. The greater $\alpha$ the more the entropy production differs from the constant reference value $\sigma_0$. This is also easy to explain since increasing values of $\alpha$ reinforce the impact of the plate temperatures, which homogenizes entropy production.

It is interesting to note that the factor $\alpha$ introduces to the entropy production two material parameters that one would intuitively expect behind dissipation, the shear viscosity and the heat conduction parameter. However, note that although the constitutive relations for the viscous stresses and for the heat flux are both linear, the corresponding parameters add a highly non-linear touch to entropy production.

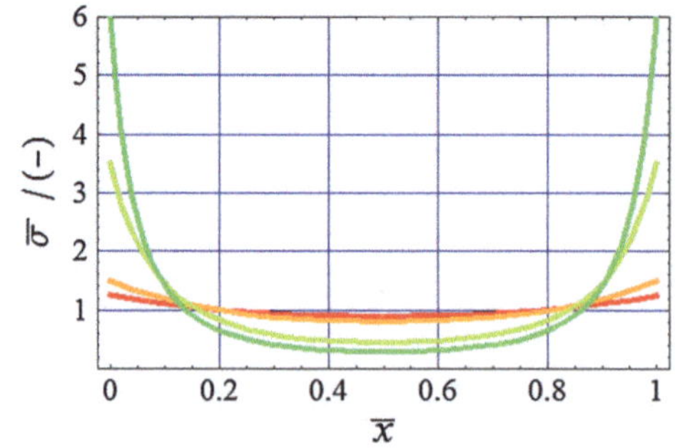


**Fig. 12.5** Local entropy production for parallel plate flow (see text)

## 12.5 Would You Like to Know More?

A first introduction to the concept of entropy from the standpoint of the Carnot cycle and engineering thermodynamics of discrete systems can, for example, be found in [6], Chapters 5 and 6, [7], Chapter 4 or [8], Chapter 4. The continuum theoretical aspects of the entropy principle are outlined in [4], Chapter 4, and in [5], Chapter 5. The latter book can also be consulted in order to learn to appreciate Carathéodory's contributions to entropy (Section 5.1.3.3), and to hear more about the notion of availability in a wider context (Section 7.2.2). The traditional ways of the chemical engineers and the physical chemists, who think in terms of minimizing Helmholtz and Gibbs functions, are outlined, for example, in [9], Chapter 3. The monograph of [10] presents a very detailed exposition of nearly all aspects of the notion of entropy, from a continuum or theoretical point-of-view, for gases, fluids, and solids. This book also provides further information on availability: Sections 7.5 / 7.6.

The principles of statistical mechanics and the entropy of the so-called great canonical ensemble are explained in the books by Münster [2] and by Tolman [11]. The latter monograph also quickly introduces aspects of quantum mechanics. The kinetic theory of gases and the corresponding notion of entropy, as introduced by Maxwell and, in particular, Boltzmann, can be studied best by consulting the bible on this topic, namely the book by Chapman and Cowling [12] including Boltzmann's famous H-Theorem. Another valuable source in the same context is the book by Becker [13], in particular Chapter 2.

The classic concepts of TIP are presented in the books by de Groot [14] or de Groot and Mazur [15]. Becker [13] gives a first introduction: Chapter 7. The methods of so-called *Rational Thermodynamics*, i.e., methods that go beyond that of TIP, can be found in the book by Truesdell [16].

## References

1. Eckart C (1940) The thermodynamics of irreversible processes I. The simple fluid. Phys Rev 58(3):267–269
2. Münster A (1969) Statistical thermodynamics, vol 1. Springer, Berlin, First English Edition
3. Haupt P (2002) Continuum mechanics and theory of materials, 2nd edn. Springer, Berlin
4. Müller I (1973) Thermodynamik. Die Grundlagen der Materialtheorie. Bertelsmann Universitätsverlag, Düsseldorf
5. Müller I (1985) Thermodynamics. Pitman Advanced Publishing Program, Boston
6. Çengel YA, Boles MA (1998) Thermodynamics: an engineering approach, 6th edn. McGraw Hill, Boston
7. Müller I (1994) Grundzüge der Thermodynamik mit historischen Anmerkungen, 1st edn. Springer, Berlin
8. Müller I, Müller WH (2009) Fundamentals of thermodynamics and applications. Springer, Berlin
9. Moore WJ (1963) Physical chemistry, 4th edn. Longmans Green and Co Ltd, London
10. Müller I, Weiss W (2005) Entropy and energy—A universal competition. Springer, Berlin

11. Tolman RC (1979) The principles of statistical mechanics. Reprint of the original edition of 1938. Dover Publications, Inc., New York
12. Chapman S, Cowling TG (1939) The mathematical theory of non-uniform gases. Cambridge at the University Press, Cambridge
13. Becker R, Leibfried G (eds) (1967) Theory of heat, 2nd edn. Springer, Berlin
14. de Groot SR (1960) Thermodynamik irreversibler Prozesse. BI Hochschultaschenbücher 18/18a. Bibliographisches Institut, Mannheim
15. de Groot SR, Mazur P (1984) Non-equilibrium thermodynamics. Dover Publications Inc., New York
16. Truesdell C (1969) Rational thermodynamics. McGraw-Hill, New York

# Chapter 13
# Fundamentals of Electromagnetic Field Theory

**Abstract** In this chapter we go beyond thermo-mechanics and extend continuum theory to electromagnetic fields. The emphasis is on a rational presentation of fundamental principles: What are the foundations of Maxwell's equations, how can the occurring fields be measured, at least in principle, and how are they linked to each other? Moreover, the question regarding frame indifference of the equations and the transformation properties of the electromagnetic fields will be posed, which had already been answered before in context with the thermo-mechanical fields. This will lead us to the beginnings of relativistic field theories.

*I am an expert of electricity.*
*My father occupied the chair of applied electricity*
*at the state prison.*

W. C. Fields

## 13.1 Preliminary Remarks

Normally engineering students are not taught the fundamentals of electrodynamics, i.e., Maxwell's equations, in the form presented in this book, one of the reasons being that neither the typical electrical engineer of daily practice nor the regular physicist are confronted with the problem to set up a *coupled electrodynamic theory of materials* in a continuum related framework. From now on our objective is not only to determine the previously established five thermo-mechanical fields, namely mass density, $\rho(\boldsymbol{x}, t)$, velocity, $\boldsymbol{v}(\boldsymbol{x}, t)$, and temperature, $T(\boldsymbol{x}, t)$, but also the (unknown) fields of electrodynamics, i.e., density of *electric charge*, $q(\boldsymbol{x}, t)$, the *current density* (*vector*), $\boldsymbol{j}(\boldsymbol{x}, t)$, in combination with the *electric field*, $\boldsymbol{E}(\boldsymbol{x}, t)$, and the *magnetic flux density* (*magnetic induction*), $\boldsymbol{B}(\boldsymbol{x}, t)$, in all points, $\boldsymbol{x}$, of a *material* continuum at all times, $t$, of its motion. Note that the focus is on the "material," i.e., on a theory of electricity for matter, and not just an electromagnetic theory of the vacuum. At that point engineers typically introduce

W. H. Müller, *An Expedition to Continuum Theory*, Solid Mechanics and Its Applications 210, DOI: 10.1007/978-94-007-7799-6_13,

very simple constitutive equations or study preferably problems for which the motion of matter is of no importance. Such cases are perfectly covered by MAXWELL's equations in a form that can be found, for example, in the renowned book by Becker [1], §53 on pg. 216, Eq. (53.2). Becker says: "... and thus we obtain the four fundamental equations, of remarkably symmetrical structure[1]:

$$\begin{aligned}&\text{(I) curl}\,\boldsymbol{H}=\frac{1}{c}\frac{\partial\boldsymbol{D}}{\partial t}+\frac{4\pi\boldsymbol{g}}{c},\ \text{(II) div}\,\boldsymbol{D}=4\pi\rho,\\&\text{(III) curl}\,\boldsymbol{E}=-\frac{1}{c}\frac{\partial\boldsymbol{B}}{\partial t},\ \text{(IV) div}\,\boldsymbol{B}=0\end{aligned}\tag{13.1.1}$$

as the final form of the Maxwell equations for media at rest." It is noteworthy that these relations hold for matter *at rest*. Of course the obvious question arises how the general equations read that hold if matter is *moving*. In what follows we attempt to clarify these issues. Moreover, we will emphasize and explain why *two* "electric" fields, i.e., $\boldsymbol{E}(\boldsymbol{x},t)$ and $\boldsymbol{D}(\boldsymbol{x},t)$, and *two* "magnetic" fields, i.e., $\boldsymbol{B}(\boldsymbol{x},t)$ and $\boldsymbol{H}(\boldsymbol{x},t)$, must be distinguish and are required in electromagnetism. In other words we will put a strong emphasis on independent measurement instructions for these four fields and all the other ones.

At this point it should already be mentioned that we will define and use the electric field $\boldsymbol{E}(\boldsymbol{x},t)$ and the magnetic induction $\boldsymbol{B}(\boldsymbol{x},t)$ as primary fields from the very beginning on. However, the fields $\boldsymbol{D}(\boldsymbol{x},t)$ and $\boldsymbol{H}(\boldsymbol{x},t)$ of Eq. (13.1.1), which are known in the pertinent literature on electrodynamics as *electric displacement* and *magnetic field*, will not be used before we talk extensively about the conservation of charges and currents. More appropriately these fields should be referred to as *charge* and *current potential in matter*. We will denote them by the symbols $\mathfrak{D}$ und $\mathfrak{H}$ in order to distinguish them from the general charge and current potentials $\boldsymbol{D}$ and $\boldsymbol{H}$, respectively.

Robert Jemison VAN DE GRAAFF was an American physicist of Dutch descent born on December 20, 1901 in Tuscaloosa, Alabama and died on January 16, 1967 in Boston, Massachusetts. He studied mechanical engineering in Tuscaloosa, University of Alabama, where he received his (first) BS and Masters degrees. After a short stay with the Alabama Power company, maybe indicating an early interest in electricity, he began wandering and performed further studies, first at the Sorbonne, and then at Oxford, where he earned a second BS 1926 and completed his PhD in 1928. He produced the first of his famous high voltage generators in Princeton, and then became a professor at MIT.

[1] Becker uses the symbols $\rho$ and $\boldsymbol{g}$ for the (true) electric charge density and the (true) current density of free charge carriers. Both symbols have been used in this book before but in a different context. In what follows we will use the symbols $q_{\mathrm{f}}$ and $\boldsymbol{j}_{\mathrm{f}}$ instead. Also note that Becker does not use SI units, which explains the factor $4\pi$ and the speed of light symbol, $c$, in his equations.

Such diversity in symbols and notions is inherent to a continuum approach to electrodynamics: From a non-atomistic, phenomenological point-of-view we must distinguish between various types of charge and currents as well. For example, there is the *density of free (or true) charges*, $\rho$, (denoted by $q_f(\boldsymbol{x}, t)$ in what follows) and $\boldsymbol{g}$ the *free (or true) current density* (further down denoted by $\boldsymbol{j}_f(\boldsymbol{x}, t)$). These were used in Maxwell's equations in the form (13.1.1). They represent the type of charge-carrying mechanisms we intuitively expect, if we think of charges collected by a VAN DER GRAAF generator or electrons migrating through a wire. However, matter may also respond to electromagnetic fields and polarize under their influence. If we cut through such a polarized medium, a surface charge will result and, what is more, we may move or deform the polarized matter which, in turn, results in an electric polarization current on a continuum scale.

Thus from the perspective of a more general materials theory it is initially much more expedient to consider the *total charge density*, $q(\boldsymbol{x}, t)$, and the *total current density*, $\boldsymbol{j}(\boldsymbol{x}, t)$, and to write MAXWELL's equations in terms of these quantities. In this context we shall use the *general* charge and current potentials, i.e., the symbols $\boldsymbol{D}$ and $\boldsymbol{H}$. Later we will split $q(\boldsymbol{x}, t)$ and $\boldsymbol{j}(\boldsymbol{x}, t)$ into several parts and realize that there is more than free charges and free currents. However, by doing so new electric fields, $\mathfrak{D}(\boldsymbol{x}, t)$ and $\mathfrak{H}(\boldsymbol{x}, t)$, will arise. This is the penalty we have to pay for our strategy. Nevertheless, it is only fair to say that such subtle distinctions are frequently not made in the literature and the same symbols for the potentials are used for an electrodynamics theory with or without matter (see, for example, [1]). This is often confusing.

Hendrik Antoon LORENTZ was born on July 18, 1853 in Arnheim and died on February 4, 1928 in Haarlem. In 1870 he turned to the University of Leiden to study mathematics and physics. After that he returned to his hometown and became a teacher in 1871. This gave him ample time to work on his dissertation, which he finished in 1875. In 1878 he finally became a professor for theoretical physics at the University of Leiden, where he stayed for the rest of his life. His scientific achievements are mostly based in theoretical physics and concern, in particular, the electromagnetism of light and of matter. For his explanation of the ZEEMAN-effect he was awarded the Nobel prize in 1902 together with ZEEMAN.

We shall later derive the connection between $\boldsymbol{D}(\boldsymbol{x}, t)$ and $\mathfrak{D}(\boldsymbol{x}, t)$ on the one hand side and $\boldsymbol{H}(\boldsymbol{x}, t)$ and $\mathfrak{H}(\boldsymbol{x}, t)$ on the other. This will allow for a conversion of one quantity into the other. Clearly, this must be possible: In the end all notations must be equivalent since there is only one theory of electromagnetism.

Moreover, we will learn that there is a connection between the fields $\boldsymbol{E}(\boldsymbol{x}, t)$ and $\boldsymbol{D}(\boldsymbol{x}, t)$ on the one hand and $\boldsymbol{B}(\boldsymbol{x}, t)$ and $\boldsymbol{H}(\boldsymbol{x}, t)$ on the other, the so-called MAXWELL-LORENTZ aether relations. Many scientists have worked on the setup of the mathematical framework preferred in this book, and we shall mention them in due course.

Finally note that we will frequently write $E_i$, $D_i$, $B_i$, $H_i$, etc. By these symbols we refer to the components of the corresponding electromagnetic vector fields in a

Cartesian frame *at rest*. This in turn indicates that we must also speak about the transformation properties of all of these fields as well as of MAXWELL's equations during change of observers.

## 13.2 The Conservation Law for the Magnetic Flux

Consider the situation illustrated in Fig. 13.1. Similarly to the general balance for a volume field density $\underset{V}{\psi}$, integrated w.r.t. a material volume *without* a singular surface as shown in Eq. (3.3.7), we present the following general balance for a vector flux density $\gamma_i$, defined for an open material surface, $S$,[2] which is *not* dissected by a singular line, $L$:

$$\frac{\mathrm{d}}{\mathrm{d}t}\iint\limits_S \gamma_i n_i \,\mathrm{d}S = -\oint\limits_{\partial S} \phi_i \tau_i \,\mathrm{d}l + \iint\limits_S (p_i + s_i)\, n_i \,\mathrm{d}S. \tag{13.2.1}$$

The vector $\phi_i$ denotes the flux of $\gamma_i$ across the closed periphery ("line") $\partial S$ in complete analogy to the nomenclature established in Sect. 3.3. Moreover $p_i$ and $s_i$ are vector densities of production and supply (per unit area) of the quantity $\gamma_i$ defined on the surface $S$.

Michael FARADAY was born on September 22, 1791 in Newington, Surrey and died on August 25, 1867 in Hampton Court Green, Middlesex. Born as a member of the lower class he first became an apprentice but quickly developed into one of the most important experimental scientists of all time in an autodidactic manner. He was both a physicist and a chemist. He studied and set up chemical reaction experiments, optical and electrical devices, recorded, interpreted, and also performed public science shows for the pleasure and entertainment of the upper class. It is fair to say that FARADAY was the experimental counterpart to MAXWELL as far as the exploration of electro-magnetic phenomena is concerned.

We will now anticipate some results from the following sections and argue in a formal manner without providing precise definitions for measuring the electromagnetic fields $E_i$ and $B_i$: Theoretical investigations, observations of natural phenomena in combination with well-defined lab experiments by diligent scientists and discoverers like FARADAY, MAXWELL, or LORENTZ have shown that in the case of the magnetic flux we must relate quantities of Eq. (13.2.1) as follows:

$$\gamma_i \rightarrow B_i,\ \phi_i \rightarrow E_i + (\boldsymbol{\upsilon} \times \boldsymbol{B})_i,\ p_i \rightarrow 0,\ s_i \rightarrow 0. \tag{13.2.2}$$

[2] For didactic reasons, which will become clearer in Sect. 13.4, we denote the open surface by the symbol $S$ and not like in Sect. 3.3 by the generic symbol for surfaces, $A$.

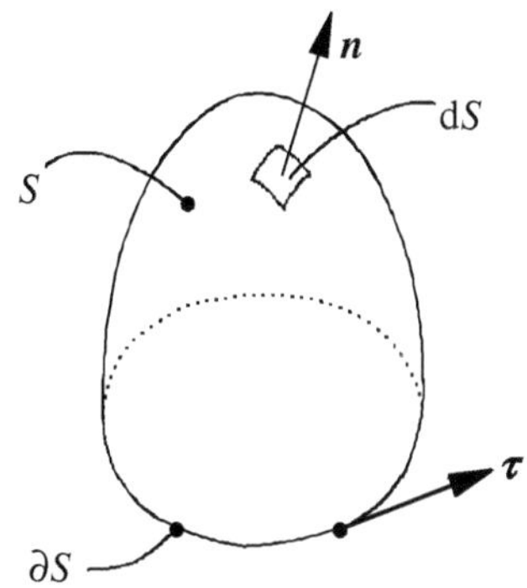


**Fig. 13.1** Open material surface without a singular line

The magnetic flux is a *conserved quantity*, which explains the vanishing production density. Moreover, the combination

$$\mathfrak{E}_i = E_i + (\boldsymbol{v} \times \boldsymbol{B})_i \tag{13.2.3}$$

is also known as the *electromotive intensity*. In honor of Lorentz the part $(\boldsymbol{v} \times \boldsymbol{B})_i$ is also called *Lorentz force density*. Thus we arrive at the following global conservation law for the magnetic flux, a.k.a. *Faraday's law of induction*:

$$\frac{\mathrm{d}}{\mathrm{d}t} \iint_S B_i n_i \,\mathrm{d}S = - \oint_{\partial S} \left[E_i + (\boldsymbol{v} \times \boldsymbol{B})_i\right] \tau_i \,\mathrm{d}l. \tag{13.2.4}$$

This result can also be applied to a closed surface $S \to \partial V$ with $\partial S \to 0$, surrounding the volume $V$:

$$\frac{\mathrm{d}}{\mathrm{d}t} \oiint_{\partial V} B_i n_i \,\mathrm{d}A = 0 \Rightarrow \oiint_{\partial V} B_i n_i \,\mathrm{d}A = \mathrm{const.}_t. \tag{13.2.5}$$

Therefore the total magnetic flux through a closed surface is conserved in time and must be equal to a constant. However, this constant must be zero, because at some point the magnetic field had to be switched on, and before that the magnetic flux was equal to zero. Thus we conclude that:

$$\oiint_{\partial V} B_i n_i \,\mathrm{d}A = 0 \tag{13.2.6}$$

and rewrite this result by means of Gauss' theorem into a volume integral. By doing so we tacitly assume that the magnetic flux densities shows no discontinuities within the volume $V$:

$$\iiint_V \frac{\partial B_i}{\partial x_i} \,\mathrm{d}V = 0. \tag{13.2.7}$$

The usual arguments lead to the corresponding local equation in regular points:

$$\frac{\partial B_i}{\partial x_i} = 0. \tag{13.2.8}$$

This is already the fourth of MAXWELL's equations (13.1.1). We now return to FARADAY's law of induction in Eq. (13.2.4). In a first step we transform the left side of the equations by means of a transport theorem for open surfaces which will be proven in the Exercise 13.2.2:

$$\frac{\mathrm{d}}{\mathrm{d}t}\iint\limits_S B_i n_i \, \mathrm{d}S = \iint\limits_S \left[\frac{\partial B_i}{\partial t} + \upsilon_i \frac{\partial B_k}{\partial x_k}\right] n_i \, \mathrm{d}S + \oint\limits_{\partial S} (\boldsymbol{B} \times \boldsymbol{\upsilon})_i \, \tau_i \, \mathrm{d}l. \tag{13.2.9}$$

The second part in the surface integral vanishes because of the previously derived MAXWELL equation (13.2.8), if we assume continuity of the magnetic flux density. Moreover the line integral cancels with the second term on the right side of Eq. (13.2.4) because of the anti-symmetry $\boldsymbol{B} \times \boldsymbol{\upsilon} = -\boldsymbol{\upsilon} \times \boldsymbol{B}$ of the vector product, so that we obtain:

$$\iint\limits_S \frac{\partial B_i}{\partial t} n_i \, \mathrm{d}S + \oint\limits_{\partial S} E_i \, \tau_i \, \mathrm{d}l = 0. \tag{13.2.10}$$

The second part is transformed into a surface integral by means of STOKES' theorem (for a proof and the requirements for its validity see the Exercise 13.2.1). By doing so we tacitly assume that the electric field $E_i$ has no singularities or discontinuities on the surface $S$:

$$\iint\limits_S \left[\frac{\partial B_i}{\partial t} + (\nabla \times \boldsymbol{E})_i\right] n_i \, \mathrm{d}S = 0. \tag{13.2.11}$$

Hence we obtain another local relation, namely the third of MAXWELL's equation from the set (13.1.1):

$$\frac{\partial B_i}{\partial t} + (\nabla \times \boldsymbol{E})_i = 0. \tag{13.2.12}$$

**Exercise 13.2.1: *STOKES' theorem***

Provide an engineering proof of STOKES' theorem, which states that we may write for continuously differentiable vector fields:

$$\oint\limits_{\partial S} \boldsymbol{\gamma} \cdot \mathrm{d}\boldsymbol{x} = \iint\limits_S \nabla \times \boldsymbol{\gamma} \cdot \mathrm{d}A. \tag{13.2.13}$$

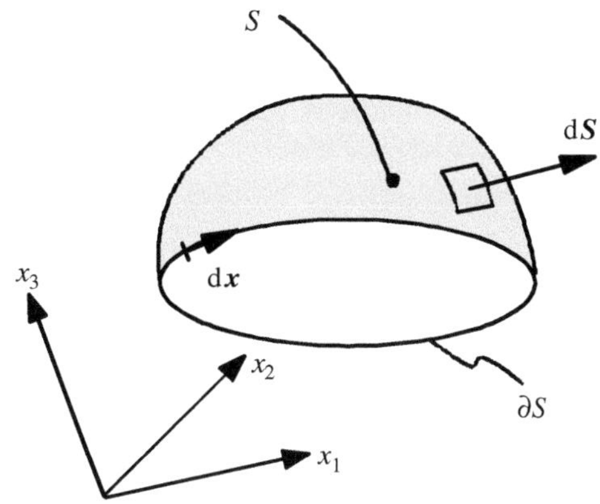


**Fig. 13.2** The scenario for STOKES's theorem

$\mathrm{d}\boldsymbol{x}$ refers to a line element along the periphery $\partial S$ of the open surface $S$ (cf., Fig. 13.2) and $\nabla$ is the del operator. Proceed as follows: First, prove the theorem for a small surface element $\Delta S = \Delta x_1 \Delta x_2$ within the $(x_1, x_2)$-plane. Then generalize the result to three dimensions and finally discuss the case of arbitrarily shaped open surfaces.

Next we will shall investigate a few criteria that are useful for characterizing a force field $\gamma$ as being conservative or not. First, recall that a conservative force can be obtained from the gradient of a potential. Now prove the following chain of arguments by using STOKES theorem: If $\nabla \times \gamma = \mathbf{0}$, then we have path independence according to $\oint_{\partial S} \gamma \cdot \mathrm{d}\boldsymbol{x} = 0$, then $\gamma \cdot \mathrm{d}\boldsymbol{x} = \gamma_1 \mathrm{d}x_1 + \gamma_2 \mathrm{d}x_2 + \gamma_3 \mathrm{d}x_3 = -\mathrm{d}\varphi(\boldsymbol{x}, t)$ , then $\gamma_i = -\partial\varphi/\partial x_i$, but then also $\nabla \times \gamma = \mathbf{0}$, which closes the logical circle. Evaluate these relations explicitly for the case of a COULOMB field of a point charge at rest, or for the gravitational field around a point mass, $\gamma = A\boldsymbol{x}/|\boldsymbol{x}|^3$, with a constant $A$ and the position vector $\boldsymbol{x} = (x_1, x_2, x_3)$ pointing away from the center of the force. What is the corresponding potential? Is it of importance that the force field has a singularity at $\boldsymbol{x} = \mathbf{0}$?

**Exercise 13.2.2:** ***Transport theorem for vector surface flux densities***

Our objective is to prove the following transport theorem for a vector surface flux density:

$$\frac{\mathrm{d}}{\mathrm{d}t} \iint_S \gamma_i n_i \,\mathrm{d}A = \iint_S \left[\frac{\partial \gamma_i}{\partial t} + \upsilon_i \frac{\partial \gamma_k}{\partial x_k}\right] n_i \,\mathrm{d}A + \oint_{\partial S} (\gamma \times \upsilon)_i \,\tau_i \,\mathrm{d}l. \qquad (13.2.14)$$

For the proof argue as follows. In a first step show that:

$$\frac{\mathrm{d}}{\mathrm{d}t}\iint\limits_{S(t)} \gamma_i(\boldsymbol{x},t)\,\mathrm{d}S_i = \iint\limits_{S} [\gamma_i(\boldsymbol{x}(\boldsymbol{X},t),t)]^{\cdot}\,\mathrm{d}S_i + \iint\limits_{S} \gamma_i(\boldsymbol{x},t)\,[\mathrm{d}S_i(\boldsymbol{x}(\boldsymbol{X},t),t)]^{\cdot} \tag{13.2.15}$$

by using the directed surface element in Lagrangian description:

$$\mathrm{d}S_i = n_i\,\mathrm{d}S,\ \boldsymbol{x} = \boldsymbol{x}(\boldsymbol{X},t). \tag{13.2.16}$$

In a second step show that:

$$[\mathrm{d}S_i(\boldsymbol{x}(\boldsymbol{X},t),t)]^{\cdot} = \left[-\frac{\partial \upsilon_j}{\partial x_i} + \frac{\partial \upsilon_m}{\partial x_m}\delta_{ij}\right]\mathrm{d}S_j. \tag{13.2.17}$$

To this end express the directed surface element by a vector product between two infinitesimal co-moving line elements $\overset{(1)}{\mathrm{d}\boldsymbol{x}}$, $\overset{(2)}{\mathrm{d}\boldsymbol{x}}$:

$$\mathrm{d}A_i = \left(\overset{(1)}{\mathrm{d}\boldsymbol{x}} \times \overset{(2)}{\mathrm{d}\boldsymbol{x}}\right)_i = \epsilon_{ijk}\mathrm{d}\overset{(1)}{x}_j\,\mathrm{d}\overset{(2)}{x}_k, \tag{13.2.18}$$

differentiate w.r.t. time and show that:

$$(\mathrm{d}x_i)^{\cdot} = \frac{\partial \upsilon_i}{\partial x_j}\,\mathrm{d}x_j. \tag{13.2.19}$$

In this context recall the peculiarities of the Lagrangian way of describing motion $\boldsymbol{x} = \boldsymbol{x}(\boldsymbol{X},t)$. Now insert the results in Eq. (13.2.15), perform some suitable algebraic transformations, finally recall that:

$$\delta_{il}\delta_{jm} - \delta_{jl}\delta_{im} = \epsilon_{ijk}\epsilon_{klm} \tag{13.2.20}$$

and Stokes' theorem from Exercise 13.2.1.

**Exercise 13.2.3:** ***Extension of the transport theorem for vector surface flux densities including singular lines***

Now prove the following extension of the transport theorem shown in Eq. (13.2.14) if the open surface contains a singular line, $L$:

$$\begin{aligned}\frac{\mathrm{d}}{\mathrm{d}t}\iint\limits_{S_-\cup S_+} \gamma_i n_i\,\mathrm{d}A &= \iint\limits_{S_-\cup S_+} \left[\frac{\partial \gamma_i}{\partial t} + \upsilon_i\frac{\partial \gamma_k}{\partial x_k}\right] n_i\,\mathrm{d}A \\ &+ \int\limits_{l_-\cup l_+} (\boldsymbol{\gamma}\times\boldsymbol{\upsilon})_i\,\tau_i\,\mathrm{d}l - \int\limits_{L} ([\![\boldsymbol{\gamma}]\!]\times\boldsymbol{w})_i\,t_i\,\mathrm{d}l.\end{aligned} \tag{13.2.21}$$

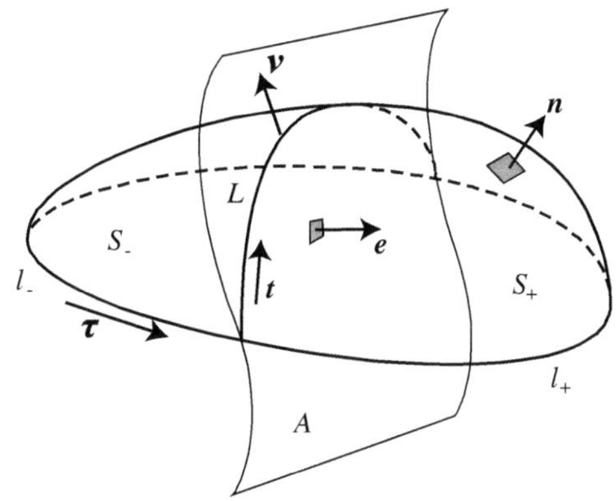


**Fig. 13.3** A singular surface intersecting with an open surface thus creating a singular line

To this end observe the notation shown in Fig. 13.3 and start the proof for an open surface $S = S_- \cup S_+ \cup L$ by application of Eq. (13.2.14) to both (regular) surfaces $S_\pm$. Assume that the singular surface $A$ moves independently of the material points of $S_\pm$ with a velocity $\boldsymbol{w}$. Explain in detail the presence of the jump term on $L$ in Eq. (13.2.21). Why is the jump proportional only to $\boldsymbol{w}$ and not to $\boldsymbol{\upsilon}$ or $\boldsymbol{w} - \boldsymbol{\upsilon}$ (say)?

Finally in this section it should be pointed out explicitly that $S$ and $\partial V$ *can* be *material* structures, moving along with matter dispersed by electromagnetic fields. Then the quantity $\boldsymbol{\upsilon}$ in the transport theorem (13.2.14) is nothing else but the local velocity of the material points. Interestingly the two local MAXWELL equations (13.2.8/13.2.12) are *independent* of this motion and, consequently, the fields $E_i$ and $B_i$ are also measurable quantities with respect to a *co-moving* observer. In what follows we will look into this more deeply.

## 13.3 Electric Charges, Currents, Electric Field Density and Magnetic Induction

Just like mass the electric charge is another fundamental, so-called *primitive* quantity, which cannot be reduced any further or related to other even more fundamental entities. Introducing this notion is useful when explaining and quantifying certain natural phenomena. This does not mean that we have truly explained them or provided the ultimate understanding. All we do is to define a rational frame, which is as far-reaching as possible, self-contained, free of contradictions, mathematically expressible, and capable of predictions which, in turn, can be verified by experiments.

In context with the fields $\boldsymbol{E}$ and $\boldsymbol{B}$ two observations must be mentioned. As we shall see they can be interpreted as *reactions to "invisible," clearly non-mechanical forces* that need to be quantified: After rubbing a Styrofoam ball with a woolen cloth it will attract small paper scraps. We explain this phenomenon as follows: Due to the rubbing negatively charged electrons are removed from the ball, so that effectively a positive charge remains. The charge generates an electric field $\boldsymbol{E}$ in the space around the sphere, which in turn polarizes the paper resulting in an attraction. We measure electric charges, $Q$, in honor of their discover in units of *Coulomb* (C). However, one Coulomb is a gigantic amount of charge, if we realize that one electron carries only a very small quantum of charge, namely $1.602 \cdot 10^{-19}$C.

Charles Augustin de COULOMB was born on June 14, 1736 in Angoulême and died on August 23, 1806 in Paris. He obtains his engineering education in Paris, joins the military engineering corps and is send for nine years to the island of Martinique, where he is in charge of civil engineering construction work. This brings him in first contact with (unsolved) problems of materials science and structural mechanics. Rumor has it that his sojourn overseas was bad for his health. Thus he "retires" in 1789 in provincial France and survives the French Revolution while performing further scientific studies. Mechanical engineers know COULOMB mostly from his law of friction. However, besides mechanics COULOMB was fascinated by electricity and magnetism, the two hot branches of physics in his days, so-to-speak. This is why he is also famous for his (experimental) discovery of the law describing attraction and repulsion between opposite or equally charged bodies, where the force is proportional to the inverse of distance. In other words, he observed the same dependence as in NEWTON's law of gravitation, which certainly served as an inspiration.

In principle we can measure the amount and the direction of an electric field generated, for example, by the Styrofoam ball, by placing a test change of known strength, $Q$, at a certain point in space and measuring the amount and the direction of the resulting force $\boldsymbol{F}$. We use this information to compute the following expression:

$$\boldsymbol{E} = \frac{\boldsymbol{F}}{Q}. \tag{13.3.1}$$

We now turn to a second electric phenomenon and consider an iron magnet at rest. In its vicinity we scatter iron filing chips and observe how these start rearranging. We interpret this behavior by saying that they follow the orientation of the field lines of the magnetic flux density, $\boldsymbol{B}$. We now investigate the behavior of an electron beam carrying a certain electric charge, $Q$, in the vicinity of a magnet. In other words a "swarm" of test charges is passing by at a certain speed. We observe that the beam does not follow a straight line. Rather it will curve and the electrons are deflected along circles away from their original straight path. The cause for this deflection must be a force and, by observation, we conclude that the greater the speed of the electrons, the greater the curvature and, thus, the greater the force.

Moreover, the force is obviously perpendicular to the velocity vector. And, finally, the force vector is perpendicular to the magnetic field lines indicated by the filing chips. Now we use all these facts and derive a rule for measuring the magnetic flux density, i.e., the amount and the direction of the vector $\boldsymbol{B}$ from the following measurable quantities, namely force vector, $\boldsymbol{F}$, amount of moving electric charge, $Q$, and their velocity vector, $\boldsymbol{\upsilon}$:

$$\frac{\boldsymbol{F}}{Q} = \boldsymbol{\upsilon} \times \boldsymbol{B}. \tag{13.3.2}$$

If an electric field $\boldsymbol{E}$ and the magnetic induction $\boldsymbol{B}$ act simultaneously, the total force per unit charge is given by:

$$\frac{\boldsymbol{F}}{Q} = \boldsymbol{E} + \boldsymbol{\upsilon} \times \boldsymbol{B} \equiv \boldsymbol{\mathfrak{E}}. \tag{13.3.3}$$

We have already used this combined quantity in Eq. (13.2.3) and interpreted it as the flux of magnetic induction across the periphery of the surface. Of course, our tests are more or less idealized *gedankenexperiments*. However, they are helpful if we want to get used to electromagnetic fields, for which we have developed no intuitive feel during our biological evolution unlike mechanical quantities, such as force or mass. Clearly, in technical practice $\boldsymbol{E}$ and $\boldsymbol{B}$ are not measured that way.

**Exercise 13.3.1:** ***Technical measurement of E and B; basic electric units***

Consult textbooks of electrical engineering and explain how engineers measure the fields $\boldsymbol{E}$ and $\boldsymbol{B}$ in daily practice. Make sure that these measurements are independent of the determination of $\boldsymbol{D}$ and $\boldsymbol{H}$ and not linked to constitutive relations. Use information from this section and derive the following units for the electric field and the magnetic induction:

$$\dim[\boldsymbol{E}] = \frac{\mathrm{N}}{\mathrm{C}}, \ \dim[\boldsymbol{B}] = \frac{\mathrm{Ns}}{\mathrm{Cm}}. \tag{13.3.4}$$

Recall the following high school experiment in preparation for calculating the energy contained in the electromagnetic field: The electric field $E$ between two electrically charged plates ("plate capacitor") is nearly homogeneous. The work, $W$, required to carry a charge of strength, $Q^+$, from the negative to the positive side over the distance $d$ is given by:

$$W = Q^+ U, \ E = \frac{U}{d}. \tag{13.3.5}$$

$U$ is known as the *electric voltage* or electric potential. Use it to motivate the following general relation, which is analogous to the notion of a conservative mechanical force, and which is in perfect harmony with Eq. $(13.3.5)_2$ for the case of homogeneous conditions:

$$E_i = -\frac{\partial U}{\partial x_i}. \tag{13.3.6}$$

The unit of the electric potential is *Volt* (V). Show by using the other results from this exercise that:

$$\dim[U] = \frac{\mathrm{J}}{\mathrm{C}} = \mathrm{V},\ \dim[\boldsymbol{E}] = \frac{\mathrm{J}}{\mathrm{Cm}} = \frac{\mathrm{V}}{\mathrm{m}},\ \dim[\boldsymbol{B}] = \frac{\mathrm{Js}}{\mathrm{Cm}^2} = \frac{\mathrm{Vs}}{\mathrm{m}^2}. \tag{13.3.7}$$

Recall that an electric current, $I$, is measured in units of *Ampere* (A). In terms of physics it is nothing else but a flow of charge per unit time. Conclude that:

$$\dim[I] = \frac{\mathrm{C}}{\mathrm{s}} = \mathrm{A},\ \dim[\boldsymbol{E}] = \frac{\mathrm{J}}{\mathrm{Cm}} = \frac{\mathrm{V}}{\mathrm{m}},\ \dim[\boldsymbol{B}] = \frac{\mathrm{J}}{\mathrm{Am}^2}. \tag{13.3.8}$$

## 13.4 Conservation of Total Charge

The second set of MAXWELL's equations are based on the conservation of charge within a material volume divided into two regions, $\partial V^+$ and $\partial V^-$, by a singular surface, $A$ (cf., Fig. 13.4). Within $V^+$ and $V^-$ the density of charge (a scalar per unit volume like mass density) is denoted by $q$ (in $\mathrm{C/m^3}$). On the singular surface the density of charge is called $\underset{A}{q}$ (per unit area in $\mathrm{C/m^2}$). Thus in the nomenclature of Eq. (3.3.6) the total charge is given by:

$$Q = \iiint\limits_{V^+\cup V^-} q(\boldsymbol{x},t)\,\mathrm{d}V + \iint\limits_{A} \underset{A}{q}(\boldsymbol{x},t)\,\mathrm{d}S \tag{13.4.1}$$

The total electric charge will change in time when electric currents leave the control volume through the surfaces $A^+ \cup A^- \cup \partial A$. We describe these directed quantities by non-convective current density vectors $\boldsymbol{j}$ (in $\mathrm{C/(m^2}s))$ and $\underset{L}{\boldsymbol{j}}$ (in $\mathrm{C/(m}s))$, which are distributed across the surface and the line, respectively. Thus we arrive at the following conservation law for charge in form of a conserved balance:

$$\begin{aligned}\frac{\mathrm{d}}{\mathrm{d}t}Q = J \Leftrightarrow \frac{\mathrm{d}}{\mathrm{d}t}\left[\iiint\limits_{V^+\cup V^-} q(\boldsymbol{x},t)\,\mathrm{d}V + \iint\limits_{A} \underset{A}{q}(\mathbf{x},t)\,\mathrm{d}S\right] \\ = -\iint\limits_{A^+\cup A^-} j_i n_i\,\mathrm{d}A - \oint\limits_{\partial A} \underset{L}{j}_i \nu_i\,\mathrm{d}l.\end{aligned} \tag{13.4.2}$$

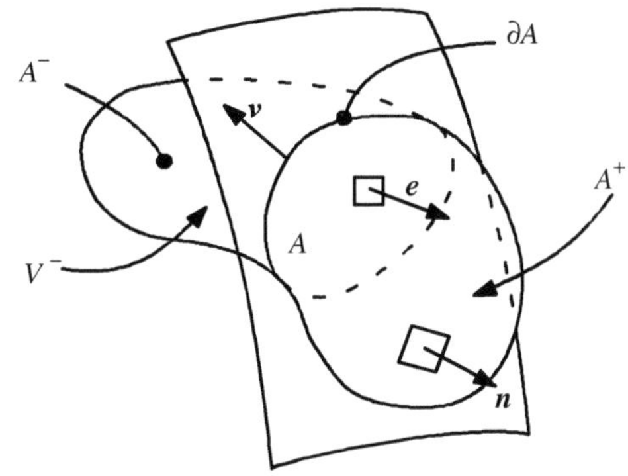


Fig. 13.4 Material volume dissected by a singular surface

The minus signs on the right side are pure convention. However, they can be motivated as follows: The total charge within a material volume will decrease if charge is removed from the inside by currents crossing the surface.

Also note that there is no (volumetric) supply or production of charge. This statement is based on experience. In mathematical terms it means (also see the corresponding remarks in Sect. 3.3) that:

$$\underset{V}{p} = 0,\ \underset{A}{p} = 0,\ \underset{V}{s} = 0,\ \underset{A}{s} = 0. \tag{13.4.3}$$

Equation (13.4.2) is now solved formally by introducing two vectors, the *charge potential* $\boldsymbol{D}$ and the *current potential* $\boldsymbol{H}$, for the *total* charge and the *total* currents as follows (cf., Fig. 13.4):

$$\oint\!\!\!\oint_{\partial V} D_i n_i \,\mathrm{d}A = \iiint_{V^+ \cup V^-} q(\boldsymbol{x}, t)\,\mathrm{d}V + \iint_A \underset{A}{q}(\boldsymbol{x}, t)\,\mathrm{d}A \tag{13.4.4}$$

and (cf., Fig. 13.5):

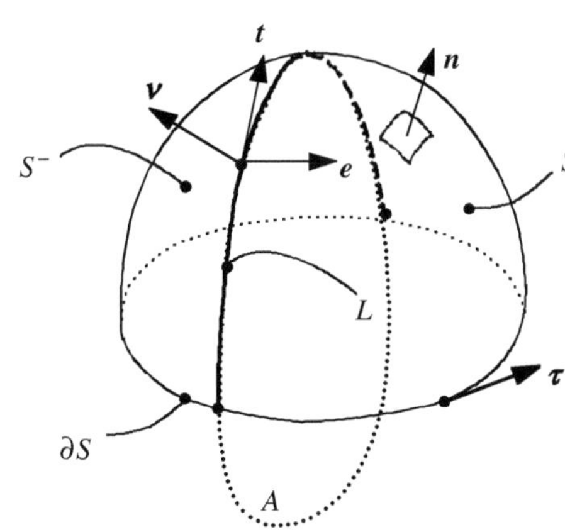


Fig. 13.5 Open surface, $S$, intercepted by a singular line, $L$

$$\frac{\mathrm{d}}{\mathrm{d}t}\iint\limits_{S} D_i n_i \,\mathrm{d}S = \oint\limits_{\partial S} \left[H_i + (\boldsymbol{D} \times \boldsymbol{v})_i\right] \tau_i \,\mathrm{d}l - \iint\limits_{S^+ \cup S^-} j_i n_i \,\mathrm{d}S - \int\limits_{L} \underset{L}{j}{}_i \, v_i \,\mathrm{d}l. \tag{13.4.5}$$

The quantities $\boldsymbol{D}$ and $\boldsymbol{H}$ are also frequently termed *dielectric displacement* and *magnetic field*. Note that according to Eq. (13.4.2) the charge $Q$ is described by two scalar densities w.r.t. the volume or the surface. It was balanced over a material volume $V = V^+ \cup V^- \cup A$. The fields of the volumetric charge density, $q$, may jump, i.e., be discontinuous across the (open) singular surface $A$. However, the current $J$ has been related to vector current densities per unit area or per unit length. The vector fields for the current per unit area behave discontinuously when crossing from $A^-$ to $A^+$, i.e., $\partial A$.

Analogous remarks hold in context with Eq. (13.4.5) and Fig. 13.5. Moreover, note that the open singular line $L$ represents a part of the periphery $\partial A$. If we now extend the open surfaces $S^+$ and $S^-$ to cover $A^+$ and $A^-$ they form a closed volume $\partial V$ together with the line $L \to \partial A$, which is then also closed. We will use this argument quite soon. Finally recall that the singular lines as well as the singular surface may have their own velocity field, $\boldsymbol{w}$, independently of the velocity of the material particles. In other words they are not necessarily material objects.

Equation (13.4.5) can also be interpreted as follows: The temporal change of the flow resulting from the charge potential, $\boldsymbol{D} \iint\limits_{S} D_i n_i \,\mathrm{d}S$, over the open surface $S$ is given by flux of this quantity, $\oint_{\partial S} \left[H_i + (\boldsymbol{D} \times \boldsymbol{v})_i\right] \tau_i \,\mathrm{d}l$ across the hull $\partial S$ as well as by supplies from current densities $\iint\limits_{S^+ \cup S^-} j_i n_i \,\mathrm{d}A$ and $\int\limits_{L} \underset{L}{j}{}_i v_i \,\mathrm{d}l$, over the open surface $S = S^+ \cup S^- \cup L$, respectively.

As previously indicated we can also specialize the second equation to the case of a *closed* surface. Then the integral $\oint_{\partial S} \left[H_i + (\boldsymbol{D} \times \boldsymbol{v})_i\right] \tau_i \,\mathrm{d}l$ vanishes because $\partial S \to 0$. Moreover we have:

$$\iint\limits_{S} \to \oiint\limits_{\partial V}, \quad \iint\limits_{S^+ \cup S^-} \to \iint\limits_{\partial V^+ \cup \partial V^-}, \quad \int\limits_{L} \to \oint\limits_{\partial A}, \tag{13.4.6}$$

and this leads us to conclude that:

$$\frac{\mathrm{d}}{\mathrm{d}t}\oiint\limits_{\partial V} D_i n_i \,\mathrm{d}A = - \iint\limits_{\partial V^+ \cup \partial V^-} j_i n_i \,\mathrm{d}A - \oint\limits_{\partial A} \underset{L}{j}{}_i v_i \,\mathrm{d}l. \tag{13.4.7}$$

Hans Christian ØRSTED was born on August 14, 1777 in Rudkøbing and died on March 9, 1851 in Copenhagen. At the age of twelve he assisted already in the pharmacy of his father. This brought him in close contact with the sciences and in particular with chemistry. Consequently, he studied chemistry and pharmacy at the University of Copenhagen where he also obtained a professorship for chemistry and physics in 1806. ØRSTED discovered various things: In 1819 he was the first to isolate piperidine, an organic solvent. Then in 1820 he discovered the magnetic effect of electric currents on compass needles. In 1825 he managed to extract aluminum for the first time in the history of mankind. He was also an active free mason and a very good friend of the fairy tale writer Hans Christian ANDERSEN.

In view of Eq. (13.4.4) we must also conclude that the balance of charge, Eq. (13.4.2), is identically satisfied. Equations (13.4.4/13.4.5) represent the first and the second of MAXWELL's equations in integral form. In terms of physics they must be interpreted as *global conservation laws for electric charge and current.* In the pertinent literature on electrodynamics Eq. (13.4.4) is also known as *COULOMB's law* (in words, the electric charge is the source of the dielectric displacement) and Eq. (13.4.5) as *ØRSTED-AMPÈRE's law*: Around 1820 ØRSTED and AMPÈRE had discovered experimentally that an electric current (more precisely the densities $\boldsymbol{j}$ and $\underset{L}{\boldsymbol{j}}$, respectively) is always accompanied by a magnetic field (more precisely the field $\boldsymbol{H}$). The corresponding rules of measurement will be investigated in the next section.

At this point we should already ask why the fields $\boldsymbol{D}$ and $\boldsymbol{H}$ were introduced to begin with and why the conservation law shown in Eq. (13.4.2) was not sufficient for further analysis. The reason is a practical one: It is possible to connect the fields $\boldsymbol{D}$ and $\boldsymbol{E}$ as well as the fields $\boldsymbol{H}$ and $\boldsymbol{B}$ by simple algebraic relations, at least in suitable frames of reference, the so-called LORENTZ systems. In contrast to that the fields $q$ and $\boldsymbol{j}$ are connected to the fields $\boldsymbol{E}$ and $\boldsymbol{B}$ by complex differential relations. We will get back to that in context with the so-called MAXWELL-LORENTZ aether relations.

Finally we will derive the *local* MAXWELL equations in *regular points* from Eqs. (13.4.4/13.4.5). If we assume that there is *no* singular surface intersecting with the volume $V$, we can apply GAUSS' theorem in its usual form rewrite Eq. (13.4.4) immediately:

$$\iiint\limits_V \frac{\partial D_i}{\partial x_i}\,\mathrm{d}V = \iiint\limits_V q(\boldsymbol{x},t)\,\mathrm{d}V \Rightarrow \frac{\partial D_i}{\partial x_i} = q. \tag{13.4.8}$$

André Marie AMPÈRE was born on January 20, 1775 in Polémieux near Lyon and died on June 10, 1836 in Marseille. Rumor has it that he did not attend school but got his education in a purely autodidactic manner. From 1799 to 1801 he served as a mathematics teacher at the École Centrale in Lyon. In 1802 he published a mathematical paper on game theory, another note on theoretical mechanics, and a treatise on partial differential equations. The latter got him a membership in the French Academy of Sciences. However, his interest in mathematics tired down and AMPÈRE turned to science instead. After a short fling with chemistry he started in 1820 to perform physical experiments with live wires and investigated their effects on magnetic needles, probably inspired by ØRSTED's experiments. He came up with the hypothesis that magnetism of any kind has its origin in electric currents and that, in fact, electric currents generate magnetic fields. As a final consequence he had to assume that the molecules of every magnet produced a small circular current.

If there is a singular surface, $A$, the pill box argument from Sect. 3.7 applies and we conclude that in a point of this surface characterized by the unit normal $\boldsymbol{e}$ (cf., Fig. 13.4) the following relation holds:

$$[\![D_i]\!]\, e_i = \underset{A}{q}\,. \tag{13.4.9}$$

Moreover, for regular open surfaces the following simplified version of Eq. (13.4.5) results:

$$\frac{\mathrm{d}}{\mathrm{d}t}\iint\limits_{S} D_i n_i\,\mathrm{d}A = \oint\limits_{\partial S}\left[H_i + (\boldsymbol{D}\times\boldsymbol{v})_i\right]\tau_i\,\mathrm{d}l - \iint\limits_{S} j_i n_i\,\mathrm{d}A. \tag{13.4.10}$$

By application of the transport theorems of Eq. (13.2.14) we find for the left side:

$$\frac{\mathrm{d}}{\mathrm{d}t}\iint\limits_{S} D_i n_i\,\mathrm{d}A = \iint\limits_{S}\left[\frac{\partial D_i}{\partial t} + \upsilon_i\frac{\partial D_k}{\partial x_k}\right] n_i\,\mathrm{d}A + \oint\limits_{\partial S}(\boldsymbol{D}\times\boldsymbol{v})_i\,\tau_i\,\mathrm{d}l. \tag{13.4.11}$$

Thus by observing Stokes' theorem in the form of Eq. (13.2.13), combining both equations, and taking Eq. $(13.4.8)_2$ into account we find the following local relation for regular points:

$$-\frac{\partial D_i}{\partial t} + (\nabla\times\boldsymbol{H})_i = j_i + q\upsilon_i. \tag{13.4.12}$$

If we use an argument analogous to the pillbox argument but for a small loop encircling the singular line, $L$, the following local equation results for singular points:

$$(\boldsymbol{e}\times[\![\boldsymbol{H}]\!])_i + [\![D_i]\!]\,\underset{A}{\upsilon_\perp} = \underset{L}{j}_i + \underset{A}{q}\,w_i, \tag{13.4.13}$$

where $w_i$ is the velocity of the singular line, which is not necessarily a material one, and the remarks and the nomenclature in context with Eq. (3.4.11) apply.

**Exercise 13.4.1:** ***Surface flux balance and local equations***

The general global balance equation for a vector flux density $\gamma_i$ across an open surface $S$ is defined in complete analogy to Eq. (3.3.6):

$$\frac{\mathrm{d}}{\mathrm{d}t}\iint\limits_{S}\gamma_i n_i\,\mathrm{d}S = -\oint\limits_{\partial S}\phi_i\,\tau_i\,\mathrm{d}l + \iint\limits_{S^+\cup S^-}(p_i + s_i)\,n_i\,\mathrm{d}S + \int\limits_{L}\left(\underset{L}{p}_i + \underset{L}{s}_i\right)\nu_i\,\mathrm{d}l. \tag{13.4.14}$$

Interpret this equation in terms of fluxes, supplies, and productions. Then recall the transport theorem from Exercise 13.2.3 and use it to confirm the following local equations:

$$\frac{\partial\gamma_i}{\partial t} + \frac{\partial\gamma_l}{\partial x_l}\upsilon_i + (\nabla\times(\gamma\times\boldsymbol{v}))_i = p_i + s_i,\ \boldsymbol{e}\times[\![\gamma\times\boldsymbol{v}+\boldsymbol{\phi}]\!] - [\![\gamma]\!]\,\underset{A}{\upsilon_\perp} = \boldsymbol{p}+\boldsymbol{s} \tag{13.4.15}$$

for regular and singular points, respectively. Note that in order to find Eq. $(13.4.15)_1$ contributions from singular lines must simply be disregarded. For a proof of Eq. $(13.4.15)_2$ consider a small closed loop around $L$ (similar to the pillbox argument presented in Chap. 3. Then recall that $\boldsymbol{v} = \boldsymbol{e} \times \boldsymbol{t}$, $\boldsymbol{e} \cdot \boldsymbol{t} = 0$, $\boldsymbol{e} \times \boldsymbol{e} = \boldsymbol{0}$, and $\underset{A}{\upsilon_\perp} = \boldsymbol{w} \cdot \boldsymbol{e}$. Make use of (and prove) the vector identity $\boldsymbol{a} \cdot (\boldsymbol{b} \times \boldsymbol{c}) = (\boldsymbol{a} \times \boldsymbol{b}) \cdot \boldsymbol{c}$. Finally, assume that the vectors of singular production and supply are tangential to the singular surface.

**Exercise 13.4.2:** ***MAXWELL's equations in regular and in singular points***

Explain in detail how Eqs. (13.4.8/13.4.9) result from the global balances. For this purpose use GAUSS' theorem with and without extension to singular surfaces (recall in this context the results from Exercise 3.4.1). Moreover, show that because of the conservation law for the magnetic flux according to Eqs. (13.2.6) the following jump condition for the magnetic induction in singular points must hold:

$$[\![B_i]\!]\, e_i = 0. \tag{13.4.16}$$

Now apply the results of Exercise 13.4.2 and identify for the case of the current potential of Eq. (13.4.5) the corresponding quantities $\gamma$, $\boldsymbol{\phi}$, etc., and prove the validity of the jump condition (13.4.13). Next show in a similar manner by means of FARADAY's law from Eq. (13.2.4) that for points of the singular line $L$ we also have:

$$(\boldsymbol{e} \times [\![\boldsymbol{E}]\!])_i - [\![B_i]\!]\, \underset{A}{\upsilon_\perp} = 0. \tag{13.4.17}$$

Finally apply REYNOLDS' transport theorem for volumes to Eq. (13.4.2) and show that in regular points:

$$\frac{\partial q}{\partial t} + \frac{\partial}{\partial x_i}(j_i + q\upsilon_i) = 0. \tag{13.4.18}$$

**Exercise 13.4.3:** ***An application of the jump condition for the electric field***

Assume that the electrons in a metal, i.e., an electrically conductive material can basically be moved freely. Consider now a charge $q$ in front of a metallic object. At which angle do the electric field lines $\boldsymbol{E}$ impinge on the surface of that object? To find the answer decompose the vector $\boldsymbol{E}$ into a normal and a tangential part as follows:

$$\boldsymbol{E} = E_n \boldsymbol{n} + E_t \boldsymbol{t}. \tag{13.4.19}$$

Recall Sect. 13.3 and interpret the vector $\boldsymbol{E}$ in terms of a force acting on the free electrons in the metal. Conclude in which direction they should move and assume that the force is too weak to pull the electrons out of the metal surface.

In a second step consider a point charge in front of an infinitely large, plane metallic surface and sketch the electric field lines in front and behind the plate. What do we conclude about the electric field behind the wall considering the jump condition (13.4.17) and the pervious result regarding the angle of entry of the electric field lines on a metallic surface? In this context explain the principle of a FARADAY cage, i.e., the complete shielding of electromagnetic fields within a closed metallic grid. Is this a way to protect yourself from poisonous, carcinogenic electro-smog or distracting cell phone calls?

## 13.5 Measuring the Charge and Current Potentials

First, we consider a point charge of intensity $Q_0$ at rest (e.g., an electron). We are interested in its charge potential. For reasons of symmetry it is reasonable to assume that the corresponding $\boldsymbol{D}$ field is time-independent and its components in spherical coordinates depend only on the radial distance $r$ from the point charge (see Fig. 13.6, left):

$$\boldsymbol{D} = D_{\langle r \rangle}(r)\, \boldsymbol{e}_r + D_{\langle \vartheta \rangle}(r)\, \boldsymbol{e}_\vartheta + D_{\langle \varphi \rangle}(r)\, \boldsymbol{e}_\varphi. \tag{13.5.1}$$

Paul Adrien Maurice DIRAC was born on August 8, 1902 in Bristol and died on October 20, 1984 in Tallahassee. He contributed fundamentally to quantum mechanics, quantum electrodynamics, to the theory of the electron, and to the properties of antimatter. Being British to the core DIRAC always kept a stiff upper lip and spoke only if absolutely required. This is confirmed by many anecdotes, for example: When DIRAC made a rare error in an equation on the blackboard during a lecture one day, a courageous student raised his hand: "Professor DIRAC," he declared, "I do not understand equation 2." When DIRAC continued writing, the student, assuming that he had not been heard, raised his hand again and repeated his remark. Again DIRAC merely continued writing ... "Professor DIRAC," another student finally interjected, "that man is asking a question." "Oh?" DIRAC replied. "I thought he was making a statement."

Due to the perfect spherical symmetry it is reasonable to assume that $D_{\langle \varphi \rangle}(r) = D_{\langle \vartheta \rangle}(r)$. This is already all the information we can get from symmetry conditions. We will now evaluate Eq. (13.4.4), i.e., the basic relation defining the charge potential. To this end we rewrite the point charge in terms of a DIRAC delta

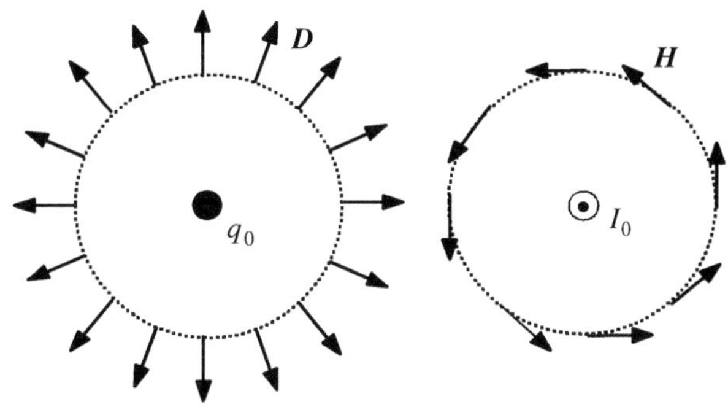


**Fig. 13.6** *D*-field around a point charge and *H*-field around a line current

function, i.e., $q = Q_0\delta(r)$. The scalar product can easily be evaluated by making use of the orthogonality of the base vectors in spherical coordinates. We obtain the following relation for the component $D_{\langle r\rangle}$ of the charge potential:

$$\oint\!\!\!\oint_{\partial V} D_i n_i \, \mathrm{d}A \equiv \oint\!\!\!\oint_{\partial V} D_{\langle r\rangle} \, \mathrm{d}A = Q_0 \iiint_V \delta(r) \, \mathrm{d}V \;\Rightarrow D_{\langle r\rangle} \oint\!\!\!\oint_{\partial V} \mathrm{d}A = Q_0$$
$$\Rightarrow \; D_{\langle r\rangle} = \frac{Q_0}{4\pi r^2}. \tag{13.5.2}$$

Thus for a known amount of charge, $Q_0$, and distance, $r$, Eq. (13.5.2) constitutes a rule of measurement for the charge potential or, more precisely, for its radial component. Note that for the time being we have no information about the two other components. However, as we shall see below they can both be rescaled to zero. Moreover, note that extended distributions of charge can, in principle, be considered as an ensemble of point charges, and we may apply the principle of superposition for vectors to obtain the corresponding total $\boldsymbol{D}$ field.

We now consider a time-independent, i.e., stationary electric current in an infinitely long wire at rest. We are interested in measuring the resulting cylindrically symmetric magnetic field $\boldsymbol{H}$ around the wire, which is also time-independent. In cylindrical coordinates we may write:

$$\boldsymbol{H} = H_{\langle r\rangle}\boldsymbol{e}_r + H_{\langle\vartheta\rangle}\boldsymbol{e}_\vartheta + H_{\langle z\rangle}\boldsymbol{e}_z. \tag{13.5.3}$$

As we shall find out below the radial component can be rescaled to zero. This corresponds to experience according to which currents may only produce rotational fields. Thus:

$$H_{\langle r\rangle} = 0. \tag{13.5.4}$$

Moreover, the line current is infinitely long. Thus the $z$ direction does not matter and we may put:

$$H_{\langle z\rangle} = 0. \tag{13.5.5}$$

Due to the cylindrical symmetry the magnetic field can only depend on the distance $r$ from the wire (see Fig. 13.6):

$$H_{\langle\vartheta\rangle} = H_{\langle\vartheta\rangle}(r). \tag{13.5.6}$$

Finally we use a DIRAC delta function to express the line current as follows:

$$\boldsymbol{j} = I_0 \delta(r)\, \boldsymbol{e}_z. \tag{13.5.7}$$

If we insert this into ØRSTED-AMPÈRE's law from Eq. (13.4.5) ($\boldsymbol{\tau} \equiv \boldsymbol{e}_\vartheta$, $\boldsymbol{n} \equiv \boldsymbol{e}_z$) we obtain:

$$0 = \oint\limits_{\partial S} H_i \tau_i \,\mathrm{d}l - \iint\limits_{S} j_i n_i \,\mathrm{d}A \Rightarrow H_{\langle\vartheta\rangle} \oint\limits_{\partial S} \mathrm{d}l = I_0 \iint\limits_{S} \delta(r)\,\mathrm{d}A \Rightarrow H_{\langle\vartheta\rangle} = \frac{I_0}{2\pi r}. \tag{13.5.8}$$

This equation constitutes a rule of measurement for the magnetic field for a given current, $I_0$, and a given distance, $r$. It allows us to determine the current potential $\boldsymbol{H}$ from other, more primitive quantities. If there are several currents the superposition principle must be applied.

**Exercise 13.5.1:** ***Technical measurement of D und H and their units***

Use the results from the previous *gedankenexperiments* and show that the units of charge and current potentials are given by:

$$\dim[\boldsymbol{D}] = \frac{\mathrm{C}}{\mathrm{m}^2} = \frac{\mathrm{As}}{\mathrm{m}^2},\ \dim[\boldsymbol{H}] = \frac{\mathrm{A}}{\mathrm{m}} = \frac{\mathrm{C}}{\mathrm{ms}}. \tag{13.5.9}$$

Moreover, consult textbooks of electrical engineering and explain how both potentials are measured in daily practice. Distinguish them from the technical rules of measurement for $\boldsymbol{E}$ and $\boldsymbol{B}$ and pay particular attention to mutual independence.

Finally the following remark should be made in this section on the principles of point wise measurement of the $\boldsymbol{D}$ and $\boldsymbol{H}$ fields. Our motivation of the spherically symmetric *ansatz* for the $\boldsymbol{D}$ field around a point charge was quite heuristically. This concerns in particular the argument that only the radial component $D_{\langle r\rangle}(r)$ is important. The same holds for the cylindrically symmetric *ansatz* for $\boldsymbol{H}$ around an infinitely long line current, where it was claimed that only the tangential component $H_{\langle\vartheta\rangle}(r)$ counts.

However, there is a more formal argument, which adds to the credibility of this heuristic approach. It turns out that the $\boldsymbol{D}$ field is only determined up to the curl of a vector field $\boldsymbol{Z}$, i.e., $\nabla \times \boldsymbol{Z}$, and the $\boldsymbol{H}$ field is only unique up to a gradient $\zeta$, i.e., $\nabla\zeta$. For a proof we write:

$$\boldsymbol{D}' = \boldsymbol{D} + \nabla \times \boldsymbol{Z}, \boldsymbol{H}' = \boldsymbol{H} + \nabla \zeta. \qquad (13.5.10)$$

If we insert these expressions in COULOMB's or in ØRSTED-AMPÈRE's law (13.4.4/13.4.5), MAXWELL's equations remain unchanged, since we may write for the corresponding expressions therein:

$$\oiint_{\partial V} \boldsymbol{D}' \cdot \boldsymbol{n} \, \mathrm{d}A = \oiint_{\partial V} \boldsymbol{D}' \cdot \boldsymbol{n} \, \mathrm{d}A + \oiint_{\partial V} \nabla \times \boldsymbol{Z} \cdot \boldsymbol{n} \, \mathrm{d}A, \qquad (13.5.11)$$

and

$$\oint_{\partial S} \boldsymbol{H}' \cdot \boldsymbol{\tau} \, \mathrm{d}l = \oint_{\partial S} \boldsymbol{H} \cdot \boldsymbol{\tau} \, \mathrm{d}l + \oint_{\partial S} \nabla \zeta \cdot \boldsymbol{\tau} \, \mathrm{d}l. \qquad (13.5.12)$$

If we now apply GAUSS' and STOKES' theorem, respectively (to this end we assume that $\nabla \times \boldsymbol{Z}$ and $\nabla \zeta$ are continuous fields) we conclude that:

$$\oiint_{\partial V} \nabla \times \boldsymbol{Z} \cdot \boldsymbol{n} \, \mathrm{d}A = \oiint_{\partial V} \epsilon_{ijk} \frac{\partial Z_k}{\partial x_j} n_i \, \mathrm{d}A = \iiint_V \epsilon_{ijk} \frac{\partial^2 Z_k}{\partial x_i \partial x_j} \mathrm{d}V \equiv 0, \qquad (13.5.13)$$

because $\epsilon_{ijk} \frac{\partial^2 Z_k}{\partial x_i \partial x_j} = 0$ and:

$$\oint_{\partial S} \nabla \zeta \cdot \boldsymbol{\tau} \, \mathrm{d}l = \oint_{\partial S} \frac{\partial \zeta}{\partial x_i} \cdot \tau_i \, \mathrm{d}l = \iint_S \epsilon_{ijk} \frac{\partial^2 \zeta}{\partial x_j \partial x_k} \mathrm{d}A_i \equiv 0, \qquad (13.5.14)$$

because $\epsilon_{ijk} \partial^2 \zeta / \partial x_j \partial x_k = 0$. Consequently we choose the fields $\boldsymbol{Z}$ and $\zeta$ such that we enforce in Eqs. (13.5.1/13.5.3) the following conditions:

$$D'_{\langle \varphi \rangle} = 0, \; D'_{\langle \vartheta \rangle} = 0, \; H'_{\langle \vartheta \rangle} = 0. \qquad (13.5.15)$$

**Exercise 13.5.2:** ***Gauge fixing of D and H***

Use the general relations for the so-called curl (i.e., $\nabla \times \boldsymbol{Z}$) of a vector field $\boldsymbol{Z}$ in spherical coordinates and for the gradient of a scalar field in cylindrical coordinates, respectively (i.e., apply the rules from Chap. 4) and show that due to the perfect spherical and cylindrical symmetry (note that $\boldsymbol{Z}$ was expressed in physical coordinates) we have:

$$\nabla \times \boldsymbol{Z} = -\boldsymbol{e}_\vartheta \frac{1}{r} \frac{\partial \left( r Z_{\langle \varphi \rangle} \right)}{\partial r} + \boldsymbol{e}_\varphi \frac{1}{r} \frac{\partial \left( r Z_{\langle \vartheta \rangle} \right)}{\partial r}, \; \nabla \zeta = \boldsymbol{e}_r \frac{\partial \zeta}{\partial r}. \qquad (13.5.16)$$

Now use the equations from above and show that the fields $\boldsymbol{Z}$ and $\zeta$ must be chosen as follows in order to guarantee that in the case of the $\boldsymbol{D}$ field around a point charge only a radial component $D_{\langle r \rangle}(r)$, and in case of the $\boldsymbol{H}$

field around an infinitely long line current only a tangential component $H_{\langle\vartheta\rangle}(r)$ survive:

$$Z_{\langle r\rangle} = \text{arbitrary},\ Z_{\langle\varphi\rangle} = \frac{1}{r}\int rD_{\langle\vartheta\rangle}\mathrm{d}r + C_\varphi,$$
$$Z_{\langle\vartheta\rangle} = -\frac{1}{r}\int rD_{\langle\varphi\rangle}\,\mathrm{d}r + C_\vartheta,\ \zeta = -\int H_{\langle r\rangle}\,\mathrm{d}r + C_r \tag{13.5.17}$$

with arbitrary constants of integration $C_\varphi$, $C_\vartheta$, and $C_r$.

## 13.6 Decomposition of the Total Charge, Polarization, Rewriting COULOMB's Law

It is customary in electrodynamics to decompose the total volumetric charge density $q$ additively into a *true* (a.k.a. *free*) charge density, $q^{\mathrm{f}}$, and a *polarization* (a.k.a. *induced*) *charge density*, $q^{\mathrm{p}}$:

$$q = q^{\mathrm{f}} + q^{\mathrm{p}}. \tag{13.6.1}$$

Polarization charges develop in so-called *dielectric* materials. These are induced in the material if it is subjected to an electric field $\boldsymbol{E}$. The polarization can be envisaged as indicated in Fig. 13.7: Due to the force action associated with the presence of an $\boldsymbol{E}$ field the electric charges within the molecules are separated and electric dipoles are formed. If we now consider within this piece of matter a material volume $V(t)$ in the continuum sense, the system boundaries $\partial V$ will "cut" through the dipoles and, speaking from a phenomenological point of view, we obtain an additional amount of electric charge on top of the true charges within the volume. In a manner of speech these charges are *fictitious*. They come into being only after the cut with the system boundaries and they must be considered as a quasi surface charge. In order to describe this contribution mathematically we define the so-called polarization vector, $\boldsymbol{P}$, (or *polarization* for short) as the product between the number density of dipoles, $n^{\mathrm{p}}$, measured in $1/\mathrm{m}^3$ with the dipole moment, $\boldsymbol{p}$. The dipole moment is nothing else but the product between the amount of charge separated in a dipole, $e$ (a positive number in units of C), and the

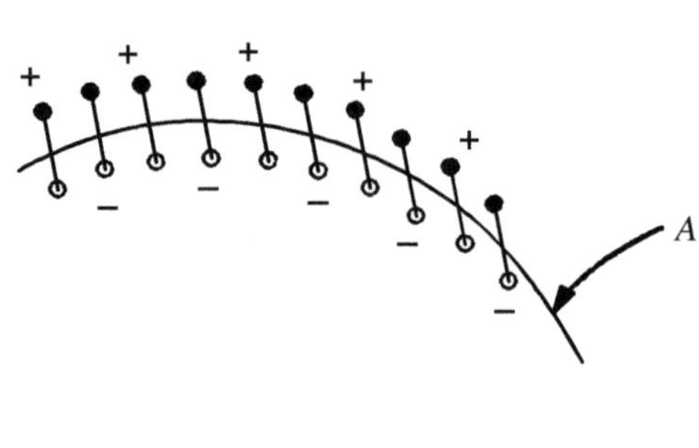


**Fig. 13.7** Cut through the surface of a polarized, dielectric material

distance vector, $\boldsymbol{\delta}$ (in units of m), from the negative to the positive centers of the dipole charges (note the analogy to the force couple known from mechanics). Thus we write:

$$\boldsymbol{P} = n^{\mathrm{p}}\boldsymbol{p} = n^{\mathrm{p}} e\, \boldsymbol{\delta}. \tag{13.6.2}$$

Consequently, the unit of polarization is $\mathrm{C/m^2}$, which corresponds perfectly to our interpretation of polarization as a surface charge density. The total charge in Eq. (13.4.1) can now be written as:

$$Q = \iiint\limits_{V^+\cup V^-} q\,\mathrm{d}V + \iint\limits_{A} q_{A}\,\mathrm{d}A = \iiint\limits_{V^+\cup V^-} q^{\mathrm{f}}\,\mathrm{d}V - \oiint\limits_{\partial V} P_i n_i\,\mathrm{d}A + \iint\limits_{A} q_{A}\,\mathrm{d}A. \tag{13.6.3}$$

The negative sign in context with the polarization vector, $\boldsymbol{P}$, can easily be interpreted (cf., Fig. 13.7): If positive dipole moments are dissected by $\partial V$, the remaining volume is negatively charged. Moreover, the scalar product $\boldsymbol{P}\cdot\boldsymbol{n}$ is what counts in the balance, since a dipole tangential to the surface gives no contribution at all. If we require that there are no singular surfaces within $V(t)$ and polarization is continuous on the surface, GAUSS theorem can be applied to Eq. (13.6.3) to reveal that:

$$Q = \iiint\limits_{V^+\cup V^-} \left(q^{\mathrm{f}} - \frac{\partial P_i}{\partial x_i}\right)\mathrm{d}V + \iint\limits_{A} q_{A}\,\mathrm{d}A. \tag{13.6.4}$$

These relations are inserted in COULOMB's law from Eq. (13.4.4). We find:

$$\oiint\limits_{\partial V} (D_i + P_i)\, n_i\,\mathrm{d}A = \iiint\limits_{V^+\cup V^-} q^{\mathrm{f}}\,\mathrm{d}V + \iint\limits_{A} q_{A}\,\mathrm{d}A \tag{13.6.5}$$

or:

$$\iiint\limits_{V} \frac{\partial}{\partial x_i}(D_i + P_i)\,\mathrm{d}V = \iiint\limits_{V^+\cup V^-} q^{\mathrm{f}}\,\mathrm{d}V + \iint\limits_{A} q_{A}\,\mathrm{d}A. \tag{13.6.6}$$

We now define the so-called *charge potential for matter*, $\mathfrak{D}(\boldsymbol{x},t)$, by:

$$\mathfrak{D}_i = D_i + P_i. \tag{13.6.7}$$

Consequently, COULOMB's law can also be written as follows:

$$\oiint\limits_{\partial V} \mathfrak{D}_i n_i\,\mathrm{d}A = \iiint\limits_{V^+\cup V^-} q^{\mathrm{f}}\,\mathrm{d}V + \iint\limits_{A} q_{A}\,\mathrm{d}A. \tag{13.6.8}$$

or:

$$\iiint\limits_{V} \frac{\partial \mathfrak{D}_i}{\partial x_i}\,\mathrm{d}V = \iiint\limits_{V^+\cup V^-} q^{\mathrm{f}}\,\mathrm{d}V + \iint\limits_{A} q_{A}\,\mathrm{d}A. \tag{13.6.9}$$

Thus we find for regular points of the volume:

$$\frac{\partial \mathfrak{D}_i}{\partial x_i} = q^{\mathrm{f}} \tag{13.6.10}$$

and for points within the singular surface $S$:

$$[\![\mathfrak{D}_i]\!]\, e_i = \underset{A}{q}\,. \tag{13.6.11}$$

Moreover, we may write because of Eqs. (13.6.4) and (13.6.1):

$$q^{\mathrm{p}} = -\frac{\partial P_i}{\partial x_i},\ q = q^{\mathrm{f}} - \frac{\partial P_i}{\partial x_i}. \tag{13.6.12}$$

## 13.7 Decomposition of the Total Currents, Magnetization, Rewriting Ørsted-Ampère's Law

Analogously to charge we now decompose the total current density into the *current density* of the *true* (or *free*) charges, *polarization current density*, and *magnetization current density*:

$$j_i = j_i^{\mathrm{f}} + j_i^{\mathrm{p}} + j_i^{\mathrm{m}}. \tag{13.7.1}$$

The current of the free charges for an open material surface (for simplicity without a singular line) is given by:

$$J^{\mathrm{f}} = \iint_S j_i^{\mathrm{f}} n_i\, \mathrm{d}S. \tag{13.7.2}$$

A polarization current results if the polarization charge on the open surface of Fig. 13.8 changes with time:

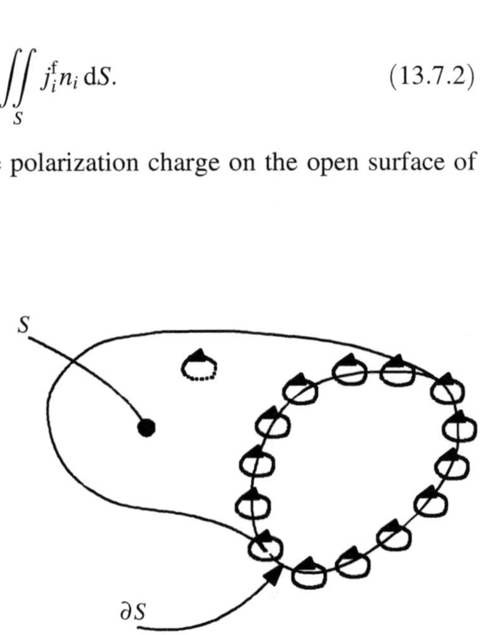


**Fig. 13.8** Magnetization current visualized by elementary current loops

$$J^{\mathrm{p}} = \iint\limits_{S} j_i^{\mathrm{p}} n_i \, \mathrm{d}S = \frac{\mathrm{d}}{\mathrm{d}t} \iint\limits_{S} P_i n_i \, \mathrm{d}S. \tag{13.7.3}$$

Finally, as illustrated in Fig. 13.8, the closed line $\partial S$ may cut through magnetic dipoles. Following an idea by AMPÈRE these magnetic dipoles may be envisaged as circular loops of electric current on an atomic level. Indeed, a circular wire carrying an electric current of intensity $I$, and enclosing a directed surface, $\boldsymbol{A}$, generates a magnetic dipole of strength $\boldsymbol{m} = I\boldsymbol{A}$. Similarly we may interpret the magnetic field generated by electrons circling around an atom as elementary magnetic dipoles. We have such atomic current loops in mind when we talk phenomenologically about magnetization currents and write:

$$J^{\mathrm{m}} = \iint\limits_{S} j_i^{\mathrm{m}} n_i \, \mathrm{d}S = \oint\limits_{\partial S} n^{\mathrm{m}} m_i \tau_i \, \mathrm{d}l = \oint\limits_{\partial S} n^{\mathrm{m}} I A_i \tau_i \, \mathrm{d}l = \oint\limits_{\partial S} M_i \tau_i \, \mathrm{d}l, \tag{13.7.4}$$

where $n^{\mathrm{m}}$ is the density of elementary current loops and $\boldsymbol{M}$ the so-called magnetization. Thus ØRSTED-AMPÈRE's law (13.4.5) can be rewritten as follows:

$$\begin{aligned} \frac{\mathrm{d}}{\mathrm{d}t} \iint\limits_{S} (D_i + P_i) \, n_i \, \mathrm{d}A = & \oint\limits_{\partial S} \left[ H_i - M_i + (\boldsymbol{D} \times \boldsymbol{\upsilon})_i \right] \tau_i \, \mathrm{d}l \\ & - \iint\limits_{S^+ \cup S^-} j_i^{\mathrm{f}} n_i \, \mathrm{d}A - \int\limits_{L} \underset{L}{j}_i \, v_i \, \mathrm{d}l. \end{aligned} \tag{13.7.5}$$

One may ask why the loop currents pierce only through the line $\partial S$ and not also through the open surface $S$. Indeed, this does happen (see Fig. 13.8). However, in that case the currents first enter *and* then leave the surface. In total there is no contribution to the magnetization current. We now use the transport theorem from Eq. (13.2.14) and obtain:

$$\begin{aligned} & \iint\limits_{S} \left[ \frac{\partial (D_i + P_i)}{\partial t} + \upsilon_i \frac{\partial (D_k + P_k)}{\partial x_k} \right] n_i \, \mathrm{d}S + \int\limits_{\partial S} \left[ (\boldsymbol{D} + \boldsymbol{P}) \times \boldsymbol{\upsilon} \right]_i \tau_i \, \mathrm{d}l \\ & - \int\limits_{L} \left( \left[\!\left[ (\boldsymbol{D} + \boldsymbol{P}) \times \boldsymbol{\upsilon} \right]\!\right] \times \boldsymbol{w} \right)_i t_i \, \mathrm{d}l = \oint\limits_{\partial S} \left[ H_i - M_i + (\boldsymbol{D} \times \boldsymbol{\upsilon})_i \right] \tau_i \, \mathrm{d}l \\ & - \iint\limits_{S^+ \cup S^-} j_i^{\mathrm{f}} n_i \, \mathrm{d}A - \int\limits_{L} \underset{L}{j}_i \, v_i \, \mathrm{d}l. \end{aligned} \tag{13.7.6}$$

We now define the current potential in matter by:

$$\mathfrak{H}_i = H_i - M_i - (\boldsymbol{P} \times \boldsymbol{\upsilon})_i \tag{13.7.7}$$

and use Eq. (13.6.10) for further transformations:

$$\iint_S \left[\frac{\partial \mathfrak{D}_i}{\partial t} + \upsilon_i q^{\mathrm{f}}\right] n_i \, \mathrm{d}S - \int_L ([\![(\mathfrak{D} + \boldsymbol{P}) \times \boldsymbol{\upsilon}]\!] \times \boldsymbol{w})_i \, t_i \, \mathrm{d}l$$
$$= \oint_{\partial S} \mathfrak{H}_i \, \tau_i \, \mathrm{d}l - \iint_{S^+ \cup S^-} j_i^{\mathrm{f}} n_i \, \mathrm{d}A - \int_L j_i^L \nu_i \, \mathrm{d}l. \tag{13.7.8}$$

Consequently, we obtain in regular points of the open surface:

$$-\frac{\partial \mathfrak{D}_i}{\partial t} + (\nabla \times \mathfrak{H})_i = j_i^{\mathrm{f}} + q^{\mathrm{f}} \upsilon_i. \tag{13.7.9}$$

This corresponds (for matter at rest) exactly to the first MAXWELL equation in (13.1.1) from Becker [1].

## 13.8 The MAXWELL-LORENTZ Aether Relations

In summary: The MAXWELL equations, e.g., in the form for regular points:

$$\nabla \cdot \boldsymbol{B} = 0, \; \frac{\partial \boldsymbol{B}}{\partial t} + \nabla \times \boldsymbol{E} = \boldsymbol{0}, \tag{13.8.1}$$

$$\nabla \cdot \boldsymbol{D} = q, \; -\frac{\partial \boldsymbol{D}}{\partial t} + \nabla \times \boldsymbol{H} = \boldsymbol{j} + q\boldsymbol{\upsilon},$$

represent two groups of partial differential equations for the fields $\boldsymbol{E}$, $\boldsymbol{B}$, $\boldsymbol{D}$, and $\boldsymbol{H}$, i.e., twelve unknowns, and four other fields, namely the charge and current densities $q$ and $\boldsymbol{j}$, respectively, or derivates thereof in form of true charges and currents, polarization, or magnetization. MAXWELL's equations consist of two vector and two scalar relations, i.e., eight equations in total. In short, the whole system is extremely underdetermined. If, in a first step, we could find a connection between the fields $\boldsymbol{E}$ and $\boldsymbol{D}$ on the one hand side and $\boldsymbol{B}$ and $\boldsymbol{H}$ on the other, the situation would improve considerably. In a second step we would "only" have to provide constitutive equations for the four unknown fields charge density (a scalar) and the current density (a vector) by relating these (for example) to $\boldsymbol{E}$ and $\boldsymbol{B}$.

Indeed, there exist such relations between the set of fields $\boldsymbol{E}$ and $\boldsymbol{D}$ and the set $\boldsymbol{B}$ and $\boldsymbol{H}$, respectively. They are known as the MAXWELL-LORENTZ aether relations. However, their specific mathematical form depends on the frame of reference. As we are about the see they can assume a rather complex algebraic form, if we choose a Euclidean observer, for example.

However, in an *inertial frame* they have a particularly simple form. Here they are simply proportional to each other:

$$\boldsymbol{D} = \varepsilon_0 \boldsymbol{E},\ \boldsymbol{H} = \frac{1}{\mu_0}\boldsymbol{B}. \tag{13.8.2}$$

The two parameters $\varepsilon_0$ and $\mu_0$ are known as the dielectric constant and the permeability of the vacuum, respectively. By means of measurements of the kind described in Sects. 13.3 and 13.5, i.e., in a terrestrial lab system at rest (which to a good approximation can be considered as an inertial system) it was found that:

$$\varepsilon_0 = 8.85 \cdot 10^{-12}\,\frac{\mathrm{As}}{\mathrm{Vm}},\ \mu_0 = 12.6 \cdot 10^{-7}\,\frac{\mathrm{Vs}}{\mathrm{Am}}. \tag{13.8.3}$$

Heinrich Rudolf HERTZ was born on February 22, 1857 in Hamburg and died on January 1, 1894 in Bonn. He was born into a family of lawyers. Nevertheless, he dedicates his life to rational thinking. In 1877 he goes to Munich and begins to study engineering. However, he soon realizes that his true love belongs to the pure sciences. Thus he turns to Berlin and studies under the famous physicists HELMHOLTZ and KIRCHHOFF. He finishes his Ph.D. in the field of electrodynamics with highest honors and becomes HELMHOLTZ' assistant in 1880. He gets interested in optical experiments, investigates the nature of NEWTONian rings and develops for this purpose a method to compute the stresses and strains of elastic bodies pressed against each other, the so-called theory of HERTZian contact.

Jean-Baptiste BIOT was born on April 21, 1774 in Paris where he also died on February 3, 1862. In 1797 he became a professor of mathematics at the École Centrale in Beauvais, in 1800 a professor of physics at the Collège de France in Paris, and in 1809 a professor of astronomy. Together with the already mentioned scientist GAY-LUSSAC he took a ride on a hydrogen balloon up to a height of 4000 m. BIOT was also interested in geodesy and improved and extended the measurements of the meridian which had been performed until then. He also discovered the effect of birefringence of mica, which explains why the mineral BIOTit was named after him.

What happens if these relations are inserted into Eq. (13.8.1), is investigated in the following exercise.

**Exercise 13.8.1:** ***The vacuum solutions of Maxwell's equations after Hertz***

Insert the Maxwell-Lorentz aether relations (13.8.2) into Eqs. (13.8.1). Specialize to the case of a vacuum (i.e., a space free charges, currents, and matter) and show that:

$$\nabla \cdot \boldsymbol{B} = 0, \ \frac{\partial \boldsymbol{B}}{\partial t} + \nabla \times \boldsymbol{E} = 0, \qquad (13.8.4)$$

$$\nabla \cdot \boldsymbol{E} = 0, \ -\frac{\partial \boldsymbol{E}}{\partial t} + c^2 \nabla \times \boldsymbol{B} = 0,$$

with $c = (\varepsilon_0 \mu_0)^{-1/2} \approx 3 \cdot 10^8 \mathrm{m/s}$ being the speed of light in vacuum. Decouple the equations by suitable differentiation and mutual insertion to arrive at the following wave equations:

$$\frac{\partial^2 \boldsymbol{E}}{\partial t^2} = c^2 \Delta \boldsymbol{E}, \ \frac{\partial^2 \boldsymbol{B}}{2} = c^2 \Delta \boldsymbol{B}, \text{ with } \Delta = \nabla \cdot \nabla. \qquad (13.8.5)$$

What kind of waves are these, in which medium do they propagate and how? Why were 19th century scientists particularly astonished about this discovery?

**Exercise 13.8.2:** ***Coulomb' law in traditional form and the definition of the unit Ampere (Biot-Savart's law)***

Use results from Sect. 13.5, specifically:

$$\boldsymbol{D} = D_{\langle r \rangle}(r)\, \boldsymbol{e}_r, \text{ with } D_{\langle r \rangle} = \frac{Q_0}{4\pi r^2}, \qquad (13.8.6)$$

the Maxwell-Lorentz aether relations, and the arguments in context with the electric field presented in Sect. 13.3 and show that a field generating charge, $Q_0$, exerts the following force on a test charge, $Q$:

$$\boldsymbol{F} = \frac{1}{4\pi\varepsilon_0} \frac{Q_0 Q}{r^2} \boldsymbol{e}_r \equiv \frac{1}{4\pi\varepsilon_0} \frac{Q_0 Q}{r^3} \boldsymbol{r}. \qquad (13.8.7)$$

Interpret this result as Coulomb's law known from high school. Now consider in an analogous manner two electric currents and the corresponding magnetic fields. Prove that the amount of force per unit length between two parallel, infinitely long wires is given by:

$$F = \frac{\mu_0}{2\pi} \frac{I_0 I}{r}. \qquad (13.8.8)$$

For this purpose follow the arguments of Sects. 13.3 and 13.5. In particular show that the quantity $I_0$ in Eq. (13.8.8) corresponds to the current of the wire generating an $\boldsymbol{H}$ field and $I$ represents the current in the wire on which a force is acting due to the resulting $\boldsymbol{B}$ field in the MAXWELL-LORENTZ aether relations. Explain how Eq. (13.8.8) is used for the definition of the SI-unit of electric currents, the Ampere. Equation (13.8.8) is sometimes a.k.a. BIOT-SAVART's law. Write the equation in vector form. In which direction does the force per unit length point?

Félix SAVART was born on June 30, 1791 in Charleville-Mézières, Ardennes and died on March 16, 1841 in Paris. He was originally a medical man. However, like most physicians he had a rational-scientific mind. His main interest in physics concerned electromagnetism and mechanical vibrations. In fact, he built a trapezoidally shaped violin and studied the peculiar sounds it made. Rumor has it that his violin was not suited as a musical instrument for the commoner, but maybe for mathematicians, to whom the eigenforms of noise might already bring comfort and joy.

## 13.9 Transformation Properties of the Electro-Magnetic Fields

During the course of their biological evolution humans have developed a certain gut feeling for the physical meaning of mechanical quantities, like mass, velocity, or force. Thus it is fair to say that the transformation rules presented in Chapter 8 have a certain natural and intuitive touch. Unfortunately this is not the case with the electrodynamic fields. Nevertheless we shall try and build upon the rules established for mechanical quantities as much as possible so that they can be extended into unknown territory. However, this is only possible to a certain degree, even if we aim "only" at intuitive clarity and in the end there is no other way but to postulate transformation properties. Of course, we shall attempt to motivate them as much as possible.

Charge is, like mass, a primitive quantity, proper to matter, which in principle can be obtained by counting elementary units. Consequently, it seems reasonable to classify charge as a Euclidean tensor of zeroth order, in other words, an objective scalar. Note that we do not treat charge like an axial scalar, because it has not been observed that reflections of coordinates have an influence on the behavior of charge. We have already explained in Sect. 13.3 that the electric field, $\boldsymbol{E}$, and the magnetic induction, $\boldsymbol{B}$, are defined in terms of measurement by the force they exert on resting or moving charges. Consequently, the question how these fields transform during a change of the frame of reference according to Euclidean transformations (see Sect. 8.2),

$$x'_i = O'_{ij} x_j + b'_i, \tag{13.9.1}$$

is easily answered, based on the transformation rule for body forces shown in Eq. (8.5.1): Charge is an objective tensor of zeroth (a Euclidean scalar) and forces are objective tensors of first order, i.e., Euclidean vectors [see also Eq. (8.5.1)]: Thus the following transformation rule applies in combination with Eq. (13.3.3):

$$\mathfrak{E}'_i = O'_{ij}\,\mathfrak{E}_j \;\Rightarrow\; E'_i + (\mathfrak{v}' \times \boldsymbol{B})_i = O'_{il}\big[E_l + (\mathfrak{v} \times \boldsymbol{B})_l\big]. \tag{13.9.2}$$

Note that the electric field, $\boldsymbol{E}$, together with the LORENTZ part, $\mathfrak{v} \times \boldsymbol{B}$, transform like a Euclidean vector. However, the individual parts do *not*. We know already that the velocity, $\mathfrak{v}$, does *not* transform like a Euclidean vector: Eq. $(8.3.18)_2$. Moreover, the vector product transforms like an *axial* tensor of third order: Eq. (8.3.27). Thus Eq. (13.9.2) leads us to conclude that the magnetic flux density must compensate the axial character. However, this is only possible if the corresponding transformation law contains the quantity $\det(\boldsymbol{O}')$. Note that this does not necessarily mean that the magnetic induction must transform like an axial Euclidean vector. The transformation rule could be more complicated, but it must contain the determinant of the rotation matrix in a suitable manner.[3] This is all information that can be obtained from Eq. (13.9.2) for the LORENTZ force.

Thus we have to motivate the transformation behavior of the magnetic induction from other sources, which only leaves MAXWELL's equations. However, we will avoid the differential form shown in Eq. (13.8.1) and try an intuitive argument instead. The magnetic flux $\mathrm{d}\phi = B_i n_i \,\mathrm{d}S$ is nothing else but the *number* of field lines piercing through the surface element $\mathrm{d}S$. Thus we postulate that it is an objective scalar. Moreover, the normal vector $n_i$ has axial character according to Eq. $(8.5.4)_2$ and, consequently, the magnetic induction should transform like an axial Euclidean vector:

$$B'_i = \det(\boldsymbol{O}')\, O'_{ij} B_j. \tag{13.9.3}$$

However, note that this argument is quite lax and rather intuitive. In the end we may say that Eq. (13.9.3) is really nothing else but a postulate. This becomes even more obvious if we start from the global relations (13.2.6/13.2.7) instead and write for the total flux $\phi$ through a closed hull:

$$\phi = \oiint_{\partial V} B_i n_i \,\mathrm{d}A = \iiint_V \frac{\partial B_i}{\partial x_i}\,\mathrm{d}V \equiv 0. \tag{13.9.4}$$

The zero on the right hand side makes it impossible to say if $\boldsymbol{B}$ has axial character or not. It provides no guidance regarding the transformation properties of $\phi$ in terms of other quantities whose transformation behavior is known. If we now turn to the left side of Eq. (13.9.4) the following can be said: Recall that $\phi$ is a

[3] Further down we shall see that the magnetic field, $\boldsymbol{H}$, has much more complicated transformation properties.

measure for the number of magnetic field lines and, consequently, should be a regular, non-axial scalar. Then the axial normal vector in the surface integral of (13.9.4) suggests that $\boldsymbol{B}$ meets the transformation rule of Eq. (13.9.3). However, the volume integral indicates non-axial transformation behavior provided we assume that the volume element behaves like a true scalar as required in Eq. (8.4.2). Also recall that we could have alternatively considered the volume element to result from a scalar triple product, which would make it axial.

Combination of the relations (13.9.2) and (13.9.3) and observing the transformation rule for the velocity shown in Eq. (8.3.18) results in the following transformation for the electric field:

$$E'_i = O'_{il}\left(E_l - \epsilon_{lbn} O'_{jb}\left[\Omega'_{jk}\left(x'_k - b'_k\right) + \dot{b}'_j\right] B_n\right) \tag{13.9.5}$$

and conclude that the electric field does *not* transform like an objective vector.

We have already mentioned that charge is an objective scalar. In the same spirit

> **Exercise 13.9.1:** ***The electric field as a non-objective vector***
>
> Show the validity of Eq. (13.9.5). To this end observe in particular Eqs. $(8.3.18)_2$ and (8.3.27).

we now postulate that the (non-convective) electric currents are objective quantities. This makes sense since they are nothing else but charges (on an atomic level) transported across a surface area. Consequently, we require in analogy to the heat flux vector of Eq. (8.6.5) that:

$$j'_i = O'_{ij} j_j. \tag{13.9.6}$$

This defines the transformation properties of the charge and of the current potentials. From Coulomb's law shown in Eq. (13.4.8) it follows from the objectivity of the charge density that the charge potential must be an objective vector:

$$D'_i = O'_{ij} D_j. \tag{13.9.7}$$

If this is inserted in Ørsted's law (13.4.12) we conclude that:

$$H'_i - (\boldsymbol{\upsilon}' \times \boldsymbol{D}')_i = \det(\boldsymbol{O}')\, O'_{il}\left[H_l - (\boldsymbol{\upsilon} \times \boldsymbol{D})_l\right]. \tag{13.9.8}$$

Thus the current potential $\boldsymbol{H}$ is *not* an objective vector. However, the combination $\boldsymbol{H} - \boldsymbol{\upsilon} \times \boldsymbol{D}$ *is* axially objective. The details are investigated in the following exercise:

**Exercise 13.9.2:** ***Transformation properties of the charge potential and of the magnetic field***

Explain the details that lead to Eq. (13.9.7). To this end use the following relation for the charge density:

$$q' = q, \tag{13.9.9}$$

the definition of Euclidean transforms (13.9.1), and the (local) form of COULOMB's law shown in Eq. $(13.4.8)_2$. Now use Eq. (13.9.6) and the following definition for the *total electric current* (i.e., the non-convective plus the convective part):

$$J_i = j_i + q\upsilon_i \tag{13.9.10}$$

in context with Eq. $(8.3.18)_2$ and show that:

$$J_i' = O_{ij}'J_j + q\left[\Omega_{ik}'\left(x_k' - b_k'\right) + \dot{b}_i'\right]. \tag{13.9.11}$$

Thus the total electric current is not objective either. Finally show by means of ØRSTED's law from Eq. (13.4.12) the validity of Eq. (13.9.7) and also of:

$$H_i' = \det(\boldsymbol{O}')\, O_{il}'\left(H_l + \epsilon_{lbn}O_{jb}'\left[\Omega_{jk}'\left(x_k' - b_k'\right) + \dot{b}_j'\right] D_n\right). \tag{13.9.12}$$

Consequently, the current potential transforms in a highly complex manner and is definitely *not* an axial objective vector. Use Eq. (13.9.5) and discuss similarities and differences to the transformation behavior of the electric field, $\boldsymbol{E}$.

Because charge is an objective scalar and the distance between two points is also objective, [see Eq. (8.3.8)] the dipole moment, $\boldsymbol{p} = e\boldsymbol{\delta}$, is another objective vector. The number density of dipoles, $n^{\mathrm{p}}$, is an objective scalar. Thus Eq. (13.6.2) leads us to conclude that the polarization vector must transform like

$$P_i' = O_{ij}'P_j. \tag{13.9.13}$$

In order to find the transformation rule for magnetization, recall AMPÈRE's idea that a magnetic dipole is given by the expression $\boldsymbol{m} = I\boldsymbol{A}$. The current $I$ is nothing else but a charge per unit of time, in other words an objective scalar. The directed surface $\boldsymbol{A}$ transforms like an axial objective vector. Hence we conclude that:

$$m_i' = \det(\boldsymbol{O}')\, O_{ij}'m_j. \tag{13.9.14}$$

Thus magnetization, $\boldsymbol{M}$, must transform the same way: It is defined by $\boldsymbol{M} = n^{\mathrm{m}}\boldsymbol{m}$, where the number density of magnetic dipoles, $n^{\mathrm{m}}$, is by its very nature an objective scalar:

$$M'_i = \det(\boldsymbol{O}')\, O'_{ij} M_j. \tag{13.9.15}$$

Then Eqs. (13.6.7) and (13.9.7/13.9.13) imply that the charge potential in matter is an objective vector:

$$\mathfrak{D}'_i = O'_{ij}\, \mathfrak{D}_j. \tag{13.9.16}$$

However, the following complicated transformation rule must hold for the current potential in matter if we only observe its definition (13.7.7) in combination with Eqs. (13.6.7), (13.9.12/13.9.13/13.9.15) and $(8.6.18)_2$:

$$\mathfrak{H}'_i = \det(\boldsymbol{O}')\, O'_{il}\left(\mathfrak{H}_l + \epsilon_{lbn} O'_{jb}\left[\Omega'_{jk}\left(x'_k - b'_k\right) + \dot{b}'_j\right] \mathfrak{D}_n\right). \tag{13.9.17}$$

Note that this is relation is completely analogous to the transformation properties of the current potential shown in Eq. (13.9.12). Thus in view of Eq. (13.9.8) and Exercise 13.9.2 we may conclude immediately that the expression

$$\mathfrak{H}'_i - (\upsilon' \times \mathfrak{D}')_i = \det(\boldsymbol{O}')\, O'_{il}\left[\mathfrak{H}_l - (\upsilon \times \mathfrak{D})_l\right] \tag{13.9.18}$$

represents an axially objective vector.

**Exercise 13.9.3:** ***Transformation property of the current potential in matter***

Prove Eqs. (13.9.17/13.9.18).

Finally it should be noted that after the transformation rules for all electromagnetic fields have been motivated and mathematically established by Eqs. (13.9.3/13.9.5–13.9.7/13.9.9/13.9.12), we have implicitly postulated that MAXWELL's equations (13.8.1) assume the same form, without any system specific terms, in a Euclidean frame moving against an inertial system. However, it is quite instructive to show this explicitly:

**Exercise 13.9.4:** ***Invariance of MAXWELL's equations***

Change from an inertial observer to a Euclidean one and use the transformation rules for the charge density, velocity, and electric current density and show that the equation for charge conservation (13.4.19) implies:

$$\frac{\partial q'}{\partial t} + \frac{\partial}{\partial x'_i}\left(j'_i + q'\upsilon'_i\right) = 0. \tag{13.9.19}$$

Note that for the transformation of the (material) time derivative it is useful to change into Lagrangian form. This has already been used effectively in Sect. 8.4 in context with the proof of form invariance of the balance of mass:

$$\frac{\partial q}{\partial t} + \upsilon_i \frac{\partial q}{\partial x_i} \equiv \left. \frac{\partial q}{\partial t} \right|_X . \tag{13.9.20}$$

Moreover, use the transformation rules for the other electromagnetic fields established in this section, start from MAXWELL's equations in an inertial frame, Eqs. (13.8.1), transform them into a Euclidean frame and show that they keep their form and system-dependent terms do not enter:

$$\frac{\partial B'_i}{\partial x'_i} = 0, \; \frac{\partial B'_i}{\partial t} + \epsilon'_{ijk} \frac{\partial E'_k}{\partial x'_j} = 0, \tag{13.9.21}$$

$$\frac{\partial D'_i}{\partial x'_i} = q', \; -\frac{\partial D'_i}{\partial t} + \epsilon'_{ijk} \frac{\partial H'_k}{\partial x'_j} = j'_i + q' \upsilon'_i .$$

Note that in this context it is also helpful to treat time derivatives in Lagrangian description, i.e.,:

$$\frac{\partial B_i}{\partial t} = \left. \frac{\partial B_i}{\partial t} \right|_X - \upsilon_k \frac{\partial B_i}{\partial x_k} \tag{13.9.22}$$

## 13.10 Transformation Properties of the MAXWELL-LORENTZ Aether Relations and MAXWELL's Equations

We now turn to the MAXWELL-LORENTZ aether relations from Eq. (13.8.2). It was mentioned before that their simple mathematical form was established in an inertial system. If we now transform them by means of Eqs. (13.9.5/13.9.7) and (13.9.3/13.9.12), respectively, into a Euclidean system we obtain:

$$D'_i = \varepsilon_0 \left( E'_i + \epsilon'_{ijp} \left[ \Omega'_{jk} (x'_k - b_k) + \dot{b}'_j \right] B'_p \right) \tag{13.10.1}$$

and:

$$\begin{aligned} H'_i = \frac{1}{\mu_0} \Big( & B'_i + \frac{1}{c^2} \epsilon'_{ijr} \left[ \Omega'_{jk} (x'_k - b'_k) + \dot{b}'_j \right] E'_r \\ & + \frac{1}{c^2} \epsilon'_{ijr} \left[ \Omega'_{jk} (x'_k - b'_k) + \dot{b}'_j \right] \epsilon'_{rst} \left[ \Omega'_{su} (x'_u - b'_u) + \dot{b}'_s \right] B'_t \Big), \end{aligned} \tag{13.10.2}$$

respectively. Obviously the MAXWELL-LORENTZ aether relations do *not* keep their simple form (13.8.2) during translational *and* rotational transformations and many system dependent terms do occur. On first glance this is not too surprising: A similar problem occurred in context with the balance of momentum, where centrifugal, CORIOLIS, EULER, and relative acceleration terms occurred after transformation to a Euclidean system all of which are system dependent.

Nevertheless there is one crucial difference. If we specialize the result for Euclidean transforms to the subgroup of Galilean transformations, i.e., rectilinear uniform types of motion with a relative velocity $V_i$ (which is *constant in time*) measured against the inertia system:

$$x_i' = O_{ij}' x_j + V_i' t \text{ with } V_i' = \dot{b}_i', \tag{13.10.3}$$

where $O_{ij}'$ is *not* time-dependent, all system-dependent terms disappear in the transformation formula for the acceleration shown in Eq. (8.3.45) and there remains:

$$a_i' = O_{ij}' a_j. \tag{13.10.4}$$

This, however, shows that the balance of momentum has the same form in Galilean systems (i.e., no system-dependent terms) as in the original inertial system. Thus nineteen century scientists concluded (prematurely) that all Galilean systems must be inertial systems.

But then electrodynamics appeared on stage: If we specialize the transformation rules for the MAXWELL-LORENTZ aether relations from Eqs. (13.10.1/13.10.2) to the special case of a Galilean transformation of Eq. (13.10.3) we obtain:

$$D_i' = \varepsilon_0 \left( E_i' + \epsilon_{ijk}' V_j' B_k' \right). \tag{13.10.5}$$

and:

$$H_i' = \frac{1}{\mu_0} \left( \left[ 1 - \frac{V'^2}{c^2} \right] \delta_{ij} + \frac{V_i' V_j'}{c^2} \right) B_j' + \varepsilon_0 \epsilon_{ijk}' V_j' E_k'. \tag{13.10.6}$$

Thus the MAXWELL-LORENTZ aether relations still contain system-dependent terms under Galilean transformations, in contrast to the momentum balance and, consequently, Galilean transformations do not lead to inertial systems.

Thus the question arises which transformations leave the MAXWELL-LORENTZ aether relations in their simple form. The answer is that this is achieved by the so-called LORENTZ transformations and, indeed, the corresponding observers qualify as inertial systems. In the next section we shall explore the exact mathematical form of LORENTZ transformations.

**Exercise 13.10.1:** ***MAXWELL-LORENTZ aether relations in Euclidean systems***

Prove Eqs. (13.10.1/13.10.2). To this end start from the MAXWELL-LORENTZ aether relations in the inertial frame, Eq. (13.8.2), insert them in the transformation rules for the charge and for the current potential according to Eqs. (13.9.7/13.9.12), use the transformation formulae for the electric field and for the magnetic induction from Eqs. (13.9.3/13.9.5), and finally Eq. (13.9.1).

**Exercise 13.10.2: *Acceleration and Maxwell-Lorentz aether relations for Galilean transformations***

First, show that for Galilean transformations Eqs. (13.9.1) and (13.10.3) imply:

$$O'_{ij} = \delta'_{ij},\ \dot{b}'_i = V'_i. \tag{13.10.7}$$

Insert this in Eqs. (8.3.46) and (13.10.1/13.10.2), and show the validity of Eqs. (13.10.4–13.10.6).

## 13.11 Four-Vector Formalism for the Electromagnetic Fields

From now on we shall add time to the group of the three spatial coordinates. In order to make sure that these four coordinates have the same dimensions we multiply time by the (constant) speed of light, $c$, and obtain in Cartesian coordinates:

$$\boldsymbol{x} = \left(x^0, x^1, x^2, x^3\right) = x^A \boldsymbol{e}_A, \quad x^0 = ct, \quad A = 0,\ 1,\ 2,\ 3. \tag{13.11.1}$$

In this equation use was made of a unit vector $\boldsymbol{e}_0$ "pointing into the future." Moreover, use is made of capital indices, e.g., $A$, and, if they appear twice in a product, automatic summation from 0 to 3 is implied. This extends the usual Einstein summation convention and combines space and time. Of course, a point in space-time can also be identified by another observer. However, the point itself is unique and, consequently, a general, invertible space-time transformation of the following form must exist:

$$x'^B = \tilde{x}'^B\left(x^A\right). \tag{13.11.2}$$

Examples of such transformations are the previously discussed Euclidean or Galilean transformations. They possess the property that the time component 0 decouples completely from the space components 1–3. As we shall see this is no longer the case during a Lorentz transformation. Moreover, it is noteworthy that the Euclidean as well as the Galilean transformation are *linear* w.r.t. the coordinates. This becomes immediately evident from Eqs. (13.9.1) and (13.10.3). We shall see that the Lorentz transformation has this property as well. However, in general a space-time transformation (13.11.2) may also be non-linear.

It is curious to note that we have employed a co-/contravariant notation (in capital letters) in context with Eqs. (13.11.1/13.11.2): Recall that a distinction between co- and contravariant in three-dimensional space was impossible. However, this is not so in 4D, where the space-time metric used to form a four-dimensional line element is no longer a unit matrix. In fact, the line element can be expressed in an inertial system by its time *and* (Cartesian) space coordinates (this extension now replaces the three-dimensional Cartesian space) as follows:

$$(\mathrm{d}s)^2 = -(\mathrm{d}ct)^2 + (\mathrm{d}x^1)^2 + (\mathrm{d}x^2)^2 + (\mathrm{d}x^3)^2 = g_{AB}\mathrm{d}x^A\mathrm{d}x^B, \tag{13.11.3}$$

where the space-time metric of the inertial system is given by[4]:

$$g_{AB} = g^{AB} = \begin{bmatrix} -1 & 0 & 0 & 0 \\ 0 & 1 & 0 & 0 \\ 0 & 0 & 1 & 0 \\ 0 & 0 & 0 & 1 \end{bmatrix}. \tag{13.11.4}$$

Further down we shall motivate why it is reasonable to introduce a four-dimensional line-element and what it can be used for. At this point all of these preliminary notes on 4D-transforms are only intended to illustrate the analogies to the 3D case. In this spirit we now introduce a so-called *world tensor*, $\boldsymbol{F}$, of $j+k$ order by its transformation behavior in terms of co-/contravariant coordinates, in complete analogy to Eq. (8.3.1):

$$F'^{A_1A_2\ldots A_j}_{\ \ B_1B_2\ldots B_k} = \left|\det\left(\frac{\partial x'}{\partial x}\right)\right|^{-w} \operatorname{sign}^p\left(\frac{\partial \boldsymbol{x}'}{\partial \boldsymbol{x}}\right) \frac{\partial x'^{A_1}}{\partial x^{C_1}} \cdots \frac{\partial x'^{A_j}}{\partial x^{C_j}} \frac{\partial x^{D_1}}{\partial x'^{B_1}} \cdots \frac{\partial x^{D_k}}{\partial x'^{B_k}} F^{C_1C_2\ldots C_j}_{\ \ D_1D_2\ldots D_k}. \tag{13.11.5}$$

Note that the functional determinant of space-time transformations shown in Eq. (13.11.2) must not be equal to $\pm 1$, in contrast to the Euclidean transformation (13.9.1) [compare also the footnote in context with Eqs. (8.3.1/8.3.2)]. The factor $p$ in this equation can assume the values 0 and 1. The exponent $w$ is a.k.a. the *weight* of the tensor. Moreover, the case of $p = 0$, $w = 0$ is referred to as an *absolute* tensor and, for $p = 1$, $w = 0$, as an *axial* tensor. For $p = 0$, $w \neq 0$ we speak of a so-called *relative* world tensor of weight $w$, and for $p = 1$, $w \neq 0$ of a relative axial tensor of weight $w$. As we shall see, the theory of electrodynamics is governed by absolute world tensors of second order and by relative world tensors of first and second order. Without going into the mathematical details it should be noted that a world tensor with a weight different from zero is also often referred to as a *tensor density*.

In a first step we now combine the electric field $E_i$ and the magnetic flux density $B_i$ in an *absolute antisymmetric covariant world tensor of second order*:

$$\varphi_{AB} = \begin{bmatrix} 0 & -E_1/c & -E_2/c & -E_3/c \\ E_1/c & 0 & B_3 & -B_2 \\ E_2/c & -B_3 & 0 & B_1 \\ E_3/c & B_2 & -B_1 & 0 \end{bmatrix}. \tag{13.11.6}$$

It transforms according to:

[4] In the older relativistic literature the stringent application of tensor calculus is avoided and the imaginary unit, $i^2 = -1$ is used in context with the definition of the time coordinate. This renders it possible to define the 4D-line element in a quasi-Pythagorean way. If we use tensors from the very beginning on we do not need this concept any more.

$$\varphi'_{CD} = \frac{\partial x^A}{\partial x'^C} \frac{\partial x^B}{\partial x'^D} \varphi_{AB}. \tag{13.11.7}$$

In a similar fashion we now combine charge and current density in the following contravariant four-vector:

$$\sigma^A = [cq, \quad j_1 + q\upsilon_1, \quad j_2 + +q\upsilon_2, \quad j_3 + q\upsilon_3]. \tag{13.11.8}$$

It transforms as follows:

$$\sigma'^B = \left|\det\left(\frac{\partial \boldsymbol{x}'}{\partial \boldsymbol{x}}\right)\right|^{-1} \frac{\partial x'^B}{\partial x^A} \sigma^A. \tag{13.11.9}$$

Thus the four-vector for the charge-current transforms like a relative contravariant world vector of weight 1. The transformation properties of the electromagnetic fields must ultimately be verified experimentally. However, they are also motivated by "intuition," namely by the measurement procedures of the *gedankenexperiments* outlined in Sects. 13.3, 13.4, 13.5 and 13.9, respectively. We are going to review the issues on transformation properties once more in the following two exercises. However, it should already be noted that this will not clarify why the four-vector from Eq. (13.11.8) is introduced as a relative and not as an absolute vector. This will not be revealed until we rewrite the Maxwell-Lorentz aether relations in four-vector notation.

**Exercise 13.11.1:** ***Four-vector notation of the electromagnetic transformation properties for Euclidean transformations***

Recall the Euclidean transformation shown in Eq. (13.9.1). Supplement it by the relation $ct' = ct$ and show that:

$$x'^B = \frac{\partial x'^B}{\partial x^A} x^A, \quad \frac{\partial x'^B}{\partial x^A} = \begin{bmatrix} 1 & 0 & 0 & 0 \\ \frac{V'_1}{c} & O'_{11} & O'_{12} & O'_{13} \\ \frac{V'_2}{c} & O'_{21} & O'_{22} & O'_{23} \\ \frac{V'_3}{c} & O'_{31} & O'_{32} & O'_{33} \end{bmatrix} \tag{13.11.10}$$

and:

$$x^A = \frac{\partial x^A}{\partial x'^C} x'^C, \quad \frac{\partial x^A}{\partial x'^C} = \begin{bmatrix} 1 & 0 & 0 & 0 \\ -\frac{V_1}{c} & O'_{11} & O'_{21} & O'_{31} \\ -\frac{V_2}{c} & O'_{12} & O'_{22} & O'_{32} \\ -\frac{V_3}{c} & O'_{13} & O'_{23} & O'_{33} \end{bmatrix}, \tag{13.11.11}$$

where:

$$\frac{\partial x'^B}{\partial x^A} \frac{\partial x^A}{\partial x'^C} = \delta^B_C, \; V'_i = \Omega'_{ij}\left(x'_j - b'_j\right) + \dot{b}'_i \text{ with } V'_i = O'_{ij} V_j \tag{13.11.12}$$

is a velocity-like quantity that, by definition, depends on the space-time coordinates of the Euclidean system. It is noteworthy that Eqs. (13.11.10/13.11.11) basically reduce to the ordinary Euclidean transformation matrix $\boldsymbol{O}'$ supplemented by the relation $t' = t$ in the limit $c \to \infty$ (or, in terms of physics, velocities small compared to the speed of light, $V_i' \ll c$). Show that Eqs. (13.11.10–13.11.12) lead to a Galilean transformation if Eq. (13.11.12) is specialized to *uniform rectilinear* motion so that *one* term cancels out (which one?). Moreover, note that Eq. (13.11.11) follow formally from Eq. (13.11.10) by applying the rules for the inversion of matrices. Show that it is also possible to use the following more physics-based rule: "The inversion of the Euclidean transformation shown in Eq. (13.11.10/13.11.12) can be accomplished by, first, replacing all dashed quantities with non-dashed ones and vice versa. Second, $V_i$ must be substituted by $-V_i$." What exactly is $O_{ij}$ under these circumstances? Moreover, show that the following transformation rules hold for rotation matrix and for the vector of angular velocity:

$$\Omega'_{ij} = -O'_{ik} O'_{jl} \Omega_{kl} \text{ with the definition } \Omega_{kl} = \dot{O}_{km} O_{lm}, \tag{13.11.13}$$

and $\omega_i' = -\det(\boldsymbol{O}')\, O'_{ik}\omega_k$ with the definition $\omega_k = \epsilon_{klm}\Omega_{lm}$.
Also observe in the same context the remarks in context with Eq. (8.2.20/8.2.21). Are the rotation matrix and the vector of angular velocity (axial) Euclidean tensors? Specialize now Eq. (13.11.7) to Euclidean transformations and show that the transformation rules for the magnetic induction and for the electric field result, i.e., Eqs. (13.9.3/13.9.5), respectively. To this end it is also necessary to apply the basic inversion rule for matrices to the rotation matrix $\boldsymbol{O}'$ [only then the expression $B_i' = \det(\boldsymbol{O}')\, O'_{ij} B_j$, i.e., Eq. (13.9.3)]. Next specialize Eq. (13.11.9) to Euclidean transformations and show that it results in the transformation rules for the charge and the current density, Eqs. (13.9.9) and (13.9.11), respectively.

**Exercise 13.11.2:** ***Maxwell's equations in four-vector notation: Pt. I***

Show by means of the definition (13.11.6) that the following four-vector equation (in Cartesian coordinates) represents Maxwell's equations (13.2.8/13.2.12):

$$\epsilon^{ABCD} \frac{\partial \varphi_{CD}}{\partial x^B} = 0. \tag{13.11.14}$$

$\epsilon^{ABCD}$ denotes the totally antisymmetric tensor of forth order defined in complete analogy to its 3D counterpart (8.3.25). It transforms like a relative axial world tensor of forth order of weight 1 (also compare Eq. (8.3.27) for the 3D case):

$$\epsilon'^{RSTU} = \det{}^{-1}\left(\frac{\partial \boldsymbol{x}'}{\partial \boldsymbol{x}}\right)\frac{\partial x'^{R}}{\partial x^{A}}\frac{\partial x'^{S}}{\partial x^{B}}\frac{\partial x'^{T}}{\partial x^{C}}\frac{\partial x'^{U}}{\partial x^{D}}\epsilon^{ABCD}. \tag{13.11.15}$$

Use this relation in combination with the transformation rule (13.11.7) and show that the first set of MAXWELL's equations is form-invariant so that we may write for an *arbitrary* coordinate system:

$$\epsilon'^{ABCD}\frac{\partial \varphi'_{CD}}{\partial x'^{B}} = 0. \tag{13.11.16}$$

The balance of electric charge (13.4.18) can be rewritten by means of the definition (13.11.9) for the four-vector as follows:

$$\frac{\partial \sigma^{A}}{\partial x^{A}} = 0. \tag{13.11.17}$$

This equation can be formally solved by introducing the antisymmetric tensor $\eta^{AB} = -\eta^{BA}$ of second order with the following property:

$$\frac{\partial \eta^{AB}}{\partial x^{B}} = \sigma^{A}. \tag{13.11.18}$$

Obviously $\eta^{AB}$ must be interpreted as a four-dimensional charge/current *potential* (note the derivative!). It is related to the well-known charge and current potentials $D_i$ and $H_i$ by:

$$\eta^{AB} = \begin{bmatrix} 0 & cD_1 & cD_2 & cD_3 \\ -cD_1 & 0 & H_3 & -H_2 \\ -cD_2 & -H_3 & 0 & H_1 \\ -cD_3 & H_2 & -H_1 & 0 \end{bmatrix}. \tag{13.11.19}$$

**Exercise 13.11.3: *MAXWELL's equations in four-vector notation*: *Pt. II***

Use Eqs. (13.11.8/13.11.17/13.11.18) and show explicitly that they result in the second set of MAXWELL's equations (13.4.8/13.4.12).

**Exercise 13.11.4:** ***Transformation rule for the four-dimensional charge-current potential***

Use the general transformation rule (13.11.5) and show starting from Eqs. (13.11.9) and (13.11.18) that the charge-current potential transforms like a relative contravariant world tensor of second order and weight $w = 1$ so that:

$$\eta'^{AB} = \left|\det\left(\frac{\partial \boldsymbol{x}'}{\partial \boldsymbol{x}}\right)\right|^{-1} \frac{\partial x'^{A}}{\partial x^{C}} \frac{\partial x'^{B}}{\partial x^{D}} \eta^{CD}, \tag{13.11.20}$$

and the second set of MAXWELL's equations has the same form in all systems:

$$\frac{\partial \eta'^{AB}}{\partial x'^{B}} = \sigma'^{A}. \tag{13.11.21}$$

Observe the following identity during the proof:

$$\frac{\partial}{\partial x^{B}}\left(\det\left(\frac{\partial \boldsymbol{x}'}{\partial \boldsymbol{x}}\right) \frac{\partial x^{B}}{\partial x'^{S}}\right) = 0. \tag{13.11.22}$$

Now prove the identity by using the LAPLACE expansion theorem for determinants (in four dimensions):

$$\left(A^{-1}\right)^{AB} = \frac{\frac{1}{6}\epsilon_{ARST}\epsilon_{BCDE}A^{CR}A^{DS}A^{ET}}{\det(\boldsymbol{A})}. \tag{13.11.23}$$

Specialize to Euclidean transformations according to Eq. (13.11.10) and show that the transformation rules (13.9.6/13.9.11) result. Finally show by means of the transformations rule (13.11.9) that the balance of charge (13.11.17) in an arbitrary system reads:

$$\frac{\partial \sigma'^{A}}{\partial x'^{A}} = 0. \tag{13.11.24}$$

Hint: Use the identity (13.11.22) again.

## 13.12 Four-Vector Notation of the MAXWELL-LORENTZ Aether Relations: The LORENTZ Transformation

Recall the so-called space-time metric $g^{AB}$ that was already introduced in Eq. (13.11.4). We require that it transforms like an *absolute* contravariant world tensor of second order:

$$g'^{AB} = \frac{\partial x'^{A}}{\partial x^{C}} \frac{\partial x'^{B}}{\partial x^{D}} g^{CD}. \tag{13.12.1}$$

For reasons that will become clear immediately we calculate the determinant of this relation and find:

$$\det\left(g'^{AB}\right) = \det^2\left(\frac{\partial x'}{\partial x}\right)\det\left(g^{CD}\right). \tag{13.12.2}$$

**Exercise 13.12.1:** ***Space-time metric for Euclidean transformations***

Use the result for the transformation matrix from an inertial to a Euclidean system, Eq. (13.11.10), in combination with the transformation rule (13.12.1) and show that the space-time metric of a Euclidean system is given by:

$$g'^{AB} = \begin{bmatrix} -1 & -\frac{V'_1}{c} & -\frac{V'_2}{c} & -\frac{V'_3}{c} \\ -\frac{V'_1}{c} & 1-\frac{V'_1V'_1}{c^2} & -\frac{V'_1V'_2}{c^2} & -\frac{V'_1V'_3}{c^2} \\ -\frac{V'_2}{c} & -\frac{V'_2V'_1}{c^2} & 1-\frac{V'_2V'_2}{c^2} & -\frac{V'_2V'_3}{c^2} \\ -\frac{V'_3}{c} & -\frac{V'_3V'_1}{c^2} & -\frac{V'_3V'_2}{c^2} & 1-\frac{V'_3V'_3}{c^2} \end{bmatrix}.. \tag{13.12.3}$$

Note that in the limit $V'_i \ll c$ Eq. (13.12.3) turns into the space-time metric of an inertial system, Eq. (13.11.4). This sounds like a contradiction since we know that a Euclidean system never represents an inertial system—not even in the limit of small velocities, unless these are truly zero. It is customary to say that every space-time metric becomes asymptotically flat in the non-relativistic case.

In fact, by using Eq. (13.11.4) for the space-time metric of an inertial system we may also put $\det(g^{CD}) = -1$. However, it is more important to note that in view of the general definition shown in Eq. (13.11.5) the determinant of the contravariant space-time metric must be a relative world tensor of zeroth order (i.e., a relative world scalar) of weight $w = -2$. Loosely speaking the determinant is a scalar density.

With this relation the MAXWELL-LORENTZ aether relations from Eq. (13.8.2) can be rewritten. They were originally verified in an inertial system, and thus:

$$\eta^{CD} = \frac{1}{\mu_0}\left(-\det g^{MN}\right)^{-\frac{1}{2}} g^{CA} g^{DB} \varphi_{AB}. \tag{13.12.4}$$

This can easily be verified by an explicit calculation (the square root $\left(-\det g^{MN}\right)^{-\frac{1}{2}} = 1$ has been omitted for convenience in the following long formula):

$$\begin{bmatrix} 0 & cD_1 & cD_2 & cD_3 \\ -cD_1 & 0 & H_3 & -H_2 \\ -cD_2 & -H_3 & 0 & H_1 \\ -cD_3 & H_2 & -H_1 & 0 \end{bmatrix} = \frac{1}{\mu_0}\begin{bmatrix} -1 & 0 & 0 & 0 \\ 0 & 1 & 0 & 0 \\ 0 & 0 & 1 & 0 \\ 0 & 0 & 0 & 1 \end{bmatrix}\begin{bmatrix} 0 & -\frac{E_1}{c} & -\frac{E_2}{c} & -\frac{E_3}{c} \\ \frac{E_1}{c} & 0 & B_3 & -B_2 \\ \frac{E_2}{c} & -B_3 & 0 & B_1 \\ \frac{E_3}{c} & B_2 & -B_1 & 0 \end{bmatrix}\begin{bmatrix} -1 & 0 & 0 & 0 \\ 0 & 1 & 0 & 0 \\ 0 & 0 & 1 & 0 \\ 0 & 0 & 0 & 1 \end{bmatrix}$$
$$= \frac{1}{\mu_0}\begin{bmatrix} 0 & \frac{E_1}{c} & \frac{E_2}{c} & \frac{E_3}{c} \\ -\frac{E_1}{c} & 0 & B_3 & -B_2 \\ -\frac{E_2}{c} & -B_3 & 0 & B_1 \\ -\frac{E_3}{c} & B_2 & -B_1 & 0 \end{bmatrix}, \quad c^2 = \frac{1}{\varepsilon_0\mu_0}. \tag{13.12.5}$$

On first glance the determinant in Eq. (13.12.4) seems rather artificial. However, it is very important because in a world tensor equation the order *and* the weight must agree. We postulate that the MAXWELL-LORENTZ aether relations in the form (13.12.4) are of the caliber of a "world equation." It is then easily confirmed that tensors of second order are present on both sides of the equation. Moreover, the weight on the left hand side is equal to one, because of Eq. (13.11.8), and so is the weight on the right hand side, because of Eqs. (13.11.7) and (13.12.1/13.12.2), respectively: $(-2)\left(-\frac{1}{2}\right) = +1$. Only then Eqs. (13.11.7/13.12.19) and (13.12.1/13.12.2) transform Eq. (13.12.4) in a form invariant manner into an arbitrary system:

$$\eta'^{LM} = \frac{1}{\mu_0}\left(-\det g'^{UV}\right)^{-\frac{1}{2}} g'^{LR} g'^{MS} \varphi'_{RS} \tag{13.12.6}$$

Our next objective is to find the transformations for which the MAXWELL-LORENTZ aether relations keep the simple form shown in Eq. (13.8.2) or (13.12.4), respectively. Then the metric $\boldsymbol{g}'$ in Eq. (13.12.6) must have exactly the same components as the metric $\boldsymbol{g}$ in Eq. (13.11.4). If we observe the transformation rules (13.12.1) we find that:

$$\begin{bmatrix} -1 & 0 & 0 & 0 \\ 0 & 1 & 0 & 0 \\ 0 & 0 & 1 & 0 \\ 0 & 0 & 0 & 1 \end{bmatrix}^{AB} = \frac{\partial x'^A}{\partial x^C}\frac{\partial x'^B}{\partial x^D}\begin{bmatrix} -1 & 0 & 0 & 0 \\ 0 & 1 & 0 & 0 \\ 0 & 0 & 1 & 0 \\ 0 & 0 & 0 & 1 \end{bmatrix}^{CD} \Leftrightarrow \begin{bmatrix} -1 & 0 & 0 & 0 \\ 0 & 1 & 0 & 0 \\ 0 & 0 & 1 & 0 \\ 0 & 0 & 0 & 1 \end{bmatrix}^{CD}$$
$$= \frac{\partial x^C}{\partial x'^A}\frac{\partial x^D}{\partial x'^B}\begin{bmatrix} -1 & 0 & 0 & 0 \\ 0 & 1 & 0 & 0 \\ 0 & 0 & 1 & 0 \\ 0 & 0 & 0 & 1 \end{bmatrix}^{AB}. \tag{13.12.7}$$

These two formulae describe the transformation from the original inertial system into the new one and *vice versa*. The second equation is particularly useful in order to derive the LORENTZ transformation in a conceptionally simple manner. We shall present the details shortly. Note that the elements of the transformation matrices $\partial \boldsymbol{x}'/\partial \boldsymbol{x}$ and $\partial \boldsymbol{x}/\partial \boldsymbol{x}'$, respectively, represent a total of sixteen unknowns. However, due to the symmetry of the matrix (13.11.4), Eq. (13.12.7) provides only ten constraints. Correspondingly, six undetermined parameter will remain in the unknown transformation. These can be interpreted as the three components of relative translational velocity, $-V_i$, between both system origins (a.k.a. *boost velocity*) as well as the three angles of rotation between the axes (i.e., the independent components of the rotation matrix). We proceed to investigate this a little further. We conclude from Eq. (13.11.4) that $\boldsymbol{g}$ as well as $\boldsymbol{g}'$ are constant and, therefore, the unknown coordinate transformation must be *linear*. Consequently, we solve the problem in two steps: First, we transform from the non-dashed system by a boost into an intermediate inertial system $\tilde{x}^B$, whose axes are not tilted w.r.t. the ones of the non-dashed system, albeit related to each other linearly:

$$x^A = \alpha^A{}_B \tilde{x}^B + \beta^A. \tag{13.12.8}$$

The four-vector $\beta^A$ is equal to zero if the origins of both coordinate systems coincide. We now assume that non-dashed observer is at rest and that the origin of the dashed one moves relatively at a *constant* speed $-V_i$.[5] Of course this speed is determined by using the yard sticks and the clocks of the original observer. Then because of $\mathrm{d}\tilde{x}^i = 0$ (the origin of the intermediate system is fixed and remains unchanged) it follows from Eq. (13.12.8):

$$\mathrm{d}x^A = \alpha^A{}_B \mathrm{d}\tilde{x}^B \Rightarrow \begin{array}{l} \mathrm{d}x^0 = \alpha^0{}_0 \mathrm{d}\tilde{x}^0 \\ \mathrm{d}x^i = \alpha^i{}_0 \mathrm{d}\tilde{x}^0. \end{array} \tag{13.12.9}$$

In order to include the velocity we divide $\mathrm{d}x^i$ by $\mathrm{d}x^0$ and observe Eq. (13.11.1):

$$-\frac{V_i}{c} \equiv \frac{\mathrm{d}x^i}{\mathrm{d}x^0} = \frac{\alpha^i{}_0}{\alpha^0{}_0}. \tag{13.12.10}$$

Another relation between $\alpha^i{}_0$ and $\alpha^0{}_0$ results if the components $A = 0$ and $B = 0$ are evaluated in Eq. $(13.12.7)_2$:

$$-1 = -\left(\alpha^0{}_0\right)^2 + \sum_{i=1}^{3} \left(\alpha^i{}_0\right)^2. \tag{13.12.11}$$

The solution to Eqs. (13.12.10/13.12.11) reads:

$$\alpha^0{}_0 = \gamma,\ \alpha^i{}_0 = -\gamma \frac{V_i}{c},\ \gamma = \frac{1}{\sqrt{1 - \boldsymbol{V}^2/c^2}}. \tag{13.12.12}$$

[5] The minus sign in the velocity is arbitrary. In fact some textbooks do not follow this convention. However, we do and this guarantees consistency with the assumed direction of the vector $\boldsymbol{b}$ shown in Fig. 8.1, the remarks in context with Eq. (13.10.3) and, finally, with Exercise 13.11.1.

The remaining equation from $(13.12.7)_2$ can be satisfied with

$$\alpha^{i}{}_{j} = \delta^{i}{}_{j} + \frac{V_i V_j}{\boldsymbol{V}^2}(\gamma - 1),\ \alpha^{0}{}_{j} = -\gamma \frac{V_j}{c}. \tag{13.12.13}$$

Thus we arrive at:

$$\begin{bmatrix} x^0 \\ x^1 \\ x^2 \\ x^3 \end{bmatrix} = \begin{bmatrix} \gamma & -\gamma \frac{V_1}{c} & -\gamma \frac{V_2}{c} & -\gamma \frac{V_3}{c} \\ -\gamma \frac{V_1}{c} & 1 + \frac{V_1 V_1}{\boldsymbol{V}^2}(\gamma - 1) & \frac{V_1 V_2}{\boldsymbol{V}^2}(\gamma - 1) & \frac{V_1 V_3}{\boldsymbol{V}^2}(\gamma - 1) \\ -\gamma \frac{V_2}{c} & \frac{V_2 V_1}{\boldsymbol{V}^2}(\gamma - 1) & 1 + \frac{V_2 V_2}{\boldsymbol{V}^2}(\gamma - 1) & \frac{V_2 V_3}{\boldsymbol{V}^2}(\gamma - 1) \\ -\gamma \frac{V_3}{c} & \frac{V_3 V_1}{\boldsymbol{V}^2}(\gamma - 1) & \frac{V_3 V_2}{\boldsymbol{V}^2}(\gamma - 1) & 1 + \frac{V_3 V_3}{\boldsymbol{V}^2}(\gamma - 1) \end{bmatrix} \begin{bmatrix} \tilde{x}^0 \\ \tilde{x}^1 \\ \tilde{x}^2 \\ \tilde{x}^3 \end{bmatrix}. \tag{13.12.14}$$

In a second step we rotate, for a fixed time coordinate, the spatial coordinates of the intermediate system so that they coincide with those of the dashed system:

$$\tilde{x}^{i} = O_{ij} x'_j \equiv O'_{ji} x'_j. \tag{13.12.15}$$

This can be rewritten in four-vector notation:

$$\begin{bmatrix} \tilde{x}^0 \\ \tilde{x}^1 \\ \tilde{x}^2 \\ \tilde{x}^3 \end{bmatrix} = \begin{bmatrix} 1 & 0 & 0 & 0 \\ 0 & O_{11} & O_{12} & O_{13} \\ 0 & O_{21} & O_{22} & O_{23} \\ 0 & O_{31} & O_{32} & O_{33} \end{bmatrix} \begin{bmatrix} x'^0 \\ x'^1 \\ x'^2 \\ x'^3 \end{bmatrix}. \tag{13.12.16}$$

This transformation can now simply be inserted in Eq. (13.12.14) and by performing the multiplication we obtain[6]:

$$\begin{aligned} x^0 &= \gamma x'^0 - \gamma \frac{V_i}{c} O_{ik} x'^k, \\ x^j &= -\gamma \frac{V_j}{c} x'^0 + \left[ \delta_{jk} + (\gamma - 1) \frac{V_j V_k}{\boldsymbol{V}^2} \right] O_{kl} x'^l. \end{aligned} \tag{13.12.17}$$

This is the famous LORENTZ transformation in its most general form, which can rarely be found in textbooks. We now apply the "rule" from Exercise 13.11.1 in an analogous manner and find for the inversion of Eq. (13.12.17) that:

[6] Note that it is important to distinguish between co- and contra variant components of the space-time vectors and must strictly be observed in the following formulae. Of course there is no difference between co- and contra variant for Cartesian non-space-time quantities, like the velocity $V_i$ or the rotation matrix $O_{ij}$. Consequently, the rule of cross-wise summation (*cf.*, the remark after Eq. (2.4.13), which holds *mutatis mutandis* also in 4D) does not hold in the subsequent formulae: Unfortunately this diminishes their beauty.

$$x'^0 = \gamma x^0 + \gamma \frac{V_k}{c} x^k,$$
$$x'^j = O'_{ji} \left\{ \gamma \frac{V_i}{c} x^0 + \left[ \delta_{ik} + (\gamma - 1) \frac{V_i V_k}{V^2} \right] x^k \right\}. \tag{13.12.18}$$

Beside the rule the following auxiliary formulae have been used:

$$V'_i = O'_{ij} V_j, \; V'_i V'_i = V_k V_k \; \Rightarrow \; \gamma' = \gamma. \tag{13.12.19}$$

Equations (13.12.17/13.12.18) allow nicely to study the transition to Galilean transformations by looking at the limit case $c \to \infty$:

$$x'^0 = x^0, \; x'^j = V'_i t + O'_{ji} x^i, \; x^j = - V_j t + O_{jl} x'^l. \tag{13.12.20}$$

These relations are in complete agreement with Eq. (13.10.3) including the signs.

**Exercise 13.12.2:** ***Further investigations of the Lorentz transformation***

Begin by verifying the solutions shown in Eqs. (13.12.11–13.12.14) and (13.12.17). Then rewrite Eqs. (13.12.17) in four-vector notation and invert the 4 × 4 matrix directly to prove Eqs. (13.12.18). Hence, reconfirm the physics based rule of inversions.

A final remark is in order. Clearly, most of the textbooks on relativity do not derive the Lorentz transformations as it was done here. However, this chapter is concerned with the foundations of an electrodynamic field theory, which explains why we have followed a non-mechanical path. Relativists argue as follows: Consider an inertial system at rest together with a second uniformly moving frame, both described in terms of Cartesian axes. The origins of both frames coincide initially and then separate at a constant boost velocity, $-V_i$, to be determined from the inertial system at rest. For simplicity the corresponding (spatial) axes of both systems are parallel to each other. In the very moment when both origins coincide a flash is ignited at the same position. The observer at rest sees a spherical light wave extending in a spherically symmetric manner at the speed of light, $c$.[7] In mathematical terms this wave can be described by an equation for a sphere of radius $ct$:

$$\left(x^1\right)^2 + \left(x^2\right)^2 + \left(x^3\right)^2 = (ct)^2. \tag{13.12.21}$$

Somewhat more formally this can also be expressed by the following space-time function:

[7] Spherically, because space is considered to behave isotropically.

$$f(\boldsymbol{x}) = -\left(x^0\right)^2 + \left(x^1\right)^2 + \left(x^2\right)^2 + \left(x^3\right)^2 \equiv 0. \tag{13.12.22}$$

One would think that a moving observer should not see a spherical wave but a somewhat deformed one instead depending on his state of motion. Intuitively we expect a certain retardation of the light wave in the direction of the velocity $-V_i$ of relative motion. However, this contradicts the result of the MICHELSON-MORLEY experiment, which we shall briefly describe in what follows. MICHELSON and MORLEY found out that the speed of light remains constant with an accuracy of $\pm 5\,\frac{\text{km}}{\text{s}}$ after turning their interferometer by 90°: Fig. 13.9.

The principle of the experiment can be illustrated as follows: We consider two equally strong swimmers (a metaphor for two light beams) of speed $c$ in a river (the aether) flowing at speed $V$. The first one swims along a distance $l_1$, first with and then against the current. The other one swims along the distance $l_2$ from one river bench to the other and then returns. Obviously the second swimmer must swim at a certain angle w.r.t. the direction of the flow of the river in order to arrive exactly opposite from the starting point. The time difference between the first and of the second swimmer is approximately given by:

$$\Delta t \approx \frac{2(l_1 - l_2)}{c} + \frac{2l_1 - l_2}{c}\left(\frac{V}{c}\right)^2, \tag{13.12.23}$$

where it has been assumed that the flow velocity of the river, $V$, is much smaller than the velocity proper, $c$, of the swimmers.

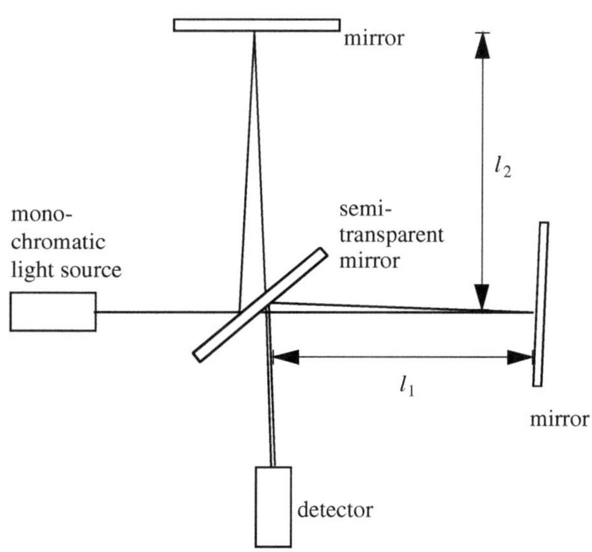


**Fig. 13.9** Sketch of the principle setup of a MICHELSON-MORLEY interferometer

Albert Abraham MICHELSON was born on December 19, 1852 in Strelno (now Poland) and died on May 9, 1931 in Pasadena, California. In 1869 he began his studies at the US-Naval Academy in Annapolis from where he graduated in 1873. He was inspired by the translation of the textbooks by the French physicist Adolphe GANOT and his remarks on a universal immaterial carrier of electromagnetic waves, the aether. As a result he dedicated the rest of his life to precise measurements of the speed of life trying to prove the existence of that "substance." In that he did not succeed. Nevertheless he was plagued by doubts to his very end that his measurements were incorrect. After a two-year scientific sojourn in Europe he left the Navy in 1881, accepted a position as a professor of physics at the Case School of Applied Science in Cleveland (Ohio) in 1883, he was appointed as a professor at Clark University in Worcester (Massachusetts) in 1889 and, finally, in 1892 as a professor and head of the physics department of the newly founded University of Chicago. In 1907 he became the first American who won the Nobel prize for physics. Rumor has it that the Cartwright family from the TV series Bonanza also had a very favorable influence on MICHELSON's scientific development. Indeed, we learn in the episode "Look to the Stars" how they help young Albert to prevail against a racist headmaster who is irritated by the Mosaic descend of his pupil.

We now apply this result to the light beams shown in Fig. 13.9. There will be a run time difference between them, potentially resulting in interference, in particular because the distances $l_1$ and $l_2$ cannot be made exactly equal, even if the mechanical parts are made with highest diligence. Consequently, the potential impact of the velocity $V$ of the aether is difficult to ascertain. However, if the apparatus is turned by 90° after the first run, so that $l_1$ and $l_2$ exchange their role, the run time difference is now:

$$\Delta\hat{t} \approx \frac{2(l_2 - l_1)}{c} + \frac{2l_2 - l_1}{c}\left(\frac{V}{c}\right)^2. \qquad (13.12.24)$$

Edward Williams MORLEY was born on January 29, 1838 in Newark, New Jersey and died on February 24, 1923 in West Hartford, Connecticut. He completed his academic studies at the Williams College in 1860. From 1869 until 1906 he was a professor of chemistry at today's Case Western Reserve University. Being a versatile experimentalist he became an ideal colleague for MICHELSON: In 1887 they conducted the famous experiment named after them. Like MICHELSON he did not really accept its negative outcome that denied the existence of the aether. MORLEY worked also on the exact chemical composition of Earth's atmosphere, the thermal expansion of solids and on measuring speed of light in magnetic fields.

Thus the total run time difference is given by $\Delta t + \Delta\hat{t} = \frac{l_1+l_2}{c}\left(\frac{V}{c}\right)^2$, which is proportional to the square of the aether drift. MICHELSON and MORLEY chose distances of $l_1 \approx l_2 = 11\,\text{m}$. Moreover, we have to keep in mind that Earth circles around the Sun at the (under terrestrial circumstances) enormous speed of roughly $30\,\frac{\text{km}}{\text{s}}$. Nevertheless, this is still small when compared to the speed of light, which is about

$3 \cdot 10^5 \frac{\text{km}}{\text{s}}$. Thus the optical distance is $\Delta d = c(\Delta t + \Delta \hat{t}) = (l_1 + l_2)\left(\frac{v}{c}\right)^2 \approx 2.2 \cdot 10^{-7}\text{m}$. This has to be related to the wave length $\lambda$ of the incoming light, which is roughly 500 nm, i.e., $\Delta d/\lambda \approx 0.44$. Thus a change of the interference pattern should clearly be visible. However, the measurements indicated a much smaller ratio of not more than 0.01. Thus scientists came to the conclusion that the speed of light in vacuum has the same (maximum) value of $3 \cdot 10^5 \frac{\text{km}}{\text{s}}$ in all frames of reference, independently of their state of motion.

The experiment has been repeated many times ever since—with the same negative outcome. It was also taken into account that Earth travels relatively to the center of our galaxy at the much greater relative speed of 200 $\frac{\text{km}}{\text{s}}$. Hence even an aether drift almost ten times as large does not seem to "impress" light waves at all. The speed of light appears to be a universal constant, independent of the observer, may he be moving or not. On the other hand, if we are honest, it would indeed be alarming if the propagation characteristics of something as fundamental as a quantum of light depended on the state of motion of the observer. This could become a real problem for the causality chain since we might be able to "overtake" events and, effectively, could no longer differentiate between the cause and the effect. The mathematical models we use to explain the physical world would suffer from inner contradictions, potentially predicting instability and chaos where no chaos has been observed.

**Exercise 13.12.3:** ***Add-on to the MICHELSON-MORLEY experiment***

Prove Eqs. (13.12.23/13.12.24).

After these remarks on the speed of light we propose a mathematical relation analogously to Eq. (13.12.22) but for the moving observer:

$$f(\boldsymbol{x}) = -\left(x'^0\right)^2 + \left(x'^1\right)^2 + \left(x'^2\right)^2 + \left(x'^3\right)^2 \equiv 0 \tag{13.12.25}$$

and conclude:

$$-\left(x'^0\right)^2 + \left(x'^1\right)^2 + \left(x'^2\right)^2 + \left(x'^3\right)^2 = -\left(x^0\right)^2 + \left(x^1\right)^2 + \left(x^2\right)^2 + \left(x^3\right)^2. \tag{13.12.26}$$

We can eliminate the common origin at $\boldsymbol{x} = \boldsymbol{0}$ and introduce a new reference position at $\boldsymbol{x} = \boldsymbol{x}_\text{r}$ as the starting point of the flash of light and write:

$$\begin{aligned} &-\left(x'^0 - x_\text{r}'^0\right)^2 + \left(x^1 - x_\text{r}'^1\right)^2 + \left(x'^2 - x_\text{r}'^2\right)^2 + \left(x'^3 - x_\text{r}'^3\right)^2 \\ &\quad = -\left(x^0 - x_\text{r}^0\right)^2 + \left(x^1 - x_\text{r}^1\right)^2 + \left(x^2 - x_\text{r}^2\right)^2 + \left(x^3 - x_\text{r}^3\right)^2. \end{aligned} \tag{13.12.27}$$

This also holds true if both points are infinitesimally close:

$$\begin{aligned} &-\left(\mathrm{d}x'^0\right)^2 + \left(\mathrm{d}x'^1\right)^2 + \left(\mathrm{d}x'^2\right)^2 + \left(\mathrm{d}x'^3\right)^2 \\ &\quad = -\left(\mathrm{d}x^0\right)^2 + \left(\mathrm{d}x^1\right)^2 + \left(\mathrm{d}x^2\right)^2 + \left(\mathrm{d}x^3\right)^2. \end{aligned} \tag{13.12.28}$$

Thus we arrived at the four-dimensional line element of Eq. (13.11.3). We conclude that the constancy of the speed of light implies that the four-dimensional line element is an absolute world scalar according to the nomenclature established in context with Eq. (13.11.5):

$$(\mathrm{d}s')^2 = (\mathrm{d}s)^2 \Leftrightarrow g'_{AB}\mathrm{d}x'^A\mathrm{d}x'^B = g_{CD}\mathrm{d}x^C\mathrm{d}x^D. \tag{13.12.29}$$

Note that the components of $g_{CD}$ as well as of $g'_{AB}$ are given by the matrix shown in Eq. (13.11.4). The space-time vector of Eq. (13.11.1) and its differentials are absolute world vectors and we conclude that:

$$\mathrm{d}x^C\mathrm{d}x^D = \frac{\partial x^C}{\partial x'^A}\frac{\partial x^D}{\partial x'^B}\mathrm{d}x'^A\mathrm{d}x'^B \tag{13.12.30}$$

Inserting this in Eq. (13.12.29) results in:

$$g'_{AB} = \frac{\partial x^C}{\partial x'^A}\frac{\partial x^D}{\partial x'^B}g_{CD}. \tag{13.12.31}$$

This is Eq. (13.12.1) in covariant form which had previously been obtained from the invariance of the MAXWELL-LORENTZ aether relations, by means of which the LORENTZ transformation was derived. In retrospect it is not surprising that the postulate of a spherical form for a flash of light in all systems arising by mutual boost $-V_i$ at equal propagation speed $c$ results in the same equations: After all light is an electromagnetic wave, and if the MAXWELL-LORENTZ aether relations in simple form hold in all these systems, Exercise 13.8.1 predicts *the same wave equation* for all of them.

## 13.13 Energy and Momentum of the Electromagnetic Field

We now want to include the effect of the electromagnetic fields on the balances of momentum and of energy. For this purpose we simply add in the balance of momentum to the (gravitational) bulk forces $\rho f_i$ the electromagnetic contribution $qE_i + [(\boldsymbol{j} + q\boldsymbol{v}) \times \boldsymbol{B}]_i$ and write in regular points:

$$\frac{\partial \rho v_i}{\partial t} + \frac{\partial}{\partial x_j}\left(\rho v_i v_j - \sigma_{ji}\right) = \rho f_i + qE_i + [(\boldsymbol{j} + q\boldsymbol{v}) \times \boldsymbol{B}]_i. \tag{13.13.1}$$

Experiments show that the power of the electromagnetic field (the so-called JOULE heating or heat production due to electric resistance) is given by the scalar product between electric current density $\boldsymbol{j} + q\boldsymbol{v}$ and the electric field $\boldsymbol{E}$. Thus the local energy balance reads:

$$\frac{\partial}{\partial t}\rho\left(u + \frac{1}{2}v^2\right) + \frac{\partial}{\partial x_j}\left[\rho\left(u + \frac{1}{2}v^2\right)v_j + q_j - \sigma_{ji}v_i\right] = \rho f_i v_i + (j_i + qv_i)E_i. \tag{13.13.2}$$

In other words, momentum and energy are no longer conserved due to *production terms* that suddenly arise. However, it is possible to rearrange MAXWELL's equations suitably and to combine them with the balances of momentum and energy so that conservation of energy and momentum is guaranteed. To this end we start from MAXWELL's equations for regular points shown in Eq. (13.8.1) and insert the MAXWELL-LORENTZ aether relations from Eq. (13.8.2):

$$\begin{aligned} &\frac{\partial B_i}{\partial x_i} = 0, \frac{\partial B_i}{\partial t} + \epsilon_{ijk}\frac{\partial E_k}{\partial x_j} = 0, \\ &\varepsilon_0\frac{\partial E_i}{\partial x_i} = q, -\varepsilon_0\frac{\partial E_i}{\partial t} + \frac{1}{\mu_0}\epsilon_{ijk}\frac{\partial B_k}{\partial x_j} = j_i + q\upsilon_i. \end{aligned} \tag{13.13.3}$$

We now multiply the second of these equations by $\frac{1}{\mu_0}B_i$ and the fourth one by $-E_i$. Both results are then added and the product rule is applied:

$$\frac{\partial}{\partial t}\frac{1}{2}\left(\boldsymbol{E}\cdot\varepsilon_0\boldsymbol{E} + \boldsymbol{B}\cdot\frac{1}{\mu_0}\boldsymbol{B}\right) + \frac{\partial}{\partial x_j}\left(\epsilon_{jki}E_k\frac{1}{\mu_0}B_i\right) = -(j_i + q\upsilon_i)E_i. \tag{13.13.4}$$

Inserting the MAXWELL-LORENTZ aether relations yields:

$$\frac{\partial}{\partial t}\frac{1}{2}(\boldsymbol{E}\cdot\boldsymbol{D} + \boldsymbol{B}\cdot\boldsymbol{H}) + \frac{\partial}{\partial x_i}(\boldsymbol{E}\times\boldsymbol{H})_i = -(j_i + q\upsilon_i)E_i. \tag{13.13.5}$$

John Henry POYNTING was born on September 9, 1852 in Monton near Manchester and died on March 30, 1914 in Birmingham. He first attended Owen's College in Manchester and obtained his B.Sc. in 1872. Then he went to Trinity College in Cambridge and graduated with a Bachelor of Arts in 1876. From 1880 until 1914 POYNTING was a professor of physics at Mason College in Birmingham. MAXWELL was one of his teachers and, not too surprising, POYNTING also worked in the field of electrodynamics. He discovered the vector named after him together with the corresponding conservation law. He also studied the interaction of solar radiation with interplanetary dust (POYNTING-ROBERT-effect). In 1888 he was elected as a Fellow of the Royal Society which awarded him the Royal Medal in 1905.

If we ignore the minus sign, the left hand side is equal to the production density of energy due to electromagnetic fields [JOULE heating, cf., Eq. (13.13.2)]. If we eliminate this term we obtain:

$$\begin{aligned} &\frac{\partial}{\partial t}\left[\rho\left(u + \frac{1}{2}\upsilon^2\right) + \frac{1}{2}(\boldsymbol{E}\cdot\boldsymbol{D} + \boldsymbol{B}\cdot\boldsymbol{H})\right] \\ &+ \frac{\partial}{\partial x_j}\left[\rho\left(u + \frac{1}{2}\upsilon^2\right)\upsilon_j + q_j - \sigma_{ji}\upsilon_i + (\boldsymbol{E}\times\boldsymbol{H})_j\right] = \rho f_i\upsilon_i. \end{aligned} \tag{13.13.6}$$

Thus the production on the right hand side of the energy balance disappears and additional terms emerge in the temporal derivative and in the divergence term. For obvious reasons the expressions $\frac{1}{2}(\boldsymbol{E}\cdot\boldsymbol{D}+\boldsymbol{B}\cdot\boldsymbol{H})$ and $\boldsymbol{E}\times\boldsymbol{H}$ are known as the *energy density* and the *energy flux of the electromagnetic field.* For historic reasons the latter is also known as the so-called *Poynting vector.*

Multiplication of Eq. $(13.13.3)_4$ by $\epsilon_{lin}B_n$ and of $(13.13.3)_3$ by $-E_l$ and summation of both results leads to:

$$\begin{aligned}\frac{\partial}{\partial t}(\epsilon_{lin}\varepsilon_0 E_i B_n) - \epsilon_{lin}\varepsilon_0 E_i\frac{\partial B_n}{\partial t} + \left(\delta_{jl}\delta_{kn} - \delta_{jn}\delta_{lk}\right)\frac{1}{\mu_0}B_n\frac{\partial B_k}{\partial x_j}\\ - \varepsilon_0 E_l\frac{\partial E_i}{\partial x_i} = -qE_l - \epsilon_{lin}(j_i + q\upsilon_i)B_n.\end{aligned} \tag{13.13.7}$$

We now eliminate the term $\partial B_n/\partial t$ by means of Eq. $(13.13.3)_2$:

$$\begin{aligned}\frac{\partial}{\partial t}(\epsilon_{lin}\varepsilon_0 E_i B_n) + \frac{\partial}{\partial x_i}\left[\frac{1}{2}\left(\boldsymbol{E}\cdot\varepsilon_0\boldsymbol{E} + \boldsymbol{B}\cdot\frac{1}{\mu_0}\boldsymbol{B}\right)\delta_{li} - E_i\varepsilon_0 E_l - B_i\frac{1}{\mu_0}B_l\right]\\ = -qE_l - \epsilon_{lin}(j_i + q\upsilon_i)B_n,\end{aligned} \tag{13.13.8}$$

or by using the Maxwell-Lorentz aether relations:

$$\begin{aligned}\frac{\partial}{\partial t}(\boldsymbol{D}\times\boldsymbol{B})_l + \frac{\partial}{\partial x_i}\left[\frac{1}{2}(\boldsymbol{E}\cdot\boldsymbol{D}+\boldsymbol{B}\cdot\boldsymbol{H})\delta_{li} - E_i D_l - B_i H_l\right]\\ = -qE_l - \left[(\boldsymbol{j}+q\boldsymbol{\upsilon})\times\boldsymbol{B}\right]_l.\end{aligned} \tag{13.13.9}$$

Clearly the left hand side of this equation is, with the exception of the sign, the production of momentum due to the electromagnetic field from Eq. (13.13.1). If we combine both equations the following conservation law is obtained:

$$\begin{aligned}&\frac{\partial}{\partial t}\left[\rho\upsilon_i + (\boldsymbol{D}\times\boldsymbol{B})_i\right] +\\ &\frac{\partial}{\partial x_j}\left(\rho\upsilon_i\upsilon_j - \sigma_{ji} + \frac{1}{2}(\boldsymbol{E}\cdot\boldsymbol{D}+\boldsymbol{B}\cdot\boldsymbol{H})\delta_{ij} - E_i D_j - B_i H_j\right) = \rho f_i.\end{aligned} \tag{13.13.10}$$

We conclude that $\boldsymbol{D}\times\boldsymbol{B}$ represents the *density of momentum of the electromagnetic field.* Moreover, in view of the mechanical stress tensor, $\sigma_{ij}$, the term $-\frac{1}{2}(\boldsymbol{E}\cdot\boldsymbol{D}+\boldsymbol{B}\cdot\boldsymbol{H})\,\delta_{ij} + E_i D_j + B_i H_j$ formally represents "electromagnetic stresses." Indeed, it is known as *Maxwell stress tensor.*

## 13.14 Simple Electrodynamic Constitutive Equations

Recall Eqs. (13.6.7) and (13.7.7), which relate the material dependent polarization $\boldsymbol{P}$ and the magnetization $\boldsymbol{M}$ with the charge and current potentials $\boldsymbol{D}$ and $\boldsymbol{H}$ to form the quantities $\mathfrak{D}$ and $\mathfrak{H}$, the charge and current potentials in matter. As their names indicate the latter are material dependent quantities:

$$\mathfrak{D} = \boldsymbol{D}+\boldsymbol{P},\ \mathfrak{H} = \boldsymbol{H}-\boldsymbol{M}-(\boldsymbol{P}\times\boldsymbol{\upsilon}). \tag{13.14.1}$$

Moreover, according to Eq. (13.7.1), there is another constitutive function, namely the density of the true currents, $\boldsymbol{j}^{\mathrm{f}}$. We must now connect all of these quantities with the primary fields $\boldsymbol{E}$ and $\boldsymbol{B}$, respectively. In general these relations are based on fundamental principles, embedded in a constitutive theory, which we will not present in full detail in this book. We rather argue empirically and start with an intuitive argument to find the simplest constitutive equation for the polarization vector: In Sect. 13.6 it was mentioned that the molecules of a dielectric material polarize under the presence of an external electric field, $\boldsymbol{E}$, under its corresponding forces. It is reasonable to describe this effect (in other words: the polarization) in a first approximation by relating it linearly to the cause (i.e., the electric field) as follows:

$$P_i = \varepsilon_0 \chi E_i. \tag{13.14.2}$$

The (temperature dependent) coefficient of proportionality, $\chi$, is known as the *dielectric susceptibility*. Recall that it was reasonable to assume in HOOKE's law that stress and strain are proportional to each other, as long as the strains are not too large and plasticity becomes an issue. Similarly the proportionality between $\boldsymbol{P}$ und $\boldsymbol{E}$ is a simple but adequate assumption for most dielectrics as long as the electric fields are not too strong. However, it must be pointed out that it holds in its simple form, Eq. (13.14.2), only for *isotropic* materials. For anisotropic materials it is necessary to use the *dielectric susceptibility tensor*, $\boldsymbol{\chi}$, instead. We write:

$$P_i = \varepsilon_0 \chi_{ij} E_j. \tag{13.14.3}$$

However, it is also fair to point out that various technical materials do show a pronounced non-linear behavior as far as polarization is concerned, e.g., Seignette salt or barium titanate. Then the loading history, in other words, the temporal development of a rising and decaying $\boldsymbol{E}$ field becomes an issue and *hysteresis* is observed. Moreover, the action of the electric field is in general coupled to the effect of mechanical stresses. We will learn more about this shortly. Before that, however, we combine Eq. $(13.14.1)_1$ with the MAXWELL-LORENTZ aether relation $(13.8.2)_1$ and obtain:

$$\mathfrak{D}_i = D_i + \varepsilon_0 \chi E_i = \varepsilon_0 (1 + \chi) E_i = \varepsilon_0 \varepsilon_{\mathrm{r}} E_i \equiv \kappa E_i, \tag{13.14.4}$$

where the so-called *relative dielectric constant* $\varepsilon_{\mathrm{r}} = 1 + \chi$ has been introduced. The dielectric constant is frequently used in high school experiments in context with plate capacitors with dielectric "fillings." Another expression used for the same quantity is the term *dielectric* or *relative permittivity* in combination with the symbols $\kappa$ or $\kappa_{ij}$ for the corresponding second order tensor used for anisotropic matter. Note that the symbol $\varepsilon_{ij}$ is sometimes used in the anisotropic case instead. Clearly this can easily be confused with the mechanical strain, especially if we consider coupled problems (see further down). For this reason we are not going to follow that convention and just warn the reader to read the instructions carefully when using other literature and programs. In summary, in the anisotropic case we replace Eq. (13.14.4) by:

$$\mathfrak{D}_i = \kappa_{ij} E_j,\ \kappa_{ij} = \varepsilon_0 (\delta_{ij} + \chi_{ij}). \tag{13.14.5}$$

In terms of the flux-force concept introduced in Sect. 12.4 we may also refer to the electric field as the "driving force" and to the polarization or to the charge potential in matter as the "fluxes." Now surely, it is only fair to refer to the electric field as a "driving force" because of the force it exerts on charges. However, if we start coupling mechanical with electric effects things are not so straightforward any more. Indeed, it is possible to use mechanical means to create a polarization or a charge potential. If we follow the arguments outlined in Sect. 12.4 strains are the drivers on the mechanical side and *not* the stresses, as one would intuitively expect. Rather the stresses are the fluxes, as we might have guessed by noting that the traction is the non-convective flux of momentum. Now if both driving forces are present, i.e., electric fields *and* mechanical strains act simultaneously, we may write within a linear theory:

$$\mathfrak{D}_i = \kappa_{ij} E_j + e_{ijk} \varepsilon_{jk} \text{ or } P_i = \varepsilon_0 \chi_{ij} E_j + e_{ijk} \varepsilon_{jk}. \tag{13.14.6}$$

Lars Onsager was born on November 27, 1903 in Kristiania (now Oslo) and died on October 5, 1976 in Coral Gables (Florida). He is well known for his work in irreversible thermodynamics and physical chemistry in particular for his discovery of the reciprocal relations in constitutive equations and the corresponding kinetic interpretation. This got him the Nobel Prize in Chemistry in 1968 which in turn resulted in envy and tirades of hate by many other thermodynamicists. He held the Gibbs Professorship of Theoretical Chemistry at Yale University and had a reputation for being an extremely bad teacher. In fact his chemistry lectures were referred to as Norwegian I and II and this had nothing to do with his mother tongue. Be that as it may, as a scientist he surely made a considerable contribution to the understanding of the nature of irreversibility.

The quantities $e_{ijk}$ are a.k.a. (*charge*) *piezoelectric constants* and they form a tensor of third order. In fact, they allow quantifying the so-called direct piezoelectric effect: If we squeeze a piezoelectric crystal the originally balanced charges will be separated and a polarization charge is generated. However, this is only half the story. Hooke's law must be extended as well in order to cover more than just the mechanical strain of Eq. (6.2.1). We write:

$$\sigma_{ij} = C_{ijkl} \varepsilon_{kl} - e_{kij} E_k. \tag{13.14.7}$$

This relation describes what is known as the *converse piezoelectric effect*: By application of an electric field the charges of the piezosensitive material are separated and a mechanical strain results which, if constrained, results in a stress.

Interestingly the *same* piezoelectric coefficients are used in Eq. (13.14.7) to relate the effects of the electric field to the mechanical stresses. This is a consequence of Onsager's *principle of reciprocity*. It can be motivated by kinetic arguments in context with Onsager's regression hypothesis and the so-called fluctuation-regression theorem, but we will not provide any details here. However, what we will do is answer how many different coefficients there are altogether in a linearized theory for a material of highest degree of anisotropy. We start with the stiffness matrix $C_{ijkl}$. Recall that the linearized strain tensor is symmetric by definition. We also assume that the stress tensor is symmetric. Thus both have six independent components and this leaves us with a total of 6×6 independent coefficients for the stiffness matrix. However, there is still more symmetry if we consider the elastically stored energy density, $w$, and its complimentary form, $w^*$. They are defined as follows:

$$w = \int_{\tilde{\boldsymbol{\varepsilon}}=\mathbf{0}}^{\tilde{\boldsymbol{\varepsilon}}=\boldsymbol{\varepsilon}} \sigma_{ij}(\tilde{\boldsymbol{\varepsilon}})\, \mathrm{d}\tilde{\varepsilon}_{ji}, \quad w^* = \int_{\tilde{\boldsymbol{\sigma}}=\mathbf{0}}^{\tilde{\boldsymbol{\sigma}}=\boldsymbol{\sigma}} \varepsilon_{ij}(\tilde{\boldsymbol{\sigma}})\, \mathrm{d}\tilde{\sigma}_{ji}. \tag{13.14.8}$$

If we now insert Hooke's law [i.e., Eq. (13.14.7) without the electric part] the integration can easily be performed and we obtain:

$$w = C_{ijkl} \int_{\tilde{\boldsymbol{\varepsilon}}=\mathbf{0}}^{\tilde{\boldsymbol{\varepsilon}}=\boldsymbol{\varepsilon}} \tilde{\varepsilon}_{kl}\, \mathrm{d}\tilde{\varepsilon}_{ij}, \quad w^* = C_{ijkl} \int_{\tilde{\boldsymbol{\varepsilon}}=\mathbf{0}}^{\tilde{\boldsymbol{\varepsilon}}=\boldsymbol{\varepsilon}} \tilde{\varepsilon}_{ij}\, \mathrm{d}\tilde{\varepsilon}_{kl}. \tag{13.14.9}$$

As indicated in Fig. 13.10 we expect for a linear elastic material both strain energies to be equal, and thus:

$$C_{ijkl} = C_{klij}. \tag{13.14.10}$$

We conclude that we may exchange index pairs and this reduces the amount of 36 independent components to $(6 \times 6 - 6)/2 + 6 = 19 = 21$. In fact this can be reduced even further, but only if we reduce the degree of anisotropy of the material. How to do this is explained (for example) in the books by Nye [2] or Tsai [3]. We will not explain this any further, and it may suffice to say that for isotropic materials we end up with two independent elastic constants, e.g., Young's modulus and Poisson's ratio or the two Lamé coefficients.

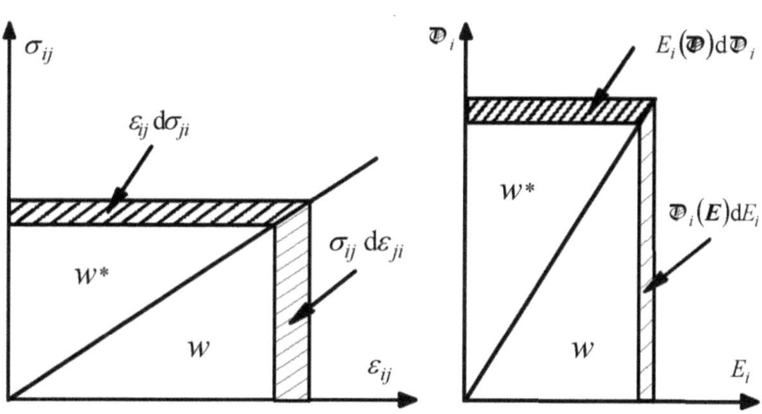


**Fig. 13.10** Strain and electric energy densities and their complements

We now turn to the susceptibility tensor or to the relative permittivity tensor, respectively. The arguments are similar as in the case of the elastic stiffness, stress, and strain. The energy density of the electric field as well as its complimentary form is given by:

$$w = \int_{\tilde{E}=0}^{\tilde{E}=E} \mathfrak{D}_i(\tilde{E})\, \mathrm{d}\tilde{E}_i,\; w^* = \int_{\tilde{\mathfrak{D}}=0}^{\tilde{\mathfrak{D}}=\mathfrak{D}} E_i(\tilde{\mathfrak{D}})\, \mathrm{d}\tilde{\mathfrak{D}}_i \tag{13.14.11}$$

We now insert Eq. $(13.14.7)_2$ without the mechanical part to find that:

$$w = \kappa_{ij} \int_{\tilde{E}=0}^{\tilde{E}=E} \tilde{E}_j\, \mathrm{d}\tilde{E}_i,\; w^* = \kappa_{ij} \int_{\tilde{E}=0}^{\tilde{E}=E} \tilde{E}_i\, \mathrm{d}\tilde{E}_j. \tag{13.14.12}$$

In the linear case both should be equal and, hence, we conclude that the susceptibility tensor and the relative permittivity tensor are symmetric and have six independent components for the highest degree of anisotropy *and* if the coordinate system does not coincide with the principal axes of the crystal:

$$\chi_{ij} = \chi_{ji},\; \kappa_{ij} = \kappa_{ji}. \tag{13.14.13}$$

The situation is similar to the case of the matrix of thermal expansion coefficients: If the principal axes of the crystal and of the coordinate system coincide the tensor is diagonal with a maximum of three different components for the three principal directions.

This leaves us with the coupling coefficients or charge piezoelectric constants $e_{ijk}$. Since the stress and strain tensors are both symmetric, i.e., consist of six independent coefficients, it follows that $e_{ijk} = e_{ikj}$ i.e., we have a maximum of $3 \times 6 = 18$ independent entries. It was Woldemar VOIGT who combined all of these material coefficients in a matrix scheme, nowadays known as VOIGT's notation. It is very popular in engineering and frequently used in finite element codes. For these reasons we will touch upon it briefly.

In VOIGT's notation symmetric index pairs $(i, j)$ are replaced by superindices $(I)$, which are easily spotted by using capital letters. For example, the symmetric stress tensor consisting of six independent components can be represented by a $6 \times 1$ matrix, i.e., a column containing these six independent components. Note that one should be careful when calling this column a vector, because it does not transform according to Eq. (2.4.1). Note that superindices are typically assigned according to the convention established in Table 13.1.

**Table 13.1** Superindex convention

| Tensor-index-pairs | 11 | 22 | 33 | 23, 32 | 31, 13 | 12, 21 |
|---|---|---|---|---|---|---|
| Superindices | 1 | 2 | 3 | 4 | 5 | 6 |

We now define the following matrices:

$$\begin{aligned} C_{IJ} &= C_{ijkl},\ e_{iJ} = e_{ikl},\ \sigma_I = \ \sigma_{ij},\ \text{and} \\ \varepsilon_I &= \left[\varepsilon_{11}, \varepsilon_{22}, \varepsilon_{33}, 2\varepsilon_{23}, 2\varepsilon_{31}, 2\varepsilon_{12}\right]^{\mathrm{T}}. \end{aligned} \tag{13.14.14}$$

Note the factors of two in the strain matrix. Their origin will become clear if we present the coupled constitutive relations as a matrix equation of the form $\boldsymbol{A}\,\boldsymbol{x} = \boldsymbol{b}$, which arises from the tensor equations (13.14.6/13.14.7). We incorporate the entries $\sigma_I$, $\mathfrak{D}_i$, $C_{IJ}$, $e_{iJ}$, $\kappa_{ij}$, $\varepsilon_I$, and $E_i$ in a giant matrix scheme as follows:

$$\begin{bmatrix} \boldsymbol{\sigma} \\ \boldsymbol{\mathfrak{D}} \end{bmatrix} = \begin{bmatrix} \boldsymbol{C} & \boldsymbol{e}^{\mathrm{T}} \\ \boldsymbol{e} & \boldsymbol{\kappa} \end{bmatrix} \begin{bmatrix} \boldsymbol{\varepsilon} \\ \boldsymbol{E} \end{bmatrix} \Leftrightarrow$$

$$\begin{bmatrix} \sigma_1 \\ \sigma_2 \\ \sigma_3 \\ \sigma_4 \\ \sigma_5 \\ \sigma_6 \\ \mathfrak{D}_1 \\ \mathfrak{D}_2 \\ \mathfrak{D}_3 \end{bmatrix} = \left[\begin{array}{cccccc|ccc} C_{11} & C_{12} & C_{13} & C_{14} & C_{15} & C_{16} & -e_{11} & -e_{21} & -e_{31} \\ C_{21} & C_{22} & C_{23} & C_{24} & C_{25} & C_{26} & -e_{12} & -e_{22} & -e_{32} \\ C_{31} & C_{32} & C_{33} & C_{34} & C_{35} & C_{36} & -e_{13} & -e_{23} & -e_{33} \\ C_{41} & C_{42} & C_{43} & C_{44} & C_{45} & C_{46} & -e_{14} & -e_{24} & -e_{34} \\ C_{51} & C_{52} & C_{53} & C_{54} & C_{55} & C_{56} & -e_{15} & -e_{25} & -e_{35} \\ C_{61} & C_{62} & C_{63} & C_{64} & C_{65} & C_{66} & -e_{16} & -e_{26} & -e_{36} \\ \hline e_{11} & e_{12} & e_{13} & e_{14} & e_{15} & e_{16} & \kappa_{11} & \kappa_{12} & \kappa_{13} \\ e_{21} & e_{22} & e_{23} & e_{24} & e_{25} & e_{26} & \kappa_{21} & \kappa_{22} & \kappa_{23} \\ e_{31} & e_{32} & e_{33} & e_{34} & e_{35} & e_{36} & \kappa_{31} & \kappa_{32} & \kappa_{33} \end{array}\right] \begin{bmatrix} \varepsilon_1 \\ \varepsilon_2 \\ \varepsilon_3 \\ \varepsilon_4 \\ \varepsilon_5 \\ \varepsilon_6 \\ E_1 \\ E_2 \\ E_3 \end{bmatrix}. \tag{13.14.15}$$

It is easy to check that this relation is equivalent to Eqs. (13.14.6/13.14.7) if the previously statements on symmetry are observed. First recall that $C_{ijkl} = C_{klij}$, hence we have $C_{IJ} = C_{JI}$ as well. Then in order to understand the factors of two all we have to do is to expand $\sigma_{ij} = C_{ijkl}\varepsilon_{kl}$ and compare it to $\sigma_I = C_{IJ}\varepsilon_J$. For example we find for the component $\sigma_6 = \sigma_{21}$:

$$\sigma_{21} = \cdots + C_{2123}\varepsilon_{23} + C_{2132}\varepsilon_{32} + \cdots = \cdots + C_{64}\varepsilon_4 + \cdots. \tag{13.14.16}$$

Similarly to Eq. (13.14.2), magnetization is linked within a first approximation linearly to the magnetic field if the term $-(\boldsymbol{P} \times \boldsymbol{\upsilon})_i$ as observed in Eq. (13.7.7) is negligibly small:

Woldemar VOIGT was born on September 2, 1850 in Leipzig and died on December 13, 1919. He was a theoretical physicist and is particularly known for his work on crystallography. He got his Ph.D. in 1875 at the University of Königsberg where he was assistant to Franz Ernst NEUMANN, whom we know from the DUHAMEL-NEUMANN extension of HOOKE's law. In fact his thesis was on the elastic behavior of rock salt. From 1875 until 1883 he worked as an assistant professor of physics in Königsberg. Then in 1883 he became a full professor for theoretical physics at the Georg-August University in Göttingen. In 1910 he published his famous "Lehrbuch der Kristallphysik," which is one of the pioneering works on theoretical crystallography. In fact, it does already contain the piezoelectric effect. VOIGT was one of the first to use the concept of tensors and he introduced the concise vector and matrix form for symmetric tensors of higher order, nowadays known as VOIGT's notation. He also anticipated the LORENTZ transformations and interpreted the outcome of the MICHELSON-MORLEY experiment. Other than tensors he loved classical music and (just like his teacher NEUMANN) was a passionate soldier, at least during his youth.

$$\boldsymbol{M} = -\frac{\chi_{\mathrm{m}}}{\mu_0}\boldsymbol{B}. \tag{13.14.17}$$

Thus the current potential in matter is given by:

$$\mu_0 \boldsymbol{\mathfrak{H}} = \mu_0(1+\chi_{\mathrm{m}})\boldsymbol{H} = \mu_{\mathrm{r}}\boldsymbol{B}. \tag{13.14.18}$$

The constant of proportionality, $\chi_{\mathrm{m}}$, is a.k.a. *magnetic susceptibility*. The combination $\mu_{\mathrm{r}} = 1 + \chi_{\mathrm{m}}$ is referred to as *relative permeability*. Note that molecular models can be used to compute the magnetic susceptibility.

To a first approximation the true current density, $\boldsymbol{j}^{\mathrm{f}}$, also obeys a linear relation:

$$\boldsymbol{j}^{\mathrm{f}} = \frac{1}{r}\boldsymbol{E}. \tag{13.14.19}$$

This makes sense because the presence of an electric field is the reason for the movement of charge carriers. Moreover, its action, i.e., the true current density, should increase if the resistance to movement, i.e., the specific electric resistance, $r$, of the corresponding material decreases. We now multiply this relation by $\mathrm{d}x_i$, where $|\mathrm{d}x_i| = \mathrm{d}l$ represents the length of an electric line current of cross-section $A$. Moreover we relate the electric field to its potential, $U$:

$$E_i = -\frac{\partial U}{\partial x_i} \tag{13.14.20}$$

and find:

$$-\frac{\partial U}{\partial x_i} A\,\mathrm{d}x_i = r j_i^{\mathrm{f}} A\,\mathrm{d}x_i \;\Rightarrow\; -\frac{\partial U}{\partial x_i} n_i A\,\mathrm{d}l = r j^{\mathrm{f}} A\,\mathrm{d}l. \tag{13.14.21}$$

By means of GAUSS' theorem we obtain after integration over the lines:

$$U_{\mathrm{out}} - U_{\mathrm{in}} = rlj^{\mathrm{f}}A = RI. \tag{13.14.22}$$

This is nothing else but OHM's law in a form known from high school physics.

Georg Simon OHM was born on March 16, 1789 in Erlangen and died on July 6, 1854 in Munich. At the young age of sixteen he began to study mathematics, physics, and philosophy at Friedrich-Alexander University in Erlangen in 1805. However, due to financial straits he had to abandon school and take a job as a mathematics teacher at a private school in Switzerland. He returned to Erlangen when he was 22, finished his Ph.D. work on light and colors in 1811, and then worked for three semesters as a private docent for mathematics. In order to survive he recommenced working as a teacher, first (1813) for a school in Bamberg, then in 1817 at the Dreikönigs Grammar School in Cologne, and in 1826 at a military school in Berlin. His main research interest belonged to electricity, a hot subject at his times. In 1833 he obtained a slightly better paid position as a docent at the Royal Polytechnic School in Nuremberg, where he became a director in 1839 and which bears his name today. In 1849 he was appointed at the University of Munich, first as an associate professor and, finally, in 1852 as a full professor for experimental physics. What a long way to go for the remaining short amount of life.

## 13.15 Would You Like to Know More?

A complete description of rational electrodynamics can be found in Chapter F of the handbook article by Truesdell and Toupin [4]. More recently the book by Kovatz [5] follows up on that subject. A discussion of the issues regarding the principle difference between the electric field and the electric displacement on the one hand side and the magnetic induction and the magnetic field on the other, their transformation properties, and the MAXWELL-LORENTZ aether relations are presented in that book in a pleasant, "digestible" manner. Moreover, the book also contains many exercises with solutions, which can lead to a more profound and sustainable understanding of the subject.

However, such books are rare. The standard literature on electrodynamics does not say much about the differences between the various fields, not to mention their transformation properties. Sometimes the presentation is so fuzzy that the beginner is inclined to believe that the form-invariance of MAXWELL's equations is limited to LORENTZ transformation, and that it cannot be understood without the theory of relativity at all. In this context the classic textbooks by Becker [1], Jackson [6], or Landau and Lifschitz [7] and Landau et al. [8] should be mentioned. They should be consulted if one is interested in further details and an alternative perspective, *after* one has gained some familiarity with the basic concepts.

Besides Einstein [9] the book by Weinberg [10] should be consulted for an in-depth understanding of the theory of relativity in field formulation. However, do not be surprised if the latter author does not mention the MAXWELL-LORENTZ aether relations in its sections on electrodynamics (Chaps. 2, 5).

More on constitutive equations in electrodynamics and on the coupling with the mechanical fields can be found in the book by Maugin [11].

In fact, this concludes our excursion to continuum theory. However, in order to quote Sir Winston CHURCHILL: "Now this is not the end. It is not even the beginning of the end. But it is, perhaps, the end of the beginning."

## References

1. Becker R (2012) Electromagnetic fields and interactions. Unabridged republication in one volume. Dover Publications, NY
2. Nye JF (1967) Physical properties of crystals: their representation by tensors and matrices, 4th edn. Oxford University Press, Oxford
3. Tsai SW (1992) Theory of composites design. Think Composites, Dayton
4. Truesdell C, Toupin R (1960) The classical field theories. In: Flügge S (ed) Encyclopedia of physics. Volume III/1 Principles of classical mechanics and field theory. Springer, Berlin
5. Kovatz A (2002) Electromagnetic theory., Oxford lecture series in mathematics and its applications. Oxford University Press, Oxford
6. Jackson JD (1975) Classical electrodynamics, 2nd edn. Wiley, New York
7. Landau LD, Lifschitz EM (1975) The classical theory of fields, vol 2, 4th edn., Course of theoretical physics. Butterworth-Heinemann, London
8. Landau LD, Lifschitz EM, Pitaevskii LP (1984) Electrodynamics of continuous media, vol 8, 2nd edn., Course of theoretical physics. Butterworth-Heinemann, London
9. Einstein A (1983) Über die spezielle und die allgemeine Relativitätstheorie. Wissenschaftliche Taschenbücher 59. 21. Ausgabe. Vieweg, Braunschweig
10. Weinberg S (1972) Gravitation and cosmology: Principles and applications of the general theory of relativity. Wiley, New York
11. Maugin GA (1988) Continuum mechanics of electromagnetic solids. North-Holland, Amsterdam

# Picture Sources

Robert Hooke (Page 2): Rita Greer, Oxford University
Claude Louis Marie Henri Navier (Page 3): Public Domain
George Gabriel Stokes (Page 2): Public Domain
Pierre Louis Dulong (Page 3): Public Domain
Alexis Thérèse Petit (Page 3): Public Domain
Euklid (Page 4): Public Domain
Carl Henry Eckart (Page 4): SIO Archives/UCSD
James Clerk Maxwell (Page 5): Public Domain
Jean Baptiste Joseph Baron de Fourier (Page 5): Public Domain
Gabriel Lamé (Page 9): Public Domain
Thomas Young (Page 9): Public Domain
Adrien Marie Legendre (Page 10): Public Domain
Jean Marie Constant Duhamel (Page 13): Public Domain
Franz Ernst Neumann (Page 13): Public Domain
Alan Arnold Griffith (Page 14): Public Domain
Pythagoras (Page 16): Wikipedia-user Galilea under CC-BY-SA 3.0
Albert Einstein (Page 17): Public Domain
Leopold Kronecker (Page 24): Public Domain
Christian Otto Mohr (Page 34): Public Domain
Richard Edler von Mises (Page 41): Smithsonian Institution
Anders Jonas Ångström (Page 48): Public Domain
Osborne Reynolds (Page 49): Public Domain
Isaac Newton (Page 51): Public Domain
Augustin Louis Cauchy (Page 52): Public Domain
Johann Carl Friedrich Gauss (Page 56): Public Domain
Joseph-Louis de Lagrange (Page 60): Public Domain
Leonard Euler (Page 61): Public Domain
Carl Gustav Jacob Jacobi (Page 62): Public Domain
Elwin Bruno Christoffel (Page 67): Public Domain
Julius Robert (von) Mayer (Page 78): Public Domain
Hermann von Helmholtz (Page 79): Public Domain
Pierre Simon de LaPlace (Page 94): Public Domain

W. H. Müller, *An Expedition to Continuum Theory*, Solid Mechanics
and Its Applications 210, DOI: 10.1007/978-94-007-7799-6,

ARCHIMEDES (Page 103): Public Domain
Siméon Denis POISSON (Page 132): Public Domain
Lorenzo Romano Amedeo Carlo AVOGADRO (Page 141): Public Domain
Ludwig BOLTZMANN (Page 142): Public Domain
Joseph-Louis GAY-LUSSAC (Page 143): Public Domain
John Prescott JOULE (Page 144): Public Domain
Eduard GRÜNEISEN (Page 148): Public Domain
Brook TAYLOR (Page 155): Public Domain
Edwin Powell HUBBLE (Page 170): Public Domain
Josiah Willard GIBBS (Page 177): Public Domain
Galileo GALILEI (Page 182): Public Domain
Tullio LEVI-CIVITA (Page 190): Public Domain
Gaspard-Gustave de CORIOLIS (Page 194): Public Domain
Jean le Rond D'ALEMBERT (Page 202): Public Domain
Ernst MACH (Page 202): Public Domain
Jean Bernard Léon FOUCAULT (Page 203): Public Domain
Hans THIRRING (Page 204): Bildarchiv der Österreichischen Nationalbibliothek, Wien
Gustav JAUMANN (Page 210): Public Domain
James Gardner OLDROYD (Page 211): Rheologica Acta, Volume 22, Number 1, 1–3, doi:10.1007/BF01679823, Springer Verlag http://www.springerlink.com/content/h8341124xu440w86/
Clifford Ambrose TRUESDELL (Page 211): Mathematisches Forschungsinstitut Oberwolfach under CC-BY-SA 2.0
Marius Sophus LIE (Page 211): Public Domain
Percy Williams BRIDGMAN (Page 228): Public Domain
Pierre Alphonse LAURENT (Page 236): Public Domain
Benjamin Olinde RODRIGUES (Page 240): Public Domain
Guillaume François Antoine Marquis DE L'HÔPITAL (Page 243): Public Domain
Johann BERNOULLI (Page 256): Public Domain
Ludwig PRANDTL (Page 298): DLR-Archiv
Endre A. REUSS (Page 298): Lehrstuhl für Technische Mechanik, Technische Universität Budapest
Constantin CARATHEODORY (Page 308): Public Domain
Nicolas CARNOT (Page 309): Public Domain
Hermann SCHWARZ (Page 311): Public Domain
Robert BOYLE (Page 312): Public Domain
Edme MARIOTTE (Page 312): Public Domain
Guillaume AMONTONS (Page 312): Sheila Terry/Science Photo Library
Bernard COLEMAN (Page 315): Springer Verlag
Walter NOLL (Page 315): personal property of Walter NOLL
Lord KELVIN (Page 318): Smithsonian Institution
Robert Jamison VANDE GRAAF (Page 332): Public Domain
Hendrik LORENTZ (Page 333): Public Domain
Michael FARADAY (Page 334): Public Domain

Charles Augustin de Coulomb (Page 340): Public Domain
Hans Christian Ørstedt (Page 344): Public Domain
André Marie Ampère (Page 345): Public Domain
Paul Dirac (Page 348): Public Domain
Heinrich Hertz (Page 357): Public Domain
Jean-Baptiste Biot (Page 357): Public Domain
Félix Savart (Page 359): Public Domain
Albert Abraham Michelson (Page 378): Public Domain
Edward Morley (Page 378): Public Domain
John Henry Poynting (Page 381): Public Domain
Lars Onsager (Page 384): http://www.chemistry.msu.edu/Portraits/images/onsagerc.jpg
Woldemar Voigt (Page 388): Public Domain
Georg Simon Ohm (Page 389): Public Domain

# Subject Index



W. H. Müller, *An Expedition to Continuum Theory*, Solid Mechanics and Its Applications 210, DOI: 10.1007/978-94-007-7799-6,

MIX
Papier aus verantwortungsvollen Quellen
Paper from responsible sources
FSC® C105338

If you have any concerns about our products,
you can contact us on
**ProductSafety@springernature.com**

In case Publisher is established outside the EU,
the EU authorized representative is:
**Springer Nature Customer Service Center GmbH**
**Europaplatz 3, 69115 Heidelberg, Germany**

Printed by Libri Plureos GmbH
in Hamburg, Germany